Sustainable Poverty Reduction in Less-favoured Areas

SUSTAINABLE POVERTY REDUCTION IN LESS-FAVOURED AREAS

Edited by

R. Ruben

Centre for International Development Issues Nijmegen (CIDIN), Radboud University, Nijmegen, The Netherlands

J. Pender

International Food Policy Research Institute (IFPRI), Washington DC, USA

and

A. Kuyvenhoven

Development Economics Group, Wageningen University, Wageningen, The Netherlands

www.cabi.org

CABI is a trading name of CAB International

CABI Head Office	CABI North American Office
Nosworthy Way	875 Massachusetts Avenue
Wallingford	7th Floor
Oxfordshire OX10 8DE	Cambridge, MA 02139
UK	USA
Tel: +44 (0)1491 832111	Tel: +1 617 395 4056
Fax: +44 (0)1491 833508	Fax: +1 617 354 6875
E-mail: cabi@cabi.org	E-mail: cabi-nao@cabi.org
Website: www.cabi.org	

A catalogue record for this book is available from the British Library, London, UK.

ISBN 978 1 84593 2770

Library of Congress Cataloging-in-Publication Data
Sustainable poverty reduction in less-favoured areas / edited by Ruerd Ruben, John Pender and Arie Kuyvenhoven.
    p cm.
  Includes bibliographical references and index.
  ISBN 978-1-84593-277-0 (alk. paper) -- ISBN 978-1-84593-278-7 (ebook)
    1. Sustainable development--Developing countries. 2. Agricultural development projects--Developing countries. 3. Rural development--Developing countries. 4. Poverty--Developing countries. I. Ruben, Ruerd. II. Pender, J. III. Kuyvenhoven, Arei. IV. Title.

HC59.72.E5S885 2007
339.4'6091724--dc22
                        2007006344

The paper used for the text pages in this book is FSC certified. The FSC (Forest Stewardship Council) is an international network to promote responsible management of the world's forests.

Typeset by Columns Design Ltd, Reading, UK
Printed and bound in the UK by Cromwell Press, Trowbridge

Contents

Contributors

Assefa Abegaz is Professor at Mekelle University, Ethiopia, and has conducted research within the framework of the Wageningen University/IFPRI research programme on less-favoured areas.

Nazneen Ahmed is a Research Fellow at the Bangladesh Institute for Development Studies (BIDS) and has conducted research within the framework of the Wageningen University/IFPRI research programme on less-favoured areas.

Ahmed Ali is a staff member of the Bangladesh Rural Advancement Committee (BRAC) and has conducted research within the framework of the Wageningen University/IFPRI research programme on less-favoured areas.

Fred Bagamba is a Researcher at the Banana Research Institute, Uganda, and has conducted research within the framework of the Wageningen University/IFPRI research programme on less-favoured areas.

Johan Brons is a Researcher at the Development Economics Group, Department of Social Sciences, Wageningen University, Netherlands.

Kees Burger is Associate Professor at the Development Economics Group, Department of Social Sciences, Wageningen University, Netherlands.

Le Chen is a Researcher from Nanjing Agricultural University, China, and has conducted research within the framework of the Wageningen University/IFPRI research programme on less-favoured areas.

Amy Damon is a Researcher at the Department of Applied Economics, University of Minnesota, USA.

Ton Dietz is Professor in Human Geography at the Amsterdam Institute for Metropolitan and International Development Studies (AMIDSt),

Amsterdam, Netherlands and Director of the Center for Resource Studies for Development (CERES) Research School.

Andrew Dorward is a Reader in Agricultural Development Economics, Centre for Environmental Policy, Imperial College, London, UK.

Shuyi Feng is Professor at the College of Land Management, Nanjing Agricultural University, China and has conducted research within the framework of the Wageningen University/IFPRI research programme on less-favoured areas.

Eleni Z. Gabre-Madhin is Programme Leader at the Development Strategy and Governance Division of the International Food Policy Research Institute (IFPRI), Washington, DC, USA.

Cornelis Gardebroek is Associate Professor at the Agricultural Economics and Rural Policy Group, Department of Social Sciences, Wageningen University, Netherlands.

Ken E. Giller is chair at the Plant Production Systems Group, Department of Plant Sciences, Wageningen University, Netherlands.

Nico Heerink is Associate Professor at the Development Economics Group, Department of Social Sciences, Wageningen University, Netherlands and Research Fellow and China Office coordinator (outposted to Beijing) at the International Food Policy Research Institute (IFPRI), Washington, DC, USA.

Moti Jaleta is Researcher at Awassa College, Ethiopia, and has conducted research within the framework of the Wageningen University/IFPRI research programme on less-favoured areas.

Hans G.P. Jansen is Research Fellow and Mesoamerica Office Coordinator (outposted at San José, Costa Rica) at the Development Strategy and Governance Division of the International Food Policy Research Institute (IFPRI), Washington, DC, USA.

Marijke Kuiper is Researcher at the International Trade and Development, Public Issues Division, Agricultural Economics Research Institute, Wageningen University, Netherlands.

Arie Kuyvenhoven is Chair of the Development Economics Group and Director of the Mansholt Graduate School for Social Sciences at Wageningen University, Netherlands.

Leslie Lipper is Researcher at the Agricultural and Development Economics Division, Food and Agriculture Organization of the United Nations (FAO), Rome, Italy.

Santiago López-Ridaura is Researcher at the Plant Production Systems Group, Department of Social Sciences, Wageningen University, Netherlands.

Aimee Milagrosa is Senior Researcher at the Centre for Development Research (ZEF), University of Bonn, Germany and has conducted research within the framework of the Wageningen University/IFPRI research programme on less-favoured areas.

Anke Niehof is Chair of the Sociology of Consumers and Households Group, Department of Social Sciences, Wageningen University, Netherlands.

Jack Peerlings is Associate Professor at the Agricultural Economics and Rural Policy Group, Department of Social Sciences, Wageningen University, Netherlands.

John Pender is Senior Research Fellow at the Environment and Production Technology Division of the International Food Policy Research Institute (IFPRI), Washington, DC, USA and Co-director of the Wageningen University/IFPRI research programme on less-favoured areas.

Prabhu Pingali is Director of the Agricultural and Development Economics Division, Food and Agriculture Organization of the United Nations (FAO), Rome, Italy.

Julieta R. Roa is Head of the Extension and Socio-economics Division of the Philippine Rootcrop Research and Training Centre based at Leyte State University, Philippines and has conducted research within the framework of the Wageningen University/IFPRI research programme on less-favoured areas.

Ruerd Ruben is Chair of Development Studies and Director of the Centre for International Development Issues (CIDIN) at Radboud University Nijmegen, Netherlands; also Director of the Wageningen University/IFPRI research programme on less-favoured areas.

Rob Schipper is Associate Professor at the Development Economics Group, Department of Social Sciences, Wageningen University, Netherlands.

Louis Slangen is Associate Professor at the Agricultural Economics and Rural Policy Group, Department of Social Sciences, Wageningen University, Netherlands.

Girmay Tesfay is Professor at Mekelle University, Ethiopia and has conducted research within the framework of the Wageningen University/IFPRI research programme on less-favoured areas.

Herman van Keulen is Professor at the Plant Production Systems Group, Department of Plant Sciences and Researcher at Plant Research International, Wageningen University, Netherlands.

Aad van Tilburg is Associate Professor at the Marketing and Consumer Behaviour Group, Department of Social Sciences, Wageningen University, Netherlands.

Willem Wielemaker is a soil scientist, formerly attached to the Atlantic Zone Programme of Wageningen University, Netherlands.

Karen Witsenburg is Researcher at the Amsterdam Institute for Metropolitan and International Development Studies (AMIDSt), Amsterdam, Netherlands.

Monika Zurek is Researcher at the Agricultural and Development Economics Division, Food and Agriculture Organization of the United Nations, Rome, Italy.

1 Sustainable Poverty Reduction in Less-favoured Areas: Problems, Options and Strategies

RUERD RUBEN,[1] JOHN PENDER[2] AND ARIE KUYVENHOVEN[3]

[1]Centre for International Development Issues Nijmegen (CIDIN), Radboud University Nijmegen, PO Box 9104, 6500 HE Nijmegen, The Netherlands; e-mail: R.Ruben@maw.ru.nl; [2]International Food Policy Research Institute (IFPRI), 2033 K St NW, Washington, DC, USA; e-mail: j.pender@cgiar.org; [3]Development Economics Group, Wageningen University, Hollandseweg 1, 6706 KN Wageningen, The Netherlands; e-mail: arie.kuyvenhoven@wur.nl

Introduction

Poor people living in less-favoured areas (LFAs) represent globally around 40% of the rural population suffering from chronic poverty. Given the limited agricultural potential and difficult access conditions in these areas, standard devices for enhancing rural development cannot appropriately address issues of poverty alleviation and sustainable natural resource management. Escaping from the downward spiral of poverty and resource degradation requires the identification of suitable pathways enabling rural households to develop production systems and livelihoods that respond to local conditions (Pender *et al.*, 2001a, b; Pender, 2004; Hazell *et al.*, 2006).

The diversity in agroecological settings and the heterogeneity amongst rural households pose particular challenges to rural development. Instead of a *one-size-fits-all* strategy, a far more targeted approach is required to exploit the comparative advantage of different resource management strategies for particular types of households and communities (Ruben and Pender, 2004). Moreover, attention needs to be paid to the incentives and governance regimes that enable farmers to adjust their production systems and livelihoods in order to guarantee both welfare and sustainability objectives. Identifying the right combination of public and private investment efforts oriented towards sustainable intensification of farming systems and rural livelihoods is of fundamental importance for attaining such win-win options.

The chapters included in this book provide an overview of research conducted within the framework of the collaborative research programme on 'Regional Food Security Policies for Natural Resource Management and Sustainable Economies' (RESPONSE). This programme has been jointly managed by the Graduate Schools for Social Sciences and Production Ecology and Resource Conservation of Wageningen University, The Netherlands, in cooperation with the International Food Policy Research Institute (IFPRI), Washington, DC, USA. The programme aimed to identify strategic options for agricultural and rural development in less-favoured areas and policy instruments that enhance rural households' investments in improved and sustainable natural resource management. Fieldwork is conducted in different LFA settings in Eastern Africa (Ethiopia, Kenya and Uganda) and South-east Asia (Bangladesh, Philippines and China) in cooperation with local partner institutes.

Development pathways for less-favoured areas demand careful adjustment of resource use strategies at field, farm-household and village level, looking for a portfolio of activities and technologies that guarantee food security and input efficiency. Given the asymmetric market access, due attention should also be given to options for reducing income inequality and resource degradation and potential pathways for escaping spatial poverty traps. Targeting of incentives towards resource-poor households may be required to guarantee both higher factor returns and improved land management. Therefore, institutional strategies for reducing transaction costs tend to be critically important for enhancing investments and enabling income diversification.

The remainder of this chapter is devoted to a discussion of: (i) the strategic interactions between natural resource management options; (ii) farm-household livelihood strategies for welfare and risk; and (iii) the surrounding market and institutional conditions for simultaneously enabling poverty reduction and sustainable land use. We start with a definition of the main characteristics of LFAs, followed by an analysis of the interactions between poverty and resource degradation in LFAs. Hereafter, the biophysical, micro- and macroeconomic dimensions of LFA development are discussed in order to provide insight in the complex interfaces between agroecological options, household drivers for change and effective incentives for resource use adjustment. We conclude with some major implications for policy and research concerning strategies for sustainable poverty reduction in less-favoured areas.

What are Less-favoured Areas?

LFAs are usually defined in terms of fragile agricultural resource base and/or limited access. An additional dimension for characterizing LFAs refers to the population dynamics. Contrary to common expectations, some LFAs – especially upland areas – have high population densities. This may be caused by historical migration (highlands as a refuge area free of malaria and other diseases and pests) or by socio-economic reasons (smallholder expulsion towards hillsides due to limited access to secure land). In addition, permanent or temporary migration of family members to other areas leads to an unbalanced population structure, characterized by high dependency rates. This is further reinforced by the decline in mortality that precedes the decline in fertility (Lipton, 2005). The demographic transition foreseen for many developing countries is likely to be delayed in LFAs.

Based on the FAO/World Bank classification of farming systems (Dixon *et al.*, 2001) and associated demographic and ecosystems data (Wood *et al.*, 1999), LFAs account for some 1.2 billion (42%) of the total 3 billion people living in the developing world. Poverty affects globally around 1.2 billion people, 75% of them living in rural areas. Of these 900 million rural poor, about 360 million live in LFAs. This is equivalent to 30% of the global poor and 40% of the total rural poor living in developing countries (see Fig. 1.1). This population is mostly con-centrated in the highlands and drylands of Asia and sub-Saharan Africa, although there are substantial numbers of poor people also in LFAs of Latin America (especially in the Andes Mountains and the hillsides of Central America).

Many disadvantaged social groups are concentrated in LFAs (Cleaver and Schreiber, 1994; Bird *et al.*, 2002). These include: (i) *women and female-headed households* facing unequal opportunities for access to land, education, employment and asset ownership, while male migration results in the feminization of agriculture and a further increase in women's work burden; (ii) *landless farmers* that depend heavily on

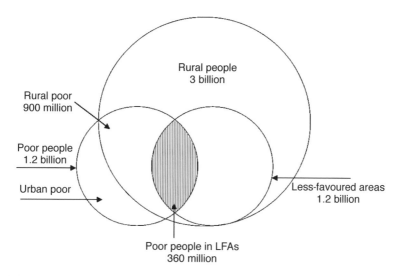

Fig. 1.1. Distribution of poor people in the developing world (from Ruben *et al.*, 2003).

poorly remunerated seasonal employment and temporary lease and sharecropping contracts; (iii) *indigenous people* who have lost their traditional land rights due to the encroachment by migrants in forest areas; (iv) *forest dwellers* in mountain areas who have become deprived of rights of collection; and (v) *fisheries communities* that derive their livelihood from capture fisheries along the coastal strips (mostly combined with cropping and off-farm activities) and are threatened by over-exploitation of fishing grounds due to increasing population pressure, undefined property rights and competition from commercial fishing.[1]

For all these reasons, LFAs are a significant part of the human development challenge facing the world community. In addition, they are also a significant part of the global environmental challenge. Land degradation concentrates in LFAs, accounting for about 40% of the total agricultural area in developing countries, including most of the areas with fragile soils. Soil erosion due to deforestation and reduced fallow periods affects some 20% of the total land area in developing countries and is heavily concentrated in LFAs. Population pressure, land fragmentation and limited access to inputs all lead to declining yields, and remoteness and the lack of services further aggravate poverty.

Deforestation is occasioning every year 14–15 million ha of forest cover loss in developing countries. Subsistence farming in LFAs is by far the most important driving factor, accounting for about 60% of total deforestation (FAO, 1997). Soil fertility mining is endemic in many LFAs, especially in areas with poor infrastructure and market access. Rainfed systems suffer from water stress and soil erosion, overgrazing and soil compaction. Agricultural production and yields are low due to very limited use of inorganic fertilizers, erratic rainfall and seasonal moisture stress. In addition, migratory livestock keeping is causing major property rights conflicts. Finally, biodiversity loss – both rare crops and traditional landraces – is loss occasioned by the expansion of traditional cropping, grazing and fuelwood collection into forests, wetlands, parks and other environmentally valued sites.

Poverty in LFAs

Most studies of rural poverty agree that, globally, there are larger numbers of poor people living in more favourable environments, simply because more people live in such environments (Ryan and Spencer, 2001; TAC, 2001; FAO, 2002). These global estimates are supported by country-level data from Kelley and Parthasarathy Rao (1995) in India, Renkow (1993) in Pakistan and Reardon *et al.* (1992) in the West African semi-arid tropics, where the absolute numbers of poor were greater in high-potential rainfed and irrigated areas.

However, there is debate about whether the incidence and severity of poverty is greater in LFAs than in more favourable environments

(Renkow, 2000; Fan and Chan-Kang, 2004). Evidence is limited on the spatial distribution of poverty in most countries, although the evidence available indicates that, in many cases, poverty is more prevalent and severe in LFAs, though this is not universal (Kelley and Byerlee, 2003).

Asia

In China, the incidence of poverty is far greater in the low-potential remote upland areas than in the coastal region (IFAD, 2002). Almost all of the 65 million people officially recognized as income-poor in the late 1990s lived in remote and mountainous rural areas (UNDP, 1997). Although the proportion of households below the national poverty line is less than 1% in the urban areas of Beijing, Shanghai, Tianjin and Guangdong, it is 20% or more in Inner Mongolia and Qinghai (De Haan and Lipton, 1998). One study found that more than 60% of the rural poor live in low-potential areas, and that the share of the national total living in such areas has increased since 1986 (Fan *et al.*, 2002).

Several studies in South-east Asia found poverty to be greater in LFAs, but also that other factors are important. For Vietnam, a household survey in the early 1990s showed that poverty was concentrated in the northern uplands and in the northern part of the central highlands, hilly areas that are far from large cities and the coast and having large ethnic minority populations (Minot, 2000). Although poverty declined rapidly between 1993 and 2002 throughout Vietnam, poverty remains highest in the north-west and central highlands (Swinkels and Turk, 2004).

There is wide variation in poverty rates within poor provinces. For example, in the northern uplands region, poverty rates range from 6% in Quang Ninh to nearly 80% in Lai Chau. The incidence and depth of poverty is greatest among ethnic minorities in Vietnam (their poverty rate was nearly 70% in 2002) (Swinkels and Turk, 2004). Other key factors contributing to poverty in these upland regions are: (i) small farm size; (ii) dependence on forest land and consequent land tenure insecurity; (iii) low education levels; (iv) limited access to health care; and (v) low public spending on investments and services in these areas (Swinkels and Turk, 2004).

In Indonesia, there is a high concentration of poverty in Java, particularly in the limestone hills of Central and East Java (IFAD, 2002). Poverty is also extremely prevalent on Madura in areas far from urban centres, and in fishing villages along the coasts of West and East Java (IFAD, 2002). In the Philippines, the incidence of poverty is 61% in the uplands compared with 50% in the lowlands (IFAD, 2002). In a study of poverty and environment linkages in Cambodia and Lao People's Democratic Republic, Dasgupta *et al.* (2003) found a positive association (correlation = 0.30) between poverty incidence and areas with steep slopes across provinces in Laos, with high incidence of poverty –

especially in the northern upland region. In Cambodia, by contrast, poverty is concentrated more in flatter lowland areas (Dasgupta *et al.*, 2003).

In South Asia, the relationship is more mixed. For India, Fan *et al.* (2003) found that the rural poor were increasingly concentrated in rainfed areas, with a relatively equal proportion in high- versus low-potential rainfed areas. There is a high degree of variability in the incidence of rural poverty across states in India, ranging from 15% in Punjab to 66% in Bihar in 1993/1994 (IFAD, 1999). There is also substantial variation within states. For example, in Maharashtra poverty incidence ranges between 24 and 38% in the coastal and western regions to 62–66% in the northern and eastern regions (IFAD, 1999). In the Himalayan belt, the largest increase in poverty between 1987/1988 and 1993/1994 occurred in West Bengal, followed by Assam Hills, Arunachel Pradesh and Manipur (IFAD, 1999). Most of these areas are dependent on rainfed agriculture, some have suffered from political unrest and many contain a large number of ethnic minorities. These findings indicate the importance of social and political as well as geographic factors in determining poverty in India.

In Pakistan, the incidence of food poverty in 1990/1991 was highest in rural areas of the South Punjab region, in contrast to low poverty rates in the neighbouring Punjab region of India (IFAD, 2002). The high level of poverty in the Punjab region of Pakistan is attributable to highly unequal access to land, indicating that poverty is affected by other factors besides agricultural potential and access to markets and infrastructure.

Recent poverty mapping work in Bangladesh found that the pockets of high poverty coincided with ecologically poor areas, including the low-lying depression area in the north-east, the drought-prone area on higher land in the north-west, several subdistricts on the fringes of major rivers and several of the south-eastern subdistricts, including the Chittagong Hill Tract (Kam *et al.*, 2005). Despite the importance of some agro-ecological conditions in explaining poverty in Bangladesh (especially the prevalence of highland, low or very low-lying land and heavy soils), socio-economic factors – especially education, but also including access to infrastructure (roads, irrigation and electricity) and landlessness are the strongest predictors of poverty (Kam *et al.*, 2005).

Africa

Several studies have also shown greater incidence and severity of poverty in LFAs of Africa. For example, the incidence of poverty is much greater in northern Uganda than in the rest of the country, where access to markets, infrastructure and biophysical conditions are more favourable (UBOS, 2003). Besides these geographic factors, however, insecurity and political marginalization are important reasons for poor performance in the northern region (Zhang, 2004).

In Kenya, income per capita is much higher in the central highlands, with their favourable agroclimatic conditions and access to the Nairobi market, than in the western highlands, which are more remote (Place *et al.*, 2006b). Household wealth (proxied by houses with a metal roof) is significantly greater in rural areas of Kenya having higher rainfall or closer to an urban area (Place *et al.*, 2006a).

In the northern Ethiopian highlands, changes in average wealth, food availability and ability to cope with drought during the 1990s were more negative in areas further from towns, in more densely populated rural areas and at higher elevations, while more favourable outcomes occurred in communities where road access had improved and where perennial production was important (Pender *et al.*, 2001a, c). Within the less-favoured northern Ethiopian region of Tigray, however, differences in household endowments of human, social and physical capital were found to be more significant determinants of household income per capita than were geographic factors such as average rainfall or access to markets (Pender and Gebremedhin, 2006; Tesfay *et al.*, Chapter 7, this volume).

Latin America

In Latin America, some studies also show greater poverty in LFAs. In the southern Andes region, rural poverty is significantly greater in the mountains than in other regions of the same countries (Walker *et al.*, 2000).[2] In Peru, for example, the incidence of rural poverty (using the nationally defined poverty level) averages about 80% in the less-favoured Sierra and Selva regions compared with less than 60% in the coastal region, while rates of chronic malnutrition are more than twice as high in the Sierra as in the coastal region (Escobal and Valdivia, 2005).

In Honduras, 92% of households in hillside areas (studied by Jansen *et al.*, Chapter 6, this volume) were found to be extremely poor (with income per capita < US$1/day) compared with the national rate of extreme poverty of 60% for all rural households. Within hillside areas, dependence on subsistence basic grains production, which is the livelihood strategy with the lowest income per capita, is greater in areas having lower rainfall, poorer road access and higher population density (Pender *et al.*, 2001b; Chapter 6, this volume). Controlling for livelihood strategies, differences in rainfall and market/road access have limited impact on incomes, and other factors – including soil fertility, ownership of physical assets and participation in agricultural training programmes – are more important determinants of income (Jansen *et al.*, 2006).

These results demonstrate not only that poverty is often more common and more severe in LFAs than in more favoured areas, but also that the social and political context and household level factors such as ethnicity, human, physical and social capital also have very important impacts on poverty. Hence, there is great diversity in the incidence and severity of poverty within LFAs, and strategies to address poverty in these areas must therefore be adequately targeted, taking this heterogeneity into account (Ruben and Pender, 2004).

Natural Resource Degradation in LFAs

Degradation of natural resources is a severe constraint to LFA development. Major problems include the folowing: (i) deforestation; (ii) soil degradation; (iii) increasing water scarcity and resulting declines in natural capital stocks and agricultural productivity; (iv) increasing poverty and vulnerability; and (v) environmental damage (losses of biodiversity, reduction in carbon sequestration in soils and plant biomass, sedimentation of rivers and reservoirs, flooding and other environmental impacts).

Land degradation in drylands[3] (or 'desertification', as defined by the United Nations Convention to Combat Desertification (UNCCD)) and in sloping highlands is a particularly severe problem in LFAs. According to the widely cited Global Assessment of Soil Degradation (GLASOD) study (Oldeman *et al.*, 1991), which used expert judgement to assess the extent and severity of soil degradation globally between 1945 and 1990, 14% of all used land globally was seriously degraded[4] during this period, but the incidence of degradation was substantially higher in Central America (31% of used land), Africa (19%) and Asia (16%) (Oldeeman *et al.*, 1991).

Soil erosion by water or wind was the major source of this degradation, causing serious degradation of 25% of used land in Central America, 16% in Africa and 15% in Asia. Other forms of soil degradation, including chemical degradation (soil nutrient depletion, salinization, acidification and pollution) and physical degradation (compaction, sealing and crusting, waterlogging and loss of organic matter) were estimated to affect substantially smaller areas in all regions, including light as well as serious degradation (6% in Central America, 5% in Africa and South America and 3% in Asia).

The results of the GLASOD study did not focus on specific agro-ecological zones, but there are good reasons to expect that the incidence and severity of soil degradation was substantially higher in the drylands and sloping highlands of Africa, Asia and Latin America than was estimated for these continents as a whole. Given the lack of vegetative cover during periods of high rainfall and high winds in drylands, and the steep slopes in many highlands, water and wind erosion are generally much more severe in these regions than in other regions.

A comprehensive study of land degradation in drylands by Dregne and Chou (1992) found even higher incidence of land degradation in drylands, estimating that more than 70% of drylands in Africa, Asia and Latin America were degraded, including 73% of rangelands, 47% of rainfed croplands and 30% of irrigated croplands. That study estimated that more than one-third of irrigated land in Asia and more than one-half of rainfed land in Africa and Asia had experienced at least a 10% loss in productive potential due to land degradation, and that over one-half of the rangelands in these regions had experienced more than 50% loss in potential productivity. Based on Dregne and Chou's estimates, Crosson (1995) estimated an average loss (weighted by value of production) for the three land uses to be about 12%, though the loss was much higher for rangelands (43%) than for irrigated or rainfed cropland.

Numerous studies have assessed soil degradation in specific regions or countries. In a study of soil degradation in South and South-east Asia that replicated the methodology of Oldeman *et al.* (1991), but using more detailed data, Van Lynden and Oldeman (1997) found that the problems of soil nutrient and organic matter depletion, salinization and waterlogging were much greater than estimated in the global assessment, though they still concluded that soil erosion was the most widespread source of degradation. They estimated that agricultural activity had caused degradation of 27% of all land in the region, while deforestation had affected 11% and overgrazing played a minor role.

Studies in China have estimated significant negative effects of land degradation (mainly erosion in sloping areas) on crop productivity, estimating that erosion reduced rice yield growth by 12% during the late 1980s and early 1990s, and reduced production of maize, wheat and cash crops by up to 20% in north China (Huang and Rozelle, 1994, 1996). For India, Sehgal and Abrol (1994) synthesized available soil survey data and concluded that more than one-half of the land was suffering from moderate to severe degradation (most of this in less favoured areas) and, on 5% of the land, degradation was so severe that the soils were unusable.

In Africa, Dregne (1990) found compelling evidence of serious land degradation problems (productivity losses of at least 20%) in subregions of 13 countries. Lal (1995) estimated erosion rates for the continent and found that about one-quarter of the land was affected by moderate to severe erosion (particularly in the highlands), estimating that the average crop yield loss in 1989 due to past erosion in Africa was about 8%, reducing annual cereal production by about 8 million tons, root and tuber production by 9 million tons and pulse production by 0.6 million tons.

Many studies have found evidence of high soil erosion levels and impacts in particular African countries, with most focusing on highland areas in eastern and southern Africa or dryland areas in southern or western Africa (FAO, 1986; Stocking, 1986; Hurni, 1988; World Bank, 1988, 1992; Bishop and Allen, 1989; Convery and Tutu, 1990; Ehui *et al.*, 1990; Bojö, 1991, 1996; Norse and Saigal, 1992; Pagiola, 1993; Sutcliffe, 1993; Grohs, 1994; McKenzie, 1994; Bishop, 1995; Bojö and Cassells, 1995; Eaton, 1996; Sonneveld, 2002). The on-site immediate produc-tivity costs of erosion estimated by these studies vary widely, from less than 1% to as high as 55% of agricultural GDP, though in most studies estimated annual losses were less than 10% (Bojö, 1996; Enters, 1998; Scherr, 1999a; Yesuf *et al.*, 2005).

Several studies have also estimated rates of soil nutrient depletion in Africa. Stoorvogel *et al.* (1993) estimated average annual nutrient losses in Africa of 22 kg/ha of nitrogen (N), 2.5 kg/ha of phosphorus (P), and 15 kg/ha of potassium (K); and much higher rates of depletion in the densely populated and erosion-prone countries of eastern and southern Africa (especially Ethiopia, Kenya, Malawi and Rwanda).

High rates of soil nutrient depletion have also been found by numerous studies conducted at lower scales in several African countries, most of these also in highland or dryland areas (e.g. Van der Pol, 1992; Smaling *et al.*, 1997; Baijukya and De Steenhuijsen Piters, 1998; Bationo *et al.*, 1998; Defoer *et al.*, 1998; De Jager *et al.*, 1998, 2004; Elias *et al.*, 1998; Folmer *et al.*, 1998; Shepherd and Soule, 1998; Van den Bosch *et al.*, 1998; Wortmann and Kaizzi, 1998; Gitari *et al.*, 1999; Onduru *et al.*, 2001; Gachimbi *et al.*, 2002, 2005; Woelcke, 2003; Nkonya *et al.*, 2004, 2005a; Abegaz, 2005), although high rates of nutrient depletion are not universal (Muchena *et al.*, 2005).

Few studies have quantified the productivity impacts of soil nutrient depletion in Africa. A long-term experimental trial in Kabete, Kenya, of 18 years of continuous maize production in absence of nutrient inputs, showed a decline of maize yields from 3 to 1 ton/ha (Bekunda *et al.*, 1997). Other long-term experimental studies in sub-Saharan Africa show that, under continuous cultivation using low external inputs, soil fertility rapidly decreases and yields decline, and that a combination of inorganic and organic sources of soil fertility is necessary to sustain crop production (Juo and Kang, 1989; Vlek, 1990; Swift *et al.*, 1994; Bationo *et al.*, 1998). A few recent studies estimated the replacement costs of lost nutrients as averaging about one-fifth of farm income in several districts of Uganda (Nkonya *et al.*, 2005a, b) and one-third of farm income in three Kenyan districts (De Jager *et al.*, 1998).

Many of the studies of land degradation in Africa and elsewhere are based on expert opinion (e.g. Oldeman *et al.*, 1991) or on assumptions and relatively few plot-level trials (Stoorvogel *et al.*, 1993). However, recent advances in remote sensing and ground survey methods have substantiated the existence of significant land degradation at landscape scale. For example, use of near infrared spectrometry to assess soil quality and land degradation over wide areas has been able to provide evidence of the extent of degradation in the Nyando River Basin of Kenya. Cohen *et al.* (2005a) found that about 56% of the land was moderately to severely degraded. Further research combining measured soil degradation with estimated effects on crop yields (Cohen *et al.*, 2005b) calculated the costs of soil erosion at the national level in Kenya to be equivalent to 3.8% of GDP. Evidence from laboratory analysis of changes in soil properties from sample plots in small farmers' fields in Uganda that were resampled 40 years after an earlier soil survey (Ssali,

2003) also supports the view that soil fertility has declined in East Africa.

For Latin America, Oldeman (1998) estimated that agricultural productivity was 37% lower in Central America and 14% lower in South America as a result of soil degradation, based on the results of Oldeman *et al.* (1991). Lutz *et al.* (1994) estimated that, over a 10-year period without conservation measures, maize yields would decline in hillsides of Honduras by 20–25%, maize and sorghum yields would decline in Haitian hillsides by 60%, bean yields would decline by 20–25% in the Dominican Republic, coffee yields would decline by 10% in the Costa Rican highlands and cocoyam yields would fall to zero in the human lowlands of Costa Rica.

Cuesta (1994) estimated that uncontrolled erosion in Costa Rica would reduce highland coffee yields by one-half within 3 years and to zero within 20 years, highland potato yields by 40% over 50 years and lowland cocoyam yields by more than one-half in the first year and to zero by the fourth year. Solórzano *et al.* (1991) estimated the replacement costs of soil nutrients lost annually due to soil erosion in Costa Rica, and estimated this cost to be in the range of 5–13% of agricultural GDP. For Mexico, McIntire (1994) estimated that soil erosion reduced maize production by an average of about 3%, but with losses as high as 12% in some states, and losses being especially high in the highlands and semi-arid regions.

Many of these estimates probably overstate the impact and cost of soil erosion, however, since they do not consider farmers' mitigation efforts. For example, Pagiola and Dixon (1997) found that farmers in hillside areas of El Salvador perceived significant erosion problems on most of their sloping fields, but also that severe, long-term productivity declines were expected by farmers on only 16% of steeply sloping fields and 5% of moderately sloping fields.

In the past decade, several critics have challenged the generality, methodology, accuracy and motivations of many studies commonly cited in the literature concerning the extent and impacts of land degradation, especially in Africa. Several studies question the extent of land degradation, providing examples of particular cases where land conditions have improved in recent history (Tiffen *et al.*, 1994; Fairhead and Leach, 1996; Leach and Mearns, 1996; McCann, 1999) or evidence that earlier land conditions (e.g. forest cover) were not as favourable as previously thought (McCann, 1999). Some studies argue that land degradation is highly context specific, acknowledging that land degradation is a problem for some farmers in some places and times, but arguing that the problem is not as universal as sometimes claimed.

Some studies critique the methods used by agronomists and others to estimate land degradation as being conceptually flawed, subject to large errors and driven by political motives (e.g. Stocking, 1996; Bassett and Crummey, 2003; Keeley and Scoones, 2003; Fairhead and Scoones, 2005). For example, the common practice of scaling-up estimates of soil erosion based on plot level measurements and models to larger national or regional scales may overstate the impacts of erosion by orders of magnitude, since most of the soil eroded from particular plots is re-deposited in nearby fields (Stocking, 1996). Several studies deconstruct and critique the 'Malthusian narrative', which predicts that land degradation is the inevitable result of population pressure and poverty and suggests that drastic action by governments is required to address it (Hoben, 1995; Leach and Mearns, 1996; Bassett and Crummey, 2003; Keeley and Scoones, 2003). Most of the authors in this tradition argue that greater appreciation of farmers' knowledge and ability to adapt and innovate is needed, as well as greater understanding of the local historical, political and sociocultural context.

Some of these criticisms are well-founded (Koning and Smaling, 2005). Land degradation is certainly not an inevitable consequence of population growth or of poverty, the relationships among these and other factors are complex and context dependent, and there are many examples of sound land management practices being practiced by small farmers in many developing countries. Nevertheless, there are many studies that document serious degradation, and some of the studies questioning the importance of land degradation also suffer from methodological flaws, such as ignoring sources of soil nutrient outflows that are difficult to quantify (Koning and Smaling, 2005).

Much of the evidence on land degradation is synthesized in the recently completed Millennium Ecosystem Assessment (MEA, 2005). The preponderance of evidence supports the view that land degradation is indeed a very serious problem, especially in the less-favoured drylands and highlands of Africa, Asia and Latin America.

Heterogeneous Resources and Mixed Farming Systems

Less-favoured areas have gained little from past agricultural successes and suffer widespread poverty and resource degradation. They include lands that are of low agricultural potential due to limited and uncertain rainfall, poor soils, steep slopes or other biophysical constraints, as well as areas that may have higher agricultural potential but which are presently under-exploited due to limited access to infrastructure and markets, low population density or other socio-economic constraints. Most LFAs are mountain regions ('uplands', see Jansen *et al.*, Chapter 6, this volume) or arid and semi-arid zones ('drylands', see Assefa and Van Keulen, Chapter 5, this volume).

They can be defined more fully on the basis of the predominant farming systems. Table 1.1 outlines the six major farming systems that encompass most of the LFAs in developing countries. These systems are defined according to resource use regimes (CGIAR/TAC, 2000) and to a recent assessment of poverty incidence in tropical farming systems by FAO/World Bank (Dixon *et al.*, 2001). These farming systems cluster into dryland and upland farming systems, and together comprise about 40%

Table 1.1. Predominant farming systems in less-favoured areas (based on FAO-World Bank, 2001).

Agro-ecological zone	Production system	Share of developing countries' rural population	Share of developing countries' agricultural land
Highlands/upland areas	Perennial/tree crops	3	2
	Shifting cultivation	2	5
	Mixed cropping	24	9
Drylands/ arid areas	Migratory herding	6	12
	Agro-pastoral	4	8
	Mixed rainfed	3	4
TOTAL		42	40

Note: estimates based on FAO expert judgements (Delphi method).

of the agricultural land area and 42% of the rural population in the developing world.

We discuss the dominant farming systems in LFAs for identifying the available options for reinforcing their productivity and sustainability (see also López-Ridaura *et al.*, Chapter 2, this volume). Development strategies for sustainable resource intensification in LFAs need a careful adjustment of resource use at field, farm-household and village levels, looking for a portfolio of activities and technologies that guarantee input efficiency and labour productivity. Major characteristics and typical constraints of each of these LFA farming systems can be illustrated as follows.

Highlands and upland areas

Perennial and tree crop systems

These systems are found in the East African highlands, Central American and Andean hillsides and South-east Asian uplands. They include crops like banana, plantain, coffee, cocoa and multi-purpose trees, interplanted with food crops like cereals and cassava, and combined with small-scale animal husbandry. Cereals, roots and tubers are the main staple, while tree products and off-farm activities provide some cash income. Major limitations are soil fertility, scarcity of good planting material and high establishment and maintenance costs. In many cases, lack of market outlets and insecure property rights constrain agroforestry development. Many indigenous peoples are dependent on these farming systems.

Upland shifting cultivation

This is found in the forest margins of East and Central Africa and South-east Asia. It is based on the annual clearing of bush fields for the cultivation of food crops (maize, sorghum and cassava) and their subsequent recovery under fallow. The slash-and-burn system is severely threatened by shortened fallow, typically driven by increasing human population and encroachment into forest margins by sedentarized cultivators, occasioning soil erosion and nutrient depletion. Recovery periods become longer and returns to agricultural activities continuously decline. Poverty is further aggravated by remoteness and the absence of service provision.

Mixed cropping system

Mixed upland cropping systems are found in the semi-humid highland areas of Southern Africa, the South-east Asian uplands and in Central America, and are based on the intercropping of cereals, in rotation with legumes, beans, tubers and pulses. Farmers use animal traction for land preparation and rely on crop residues and mulching for soil fertility management. Upland rice systems increasingly suffer from decreasing water efficiency and salinity problems. Other major constraints are low soil fertility and high seasonal labour demands. Distance is prohibitive for further market integration.

Drylands and (semi-)arid areas

Migratory herding

This is found mainly in the arid areas of sub-Saharan Africa, the Middle East, North Africa and Central Asia. It is based on transhumant pastoralism with mixed herds of cattle, camels, sheep and goats that depend on the availability of grass, water and crop residues in neighbouring arable systems. Critical management issues relate to animal health care and scarce feed availability, especially during periodic droughts. Socio-economic differentiation is considerable and many herders lose their stock to drought or theft. Opportunistic grazing by sometimes externally controlled herds leads to conflicts over land and water rights with agricultural communities.

Agro-pastoral systems

These are found in the semi-arid regions of sub-Saharan Africa, the Middle East, North Africa and South-east Asia. They are based on the integration of cropping and livestock activities. Major components include linear agroforestry arrangements, production of fodder crops, manure recycling and use of animals for land preparation and transport. Sections of the animal herds may seasonally migrate to semi-humid areas. Farmers use early-maturing and drought-resistant crop varieties and grain storage for overcoming the dry period. The system is vulnerable to frequent crop failure, shortage of animal feed, large grain price variations and the periodic collapse of livestock prices.

Rainfed mixed cropping

This is found in Central and Southern Africa, South Asia, the coastal part of North Africa, north-east Brazil and the Yucatan peninsula in Mexico. It is based on seasonal cultivation of food and cash crops, using locally available resources. Sorghum, millet, maize and barley are major food crops, providing stubble grazing by animals after the harvest; cattle, sheep and goats provide the major part of cash income. Water shortages and soil fertility decline limit yields. Crop harvests are also seriously affected by weed infestation (*Striga*). Problems of unequal land distribution and poor market linkages lead to chronic poverty.

Farming systems development in LFAs

Local experiences with sustainable agricultural intensification offer some promising perspectives for simultaneously improving resource management and generating stable income streams to rural households living in LFAs. We outline six major strategies for pro-poor sustainable intensification in highland and dryland areas, focusing on their potential for adoption by poor farmers in different LFA settings (see Table 1.2). Particular attention is given to the identification of local economic incentives and the knowledge infrastructure that should be in place to facilitate the process of upscaling of sustainable agricultural practices in LFAs.

Agroforestry

Agroforestry systems supply rural households with a wide range of products for domestic use and sales, including food, fruits, medicine,

Table 1.2. Technology options for LFA production systems intensification (from Hazell *et al.*, 2006).

Farm management practices / Farming systems	Agroforestry	Nutrient management	Plant protection	Water management	Livestock and pastures	Seed and biotech
Highland and upland areas						
Perennial tree crops	■					■
Shifting cultivation	■	■	■		■	
Mixed cropping		■	■	■	■	■
Semi-arid and dryland areas						
Migratory herding					■	
Agro-pastoral	■			■	■	■
Mixed rainfed		■	■	■		■

Source: Hazell *et al.* (2006)

feed and fodder, fuelwood and timber, and also provide environmental services, such as erosion control and moisture conservation. Prospects for establishing agroforestry systems are particularly high in fragile hillsides and semi-arid lowlands where the benefits of annual cropping are low and risky (Neupane and Thapa, 2001).

Agroforestry can offer attractive returns to poor farmers. In the Dhading district of Nepal, where planting of multipurpose trees almost doubled net returns to farming at only slightly higher costs, including the additional income from fodder, fuelwood and timber. Short-rotation improved fallow with *Sesbania* species as green manure crop in densely populated hillside areas of Western Kenya had little effect on cereal yields but led to savings in weeding labour, thus offering farmers options for engagement in off-farm employment (Swinkels and Franzel, 1997). Other benefits from agroforestry systems include the added value from woody and non-timber forest products (NTFPs). For Cameroon, the commercial value of NTFP almost doubles farmers' incomes (Leaky and Tchoundjeu, 2001). Especially women in homestead and community areas frequently manage mixed-tree systems with fruit and woody species.

Soil conservation programmes in dryland areas rely on linear forestry arrangements (windbreaks and shelterbelts), while contour hedgerows are applied on moderate slopes, and agroforestry and mixed trees on steeper slopes. Soil fertility and nutrient uptake in mixed cropping systems benefit greatly from intercropping with leguminous trees. In sub-Saharan Africa, the combination of agroforestry with phosphate rock applications substantially improves nitrogen uptake and crop yields (Sanchez and Jama, 2002). Land reclamation through agroforestry is done with multipurpose trees, fodder shrubs and grasses, usually assisted by biomass transfers and specific soil conservation measures (ridge tillage) on sloping lands. In Tobora district in Tanzania, rotational woodlots assist farmers in generating substantial income while at the same time conserving large forest areas (Ramadhani *et al.*, 2002). Agroforestry contributes to the reduction in the pressure of shifting cultivation and the control of erosion through more permanent land cover.

Agroforestry systems that are particularly interesting to poor farmers in LFAs include those that have low establishment and maintenance costs, face limited competition with other activities and exhibit high synergy effects for enhancing soil fertility and water storage capacity. Moreover, expected returns should become available after a short period, and stable market outlets need to be accessible when a marketable surplus is produced.

Although most agroforestry research has focused on semi-humid hillside regions, tree crops that have a high drought tolerance also represent an interesting component of agro-pastoral development in dryland areas. Trees that enhance synergy with arable or pastoral activities can contribute to both sustainability and profitability of mixed farming systems. Current initiatives in agroforestry seek the domestication of trees in order to integrate indigenous species into tropical farming systems (Wiersum, 1997; Leaky and Tchoundjeu, 2001).

Most promising results in agroforestry are reached with fast-growing tree species (*Tephrosia, Sesbania*) incorporated in improved fallow systems for maize and vegetables (Western Kenya and Eastern Zambia), drought-resistant trees (pistachio, almond) used for soil moisture upgrading and fodder provision in the semi-arid regions of Tunisia, living fences used in the marginal drylands of the Sahel and Central America and contour farming on acid soils in the uplands of the Philippines. Techniques like alley cropping, hedgerow intercropping and community woodlots are less easily adopted by smallholder farmers due to their high labour or input costs.

The adoption of agroforestry technologies typically depends on local resource endowments, the opportunity costs of land and labour, and market and institutional conditions. Where population density is high, fallow periods are decreasing and, when farmers perceive a decline in soil fertility, improved tree fallow has a great potential (Franzel, 1999). In more densely populated areas, forestry is feasible only with rapidly growing varieties and high-value by-products. Adoption of hedgerow technologies has been rather limited because short-term returns are too small and rather uncertain (Nelson *et al.*, 1998). Contour farming with natural vegetative strips – although less effective for reducing soil erosion – can be a better alternative due to its lower establishment costs and reduced maintenance requirements (Garrity, 2002).

Soil nutrient management

Farmers' soil fertility management practices include a wide variety of agronomic, biological and mechanical measures to reduce soil and water erosion, reinforce soil fertility and soil structure and to safeguard soil biological processes. These measures enhance the availability and efficient uptake of soil nutrients, and reduce water constraints. Strategies for Integrated Nutrient Management (INM) rely on the combination of appropriate organic and inorganic fertilizer applications, soil and water management practices, and agronomic and soil conservation measures to increase yields and maintain the ecosystem stability of the environment (Vanlauwe *et al.*, 2002).

High nutrient deficits are registered in subsistence-oriented rainfed cropping systems that use almost no fertilizers and are located in more remote areas. Agricultural yields in sub-Saharan Africa are severely limited due to very low inorganic fertilizer use (on average only 8 kg/ha compared with 107 kg/ha in all developing countries). Sustained treatments with only inputs of organic matter (green or animal manure, crop residues) are usually not sufficient to halt declining yields. Soil replenishment with phosphate rock in West Africa could increase crop yields by 20–30% (Diop, 2001). Farmers are somewhat reluctant to use phosphate rock given the high investment costs and detrimental effects on health (Kuyvenhoven *et al.*, 1998).

The impact of soil and water conservation measures on farmers' income varies strongly between different settings. In the Sahelian countries, simple and low-cost technologies (e.g. earth bunds, vegetation strips, windshields) that retain soil nutrients and reduce erosion contribute to slightly higher (and more stable) yields and higher income (De Graaff, 1996; Reij and Steeds, 2003). In Northern Africa and in the East African Highlands regions, land management is more focused on soil and water conservation, and agronomic measures (intercropping, agroforestry) proved to be able to provide win-win solutions with reduced erosion and increased productivity (Shiferaw and Holden, 1999).

On the steep hillsides of the Chiapas region in Mexico, the combination of conservation tillage and crop mulching increased net returns to land and labour by 13 and 28%, respectively (Erenstein, 1999). In the Central American hillsides, average smallholder maize yields were three to nine times higher after a period of 10–22 years relying on cover crops and green manure (velvet beans) as a method for improving soil fertility (Bunch, 2002). Intensive training and support for these programmes has been provided by local NGOs, and diffusion took place through a 'farmer to farmer' methodology. Recent studies point, however, to dis-adoption of green manuring due to plant diseases and difficulties in adapting to changing agroecological conditions.

Population density, rainfall and market orientation influence the scope and feasibility of specific soil fertility measures (Scoones and Toulmin, 1999). Crop diversification, terracing and INM practices are used mainly in high population density highland regions like western Kenya and north-east Nigeria. Grass-strips, contour bunds, composting and manure are applied in more remote locations in the south-west Ethiopian highlands and the central region of Malawi. Mixed farming systems with strong crop–livestock integration become a feasible option in rainfed regions with a regular population density. In the semi-arid regions with lower population densities and variable rainfall, labour-extensive silvopastoral and improved fallow systems offer feasible alternatives.

Even within the same region, adoption of soil and water conservation (SWC) measures is unequally distributed amongst households. In Atacora district in north-west Benin, larger farmers with cattle and commercial crops used improved fallow, chemical fertilizer and manure for soil fertility management, while small farmers with more food crops used more labour-intensive crop residue management options to maintain soil fertility (Mulder, 2001). Less-endowed households benefit substantially from programmes for indigenous SWC technologies, like water harvesting (*tassa*) in Niger, rectangular ridges (*sagan*) in Nigeria and bench terraces in Togo (Reij *et al.*, 1996).

Plant protection

Pests and diseases lead to high crop loss at different stages of the production process. Postharvest losses in farm level storage typically represent 2–8% of the production weight, while losses in storage and transport are about 3–7% (Boxall, 2001). In tropical areas, plant diseases and pest infestation strongly limit the harvested output until reaching complete crop failure. Post-harvest losses can be equally large. Pest incidence in LFAs is most relevant in semi-intensive cropping systems that face water stress or nutrient limitations typical for poor farmers, and is best controlled through cultural practices that maintain pests at a controllable level.

Conventional synthetic pesticides tend gradually to lose their effectiveness and concerns are growing regarding the detrimental health and environmental effects. Integrated pest management (IPM) strategies for cereals and cassava crops have been developed that use a combination of biological, cultural, genetic and chemical techniques to maintain pest populations below an economically damaging level (Swinton and Williams, 1998). IPM programmes that reduce pest problems while minimizing environmental damage are potentially win-win strategies.

Plant protection measures by poor farmers in LFAs are mostly based on local techniques. Due to high variability, site-specific pest management techniques are needed. Crop rotation and intercropping are the most commonly used plant protection devices. Labour shortages that lead to low-intensity weeding may increase the vulnerability to pests and diseases. Most recommended IPM strategies for combating pests in maize, beans and pigeon pea in southern Malawi, like inorganic fertilization and weeding for *Striga* control, and mulching for controlling bean stem maggot, require considerable labour or cash inputs and are thus less accessible to poor farmers. Poor farmers tend to select simple and low-cost IPM methods (Chaves and Riley, 2001), and measures based on varietal resistance, botanical seed dressing, biological control with fungi and trap crops have a large potential for being adopted by resource-poor farmers (Orr and Jere, 1999).

Water management

Poor farmers in semi-arid areas suffer from water shortages and declining water quality. Competition for water by different stakeholders asks for improved efficiency in water use through appropriate systems for water allocation and management. Participatory planning of resource use at watershed level enables the selection of appropriate water conservation practices and makes significant contributions to agricultural productivity, natural resource conservation and poverty alleviation (Kerr *et al.*, 2001).

Water-harvesting techniques like small dunes and planting pits (*tassa*) are widely adopted by smallholders in Niger. Within the framework of the IFAD-funded Soil and Water Conservation Project, within 4 years 46% of the farmers had applied the new techniques in small areas of 0.9–1.4 ha. Training and extension, and food-for-work rations (in dry years), tools and community infrastructure have supported rapid dissemination (Hassane *et al.*, 2000). Surface drainage with broad-beds and furrows is widely used in the central highlands of Ethiopia to protect crops from waterlogging (Deckers, 2002). Given the climate change projections, rainfall variability in sub-Saharan Africa tends to increase, and cropping periods may become shorter (Dietz *et al.*, 2004). Consequently, demand for micro-irrigation and further improvement of water harvesting strategies is of primary importance (Rosegrant and Perez, 1997).

Small-scale, farmer-controlled irrigation programmes that use simple and low-cost technologies of river diversion, lifting with small (hand or rope-)pumps from shallow groundwater or rivers, or seasonal flooding, are successful in Africa, Central America and the Andes region. In Zimbabwe, low-cost indigenous water management systems for dambo gardens managed by individual farmers on land allocated by local communities allow flexible water management, based on shallow wells and water channels between beds. Initial investments (US$500/ha) are four to 20 times lower than for conventional gravity irrigation, while returns are twice as high because high-value horticulture crops are grown instead of grains (Rukuni *et al.*, 1994). Dambo gardens are about ten times more productive than dryland farming, prevent erosion and protect downstream water flows, and can relieve pressure on upland resources. Dambo gardens now cover about 15,000–20,000 ha in Zimbabwe (compared with 150,000 ha of formal irrigation) and there is a potential to develop another 60,000 ha.

Programmes for integrated aquaculture require only limited amounts of land and show highly positive returns, but require access to stable feed sources from animal or chicken manure, garden and kitchen waste. In Malawi, the vegetable–garden–pond system generates annually US$14/100 m², compared with US$1–2 for cropping activities (Brummett, 2002). In addition, income from fishponds is more stable and compensates in adverse years for losses in crop income.

Livestock and pasture management

Livestock production is considered as an attractive device for supporting simultaneously objectives of poverty reduction, food security and environmental sustainability (De Haan *et al.*, 2001). In LFAs, the dual function of livestock as a production and a stock-keeping activity contributes to both income and wealth. The latter function is especially important in parts of sub-Saharan Africa, providing a source of savings and a buffer against calamities (Fafchamps *et al.*, 1998). Livestock provides a relative high share of income to the rural poor, including landless farmers (Delgado *et al.*, 1999). Proteins and calories from animal products are a critical supplement to the rural diet. Women play a predominant role in the marketing of livestock products. Given the low labour requirements of livestock keeping, it can be easily combined with other income-generating activities. Moreover, economies of scale are relatively small in primary livestock production, but become more important in processing and marketing.

In LFAs, most widespread constraints to livestock intensification are the availability of feed, forage and fodder from different sources (McIntyre *et al.*, 1992). Fertility and mortality rates, gestation periods and inter-calving periods are influenced by feed intake adequacy, sanitary measures and infrastructure provisions (stalls), and veterinary care activities (Hengsdijk, 2002). Alternatives have been developed to reduce feed and water constraints, based on improved pasture management (area rotation, silvo-pastoral systems), production of leguminous fodder crops and the use of crop residues and industrial sub-products (e.g. feedblocks in Northern Africa, cottonseed in West Africa).

Success has been reached mostly in areas where increasing stocking rates – associated with higher population density and land scarcity – and more commercially oriented livestock production provoke a better delineation of grazing and watering rights. Moreover, in closed settled zones, livestock and crop systems become more integrated and land use intensity, as well as labour input, increases with population densities. Hoffman *et al.* (2001) found that, in remote areas of north-west Nigeria, indigenous strategies to exchange manure for crop residues with transhuman herders were still effective. Cattle played a key role in nutrient recycling, thus taking advantage of spatial and temporal variability.

In a similar vein, in intensive smallholder systems of upland Java, cattle are permanently maintained in backyards and fed with indigenous forage cut from field margins and roadsides (Tanner *et al.*, 2001). Intensive cropping cycles, high population densities and high livestock densities leave little land for grazing. Although cut-and-carry feeding is labour intensive, it is surprising that farmers collect quantities greatly in excess of the requirements of their livestock. The refused feeds associated with this 'excess feeding' are composted with animal manure for subsequent use on surrounding fields. The compost contributed to increased value by improving soil structure and water-holding capacity, in addition to providing soil nutrients. Poor, landless households participate in the cut-and-carry compost production system and thus gain an important supplementary income.

Direct income effects of livestock development projects tend to be attractive. Farmers participating in the Sichuan livestock development project in China (with 68,000 direct beneficiaries) saw their average net income rise by almost 50%. In a similar vein, farmers in the Mahreq and Maghreb programme in the Middle East reduced input costs for feeding by more than 50%. In Indonesia, the IAF/WB Smallholder Cattle Development Programme involved more than 70,000 smallholders in livestock keeping and showed an economic rate of return of 16% (Afifi-Affat, 1998).

Integration of cropping and livestock activities represents in semi-arid areas a main strategy for income diversification, asset creation and risk management. Livestock can reinforce arable cropping activities through manure provision and animal traction. Household welfare proved to be strongly dependent on the availability of both animal traction and simple implements for timely land preparation (Berckmoes *et al.*, 1990). Access to better-quality feed and fodder reduces the exclusive reliance on pastures and permits an increase in stocking rates. Public and private financial and extension institutions played a catalytic role in promoting sustainable intensification of mixed and integrated crop–livestock systems in Burkina Faso, Mali and Tanzania (Williams *et al.*, 1999). Subsidized feeding programmes may, however, easily lead to overstocking, and input subsidies artificially impose economies of scale.

Seed, breeding and biotechnology

The development of agroforestry and tree systems is strongly dependent on the availability of high-quality germoplasm and planting material of appropriate species that permit adequate planting intervals for specific agroecosystems. Agro-pastoral and mixed rainfed cropping systems could greatly benefit from new breeding technologies that increase crop tolerance to drought and extreme temperatures, and improve pests and disease resistance. Conservation tillage greatly benefits from herbicide-resistant varieties. Breeding for resistant varieties of maize and beans that are suitable for direct seeding holds potential environmental benefits for developing countries, where these crops are often grown on erosion-prone hillsides in LFAs.

Other prospects for genetically modified (GM) techniques are found in the control of major diseases in livestock, nitrogen fixation in cereals and new types of processed foods, etc. Cross-breeding and selection of native breeds are used as procedures for genetic improvement and gradual upgrading in animal systems, but artificial insemination and imported breeds can be applied to speed up the process. The performance of improved breeds tends to suffer from feeding and management problems, and needs to be accompanied by strong training and extension activities.

GM techniques are an increasingly important part of crop improvement strategies and can be potentially very useful for addressing some location-specific production constraints faced by the poor in marginal areas, for reducing environmental damage as well as for producing more nutritious foods that are of particular importance for addressing malnutrition in LFAs. Since it is unlikely that the private sector is willing to invest in research for areas with limited financial resources, higher public investment in agricultural research is needed. The same holds for marker-assisted breeding (not involving gene transfer) in combination with Information and Communication Technology (ICT), offering substantial promises for speeding up on-farm participatory breeding to evaluate the potential of different varieties for LFAs.

Biotechnology also brings new risks and problems. Most current agricultural biotechnology research is being undertaken by multinational companies and caters to the problems of rich farmers and developed-country consumers. Few outputs from this research will be appropriate for farmers in LFAs. Crop varieties with built-in herbicide resistance require much greater reliance on herbicides than is common in developing countries, where most weeding is still done by hand. Crop varieties that incorporate *Bt* genes for insect pest resistance need to be surrounded by refuge areas of non-*Bt* varieties if insects are not to become resistant. This may be hard to enforce in most developing countries. Biotechnology might also bring environmental risks associated with the release of genetically modified material (e.g. gene jumping, new pests) and from the consumption of genetically modified foods (e.g. allergic reactions, toxins). These risks are not yet fully understood and provoke a great deal of anxiety among some segments of the public. National institutions must have the capacity to evaluate these risks, and to implement and rigorously enforce appropriate regulatory systems for biosafety (Ruben *et al.*, 2003).

Livelihood Strategies and Development Pathways in LFAs

In recent years, development practitioners and researchers have emphasized the need to take into account the diverse set of activities that individuals, households and communities undertake to sustain and improve their livelihoods, and the underlying driving and conditioning factors that promote or hinder livelihood improvement, in order to more effectively address rural poverty, food insecurity and natural resource degradation in developing countries (see Brons *et al.*, Chapter 3, this volume).

The sustainable livelihoods framework (SLF), promoted by the Department for International Development of the United Kingdom (DFID), emphasizes the role of the vulnerability context and household assets (broadly defined to include physical, human, natural, social and financial assets) in determining the livelihood strategies of individuals and households (Ashley and Carney, 1999; DFID, 1999; see Fig. 1.2). In the SLF, livelihood strategies are defined as the range and combination of activities and choices that people make/undertake in order to achieve their livelihood goals (including productive activities, investment strategies, reproductive choices, etc.).

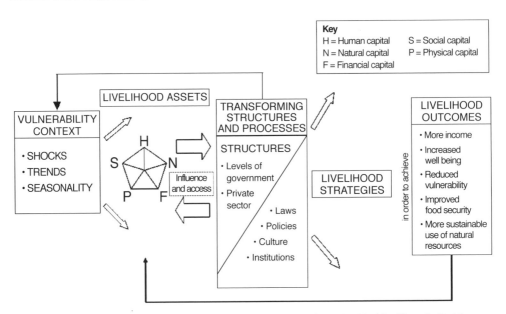

Fig. 1.2. Sustainable Livelihoods Framework (SLF) (from Sustainable Livelihoods Guidance Sheets (DFID), 1999).

Livelihood strategies are conditioned by transforming structures and processes (the institutions, organizations and policies that shape livelihoods by determining access to different types of capital, livelihood strategies and decision-making bodies, and by affecting the terms of exchange for and returns to investment in different types of capital) and affect outcomes for individuals and households (such as changes in income, well-being, vulnerability, food security and natural resource conditions), which feed back to affect their asset endowments over time.

Households (and individuals) may change their livelihood strategies over time as a result of changing opportunities and constraints. Pender (2004) defines a development pathway as a common pattern of change in livelihood strategies, focusing on community-level development pathways.[5] Examples of common development pathways found in the work of Pender and colleagues in Central America and East Africa include intensification of mixed food grains and livestock production, adoption or expansion of perishable horticultural cash crops, expansion of perennial cash crops and increased off-farm employment or non-farm activity in combination with continued food crop production (Pender, 2004).[6]

As with the SLF, this work emphasizes a complex set of factors influencing livelihood strategies at a given point in time and dynamic processes of change in livelihood strategies, affected by national level driving forces such as: (i) population growth, changes in technologies and market prices; (ii) local conditioning factors such as local population density, market access, agro-ecological conditions, local market development and local institutions and organizations; (iii) household level assets; and (iv) government policies, programmes and institutions affecting these (see Fig. 1.3). These factors influence households' choice of income strategy and natural resource management decisions (both part of livelihood strategies) which, in turn, affect outcomes including agricultural production, natural resource conditions and income and welfare, with feedback effects on the causal factors.

Although similar to the SLF, an important difference in this framework of relevance to the study of LFAs is its emphasis on local differences in geographical factors determining comparative advantage, such as agricultural potential and access to markets and infrastructure, in determining livelihood strategies. Based on these considerations, hypotheses can be derived concerning which types of livelihood strategies and development pathways have more potential and are likely to be pursued in different development domains, including LFAs of different types (Pender *et al.*, 2006a).

Key issues

These concepts and frameworks emphasize the heterogeneity of situations in rural areas of developing countries, including LFAs, and the complex set of factors that may influence livelihood strategies and development pathways and their outcomes. Several key issues arise from

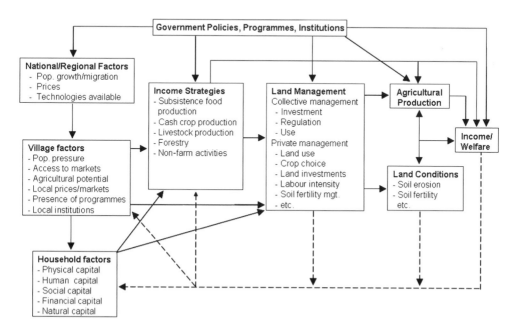

Fig. 1.3. Factors affecting income strategies, land management and their implications (from Nkonya *et al.*, 2004).

these frameworks as guides for policy-relevant empirical research in LFAs:

1. What livelihood strategies and development pathways are being/have been pursued by individuals, households and communities in LFAs?
2. What are the impacts of these strategies and pathways on important outcomes such as household income, food security, vulnerability and natural resource degradation?
3. What key factors determine which livelihood strategies and development pathways have been/are being pursued by individuals, households and communities in different contexts of LFAs? What has been the role of policies, institutions, programmes and organizations in promoting or inhibiting development pathways that lead to better outcomes?
4. What opportunities are there to pursue development pathways in the future that will lead to better outcomes? What are the key constraints to realizing these opportunities, and what is the role of policies, institutions, programmes and organizations in facilitating this?
5. What trade-offs or complementarities between different outcomes and across different target groups are likely to result from changes in policies, institutions, programmes and organizations to promote alternative development pathways?

Development pathways for LFAs

A substantial amount of empirical research has been conducted on many of these issues. We do not attempt a full review of all relevant research to these issues, but will highlight some key findings from recent literature and from some of the studies included in this volume (see Chapters 3, 8, 9 and 10, this volume).

The empirical literature has shown that a wide variety of livelihood strategies and development pathways are being pursued in LFAs (focusing particularly on income strategies). These usually involve staple food production as a primary or secondary activity, combined with an array of other types of activities, usually including livestock production and often including production of cash crops, off-farm employment and non-farm activities (Ellis, 2000; Nkonya *et al.*, 2004; Pender, 2004; Jansen *et al.*, 2006; Pender and Gebremedhin, 2006; Pender *et al.*, 2006b). Rural non-farm activities and off-farm employment are very important components of household income in many rural areas of developing countries (Reardon, 1997; Ellis, 2000; Barrett *et al.*, 2001; Reardon *et al.*, 2001; Ruben *et al.*, 2001; Stroosnijder and Van Rheenen, 2001; Ellis *et al.*, 2003; Ellis and Bahiigwa, 2003; Ellis and Mdoe, 2003; Nkonya *et al.*, 2004; Jansen *et al.*, 2006; Pender and Gebremedhin, 2006).

However, in many LFAs, such opportunities are limited due to limited access to opportunities in urban areas and limited opportunities in local markets resulting from low agricultural incomes (Reardon, 1997; Barrett *et al.*, 2001; Ruben and Pender, 2004). In addition, entry into more highly remunerated segments of the labour market is often hindered by the costs of lumpy investments in training, relocation and/or equipment that are usually required (Ruben and Pender, 2004). Given limited off-farm and non-farm employment opportunities and low agricultural productivity in many LFAs, dependence on food for work (FFW), cash for work (CFW) or other employment-generation or -assistance schemes can be quite high in some LFAs, such as drought-prone areas of Ethiopia (Pender and Gebremedhin, 2006). In such circumstances, income from migration and remittances is often quite important. Other natural resource-based activities, including forestry and fishing, are important in LFAs where such resources are available.

The outcomes of these different livelihood strategies vary across different contexts, although livelihood strategies based primarily on food production generally result in lower incomes and welfare conditions than more diversified livelihoods, especially where there are opportunities for cash crops, higher-value livestock production or non-farm activities (Ellis, 2000; Holden *et al.*, 2004; Nkonya *et al.*, 2004; Pender, 2004; Jansen *et al.*, 2006; Pender and Gebremedhin, 2006; Place *et al.*, 2006b). Consistent with the prior literature, Roa (Chapter 8, this volume) finds that non-farm income-earning opportunities (through handicrafts production) are critical to the welfare of households in a LFA in the hillsides of the Philippines, and are a key factor differentiating livelihoods and food security in the two communities studied. These findings strengthen the case that expanding the opportunities for non-farm income are critical for reducing poverty in LFAs.

Available evidence on the impacts of different livelihood strategies on natural resource management and degradation suggests that these impacts are highly context- and resource-specific. Pender *et al.* (2001b, d) found that adoption of improved land management practices and improvements in cropland quality and forest availability were greater, but that water availability was also more constrained, in communities where horticultural expansion was occurring than where basic grains production continued to dominate, consistent with the predictions of a bio-economic model developed for a community in central Honduras by Barbier and Bergeron (2001). By contrast, in Uganda, Pender *et al.* (2004) found little impact of increased horticultural production on most land management practices and indicators of changes in natural resource conditions; while in northern Ethiopia, Pender *et al.* (2001c) found that horticultural production was associated with improvements in soil fertility in croplands but reduced forest availability.

Production of perennial cash crops such as coffee and bananas is associated with adoption of several improved land management practices (particularly organic practices) in Uganda (Nkonya *et al.*, 2004; Pender *et al.*, 2004), the highlands of Kenya (Place *et al.*, 2006b) and northern Ethiopia (Pender *et al.*, 2001a, c), but with less adoption of several land management practices in Honduras (Jansen *et al.*, 2006). Nevertheless, soil fertility depletion is more rapid in perennial production (especially of bananas) in Uganda (Nkonya *et al.*, 2005b).

Off-farm and non-farm activities are associated with greater adoption of purchased inputs such as fertilizer and improved seeds in northern Ethiopia, but with less use of manure and compost, probably due to labour constraints (Pender and Gebremedhin, 2006). Similar results were found in central Honduras, where communities and households pursuing non-farm or off-farm activities were more likely to use agricultural chemicals but less likely to use mulching (Pender *et al.*, 2001b, e).

Access to off-farm employment has theoretically ambiguous impacts on farmers' investments in soil and water conservation (SWC) measures: it can reduce investment by increasing the opportunity cost of labour but can increase it by increasing farmers' ability to finance labour and other costs. The findings of Pender and Kerr (1998) for a village in semi-arid India and of Clay *et al.* (1998) for Rwanda support the argument that off-farm activities increase SWC investment. Kuiper and Ruben (Chapter 17, this volume) predict that increased access to employment opportunities through migration or CFW programmes will reduce erosion through a different mechanism; i.e. by reducing the intensity of agricultural production.

By contrast, the predictions of a bio-economic model for a community in northern Ethiopia support the expectation that off-farm income reduces investment in SWC and increases erosion (Holden *et al.*, 2004). For Uganda, non-farm activities are associated with increased fallowing and reduced labour use in one recent study (Nkonya *et al.*, 2005b), but have limited impact on land management or labour use in two prior studies (Nkonya *et al.*, 2004; Pender *et al.*, 2004). However, Pender *et al.* (2004) found that increase in non-farm activities was associated with community perceptions of reduced soil erosion and improved water availability and quality in Uganda, while Nkonya *et al.* (2004) found that dependence on off-farm income improved soil nutrient balances in eastern Uganda. More evidence is needed on such issues in different contexts to better identify the conditions under which non-farm or off-farm opportunities will reduce or increase land degradation.

As with the outcomes of livelihood strategies, the factors determining livelihood strategies are also context dependent, although some generalizations appear to follow from the available literature. For example, several studies using different methods in different countries find that higher-value cash crop production is more common in areas of higher agricultural potential and better access to markets and infrastructure (Pender *et al.*, 2001c, d, 2004; Nkonya *et al.*, 2004; Jansen *et al.*, 2006; Kruseman *et al.*, 2006; Place *et al.*, 2006a, b). Improved dairy production has also been found to be promoted by favourable agroclimatic conditions and market access in several studies in Kenya (Staal *et al.*, 2002; Place *et al.*, 2006a, b), although results from other East African countries are less clear (Pender *et al.*, 2006b).

Household-level factors, particularly human capital, also influence livelihood strategies. Education of the household head is particularly important in determining whether more highly remunerated off-farm salary employment or rural non-farm activities can be pursued (Barrett *et al.*, 2001; Reardon *et al.*, 2001; Nkonya *et al.*, 2004). Gender also is a very important determinant of livelihood strategies. In Ethiopia, for example, female-headed households are prevented by a cultural taboo from using oxen, which limits their ability to farm, and often results in such households sharecropping out their land (Benin, 2006; Pender and Gebremedhin, 2006; Tesfaye, Chapter 7, this volume).

In western Kenya, women are often left as land managers, while their husbands migrate for off-farm employment, but lack decision-making authority, which can limit their ability to invest in land improvements or make other agricultural investments (Place *et al.*, 2006b). In Uganda, female-headed households are more likely than male-headed households to use fertilizer, while households with more males use more of some labour-intensive land management practices (Jagger and Pender, 2006). Nevertheless, the difference in crop production between male- and female-headed households is insignificant in Uganda (Nkonya *et al.*, 2004), suggesting that female-headed households are able to overcome labour shortages in agricultural production by using other inputs. Over the longer term, however, women's lack of inheritance rights to land can undermine their ability and incentive to invest in land improvements.

In Honduras, female-headed households are less likely to be involved in off-farm employment and more likely to be focused on basic grain production as their income strategy, increasing the likelihood that such households will remain poor. Consistent with these and many other studies, Ali and Niehof (Chapter 9, this volume) find that gender inequality contributes to lower incomes of female-headed households and poor nutrition of women in Bangladesh, although there has been improvement in women's status as a result of many government and NGO programmes promoting improved female education, alternative livelihoods for women and recovery after a major flood. Although such social changes have not eliminated the gender gap, it appears to be narrowing. These findings suggest that it is possible to address problems caused by gender inequality through targeted programmes for women, although a long-term effort is likely to be needed.

Social capital also has a strong influence on livelihood strategies. In Uganda, differences in ethnicity are a major factor explaining differences in livelihood strategies, and this may be in part due to differences in social capital and experience as well as differences in consumer preferences across ethnic groups (Nkonya *et al.*, 2004). The success of horticultural development for export markets in central Kenya depended to an important extent on the presence of a local merchant class with considerable international trading experience, and upon development of long-term relationships (often personalized) between Kenyan exporters and overseas buyers and distributors (Jaffee, 1995). Dairy production in central Kenya depended heavily upon the organization of effective dairy cooperatives, supported by public policies (Staal, 1995). The absence of a vibrant cooperative sector in Ethiopia is one of the important factors explaining the more limited development of the dairy sector around Addis Ababa (Staal, 1995).

Milagrosa and Slangen (Chapter 10, this volume) contribute in an important way to the literature on social capital and its impacts on livelihood strategies by analysing the nature of social capital, its determinants and impacts on vegetable markets in a LFA in the Philippines. They identify several types of social capital in their study area, and find that bonding social capital is more important for vegetable farmers and that bridging social capital is more important for traders. They find several factors that are associated with the development of social capital, including gender, education, religion, age and ethnicity of the farmer, and argue that social capital is essential for the development of vegetable markets. These findings could be helpful in guiding efforts to target promotion of social capital in the Philippines and elsewhere.

Concerning opportunities to promote improved development pathways in LFAs, the means to facilitate this and potential trade-offs or complementarities among outcomes, several studies have addressed these issues in different LFA contexts using bio-economic modelling approaches. Barbier and Bergeron (2001) considered alternative policies and technologies to promote more profitable and sustainable land use in a micro-watershed in central Honduras. They found that infrastructure development (including roads and low-cost irrigation) and dissemination of improved technologies were critical contributors to the development of horticultural production in this community, which was helping to overcome the negative impacts of population growth and declining maize prices.

For a community in northern Ethiopia, Holden *et al.* (2005) considered alternative policies and programmes to promote sustainable development, and found that the most promising option was to promote tree planting on marginal lands combined with FFW programmes linked to investments in soil and water conservation, which could substantially increase household incomes while contributing to reduced land degradation. Other approaches, such as fertilizer credit and promotion of off-farm employment, had either limited or negative impacts on productivity and were predicted to reduce farmers' investments in SWC and increase erosion, leading to trade-offs between income and sustainability objectives.

Okumu *et al.* (2002) reached similar conclusions based on their bio-economic model of a community in central Ethiopia, i.e. that by combining tree planting on marginal lands with technologies to promote more productive crop production, substantial increases in productivity and farm income were possible whilst reducing land degradation. Woelcke *et al.* (2006) reached less optimistic conclusions using their bio-economic model of a maize-producing community in eastern Uganda, finding that soil nutrient depletion was likely to continue at a rapid pace even with very large subsidies on input or output prices. This is due largely to the low response of maize to various technological options tested in the study villages, leading to low profitability of efforts to promote such options for maize production.

By contrast, Woelcke *et al.* (2006) found that significant improvements in incomes and nutrient balances are possible by combining improvements in the efficiency of input and output markets with introduction of improved technologies, provision of credit and promotion of labour exchange to relax labour constraints, although nutrient depletion is likely to continue for most households. These findings highlight the importance of providing opportunities for higher-value crops and other livelihood opportunities to increase household welfare, but also demonstrate that such shifts in livelihood strategies may not be sufficient to ensure sustainable use of natural resources. Targeted efforts to address such problems, such as the interlinkage of FFW with conservation objectives, as investigated by Holden *et al.* (2005), may be necessary.

The findings of Okumu *et al.* (2002), Holden *et al.* (2005) and Woelcke *et al.* (2006) concerning the impacts of multiple interventions suggest (but do not demonstrate) that there may be synergies among different policies if combined. Kuiper and Ruben (Chapter 17, this volume) make an important contribution to the literature on this issue by investigating the impacts of alternative policies or programmes separately and then jointly, and comparing the outcomes. They find that, in most cases studied, there appear to be synergies among policies. For example, combining a cash for work programme with investments in infrastructure to reduce transaction costs leads to a substantially greater reduction in poverty than the sum of the impacts of these individual policies if implemented singly, because the income from the CFW programme helps poorer households to overcome asset limitations that otherwise constrain their ability to take advantage of new market options resulting from reduced transaction costs. This is an important insight, and the method developed to demonstrate such complementarities should be applied in future modelling efforts.

Market Structures and Institutional Development

Rural households in LFAs are embedded in a socio-economic environment characterized by market imperfections and institutional failure. Analysing the relationship between market access, potential agricultural productivity and development, Dorward (Chapter 19, this volume) envisages three broad pathways for escaping from poor, high-risk and thin markets. First, dramatic investment in market access infrastructure may create pathways through, e.g. extensive (in more remote areas) or intensive (in more accessible areas) livestock systems.

Such pathways are likely to be least demanding in terms of productivity potential while generating substantial marketable surpluses. A second type of pathway reflects a sudden shift in productivity potential, e.g. through investment in irrigation and/or the adoption of better technologies. To generate a marketable surplus, staple food production may be least demanding of market access (and productivity), but conventional cash crops will certainly be more demanding of both.

The third and probably most common type of pathway is characterized by more incremental technical, institutional and infrastructure investment and is positioned between the two other types. It most likely describes the potential for gradual livelihood improvements in many LFAs without creating major surpluses, but the development of horticultural crops for local and international markets is likely to fit this pathway as well, although conditional on substantial market access and marketing effort. In general, intensification of farming activities, diversification into higher-value crops and farm size (and structure) improvement need to be matched by improved access to markets in order to absorb surplus produce.

Similar arguments for market access apply for the diversification out of agriculture as part of a household's livelihood strategy (see also Niehof, 2004; Brons *et al.*, Chapter 3, this volume). These strategies or pathways can take two forms: (i) local reliance on non-agricultural activities, initially through labour diversification, but in case of self-employment requiring a certain amount of start-up capital; and (ii) migration to other areas to find employment in urban areas or more productive agricultural activities elsewhere. The migration strategy usually assumes relatively high capital endowments to finance new activities elsewhere (in addition to some education, communication and information networks). Sources of capital can be accumulated from past remittances, savings from agricultural surpluses or other sources of local wealth. In both strategies, overall labour mobility is likely to be constrained by initial inequalities in wealth and human capital coupled with an imperfect capital market.

Interestingly, all pathways for LFA development imply improved access to, or functioning of, product and factor markets, as documented in a number of case studies in this volume (see Chapters 12–15). Although the case studies usually focus on particular elements of a livelihood strategy, in reality a multitude of pathways or strategies can be observed in most regions, reflecting unique local comparative advantages (Pender, 2004). This phenomenon strengthens the argument for market development as a prerequisite that is largely independent of the particular pathway chosen.

The general need for market development also reduces somewhat the importance attached by different authors to appropriate strategies for LFAs. Thus, FAO/World Bank (2001) emphasizes exit from agriculture (migration), off-farm activities and on-farm diversification (in that order) as strategies for low-potential areas (defined as mainly rain-fed highlands and drylands). Dorward, Chapter 19, this volume, argues that (sustainable) intensification ('stepping up') migration and non-agricultural activities ('stepping out') are promising strategies. Whatever combination of strategies is adopted, the realization of an improvement of livelihoods in LFAs crucially depends on a better functioning product and factor markets.

Market imperfections

Markets are one particular form of exchange mechanisms, and the extent to which they tend to function better than other ways of coordination or exchange (formal and informal contracts, vertical integration, social networks) depends on a variety of factors. Ahmed and Peerlings, Chapter 16, this volume, mention transaction cost, information, uncertainty, specificity of transaction investment, frequency of exchange, number of actors, entry and exit cost, product heterogeneity and (natural resource) externalities as factors explaining market imperfections. With high transaction cost and limited information, product as well as factor markets may be thin or even missing. Poor infrastructure may cause shallow markets that are regionally dispersed and show high price fluctuations. Absence of credit facilities, e.g. because land markets are missing so that land has no collateral function, can lead to interlinked markets for, e.g. input supply, output and credit. Asymmetric information on the side of buyers (traders) notoriously results in imperfect competition favouring traders. Government interventions can seriously distort markets (and create opportunities for rent seeking).

As many of the case studies show, such market imperfections tend to be more frequent in and/or harmful for LFAs. As a consequence, LFAs suffer substantial efficiency losses compared with other regions, where a more favourable resource position has often enabled earlier market development. Studies by Fan and Chan-Kang (2004) show how investment in rural infrastructure can redress such inefficiencies by reducing transaction cost and capturing externalities. Their work points to high returns on investment and large contributions to poverty alleviation, relative to better situated areas, of roads and public utilities, education, research and development and, to a lesser extent, irrigation.

Product markets

Most of the cases focus mainly on product markets: (i) food staples (rice in China, Chen Le and Peerlings, Chapter 15, this volume); (ii) vegetables in the Philippines (Milagrosa and Slangen, Chapter 10, this volume); (iii) food and cash crops in Ethiopia (Jaleta and Gardebroek, Chapter 14, this volume); and (iv) apparel exports from Bangladesh (Ahmed and Peerlings, Chapter 16, this volume).

The transition in rice marketing from formerly state-owned trading companies towards a system with private traders in three villages in Jiangxi Province, south-east China, illustrates a variety of market imperfections due to unfavourable natural conditions and high transportation costs. After the reforms, markets remain thin, distorted and oligopsonistic, with traders suffering from a missing credit market. Moreover, local farmers are isolated from changes in the main consumption centres so that price transmission is incomplete. Trust and informal relationships (networks) therefore continue to play an important role in exchange between farmers and traders. However, communication by cell phones has reduced search cost and strengthened the position of surplus farmers. Overall, and depending on the nature of the initial market imperfections, those farmers situated closest to consumption centres who are able to offer larger surpluses and have good market access are expected to gain most from further liberalization and deregulation.

Vegetable farming (carrots, potatoes and cabbage) in the province of Benguet in the northern Philippines, an important production area, is characterized by poor market access due to low-quality roads and risk of landslides, and by limited agricultural potential as a result of soil depletion and lack of irrigation facilities. The area provides the bulk of vegetables for the national market, but local trading posts are incapable of handling harvest overflows (shallow markets), postharvest losses are high and physical market infrastructure for cold storage, handling and communication is insufficient. Despite strong interpersonal networks, trust within the farming community is low, and farmers are suspicious of traders in matters of weight, grade and price of vegetables, resulting in high transaction cost. The latter is related to asymmetric information, giving traders power over an individual farmer. Produce is normally sold using spot markets, but farmers prefer to deal with well-known traders with whom they are acquainted. These traders also supply them with inputs, cash advances and gifts, thereby creating interlinked transactions.

The Ethiopian study focuses on farm household land and labour decision making under imperfect factor and product markets in central and eastern Ethiopia. High transaction cost to enter factor or products markets causes the production and consumption decisions in a farm household to become non-separable. Households satisfy their own consumption needs first before considering allocating resources to high-value cash crops, the price of which can bear high transaction cost. Risk in those circumstances may further aggravate the wedge between buying and selling prices, making households in these areas almost autarkic.

The empirical results show that, in this environment of risk and high transaction cost, it is only those households who have access to their own farm capital (in this case a motor pump for irrigation) and non-farm income sources, that can afford to diversify into high-value crops. The resulting higher land and labour productivity encourages them subsequently to rent more land and employ more labour. The shift towards labour-intensive vegetables, in turn, contributes to higher incomes for landless workers.

The rapid increase of apparel exports by Bangladesh under the former (and discriminatory) Multi-Fibre Arrangement (MFA) illustrates the major advantages enjoyed by an exit-out-of-agriculture strategy for LFAs. Under the MFA, import quotas for apparel in the markets of Europe and North America were less restrictive for Bangladesh than for its competitors. As a result, Bangladesh enjoyed import tariff and import quota preferences that gave the country an (artificial) competitive edge over potentially stronger producers like India and China, as a result of an (intended) market distortion.

Simulations using a GTAP general equilibrium approach of different liberalization scenarios confirm the substantial welfare gains Bangladesh enjoyed under the MFA regime. Almost 2 million workers, more than 90% female, are currently employed by the export industry, of whom three-quarters are migrants from poor rural areas that are able to send substantial remittances back home. Better infrastructure and communications, as well as product diversification, are needed to improve the competitiveness of the sector and retain the past achievements under less-distorted market conditions.

Interlinked factor and product markets

Market configurations in many LFAs are characterized by strong imperfections at major factor markets, as well as interlinkages between factor and product markets, both for reducing transaction costs and for managing risk. The Ethiopia study (Jaleta and Gardebroek, Chapter 14, this volume) on decision making regarding food and cash crops shows how imperfections in the product market affect land and labour allocation within the household, driving a wedge between the factor's (lower) internal marginal value product and the possible remuneration elsewhere. Farmers are aware of this discrepancy, but high transaction costs in the product market justify the fulfilment of consumption needs from own resources, even if this means forgoing higher returns outside the household. Such interlinkages between different markets are typical of many LFAs, and have been illustrated in two more cases: (i) cooking banana, a staple and cash crop in Uganda (Bagamba *et al.*, Chapter 12, this volume); and (ii) participation in factor markets in the previously mentioned rice production area in Jangxi Province, China (Feng *et al.*, Chapter 13, this volume).

Cooking banana is a key staple food in Uganda and is produced mainly for own consumption in low-elevation areas and for both consumption and sale at higher levels of elevation. It has one of the best functioning commodity markets in the country. In response to poorly functioning financial and insurance markets, farmers react by diversifying their crops at the expense of banana, giving preference to subsistence crops, engaging in off-farm employment and limiting the extent to which they can grow labour-intensive (cash) crops. It is this behaviour that explains why, in areas suitable for banana production, farmers allocate more labour and land to sweet potato and cassava than to banana.

For the same reason, remote households allocate more resources to the production of coffee (to obtain cash) and maize (for home consumption), despite their low-value marginal products. As a result, farmers close to consumption centres and in competition with off-farm activities devote more of their household labour to low-value subsistence crops rather than to higher-value bananas for sale. In more remote areas, with limited off-farm opportunities, considerably more labour is used for banana production. It is therefore the functioning of other markets that enters into the explanation of factor allocation in banana production in different regions.

In a sequel to the earlier Chinese study, an analysis was made of village household participation in the land and labour market, showing that these decisions are taken simultaneously, thereby linking the two markets. To explain this phenomenon, the size of a farmer's land holding plays an important role. Households with small land endowments may not be wealthy enough to utilize off-farm employment, whereas those with larger land holdings are likely to have difficulties renting out their land and will prefer to work on-farm instead.

In general, the likelihood of being involved in migration shows an inverted U-shaped relationship with the average age of adults in the household. The turning point comes earliest for those that rent in land (and need labour), and comes latest for households renting out. The larger the household size and the fewer the number of dependents the more likely is migration, but not local, off-farm employment (probably because food requirements remain the same for local employment while having less labour on the farm). Durable household assets, possession of a land contract and access to a migration network all favour out-migration. As such, this case emphasizes the important role of a (temporary) 'stepping out' strategy to improve the livelihood of LFA households.

Other options: looking beyond agriculture

To improve living conditions in LFAs many development pathways are conceivable, as shown in this volume. The gradual conversion of low-productivity systems through suitable technologies, economic incentives (largely through well-functioning markets) and support services and institutional innovation, emphasizes the important role external factors can play. In general, what happens outside agriculture can be an important force for LFA development, as Lipper *et al.* (Chapter 18, this volume) argue. Using an induced innovation-type framework, they show for different categories of dry-land farming how economic development and trade outside an LFA, coupled with local investment in irrigation and a good market infrastructure, have transformed these farming systems into highly developed and competitive systems in many parts of the world.

Such transformations can currently be observed in some LFAs (and other rural areas for that matter) where the rapid development of the non-farm rural economy, with newly emerging markets and support services, pulls primary agriculture out of a state of low productivity (Hazell *et al.*, 2006). The rapid rise of supermarkets, changes in consumer preferences away from 'bulk' quality and the concurrent changes in food chain management and (international) outsourcing, have created new opportunities for those agricultural suppliers able to meet the technical, sanitary and phytosanitary standards of processing firms and exporters.

New challenges for LFA producers are emerging here, as the exchange mechanisms they face can be either contract farming or spot markets and are characterized by considerable product differentiation. Both institutional and market support have proved indispensable to farmers in order to satisfy successfully the requirements posed by modern food chains (Ruben *et al.*, 2006).

Another relatively new opportunity for LFA development is the provision of ecosystem services, as put forward in Lipper *et al.* (Chapter 18, this volume). To make a lasting contribution to the livelihoods of poor households, the provision of such environmental services needs to be integrated in the development pathways for specific LFAs, and funding based on willingness to pay needs to be mobilized to actually pay for these services as an additional source of income. In principle, such schemes internalize positive externalities thus far provided for free (and usually known to both private actors and local authorities). To the extent that property rights to land and water can be recognized and assigned to households, a market for these services can be created where previously none existed.

Public international demand for ecoservices like climate change mitigation, biodiversity conservation and the management of water resources is emerging (for instance, through the Global Environmental Fund) and, in some countries, national public demand is institutionalized (China's programme to reduce soil erosion and water pollution; Brazil's biodiversity conservation programme). Private sector purchase of ecoservices is developing as well, as are market-based implementation systems. What is currently lacking is knowledge about the type of institutions that can facilitate the exchange of environmental services for payments at low transaction cost and certified product quality. Certain NGOs provide these mechanisms, but the transaction cost remains prohibitively high for successful upscaling.

Implications for Policy and Research

Based on our review of the options for farming systems intensification and sustainable livelihoods that are available for rural households in LFAs, we can discuss the effectiveness of policy instruments that may support the resilience of LFA farming systems and their contributions for escaping spatial poverty traps in LFAs. Targeting of incentives towards resource-poor households and remote regions will be required to guarantee simultaneous increase in the returns to land and labour. Otherwise, institutional strategies for enhancing collective action and reducing transaction costs tend to be critically important for enhancing farmers' investments and enabling them to exploit their comparative advantage.

Policy incentives

Public investments in LFAs have the potential to generate competitive, if not greater, agricultural growth on the margin than comparable investments in many high-potential areas, with a greater impact on the poverty and environmental problems. Recent studies on India (Fan *et al.*, 2000), China (Fan and Hazell, 2001) and Uganda (Fan and Chan-Kang, 2004) have shown that many investments in LFAs now give comparable or higher returns than investments in irrigated and high-potential rainfed areas, having a greater impact on poverty as well. Targeting investments in roads, agricultural research and education in LFAs seems to offer good prospects for increasing productivity and contributes most to poverty reduction.

In Uganda, sizeable differences are also observed among investments across regions. Agricultural R&D and feeder roads appeared to be the most profitable investments. In terms of poverty reduction, all types of investments apart from health care generated the highest returns in the northern region, where most of the poor in Uganda live, whereas relatively small poverty impacts occurred in the most developed central region. In all regions, agricultural R&D was the most effective investment for cutting poverty. Feeder roads proved the second most effective investment whereas health had, overall, a small impact on poverty reduction.

On the whole, returns – in terms of agricultural productivity and poverty – are still quite sizeable in Uganda's most developed regions. This suggests that, in eastern Africa in some of the high-potential areas, returns to public investment are still high. These results contrast markedly with those for Asian countries, where for most investments the estimated returns in the less-developed areas are substantially higher than those in the high-potential areas.

The transformation of production systems and livelihood strategies in LFAs also requires a set of local incentives to enhance the responsive capacity of communities and households (Vitale and Sanders, 2002). Input support is frequently mentioned as an incentive for enhancing soil and water conservation investment and crop diversification in mixed cropping systems, and for pasture and tree crop improvement.

Results of input support measures are, however, rather mixed, especially when implicit subsidies are involved. Even while farmers have clear perceptions of the causes and effects of soil degradation, general adoption of land conservation measures rarely takes place.[7] Effective incentive regimes for sustainable land management and small-scale irrigation programmes offer support for covering sunk costs and, furthermore, focus on reinforcing the marketing structure (higher-value crops, crop diversification) as a device for enabling farmers to make the required investments. Furthermore, soil conservation and fertility strategies that effectively combine short- and long-term interests of farmers are likely to be better accepted by farmers (Erenstein, 1999).

Rural financial institutions, traditional rotating savings and credit schemes (ROSCAs) have proved successful in providing access to resources for agricultural intensification and income diversification. Local group-lending schemes can offer a cost-efficient system of insurance against risk, but tend to be biased in favour of wealthier farmers (Udry, 1990; Dercon, 1998). Institutional alternatives of area-based insurance that offer poor rural households a suitable risk management option have now been developed (Skees *et al.*, 1999).[8]

Animal systems usually require credit services with a long grace period. Drought assistance schemes are used to support down-stocking of animals in order to reduce stocking rates and to maintain cattle prices during periods of high market supply.[9] In Indonesia, livestock acquisition for landless farmers is promoted through in-kind credit facilities. Smallholder dairy credit in Kenya and Bangladesh provided through local NGOs was successful in increasing productivity and income, focusing on simultaneous improvements in genetics, milking equipment, feed supplements, traction implements, marketing and processing (De Haan *et al.*, 2001).

Water charges can be introduced as a step towards efficient water distribution according to the real opportunity costs. In perspective, tradeable water rights are considered an effective mechanism for optimizing water allocation, although measurement problems still inhibit a more general application (Pingali and Rosegrant, 2001). Watershed protection programmes that focus attention on hillside and upland conservation and sediment control may offer the additional benefits of hydro-energy production and wildlife and biodiversity conservation.

Food-for-work activities are frequently used for rural infrastructure construction and the creation of soil and water conservation structures. Drought-relief interventions that rely on food-for-work programmes can be helpful in relieving the pressure on natural resources. The total costs of food-for-work programmes are, however, extremely high (transport costs commonly take half of the resources) and participants typically accept a lower cash wage to do the same work (Barrett and Maxwell, 2005).

Perennial tree crop and agro-pastoral systems can be stimulated through sales of environmental services. Agroforestry and improved pastures reduce the build-up of carbon dioxide and other greenhouse gases (Dixon, 1995). Widespread adoption of agroforestry and other methods of improved soil management in the tropics could lead to a relative increase in soil carbon storage of about 1%. Indirect effects can be much greater, reducing carbon dioxide emissions caused by forest clearance. In a similar vein, mixed cropping systems in highland and rainfed areas could generate local markets for environmental services for their contributions to water management at watershed level.

Promotion of integrated pest (IPM) and nutrient management (INM) practices in mixed highland and rainfed systems requires a thorough training and participatory involvement of farmers to create sufficient knowledge about biological and agronomic control practices. Farmers field schools (FFS) that have offered such training in various Asian countries provided a useful framework for experimental learning based on field trials with different management practices, followed by joint damage or impact assessment. Recent evidence indicates, however, that poor upscaling has seriously reduced the impact of the FFS programme in Indonesia.

Market development based on the domestication of trees in agroforestry systems offers some prospects for non-timber forest products, like fruit, flesh, kernels and seedoils. ICRAF identified promising perspectives for particular products in the humid and semi-arid lowlands of West Africa, the southern African plateau and in the Amazon region (ICRAF, 1999). Success depends on further improvement of plant traits (clonal forestry), increasing the length of the productive season, reducing tree height and improving yield and product quality. More importantly, access to export markets requires close collaboration with the food industry engaged in the development of novel foods (e.g. antimicrobial properties of some kernels for dental toothpaste, fruit pulp applications in processed foods, etc.).

Finally, improved chain integration can be helpful in creating prospects for sustainable resource management for tropical fruit, fish and vegetables based on stable access to markets and information that enables additional investment in quality management for increasing value added (Kuyvenhoven and Bigman, 2002). While public market information systems formerly received much attention, reliable private marketing arrangements are now considered more important for enhancing stable market access. Local processing also creates major rural, non-farm employment opportunities. New procedures and practices for organizing food supply networks – with direct contractual ties between primary producers, processors and retailers – have emerged to cope with food quality, safety and health demands (Glover, 1990; Key and Runsten,

1999). These practices require the development of grades and standards, together with agreements on best practices.

Property rights and community organization

Supporting the process of sustainable rural development and agricultural intensification in highland and dryland areas requires institutional structures that guarantee local stakeholders equitable access and legally secure entitlements to assets, knowledge and information (Knox *et al.*, 2002). This would create the necessary conditions for demand-driven and participatory processes of local technology development that are responsive to the needs of poor people. Community social capital is of primary importance for gaining access to markets and exchange networks. Land use intensification in LFAs will become more market oriented when local farmers' groups are able to achieve better market integration for finding stable and rewarding outlets.

When the property rights of the rural poor to critical assets like land, water, trees and pastures are sufficiently ensured and local communities are able to exercise control over their resources, economic incentives for initiating an endogenous process of technological innovations can be set in motion. Improved resource management strategies for LFAs require effective enforcement of property rights and a high degree of collective action. Agroforestry and perennial tree crops are long-term investments, and individual farmers will only plant trees if they have secure (land or tree) property rights or leasehold arrangements enabling them capture the future returns from their investment.

Improving shifting cultivation in highland areas relies on improved fallow, with investments in contour ridges, formation terraces and fast-growing, multi-purpose tree species that provide early returns. Control of traditional slash-and-burn agriculture within community-based watershed development and integrated nutrient and pest management programmes puts a high demand on collective action. Migratory herding and transhumant pastoralism in densely populated semi-arid areas increasingly face land conflicts with rural communities regarding grazing and water rights, and grazing codes are put forward for conflict resolution. Agro-pastoral systems that use simple erosion control measures, water-harvesting technologies and drought-resistant varieties can be adopted by individual farmers as long as stable use rights are in place. Mixed rainfed cropping in (semi-)arid areas relying on soil and water conservation structures and weed control measures requires the consolidation of fragmented holdings and regulatory measures regarding access to land and water at community level.

In highland LFAs, property rights and tenure security strongly influence prospects for agroforestry development. However, customary land tenure systems in East Africa and Asia need not to be an impediment for agroforestry investment (Bruce and Migot-Adholla,

1993). In practice, land rights have become more individualized after investments in tree planting took place. Community forestry systems are gradually evolving towards individual tenure regimes (Otsuka and Place, 2001). Migrant herders rely on mobility of cattle over long distances as a strategy to take advantage of spatial ecosystem variability. Recognition of customary rights (pastoral codes) can be an effective way of reducing overgrazing, but fiscal and price instruments (pasture fees) are less effective in controlling stocking rates. In Mongolia, the legal recognition of customary forms of pasture tenure provides better security to nomads. Similarly, artisanal fisheries communities define territorial use rights for matching catching capacity with resource productivity, guaranteeing community members sequential access to different fisheries areas (Allison and Ellis, 2001).

Pastoral organizations have also been successful in mobilizing producers around input provision, but proved less effective in establishing rangeland management procedures; customary institutions show considerably better performance in this respect. Grazing and water fees have been faced with implementation difficulties, and long-term leasehold contracts proved to be a more effective means for controlling pasture degradation. Upgrading of feed and water resources is most successful in livestock programmes that focus on marketing and processing in areas with increasing land scarcity. Forward market linkages thus seem to be an important driver for technology adoption.

The reversal of rangeland degradation with agro-pastoral practices requires clear regimes of property or grazing rights. Common property of arid rangelands with well-identified membership, entry boundaries and management rules and regulations can be an effective vehicle for establishing coordination between independent livestock owners (McCarthy *et al.*, 1999). In addition, insurance systems that are able to reduce risk support the adjustment of stocking levels and the intensification of pasture management, although at the cost of greater heterogeneity amongst herders and possibly reduced willingness for cooperation (Hazell, 1999).

Secure tenure arrangements are especially important in encouraging farmers to make major investments in soil fertility management and watershed conservation. Well-defined (individual or community) property rights provide incentives to farmers to control land degradation and to enhance their efforts in appropriate soil and water conservation measures. In Niger and Burkina Faso, secure family fields receive twice as much manure and fertilizers as village fields (Stroosnijder and Van Rheenen, 2001). In South-east Asia, short-term sharecropping arrangements may hinder investments in soil and water conservation, but long-term leasehold contracts provide substantial positive incentives for improved soil fertility management (Hayami and Otsuka, 1993).

Local organizations have been particularly successful in empowering communities in overcoming social and institutional constraints. Improving resource management in LFAs using participatory research methods can build on farmers' own knowledge and experiences (Goma *et al.*, 2001; Zurayek *et al.*, 2001). Watershed protection programmes that focus attention on hillside and upland conservation and sediment control ask for a participatory approach involving a broad coalition of all stakeholders, as well as a clear recognition of the 'public goods' character of watershed services (De Graaff, 1996; McNeely, 2001). Decentralization of management and community participation contributed to the success of watershed programmes in India (Kerr *et al.*, 2001). Additional financial resources could be mobilized through the generation of hydro-energy, tourism and the sale of environmental services like wildlife, biodiversity conservation and carbon fixation.

Successful implementation of water management programmes that benefit poor households is highly dependent on the regulation of water rights and the procedures for water distribution. The social capital of village institutions in Asia provides a wide basis for putting user organizations in charge of water management and distribution (Meinzen-Dick, 1997). Clear rules and regulations, as well as proper implementation and enforcement strategies, are needed to guarantee user participation. In a similar vein, control of soil erosion requires an institutional framework for nutrient management at regional level in order to control off-site effects. Strong community organization amongst farmers is also required for controlling plant diseases that are easily spread, and therefore require collective action. Early detection and understanding of the field ecology enables farmers to determine critical levels of infestation and the appropriate moment when interventions are required.

Institutional development of service provision, extension and training are essential components for sustainable intensification. Decentralization of authority and empowerment of local communities are of key importance in overcoming resource constraints and contributing to capacity building. Private delivery of inputs and services can be effective for reaching poor households, although sanitary services related to disease surveillance and food safety surveillance maintain a public goods character (Umali *et al.*, 1992). Community social capital is often of primary importance in gaining access to markets and exchange networks. Farming systems intensification in LFAs can become more market driven when local farmers groups are able to establish stable and rewarding outlets. However, for those services where missing markets persist (e.g. for carbon fixation, food safety surveillance and insurance), collective action, e.g. through state intervention, is still required.

Policy targeting

The identification of particular geographical areas for targeting LFA development efforts is based on the assumption that location is a prime determinant for poverty.[10] Whereas large inequalities in living standards and poverty incidence are registered between different geographical settings, focusing attention on such 'backward areas' can substantially improve targeting efficiency (Bigman and Fofack, 2000).

Empirical evidence derived from several studies on geographical poverty traps indicates that geographic factors (residence) have a strong and significant effect on household wealth and consumption (Ravallion and Jalan, 1996; Minot, 2000). Initial conditions consistently reduce returns to private investment, and the stock of community capital has a strong negative effect on the productivity of private investment. In addition, health status is usually lower and educational levels are limited, thus further reducing the returns to labour. This suggests that households living in LFAs face critical geographical constraints and meet consumption levels that are substantially lower than those of otherwise identical households in better areas.

Whereas broad spatial targeting based on access and distance (road density) criteria can be effective for reducing poverty due to adverse (man-made and natural) geographic variables, differences in the distribution of private assets may negatively influence income distribution (Escobal, 2005). Therefore, complementarities between social and physical infrastructure and the role of local institutions need to be seriously considered. Other studies suggest that substantial disaggregation is required and that attention should be focused on relatively small administrative units (districts, villages), since a large part of the variation in household income can be attributed to within-village differences in resource endowments (Baker and Grosh, 1994; Jayne et al., 2003; Elbers et al., 2004).

Within the framework of livelihood studies, far more attention is given to social heterogeneity and the identification of 'people at risk'. Vulnerable population groups are identified according to criteria of gender, age, ethnicity, household size and education. It is noted, however, that some of these individual characteristics tend to be geographically correlated, since disadvantaged groups are likely to be concentrated in remote and less productive areas (Van der Walle and Gunewardena, 2001). Given equal endowments and location, disadvantaged and minority groups still receive lower returns to given individual and household characteristics. Lower returns may, however, be contested by behavioural responses that partly compensate for geographical disadvantages. Even while the absolute advantage of poor areas is limited, a wide diversity of development pathways can be identified based on specific local, and individual relative, advantage (Pender, 2004).

Frontiers for future research[11]

Many of the studies demonstrate the primary importance of identifying profitable options for farmers in LFAs, if poverty is to be reduced. Despite the importance of profitability of livelihood options, there is still a dearth of information about this. Recent literature has shed light upon the profitability of some options in particular circumstances, but much more needs to be known in order to develop effective targeted interventions. There is still limited systematic and reliable information collected on a regular basis about the profitability of different crop, livestock or forestry combinations in different types of LFAs, or the profitability of land management practices linked to these different livelihood activities. Beyond estimating private profitability, information on the social profitability of alternative activities in different domains is also needed, taking into account externalities, market price distortions and non-marketed inputs and outputs.

In addition, the social profitability of alternative public programmes and investments also needs to be better understood, to help guide development investors and governments as to where the highest returns can be expected. The returns, costs, risks and social and environmental impacts of other public investments – such as investments in infrastructure, education, agricultural research and extension, and others – are also not well quantified. More research is needed to estimate the costs, risks and social and environmental impacts of alternative investments.

To better assess such impacts, more long-term research with panel data sets and dynamic models is needed to better understand the dynamic relationships between: (i) policy and programme interventions; (ii) local institutions and endowments of physical, human, natural, financial and social capital; (iii) community and household responses in terms of collective action, livelihood strategies and land management practices; (iv) changes in production, income, land degradation and other outcomes; and (v) the feedback effects of these responses and outcomes on interventions, local institutions and endowments, and future responses. It is difficult to know the extent to which communities and households are trapped in a downward spiral or poverty trap, stagnation, a virtuous upward spiral or other kind of dynamic development path, or what the most effective interventions will be to promote sustainable development, without better diagnosis of the problems and the key causal factors and feedback relationships that are driving them.

For example, some communities may be falling deeper into poverty and depleting all of their endowments as a result of a lack of sufficiently profitable investment opportunities for any type of capital. Unless profitable investments of some kind can be identified, a sustainable development solution may not be possible without promoting large-scale emigration out of such areas. In other cases, communities and households may be depleting their natural capital but investing in other forms of capital that are yielding higher returns (Pender, 1998). Such a development path may be sustainable as long as households are aware of the depletion of natural capital, and will eventually address it adequately as the returns from investing in natural capital increase relative to the returns from investing in other types of capital (Pender, 1998). Alternatively, they may not be sufficiently aware of the depletion, may not have adequate incentive or ability to address it because of externalities or other market failures, or may be crossing a threshold into a poverty–degradation trap in which the costs are too high or the marginal returns too low to maintain or restore the natural capital stock (Pender, 1998; Barrett *et al.*, 2002).

In order to prescribe effective actions, it is essential to diagnose the problem correctly. If there is sufficient awareness and no major market failure, the problem is likely to take care of itself as the relative returns to investment in different types of capital adjust (Pender, 1998). If there is insufficient awareness of the degradation problem, educational and technical assistance approaches may be sufficient to solve it. If the problem is due to market failures or a degradation trap, more intervention will be necessary to address these causes. Without more information to diagnose what kind of dynamic situations that communities and households are facing, it will be difficult to prescribe effective remedies.

Even without dynamic information, however, it would be very useful to identify areas and household types for which profitable livelihoods and land management practices are feasible, but are not being pursued. Where such untapped potentials exist, it is useful to investigate the reasons why, and identify the extent to which policies, public investments and programmes could facilitate fulfilment of these potentials.

Investigation of synergies among different policies using a modelling approach, as demonstrated by Kuiper and Ruben (Chapter 17, this volume), can be very useful for this. Past research and the research in this book has identified some examples of such potentials, such as production of high-value commodities in areas of high potential close to urban markets and tree-planting activities in many other areas, and has also provided some insights into the reasons why such potentials are not being more widely exploited. Further case study research into these and other promising livelihood options could yield valuable insights.

More historical case study research investigating the dynamics of changes in income strategies, land use, land management, land degradation, productivity and welfare outcomes – such as the influential case study of Machakos by Tiffen *et al.* (1994) – would also be valuable. Such long-term historical studies can yield a wealth of insights into the processes of land degradation or improvement, and also into key driving forces and responses that are not achievable using only cross-sectional surveys of the type emphasized by many of the studies in this book. However, the conclusions of such a focused case study can easily be over-generalized.

Similar studies are needed in different development domains and different historical, political and social contexts to draw more robust and generalizable conclusions about the dynamics of livelihoods and resource degradation, causes and responses in different LFAs. A combination of quantitative survey and qualitative case study research methods, building on the strengths and addressing the weaknesses of each approach, is more likely to produce clear and robust conclusions than reliance on any single approach.

Finally, the scale of interventions and their impacts also need to be better understood, and have been addressed in several studies (see Lopez Ridaura *et al.*, Chapter 2, and Dorward, Chapter 19, this volume). Interventions that are able to increase production and household income when pursued on a small scale may lead to quite different impacts when implemented on a large scale.[12]

Research tracing impacts across scales is needed, from the assessment of impacts of policy and programme interventions on adoption decisions at the plot – and household scale and their implications for local natural resource conditions – to the impacts on prices and other outcomes at the community, national and regional scales. The feedback effects occurring between these scales must be better understood and accounted for in planning interventions, if the benefits of such interventions are to be maximized and unintended negative impacts are to be minimized. The use of integrated bio-economic models at farm-household, community and higher scales (as illustrated by Jansen *et al.*, Chapter 6 and Kuiper and Ruben, Chapter 17, this volume) is likely to be essential for a better understanding of these impacts.

Endnotes

[1] LFAs also carry a disproportionate burden with vector-borne diseases (e.g. malaria, plague) because of lack of preventive and curative care, and are affected by the threats of HIV/AIDS, in part because return migrants bring the disease back home. The associated decline in the available labour force of prime working age is leading to rising dependency ratios, and available labour for maintaining and improving natural resources becomes strongly reduced (Niehof, 2004). Income shortfalls affect food consumption, while seasonal labour shortages lead to the removal of children from school and the decline of gross primary enrolment rates. Rising health expenditures for medical treatment force households to deplete their savings and eventually sell assets. In addition, agricultural knowledge and farm management skills get lost and traditional social security systems may become disrupted.

[2] Walker *et al.* (2000) assert that this is not the case in the northern Andes region, where agro-ecological and market conditions are more favourable in the Andes mountains.

[3] The UNCCD defines drylands as arid (with average length of growing period (LGP) < 60 days), semi-arid (LGP 60–119 days) or dry sub-humid (LGP 120–179 days) areas.

[4] The figures for shares of land seriously degraded are from Scherr (1999), who combined the GLASOD categories of 'moderately degraded', 'strongly degraded' and 'extremely degraded' into the class of 'seriously degraded'. Oldeman *et al.* (1991) defined moderately degraded soils as having suffered greatly reduced productivity, but which were still suitable for use in local farming systems; strongly degraded

soils as soils in which productivity is virtually lost and are not suitable for agricultural use without major restoration investments; and extremely degraded soils as 'human-induced wasteland' beyond restoration.

5   Both livelihood strategies and development pathways may be studied at different scales, such as at community, household or individual level.

6   The empirical work of Pender and colleagues focuses on changes in income strategies and natural resource management, which are subsets of the more encompassing notion of livelihood strategies as defined by DFID (1999). Similar concepts have been introduced by other researchers. For example, Scoones and Wolmer (2002) refer to heterogeneous pathways of change in crop–livestock systems in different contexts of sub-Saharan Africa, including: (i) development of mixed crop–livestock integrated systems; (ii) integration of communal rangelands and individualized crop production; (iii) specialization and separation of extensive livestock and crop production; (iv) separate intensification of crop and livestock production; and (v) abandonment of cattle production with intensification of small-farm garden agriculture and off-farm income. Dixon *et al.* (2001) refer to household poverty reduction strategies, including agricultural intensification, diversification, increased farm size and increased off-farm income. While there are differences among these concepts, they are similar in emphasizing dynamic processes of change in livelihoods.

7   Local projects rely on direct and indirect incentives (input subsidies, free seed provision, food-for-work programmes, etc.) to stimulate adoption, but maintenance of SWC structures is easily abandoned once these facilities are phased out (Feder *et al.*, 1985).

8   Such schemes, already used in Mexico and piloted in India, index contracts to measurements at local weather stations rather than individual experiences, and can be provided by the private sector without the need for subsidies.

9   Early interventions of this kind in the Isiolo district Kenya proved to be rather effective for maintaining purchasing power of pastoralists and stabilizing the local livestock sector. After drought, credit for new foundation stocks is provided to farmers in order to re-establish their activities (De Haan *et al.*, 2001).

10 Areas can be defined according to place and space characteristics. *Place* refers to climate and soil conditions that limit returns to agricultural production and render yields highly uncertain, especially under conditions of high population density. *Space* refers mainly to distance to markets and services, occasioning high transaction costs. In both settings, barriers to migration – either because moving is costly and risky or because people cannot move due to local patronage systems – could easily lead to spatial poverty traps. Thin land markets and barriers to borrowing further reduce the prospects for escaping from poverty.

11 This subsection draws heavily from Pender *et al.* (2006b).

12 For example, rapid adoption of improved maize varieties and fertilizer in Ethiopia in the early 2000s led to a dramatic fall in maize prices and farmers' disillusionment with the technology.

References

Abegaz, A. (2005) Farm management in mixed crop–livestock systems in the Northern Highlands of Ethiopia. Tropical Resource Management Papers No. 70, PhD dissertation, Wageningen University and Research Center, The Netherlands.

Afifi-Affat, K.A. (1998) Heifer in trust: a model for sustainable livestock development? *World Animal Review* 91, 13–20.

Allison, E.H. and Ellis, F. (2001) The livelihoods approach and management of small-scale fisheries. *Marine Policy* 25, 377–388.

Ashley, C. and Carney, D. (1999) *Sustainable Livelihoods: Lessons from Early Experience.* Department for International Development, London.

Baijukya, F.P. and De Steenhuijsen Piters, B. (1998) Nutrient balances and their consequences in the banana-based land use systems of Bukoba District, northwest Tanzania. *Agriculture Ecosystems and Environment* 71, 149–160.

Baker, J.L. and Grosh, M.E. (1994) Poverty reduction through geographical targeting: how well does it work? *World Development* 22, 983–995.

Barbier, B. and Bergeron, G. (2001) *Natural Resource Management in the Hillsides of Honduras: Bioeconomic Modeling at the Microwatershed Level.* Research Report No. 123, International Food Policy Research Institute, Washington, DC.

Barrett, C.B. and Maxwell, D.G. (2005) *Food Aid after Fifty Years: Recasting its Role.* Routledge, London.

Barrett, C.B., Reardon, T. and Webb, P. (2001) Non-income diversification and household livelihood strategies in rural Africa: concepts, dynamics and policy implications. *Food Policy* 26, 315–331.

Barrett, C.B., Place, F. and Aboud, A. (2002) The challenges of stimulating adoption of improved natural resource management practices in African agriculture. In: Barrett, C.B., Place, F. and Aboud, A.A. (eds) *Natural Resources Management in African Agriculture.* ICRAF and CABI, Nairobi, Kenya.

Bassett, T.J. and Crummey, D. (2003) Contested images, contested realities: environment and society in African savannas. In: Bassett, T.J. and Crummey, D. (eds) *African Savannas: Global Narratives and Local Knowledge of Environmental Change.* James Currey, Oxford.

Bationo, A., Lompo, F. and Koala, S. (1998) Research on nutrient flows and balances in West Africa: state-of-the-art. *Agriculture Ecosystems and Environment* 71, 19–35.

Bekunda, M.A., Bationo, A. and Ssali, H. (1997) Soil fertility management in Africa: a review of selected research trials. In: Buresh, R.J., Sanchez, P.A. and Calhoun, F. (eds) *Replenishing Soil Fertility in Africa.* Soil Science Society of America, Madison, Wisconsin.

Benin, S. (2006) Policies and programs affecting land management practices, input use and productivity in the highlands of Amhara region, Ethiopia. In: Pender, J., Place, F., and Ehui, S. (eds) *Strategies for Sustainable Land Management in the East African Highlands.* IFPRI, Washington, DC.

Berckmoes, W.M.L., De Jager, E.J. and Kone, Y. (1990) *L'Intensification Agricole au Mali–Sud: Souhait ou Réalité?* Bulletin no. 318, Royal Tropical Institute, Amsterdam.

Bigman, D. and Fofack, H. (2000) *Geographical Targeting for Poverty Alleviation: Methodology and Applications.* World Bank, Washington, DC.

Bird, K., Hulme, B., Moore, K. and Sheperd, A. (2002) *Chronic Poverty and Remote Rural Areas.* CPRC Working Paper No. 13, Chronic Poverty Research Centre, Birmingham and Manchester, UK.

Bishop, J. (1995) *The Economics of Soil Degradation: an Illustration of the Change in Productivity Approach to Valuation in Mali and Malawi.* LEEC paper DP 95-02, International Institute for Environment and Development, London.

Bishop, J. and Allen, J. (1989) *The On-site Costs of Soil Erosion in Mali.* Environment Working Paper No. 21, Environment Department. The World Bank, Washington, DC.

Bojö, J. (1991) *The Economics of Land Degradation: Theory and Applications to Lesotho.* The Stockholm School of Economics, Stockholm.

Bojö, J. (1996) The costs of land degradation in Sub-Saharan Africa. *Ecological Economics* 16, 161–173.

Bojö, J. and Cassells, D. (1995) *Land Degradation and Rehabilitation in Ethiopia: a Reassessment.* AFTES Working Paper No. 17, World Bank, Washington, DC.

Boxall, R.A. (2001) Postharvest losses to insects: a world overview. *International Biodeterioration and Biodegradation* 48, 137–152.

Bruce, J. and Migot-Adholla, S. (eds) (1993) *Searching for Land Tenure Security in Africa.* Kendall/Hunt Publishing Company, Dubuque, Iowa.

Brummett, R.E. (2002) Realizing the potential of integrated aquaculture: evidence from Malawi. In: Uphoff, N. (ed.) *Agroecological Innovations: Increasing Food Production with Participatory Development.* Earthscan, London.

Bunch, R. (2002) Increasing productivity through agro-ecological approaches in Central America: experiences from hillside agriculture. In: Uphoff, N. (ed.) *Agro-ecological Innovations: Increasing Food Production with Participatory Development.* Earthscan, London.

CGIAR/TAC (2000) *CGIAR Research Priorities for Marginal Lands.* Consultative Group on International Agricultural Research/Technical Advisory Committee, TAC Secretariat, Food and Agriculture Organization of the United Nations (FAO), Rome.

Chaves, B. and Riley, J. (2001) Determination of factors influencing integrated pest management adoption in coffee berry borer in Colombian farms. *Agriculture, Ecosystems and Environment* 87, 159–177.

Clay, D.C., Reardon, T. and Kangasniemi, J. (1998) Sustainable intensification in the highland tropics: Rwandan farmers' investments in land conservation and soil fertility. *Economic Development and Cultural Change* 46, 351–378.

Cleaver, K. and Schreiber, G. (1994) *Reversing the Spiral: the Population Agriculture, and Environment Nexus in sub-Saharan Africa.* World Bank, Washington, DC.

Cohen, M., Shepherd, K. and Walsh, M. (2005a) Empirical reformulation of the universal soil loss equation for erosion risk assessment in a tropical watershed. *Geoderma* 124, 235–252.

Cohen, M.J., Brown, M.T. and Shepherd, K.D. (2005b) *Estimating the Environmental Costs of Soil Erosion at Multiple Scales in Kenya using Energy Synthesis.* Mimeo, International Centre for Research in Agroforestry, Nairobi.

Convery, F. and Tutu, K. (1990) *Evaluating the Costs of Environmental Degradation: Ghana – Applications of Economics in the Environmental Action Planning Process in Africa.* University College Dublin, Environmental Institute, Dublin.

Crosson, P.R. (1995) *Soil Erosion and its On-farm Productivity Consequences: What do we Know?* Resources for the Future Discussion Paper 95-29, Resources for the Future, Washington, DC.

Cuesta, M.D. (1994) Economic Analysis of Soil Conservation Projects in Costa Rica. In: Lutz, E., Pagiola, S. and Reiche, C. (eds) *Economic and Institutional Analyses of Soil Conservation Projects in Central America and the Caribbean.* A CATIE-World Bank Project, World Bank Environment Paper 8, The World Bank, Washington, DC.

Dasgupta, S., Deichmann, U., Meisner, C. and Wheeler, D. (2003) *The Poverty/ Environment Nexus in Cambodia and Lao People's Democratic Republic.* World Bank Policy Research Working Paper 2960, World Bank, Washington, DC.

Deckers, J. (2002) A systems approach to target balanced nutrient management in soilscapes of Sub-Saharan Africa. In: Vanlauwe, B., Diels, J., Sanminga, N. and Merckx, R. (eds) *Integrated Plant Nutrient Management in Sub-Saharan Africa: from Concept to Practice.* CAB International, Wallingford, UK.

Defoer, T., De Groote, H., Hilhorst, T., Kanté, S. and Budelman, A. (1998) Participatory action research and quantitative analysis for nutrient management in southern Mali. *Agriculture Ecosystems and Environment* 71.

De Graaff, J. (1996) *The Price of Soil Erosion. An Economic Evaluation of Soil Conservation and Watershed Development*. Mansholt Studies No. 3, Wageningen Agricultural University, Wageningen, The Netherlands.

De Haan, A. and Lipton, M. (1998) Poverty in Emerging Asia: progress, setbacks, and logjams. *Asian Development Review* 16, 2.

De Haan, C., Van Veen, T.-S., Brandenburg, B., Gauthier, J., Le Gall, F., Mearns, R. and Simeon, M. (2001) *Livestock Development: Implications for Rural Poverty, the Environment and Global Food Security*. World Bank, Washington, DC.

De Jager, A., Kariuki, I., Matiri, F.M., Odendo, M. and Wanyama, J.M. (1998) Monitoring nutrient flows and economic performance African farming systems (NUTMON). IV. Linking nutrient balances and economic performance in three districts in Kenya. *Agriculture Ecosystems and Environment* 71, 81–92.

De Jager, A., Onduru, D. and Walaga, C. (2004) Facilitated learning in soil fertility management: assessing potentials of low-external-input technologies in East African farming systems. *Agricultural Systems* 79, 205–223.

Delgado, C., Rosegrant, M., Steinfeld, H., Ehui, S. and Courbios, C. (1999) *Livestock to 2020: The Next Food Revolution*. Food, Agriculture and Environment Discussion Paper No. 28, International Food Policy Research Institute, Washington, DC.

Dercon, S. (1998) Wealth, risk and activity choice: cattle in Western Tanzania. *Journal of Development Economics* 55, 1–42.

DFID (1999) *Sustainable Livelihoods Guidance Sheets*. Department for International Development, UK (http://www.livelihoods.org).

Dietz, T., Verhagen, J. and Ruben, R. (eds) (2004) *Impact of Climate Change on Drylands, with a Focus on West Africa*. Kluwer Academic Publishers, Dordrecht, The Netherlands/Boston, Massachusetts.

Diop (2001) Management of organic inputs to increase food production in Senegal. In: Uphoff, N. (ed.) *Agroecological Innovations: Increasing Food Production with Participatory Development*. Earthscan, London.

Dixon, J., Gulliver, A. and Gibbon, D. (2001) *Farming Systems and Poverty: Improving Farmers' Livelihoods in a Changing World*. FAO, Rome and World Bank, Washington, DC.

Dixon, R.K. (1995) Agroforestry systems: sources of sink or greenhouse gasses? *Agroforestry Systems* 31, 99–116.

Dregne, H.E. (1990) Erosion and soil productivity in Africa. *Journal of Soil and Water Conservation* 45, 432–436.

Dregne, H.E. and Chou, N.T. (1992) Global desertification dimensions and costs. In: Dregne, H.E. (ed.) *Degradation and Restoration of Arid Lands*. Texas Tech University, Lubbock, Texas.

Eaton, D. (1996) *The Economics of Soil Erosion: a Model of Farm Decision Making*. Discussion paper 96-01, Environmental Economics Programme, International Institute for Environment and Development, London.

Ehui, S., Hertel, T. and Preckel, P. (1990) Forest resource depletion, soil dynamics, and agricultural productivity in the tropics. *Journal of Environmental Economics and Management* 18, 136–154.

Elbers, C., Fujii, T., Lanjouw, P., Ozler, B. and Yin, W. (2004) *Poverty Alleviation through Geographic Targeting: how much does Aggregation Help?* Mimeo, Vrije Universiteit, Amsterdam.

Elias, E., Morse, S. and Belshaw, D.G.R. (1998) The nitrogen and phosphorus balances of some Kindo Koisha farms in southern Ethiopia. *Agriculture Ecosystems and Environment* 71, 93–113.

Ellis, F. (2000) *Rural Livelihoods and Diversity in Developing Countries.* Oxford University Press, Oxford, UK.

Ellis, F. and Bahiigwa, G. (2003) Livelihoods and rural poverty reduction in Uganda. *World Development* 31, 997–1013.

Ellis, F. and Mdoe, N. (2003) Livelihoods and rural poverty reduction in Tanzania. *World Development* 31, 1367–1384.

Ellis, F., Kutengule, M. and Nyasulu, A. (2003) Livelihoods and rural poverty reduction in Malawi. *World Development* 31, 1495–1510.

Enters, T. (1998) Method for economic assessment of the on- and off-site impacts of soil erosion. *Issues in Sustainable Land Management* 2.

Erenstein, O.C.A. (1999) The economics of soil conservation in developing countries: the case study of crop residue mulching. PhD Thesis, Wageningen University, Wageningen, The Netherlands.

Escobal, J. (2005) The role of public infrastructure in market development in rural Peru. PhD Thesis. Wageningen University, Wageningen, The Netherlands.

Escobal, J. and Valdivia, M. (2005) *Characterizing Diversity in the Peruvian Rural Sierra: a First Trial towards a Typology.* Mimeo GRADE, Lima, Peru and International Food Policy Research Institute, Washington, DC.

Fafchamps, M., Udry, C. and Czukas, K. (1998) Drought and savings in West Africa: are livestock a buffer stock? *Journal of Development Economics* 55, 273–305.

Fairhead, J. and Leach, M. (1996) *Misreading the African Landscape: Society and Ecology in a Forest-Savanna Mosaic.* Cambridge University Press, Cambridge, UK.

Fairhead, J. and Scoones, I. (2005) Local knowledge and the social shaping of soil investments: critical perspectives on the assessment of soil degradation in Africa. *Land Use Policy* 22, 33–42.

Fan, S. and Chan-Kang, C. (2004) Returns to investment in less-favored areas in developing countries: a synthesis of evidence and implications for Africa. *Food Policy* 29, 431–444.

Fan, S. and Hazell, P. (2001) Returns to public investments in the less-favored areas of India and China. *American Journal of Agricultural Economics* 85, 1217–1222.

Fan, S., Hazell, P. and Haque, T. (2000) Targeting public investments by agro-ecological zone to achieve growth and poverty alleviation goals in rural India. *Food Policy* 25, 411–428.

Fan, S., Zhang, L. and Zhang, X. (2002) *Growth, Inequality, and Poverty in Rural China.* Research Report 125, International Food Policy Research Institute, Washington, DC.

Fan, S., Thorat, S. and Rao, N. (2003) *Investment, Subsidies, and Pro-poor Growth in Rural India.* Draft report submitted to DFID. International Food Policy Research Institute, Washington, DC.

FAO (1986) *Highlands Reclamation Study: Ethiopia.* Final Report, vols I and II, Food and Agriculture Organization of the United Nations, Rome.

FAO (1997) *State of the World's Forest.* Food and Agriculture Organization of the United Nations, Rome.

FAO (2002) *Food Insecurity, Poverty, and Agriculture: a Concept Paper.* Food and Agriculture Organization of the United Nations, Rome.

FAO/World Bank (2001) *Farming Systems and Poverty: Improving Farmers' Livelihoods in a Changing World.* FAO, Rome and World Bank, Washington, DC.

Feder, G., Just, R.E. and Zilberman, D. (1985) Adoption of agricultural innovations in developing countries: a survey. *Economic Development and Cultural Change* 33, 255–298.

Folmer, E.C.R., Geurts, P.M.H. and Francisco, J.R. (1998) Assessment of soil fertility depletion in Mozambique. *Agriculture, Ecosystems and Environment* 71, 159–167.

Franzel, S. (1999) Socio-economic factors affecting the adoption potential of improved tree fallows in Africa. *Agroforestry Systems* 47, 305–321.

Gachimbi, L.N., De Jager, A., Van Keulen, H., Thuranira, E.G. and Nandwa, S.M. (2002) *Participatory Diagnosis of Soil Nutrient Depletion in Semi-arid Areas of Kenya*. Managing Africa's Soils No. 26, International Institute for Environment and Development, London.

Gachimbi, L.N., Van Keulen, H., Thuranira, E.G., Karuku, A.M., De Jager, A., Nguluu, S., Ikombo, B.M., Kinama, J.M., Itabari, J.K. and Nandwa, S.M. (2005) Nutrient balances at farm level in Machakos (Kenya), using a participatory nutrient monitoring (NUT-MON) approach. *Land Use Policy* 22, 13–22.

Garrity, D. (2002) Increasing the scope for food production on sloping lands in Asia: contour farming with natural vegetation strips in the Philippines. In: Uphoff, N. (ed.) *Agro-ecological Innovations: Increasing Food Production with Participatory Development*. Earthscan, London.

Gitari, J.N., Matiri, F.M., Kariuki, I.W., Muriithi, C.W. and Gachanja, S.P. (1999) Nutrient and cash flow monitoring in farming systems on the eastern slopes of Mount Kenya. In: Smaling, E.M.A., Oenema, O. and Fresco, L.O. (eds) *Nutrient Disequilibria in Agro-ecosystems*. CAB International, Wallingford, UK.

Glover, D. (1990) Contract farming and outgrower schemes in East and Southern Africa. *Journal of Agricultural Economics* 41, 303–315.

Goma, H.C., Rahim, K., Nangendo, G., Riley, J. and Stein, A. (2001) Participatory studies for agro-ecosystem evaluation. *Agriculture, Ecosystems and Environment* 87, 179–190.

Grohs, F. (1994) *Economics of Soil Degradation, Erosion, and Conservation: a Case Study of Zimbabwe*. Arbiten zur Agrarwirtschaft in Entwicklungsländern, Wissenschafts-verlag Vauk KG, Kiel, Germany.

Hassane, A., Martin, P. and Reij, C. (2000) *Water Harvesting, Land Rehabilitation and Household Food Security in Niger*. VU/IFAD, Amsterdam/Rome.

Hayami, Y. and Otsuka, K. (1993) *The Economics of Contract Choice: an Agrarian Perspective*. Clarendon Press, Oxford, UK.

Hazell, P. (1999) Public policy and drought management in agropastoral systems. In: McCarthy, N., Swallow, B., Kirk, M. and Hazell, P. (eds) *Property Rights, Risk and Livestock Development in Africa*. International Food Policy Research Institute, Washington, DC and International Livestock Research Institute, Nairobi.

Hazell, P., Ruben, R., Kuyvenhoven, A. and Jansen, H. (2006) *Investing in Poor People in Less-favored Areas*. International Food Policy Research Institute (IFPRI) and WUR research report for IFAD, Rome.

Hengsdijk, H. (2002) Formalising agro-ecological knowledge for future-oriented land use studies. Doctoral thesis. Wageningen University, Wageningen, The Netherlands.

Hoben, A. (1995) Paradigms and politics: the cultural construction of environmental policy in Ethiopia. *World Development* 23, 1007–1021.

Hoffmann, I., Gerling, D., Kyiogwom, U.B. and Mané-Bielfeldt, A. (2001) Farmers' management strategies to maintain soil fertility in remote areas in northwest Nigeria. *Agriculture, Ecosystems and Environment* 86, 263–275.

Holden, S., Shiferaw, B. and Pender, J. (2004) Non-farm income, household welfare, and sustainable land management in a less-favoured area in the Ethiopian highlands. *Food Policy* 29, 369–392.

Holden, S., Shiferaw, B. and Pender, J. (2005) *Policy Analysis for Sustainable Land Management and Food Security: a Bio-economic Model with Market Imperfections*. International Food Policy Research Institute Research Report No. 140, Washington, DC.

Huang, J. and Rozelle, S. (1994) Environmental stress and grain yields in China. *American Journal of Agricultural Economics* 77, 246–256.

Huang, J. and Rozelle, S. (1996) Technological change: rediscovering the engine of productivity growth in China's rural economy. *Journal of Development Economics* 49, 337–369.

Hurni, H. (1988) Degradation and conservation of the resources in the Ethiopian highlands. *Mountain Research and Development* 8, 123–130.

ICRAF (1999) *Sahelian Programme: Caring for the Sahel through Agroforestry. Concept Note.* International Center for Research in Agroforestry, Nairobi.

IFAD (1999) *India: Country Strategic Opportunities Paper.* Report No. 896-IN, IFAD, Rome.

IFAD (2002) *Assessment of Rural Poverty: Asia and the Pacific.* International Fund for Agricultural Development, Asia and the Pacific Division, Project Management Department. Rome.

Jaffee, S. (1995) The many faces of success: the development of Kenyan horticultural exports. In: Jaffee, S. and Morton, J. (eds) *Marketing Africa's High-value Foods: Comparative Experiences of an Emergent Private Sector.* The World Bank, Washington, DC.

Jagger, P. and Pender, J. (2006) Impacts of programs and organizations on the adoption of sustainable land management technologies in Uganda. In: Pender, J., Place, F. and Ehui, S. (eds) *Strategies for Sustainable Land Management in the East African Highlands.* IFPRI, Washington, DC.

Jansen, H.G.P., Pender, J., Damon, A. and Schipper, R. (2006) *Rural Development Policies and Sustainable Land Use in the Hillside Areas of Honduras: a Quantitative Livelihoods Approach.* Research Report No. 144, International Food Policy Research Institute, Washington, DC.

Jayne, T.S., Yamano, T., Weber, M.T., Tschirley, D., Benfica, R., Chapoto, A. and Zulu, B. (2003) Smallholder income and land distribution in Africa: implications for poverty reduction strategies. *Food Policy* 28, 253–275.

Juo, A.S.R. and Kang, B.T. (1989) Nutrient effects of modification of shifting cultivation in West Africa. In: Proctor, J. (ed.) *Mineral Nutrients in Tropical Forest and Savanna Ecosystems.* Blackwell, Oxford, UK.

Kam, S., Hossain, M., Bose, M. and Villano, L.S. (2005) Spatial patterns of rural poverty and their relationships with welfare influencing factors in Bangladesh. *Food Policy* 30 (5/6), 551–567.

Keeley, J. and Scoones, I. (2003) *Understanding Environmental Policy Processes: Cases from Africa.* Earthscan, London.

Kelley, T.G. and Byerlee, D. (2003) *Surviving on the Margin: Agricultural Research and Development Strategies for Poverty Reduction in Marginal Areas.* Mimeo CGIAR, Science Council Secretariat, Rome and World Bank, Washington, DC.

Kelley, T.G. and Parthasarathy Rao, P. (1995) Marginal environments and the poor: evidence from India, *Economic and Political Weekly* 30, 2494–2495.

Kerr, J., Pangare, G. and Pangare, V.L. (2001) *The Role of Watershed Projects in Developing Rainfed Agriculture in India.* Research Report, International Food Policy Research Institute, Washington, DC.

Key, N. and Runsten, D. (1999) Contract farming, smallholders and rural development in Latin America: the organization of agroprocessing firms and the scale of outgrower production. *World Development* 27, 381–401.

Knox, A., Meinzen-Dick, R. and Hazell, P. (2002) Property rights, collective action, and technologies for natural resource management: a conceptual framework. In: Meinzen-Dick, R., Knox, A., Place, F. and Swallow, B. (eds) *Innovation in Natural Resource*

Management; the Role of Property Rights and Collective Action in Developing Countries. IFPRI, Johns Hopkins University Press, Baltimore, Maryland.

Koning, N. and Smaling, E. (2005) Environmental crisis or 'lie of the land'? The debate on soil degradation in Africa. *Land Use Policy* 22, 3–12.

Kruseman, G., Ruben, R. and Tesfay, G. (2006) Village stratification for policy analysis: multiple development domains in the Ethiopian highlands of Tigray. In: Pender, J., Place, F. and Ehui, S. (eds) *Strategies for Sustainable Land Management in the East African Highlands.* IFPRI, Washington, DC.

Kuyvenhoven, A. and Bigman, D. (2002) Technical standards in a liberalised agri-food system: institutional implications for developing countries. *Tijdschrift voor Sociaalwetenschappelijk Onderzoek van de Landbouw* 17, 51–62.

Kuyvenhoven, A., Becht, J.A. and Ruben, R. (1998) Financial and economic evaluation of phosphate rock use to enhance soil fertility in West Africa. In: Wossink, G.A.A., Van Kooten, G.C. and Peters, G.H. (eds) *Economics of Agro-chemicals.* Ashgate, Aldershot, UK, pp. 249–261.

Lal, R. (1995) Erosion-crop productivity relationships for soil of Africa. *Soil Science Society of America Journal* 59, 661–667.

Leach, M. and Mearns, R. (1996) Environmental change & policy: challenging received wisdom in Africa. In: Leach, M. and Mearns, R. (eds) *The Lie of the Land: Challenging Received Wisdom on the African Environment.* The International African Institute in association with James Currey, London.

Leaky, R.R.B. and Tchoundjeu, Z. (2001) Diversification of tree crops: domestication of companion crops for poverty reduction and environmental services. *Exploratory Agriculture* 37, 270–296.

Lipton, M. (2005) *The Family Farm in a Globalizing World: the Role of Crop Science in Alleviating Poverty.* 2020 Discussion Paper No. 40, International Food Policy Research Institute, Washington, DC.

Lutz, E., Pagiola, S. and Reiche, C. (eds) (1994) *Economic and Institutional Analyses of Soil Conservation Projects in Central America and the Caribbean.* A CATIE–World Bank Project, World Bank Environment Paper 8, The World Bank, Washington, DC.

McCann, J.C. (1999) *Green Land, Brown Land, Black Land: an environmental history of Africa, 1800–1990.* James Currey, Oxford, UK.

McCarthy, N., Swallow, B., Kirk, M. and Hazell, P. (eds) (1999) *Property Rights, Risk and Livestock Development in Africa.* IFPRI, Washington, DC and ILRI, Nairobi.

McIntire, J. (1994) A review of the soil conservation sector in Mexico. In: Lutz, E., Pagiola, S. and Reiche, C. (eds) *Economic and Institutional Analyses of Soil Conservation Projects in Central America and the Caribbean.* A CATIE–World Bank Project, World Bank Environment Paper 8, The World Bank, Washington, DC.

McIntyre, J., Bourzat, D. and Pingali, P. (1992) *Crop–Livestock Interaction in Sub-Saharan Africa.* World Bank, Washington, DC.

McKenzie, C. (1994) Degradation of arable land resources: policy options and considerations within the context of rural restructuring in South Africa. Paper prepared for the *LAPC Workshop,* Johannesburg, South Africa.

McNeely, J.A. (ed.) (2001) *The Great Reshuffling: Human Dimensions of Invasive Alien Species.* International Union for the Conservation of Nature, Gland, Switzerland.

MEA (2005) *Millennium Ecosystem Assessment Synthesis Report.*

Meinzen-Dick, R.S. (1997) Farmer participation in irrigation: 20 years of experience and lessons for the future. *Irrigation and Drainage Systems* 11, 103–118.

Minot, N. (2000) Generating disaggregated poverty maps: an application to Viet Nam. *World Development* 28, 319–331.

Muchena, F.N., Onduru, D.D., Gachini, G.N. and De Jager, A. (2005) Turning the tides of soil degradation in Africa: capturing the reality and exploring opportunities. *Land Use Policy* 22, 23–32.

Mulder, I. (2001) Soil degradation in Benin: farmers' perceptions and responses. Doctoral thesis, Thela Research series No. 240, Tinbergen Institute, Amsterdam.

Nelson, R., Cramb, R.A., Manz, K.M. and Mamicpic, M.A. (1998) Bio-economic modelling of alternative forms of hedgerow intercropping in the Philippine uplands. *Agroforestry Systems* 30, 241–262.

Neupane, R. and Thapa, G.B. (2001) Impact of agroforestry intervention on soil fertility and farm income under the subsistence farming system of the middle hills, Nepal. *Agriculture, Ecosystems and Environment* 84, 157–167.

Niehof, A. (2004) The significance of diversification for rural livelihood systems. *Food Policy* 29, 321–338.

Nkonya, P., Pender, J., Jagger, P., Sserunkuuma, D., Kaizzi, C.K. and Ssali, H. (2004) *Strategies for Sustainable Land Management and Poverty Reduction in Uganda.* Research Report No. 133, International Food Policy Research Institute, Washington, DC.

Nkonya, E., Kaizzi, C.K. and Pender, J. (2005a) Determinants of nutrient balances in maize farming systems in eastern Uganda. *Agricultural Systems* 85, 155–182.

Nkonya, E., Pender, J., Kaizzi, C., Edward, K. and Mugarura, S. (2005b) *Policy Options for Increasing Crop Productivity and Reducing Soil Nutrient Depletion and Poverty in Uganda.* Environment and Production Technology Division Discussion Paper No. 134, International Food Policy Research Institute, Washington, DC.

Norse, D. and Saigal, R. (1992) National economic cost of soil erosion: the case of Zimbabwe. Paper prepared for the *CIDIE Workshop on Environmental Management and Natural Resource Management in Developing Countries*, 22–24 January, World Bank, Washington, DC.

Okumu, B.N., Jabbar, M.A., Colman, D. and Russell, N. (2002) A bio-economic model of integrated crop–livestock farming systems: the case of the Ginchi watershed in Ethiopia. In: Barrett, C.B., Place, F. and Aboud, A.A. (eds) *Natural Resources Management in African Agriculture: Understanding and Improving Current Practices.* CAB International, Wallingford, UK.

Oldeman, L.R. (1998) *Soil Degradation: a Threat to Food Security?* Report 98/01, International Soil Reference and Information Centre, Wageningen, The Netherlands.

Oldeman, L.R., Hakkeling, R.T.A. and Sombroek, W.G. (1991) *World Map of the Status of Human-induced Soil Degradation: an Explanatory Note.* International Soil Reference and Information Centre, Wageningen, The Netherlands and United Nations Environment Programme, Nairobi.

Onduru, D.D., De Jager, A., Gachini, G.N. and Diop, J.-M. (2001) *Exploring New Pathways for Innovative Soil Fertility Management in Kenya.* Managing Africa's Soils No. 25, International Institute for Environment and Development, London.

Orr, A. and Jere, P. (1999) Identifying smallholder target groups for IPM in southern Malawi. *International Journal of Pest Management* 45, 179–187.

Otsuka, K. and Place, F. (2001) *Land Tenure and Natural Resource Management: a Comparative Study of Agrarian Communities in Asia and Africa.* The Johns Hopkins University Press, Baltimore, Maryland and IFPRI, Washington, DC.

Pagiola, S. (1993) Soil conservation and the sustainability of agricultural production. PhD dissertation, Stanford University, California.

Pagiola, S. and Dixon, J. (1997) *Land Degradation Problems in El Salvador, Annex 7 – El Salvador.* Rural Development Study Report No. 16253-ES, World Bank, Washington, DC.

Pender, J. (1998) Population growth, agricultural intensification, induced innovation and natural resource sustainability: an application of neoclassical growth theory. *Agricultural Economics* 19, 99–112.

Pender, J. (2004) Development pathways for hillsides and highlands: some lessons from Central America and East Africa. *Food Policy* 29, 339–367.

Pender, J. and Gebremedhin, B. (2006) Land management, crop production and household income in the highlands of Tigray, northern Ethiopia. In: Pender, J., Place, F. and Ehui, S. (eds) *Strategies for Sustainable Land Management in the East African Highlands.* International Food Policy Research Institute, Washington, DC.

Pender, J. and Kerr, J. (1998) Determinants of farmers' indigenous soil and water conservation investments in India's semi-arid tropics. *Agricultural Economics* 19, 113–125.

Pender, J., Gebremedhin, B., Benin, S. and Ehui, S. (2001a) Strategies for sustainable development in the Ethiopian highlands. *American Journal of Agricultural Economics* 83, 1231–40.

Pender, J., Scherr, S.J. and Durón, G. (2001b) Pathways of development in the hillsides of Honduras: causes and implications for agricultural production, poverty, and sustainable resource use. In: Lee, D.R. and Barrett, C.B. (eds) *Tradeoffs or Synergies? Agricultural Intensification, Economic Development and the Environment.* CAB International, Wallingford, UK.

Pender, J., Gebremedhin, B., Benin, S. and Ehui, S. (2001c) *Strategies for Sustainable Agricultural Development in the Ethiopian Highlands.* Environment and Production Technology Discussion Paper No. 77, Environment and Production Technology Division, International Food Policy Research Institute, Washington, DC.

Pender, J., Jagger, P., Nkonya, E. and Sserunkuuma, D. (2001d) *Development Pathways and Land Management in Uganda: Causes and Implications.* Environment Production and Technology, Discussion Paper 85, International Food Policy Research Institute, Washington, DC.

Pender, J., Scherr, S. and Durón, G. (2001e) Pathways of development in the hillside areas of Honduras: causes and implications for agricultural production, poverty and sustainable resource use. In: Lee, D.R. and Barrett, C.B. (eds) *Tradeoffs or Synergies? Agricultural Intensification, Development and the Environment.* CAB International, Wallingford, UK.

Pender, J., Jagger, P., Nkonya, E. and Sserunkuuma, D. (2004) Development pathways and land management in Uganda. *World Development* 32, 767–792.

Pender, J., Ehui, S. and Place, F. (2006a) Conceptual framework and hypotheses. In: Pender, J., Place, F. and Ehui, S. (eds) *Strategies for Sustainable Land Management in the East African Highlands.* International Food Policy Research Institute, Washington, DC.

Pender, J., Place, F. and Ehui, S. (2006b) Strategies for sustainable land management in the East African highlands: conclusions and implications. In: Pender, J., Place, F. and Ehui, S. (eds) *Strategies for Sustainable Land Management in the East African Highlands.* International Food Policy Research Institute, Washington, DC.

Pingali, P.L. and Rosegrant, M. (2001) Intensive food systems in Asia: can the degradation problems be reversed? In: Lee, D.R. and Barrett, C.B. (eds) *Tradeoffs or Synergies? Agricultural Intensification, Economic Development and Environment.* CABI Publishing, Wallingford, UK, pp. 383–397.

Place, F., Kristjanson, P., Staal, S., Kruska, R., DeWolff, T., Zomer, R. and Njuguna, E.C. (2006a) Development pathways in medium to high-potential Kenya: a meso-level analysis of agricultural patterns and determinants. In: Pender, J., Place, F. and Ehui, S. (eds) *Strategies for Sustainable Land Management in the East African Highlands.* International Food Policy Research Institute, Washington, DC.

Place, F., Njuki, J., Murithi, F. and Mugo, F. (2006b) Agricultural enterprise and land management in the highlands of Kenya. In: Pender, J., Place, F. and Ehui, S. (eds) *Strategies for Sustainable Land Management in the East African Highlands.* International Food Policy Research Institute, Washington, DC.

Ramadhani, T., Otsyina, R. and Franzel, S. (2002) Improving household incomes and reducing deforestation using rotational woodlots in Tabora district, Tanzania. *Agriculture, Ecosystems and Environment* 89, 229–239.

Ravallion, M. and Jalan, J. (1996) Growth differences due to spatial externalities. *Economic Letters* 53, 227–232.

Reardon, T. (1995) Sustainability issues for agricultural research strategies in the semi-arid tropics: focus on the Sahel. *Agricultural Systems* 48, 345–360.

Reardon, T. (1997) Using evidence of household income diversification to inform study of the rural nonfarm labor market in Africa. *World Development* 25, 735–748.

Reardon, T., Delgado, C. and Matlon, P. (1992) Determinants and effects of income diversification amongst farm households in Burkina Faso. *Journal of Development Studies* 28, 264–296.

Reardon, T., Berdegué, J. and Escobar, G. (2001) Rural nonfarm employment and incomes in Latin America: overview and policy implications. *World Development* 29, 395–409.

Reij, C. and Steeds, D. (2003) *Success Stories in African Drylands: Supporting Advocates and Assessing Sceptics.* Paper commissioned by Global Mechanisms of the Convention to Combat Desertification, Centre for International Cooperation (CIS-VU), Amsterdam.

Reij, C., Scoones, I. and Toulmin, C. (1996) *Sustaining the Soil: Indigenous Soil and Water Conservation in Africa.* Earthscan, London.

Renkow, M. (1993) Differential technology adoption and income distribution in Pakistan: implications for research resource allocation. *American Journal of Agricultural Economics* 75, 33–43.

Renkow, M. (2000) Poverty, productivity and production environment: a review of the evidence. *Food Policy* 25, 463–478.

Rosegrant, M.W. and Perez, N.D. (1997) *Water Resources Development in Africa: a Review and Synthesis of Issues, Potentials, and Strategies for the Future.* Environment and Production Technology, Discussion Paper 28, International Food Policy Research Institute, Washington, DC.

Ruben, R. and Pender, J. (2004) Rural diversity and heterogeneity in less-favoured areas: the quest for policy targeting. *Food Policy* 29, 303–320.

Ruben, R., Kruseman, G., Hengsdijk, H. and Kuyvenhoven, A. (1996) The impact of agrarian policies on sustainable land use. In: Teng, P.S., Kropff, M.J., Ten Berge, H.F.M., Dent, J.B., Lansigan, F.P. and Van Laar, H.H. (eds) Applications of systems approaches at the farm and regional levels. *Proceedings of the 2nd Symposium on System Approaches for Agricultural Development*, Kluwer, Dordrecht, The Netherlands, pp. 65–82.

Ruben, R., Kruseman, G. and Kuyvenhoven, A. (2006) Strategies for sustainable intensification in East African highlands: labour use and input efficiency. *Agricultural Economics* 34, 167–181.

Ruben, R., Kuyvenhoven, A., Kruseman, G. (2001) Bio-economic models for eco-regional development: policy instruments for sustainable intensification. In: Lee, D.R. and Barrett, C.B. (eds) *Tradeoffs or Synergies? Agricultural Intensification, Economic Development and the Environment in Developing Countries.* CAB International, Wallingford, UK, pp. 115–134.

Ruben, R., Kuyvenhoven, A. and Hazell, P. (2003) Investing in poor people in poor lands. Contributed paper at the *International Conference 'Staying Poor: Chronic Poverty and Development Policy'*, 7–9 April, Manchester, UK.

Rukuni, M., Svendsen, M., Meinzen-Dick, R. and Makombe, G. (1994) *Irrigation Performance in Zimbabwe*. Faculty of Agriculture, University of Zimbabwe, Harare.

Ryan, J.G. and Spencer, D.S.C. (2001) *Future Challenges and Opportunities for Agricultural R&D in the Semi-arid Tropics*. International Crops Research Institute for the Semi-arid Tropics, Patancheru, Andhra Pradesh, India.

Sanchez, P.A. and Jama, B.A. (2002) Soil fertility replenishment takes off in east and southern Africa. In: Vanlauwe, B., Diels, J., Sanginga, N. and Merckx, R. (2002) *Integrated Plant Nutrient Management in Sub-Saharan Africa: from Concept to Practice*. CAB International, Wallingford, UK with IITA, pp. 23–45.

Scherr, S.J. (1999a) *Soil Degradation: a Threat to Developing Country Food Security by 2020?* Food, Agriculture and Environment Discussion Paper 27, International Food Policy Research Institute, Washington, DC.

Scherr, S.J. (1999b) *Poverty–Environment Interactions in Agriculture: Key Factors and Policy Implications*. Poverty Environment Initiative Paper No. 3, UNDP/EC, Washington, DC.

Scoones, I. and Toulmin, C. (1999) *Policies for Soil Fertility Management in Africa*. IIED/IDS, London and Brighton, UK, for DFID.

Scoones, I. and Wolmer, W. (2002) Pathways of change: crop–livestock integration in Africa. In: Scoones, I. and Wolmer, W. (eds) *Pathways of Change in Africa: Crops Livestock and Livelihoods in Mali, Ethiopia and Zimbabwe*. James Curry Ltd., Oxford, UK.

Sehgal, J. and Abrol, I.P. (1994) *Soil Degradation in India: Status and Impact*. Oxford University Press and IBH, New Delhi, India.

Shepherd, K.D. and Soule, M.J. (1998) Soil fertility management in West Kenya: dynamic simulation of productivity, profitability and sustainability at different resource endowment levels. *Agriculture Ecosystems and Environment* 71, 131–145.

Shiferaw, B. and Holden, S. (1999) Soil erosion and smallholders' conservation decisions in the highlands of Ethiopia. *World Development* 27, 739–752.

Skees, J., Hazell, P. and Miranda, M. (1999) *New Approaches to Crop Yield Insurance in Developing Countries*. EPTD Discussion Paper No. 55, International Food Policy Research Institute, Washington, DC.

Smaling, E.M.A., Nandwa, S.M. and Janssen, B.H. (1997) Soil fertility in Africa is at stake. In: Buresh, R.J., Sanchez, P.A. and Calhoun, F. (eds) *Replenishing Soil Fertility in Africa*. Soil Science Society of America and American Society of Agronomy, SSSA Spec. Publ. 51. SSA, Madison, Wisconsin, pp. 151–192.

Solórzano, R., De Camino, R. and Woodward, R. (1991) *Accounts Overdue: Natural Resource Depreciation in Costa Rica*. World Resources Institute, Washington, DC.

Sonneveld, B.G.J.S. (2002) Land under pressure: the impact of water erosion on food production in Ethiopia. PhD dissertation. Shaker Publishing, The Netherlands.

Ssali, H. (2003) Soil organic matter and its relationship to soil fertility changes in Uganda. In: Benin, S., Pender, J. and Ehui, S. (eds) *Policies for Sustainable Land Management in the East African highlands*. Summary of papers and proceedings of the conference held at the United Nations Economic Commission for Africa, Addis Ababa, Ethiopia, 24–26 April. Environment and Production Technology Division Workshop Summary Paper No. 13, International Food Policy Research Institute, Washington, DC.

Staal, S.J. (1995) Periurban dairying and public policy in Ethiopia and Kenya: a comparative economic and institutional analysis. PhD dissertation, Department of Food and Resource Economics, University of Florida, Gainesville, Florida.

Staal, S.J., Baltenweck, I., Waithaka, M.M., deWolff, T. and Njoroge, L. (2002) Location and uptake: integrated household and GIS analysis of technology adoption and land use, with application to smallholder dairy farms in Kenya. *Agricultural Economics* 27, 295–315.

Stocking, M. (1986) *The Cost of Soil Erosion in Zimbabwe in Terms of the Loss of Three Major Nutrients.* Consultant's Working Paper No. 3, Soil Conservation Programme, Land and Water Development Division, AGLS, FAO, Rome.

Stocking, M. (1996) Soil erosion: breaking new ground. In: Leach, M. and Mearns, R. (eds) *The Lie of the Land: Challenging Received Wisdom on the African Environment.* The International African Institute in association with James Currey, London.

Stoorvogel, J.J., Smaling, E.M.A. and Janssen, B.H. (1993) Calculating soil nutrient balances at different scales. *Fertilizer Research* 35, 227–235.

Stroosnijder, L. and Van Rheenen, T. (2001) *Agro-silvo-pastoral Land Use in Sahelian Villages.* Advances in Geology 33, Catena Verlag, Reiskirchen, Germany.

Sutcliffe, J.P. (1993) *Economic Assessment of Land Degradation in the Ethiopian highlands: a Case Study.* National Conservation Strategy Secretariat, Ministry of Planning and Economic Development, Transitional Government of Ethiopia, Addis Ababa.

Swift, M.J., Seward, P.D., Frost, P.G.H., Qureshi, J.N. and Muchena, F.N. (1994) Long-term experiments in Africa: developing a database for sustainable land use under global change. In: Leigh, F.A. and Johnston, A.E. (eds) *Long-term Experiments in Agricultural and Ecological Sciences.* CAB International, Wallingford, UK.

Swinkels, R. and Franzel, S. (1997) Adoption potential of hedgerow intercropping systems in the highlands of western Kenya. Part II. Economic and farmers' evaluation. *Experimental Agriculture* 33, 211–223.

Swinkels, R. and Turk, C. (2004) Poverty and remote areas: evidence from new data and questions for the future. Background paper for the *PAC Conference*, Hanoi, Vietnam, 24–26 November, World Bank, Vietnam.

Swinton, S.M. and Williams, M.B. (1998) *Assessing the Economic Impact of Integrated Pest Management: Lessons from the Past, Direction for the Future.* Staff Paper No. 98-12, Michigan State University, East Lansing, Michigan.

TAC (2001) CGIAR *Research Priorities for Marginal Lands.* SDR/TAC: IAR/99/12, Technical Advisory Committee of the CGIAR, Washington, DC.

Tanner, J.C., Holden, S.J., Owen, E., Winugroho, M. and Gill, M. (2001) Livestock sustaining intensive smallholder crop production through traditional feeding practices for generating high quality manure-compost in upland Java. *Agriculture, Ecosystems and Environment* 84, 21–30.

Tiffen, M., Mortimore, M. and Gichuki, F. (1994) *More People – Less Erosion: Environmental Recovery in Kenya.* Wiley and Sons, London.

UBOS (2003) *Uganda National Household Survey 2002/2003.* Report on the socio-economic survey, Uganda Bureau of Statistics, Entebbe, Uganda.

Udry, C. (1990) Credit markets in Northern Nigeria: credit as insurance in a rural economy. *World Bank Economic Review* 4, 251–270.

Umali, D., Feder, G. and De Haan, C. (1992) *The Balance between Public and Private Sector Activity in the Delivery of Livestock Services.* World Bank Discussion Paper No. 163, World Bank, Washington, DC.

UNDP (1997) *Human Development Report 1997.* United Nations Development Programme, Oxford University Press, New York.

Van den Bosch, H., Gitari, J.N., Ogaro, V.N., Maobe, S. and Vlaming, J. (1998) Monitoring nutrient flows and economic performance in African farming systems (NUTMON). Monitoring nutrient flows and balances in three districts in Kenya. *Agriculture, Ecosystems and Environment* 71, 63–80.

Van der Pol, F. (1992) *Soil Mining: an Unseen Contributor to Farm Income in Southern Mali.* Bulletin 325, Royal Tropical Institute, Amsterdam.

Van der Walle, D. and Gunewardena, D. (2001) Sources of ethnic inequality in Vietnam. *Journal of Development Economics* 65, 177–207.

Vanlauwe, B., Diels, J., Sanginga, N. and Merckx, R. (2002) *Integrated Plant Nutrient Management in Sub-Saharan Africa: From Concept to Practice*. CAB International, Wallingford, UK with IITA.

Van Lynden, G. and Oldeman, L.R. (1997) *Soil Degradation in South and Southeast Asia*. International Soil Reference and Information Centre, Wageningen, The Netherlands.

Vitale, J.D. and Sanders, J.H. (2002) *Demand-driven Technological Change and the Traditional Cereals in Sub-Saharan Africa: The Malian Case*. Discussion Paper, Purdue University, West Lafayette, Indiana.

Vlek, P.L.G. (1990) The role of fertilizers in sustaining agriculture in sub-Saharan Africa. *Fertiliser Research* 26, 327–339.

Walker, T., Swinton, S., Hijmans, R., Quiroz, R., Valdivia, R., Holle, M., León-Velarde, C. and Posner, J. (2000) Technologies for the tropical Andes, Brief 3. In: Pender, J. and Hazell, P. (eds) *Promoting Sustainable Development in Less-favored Areas*. 2020 Vision Focus 4, International Food Policy Research Institute, Washington, DC.

Wiersum, K.F. (1997) From natural forest to tree crops, co-domestication of forests and tree species. *Netherlands Journal of Agricultural Research* 45, 425–438.

Williams, T.O., Hiernaux, P. and Fernandez-Riviera, S. (1999) Crop–livestock systems in Sub-Saharan Africa: determinants and intensification pathways. In: McCarthy, N., Swallow, B., Kirk, M. and Hazell, P. (eds) *Property Rights, Risk and Livestock Development in Africa*. International Food Policy Research Institute, Washington, DC and International Livestock Research Institute, Nairobi.

Woelcke, J. (2003) Bio-economics of sustainable land management in Uganda. In: Heidhues, F. and Von Braun, J. (eds) Series *Developing Economics and Policy*. Peter Lang, Frankfurt am Main, Germany, Berlin, Berne, Switzerland, Brussels, New York, Oxford.

Woelcke, J., Berger, T. and Park, S. (2006) Sustainable land management and technology adoption in eastern Uganda. In: Pender, J., Place, F. and Ehui, S. (eds) *Strategies for Sustainable Land Management in the East African Highlands*. International Food Policy Research Institute, Washington, DC.

Wood, S., Sebastian, K., Nachtergaele, F., Nielsen, D. and Dai, A. (1999) *Spatial Aspects of the Design and Targeting of Agricultural Development Strategies*. Environment and Production Technology, Discussion Paper No. 44, International Food Policy Research Institute, Washington, DC.

World Bank (1988) *Madagascar – Environmental Action Plan, Vol. 1*. In cooperation with USAID, Swiss Cooperation, UNESCO, UNDP and WWF, Washington, DC.

World Bank (1992) *Malawi – Economic Report on Environmental Policy, Vols I and II*. Report No. 9888-MAI. Washington, DC.

Wortmann, C.S. and Kaizzi, C.K. (1998) Nutrient balances and expected effects of alternative practices in farming systems of Uganda. *Agriculture, Ecosystems and Environment* 71, 115–130.

Yesuf, M., Mekonnen, A., Kassie, M. and Pender, J. (2005) *Costs of Land Degradation in Ethiopia: a Critical Review of Past Studies*. Mimeo Environmental Economics Policy Forum in Ethiopia, Addis Ababa and International Food Policy Research Institute, Washington, DC.

Zhang, X. (2004) *Security is Like Oxygen: Evidence from Uganda*. Development Strategy and Governance Division Discussion Paper No. 6, International Food Policy Research Institute, Washington, DC.

Zurayk, R., El-Awar, F., Hamadeh, F., Talhouk, S., Sayegh, C., Chehab, A.G. and Al Shab, K. (2001) Using indigenous knowledge in land use investigations: a participatory study on semi-arid mountainous region of Lebanon. *Agriculture, Ecosystems and Environment* 86, 247–262.

I Development Strategies for Poor People in Less-favoured Areas

2

Designing and Evaluating Alternatives for More Sustainable Natural Resource Management in Less-favoured Areas

SANTIAGO LÓPEZ-RIDAURA,[1] HERMAN VAN KEULEN[1, 2,*] AND KEN E. GILLER[1,**]

[1]Plant Production Systems, Department of Plant Sciences, Wageningen University and Research Centre, Wageningen, The Netherlands; *e-mails: herman.vankeulen@wur.nl; **ken.giller@wur.nl; [2]Plant Research International, Plant Sciences Group, Wageningen University and Research Centre, Wageningen, The Netherlands.

Abstract

Less-favoured Areas (LFAs) are defined as regions with low agricultural potential because of limited and uncertain rainfall, poor soils, steep slopes and/or other biophysical constraints, as well as regions that may have higher agricultural potential, but with poor infrastructure and limited access to markets, low population density and/or other socio-economic constraints. About 800 million people live in LFAs, mostly in the semi-arid tropics of Africa and South Asia, mountain and hillside areas in Africa, Latin America and South-east Asia, as well as in large parts of the humid (sub-)tropics of Africa and Latin America.

The objective of this chapter is to contribute to the design and evaluation of options for more sustainable NRM (Natural Resource Management) in LFAs. First, we deal with some basic issues in the design of alternatives for LFAs, including their definition, their biophysical and socio-economic diversity and heterogeneity and the main development pathways in which alternative NRM strategies may play a role. As LFAs are characterized by strong resource limitations at different scales, relevant issues on productivity and resource use efficiency in LFAs are discussed and a generic framework for the evaluation and design of alternatives is presented.

In the search for options for more sustainable NRM in LFAs (technically feasible, economically viable, ecologically maintainable, socially acceptable), alternatives at different scales must be designed (including technological innovations and policy measures), aiming at maximum marginal resource use efficiency for the most limiting resource, and maximum absolute resource use efficiency for resources locally in more abundant supply. Essential for the success of those alternatives is strong stakeholder participation in both, their design and evaluation, using a multi-scale approach and taking into account the biophysical and socio-economic diversity and heterogeneity of NRM in LFAs.

Introduction

The contribution of the 'Green Revolution' to the increase in food production and alleviation of world hunger is indisputable: two- to four-fold yield increases in main grain crops were achieved following adoption of high-yielding varieties and crop management based on the use of external inputs (irrigation, machinery, fertilizers, plant protection agents) (Conway and Barbier, 1999; Borlaug, 2003; Dalgaard *et al.*, 2003; Pretty *et al.*, 2003). Over the past 40 years, adoption of these technologies has allowed food prices for the growing urban population to remain low and the proportion of hungry people worldwide to decline from 37 to 17% (Sanderson, 2005).

Green Revolution technologies however, have been severely criticized on two accounts, potential environmental consequences and inappropriate technologies.

Environmental consequences

Possible environmental consequences, when mismanaged, include: (i) the loss of (agro-)biodiversity and the local gene pool; (ii) the reliance on non-renewable resources; (iii) soil degradation (e.g. acidification, salinization, compaction, erosion); and (iv) emission of compounds harmful to environmental and human health (e.g. pesticides, greenhouse gases, nitrates). Such environmental concerns were voiced in the United Nations Conference on Environment and Development (UNCED), the so-called 'Earth Summit', in 1992, combined with a plea for more sustainable, environmentally friendly management of natural resources.

Inappropriate technologies

Such technologies were not suitable for, and therefore not adopted by, small-scale farmers operating in resource-poor, diverse and 'fragile' environments with limited infrastructure and market development. About 20% of the population in developing countries (*c.*800 million people) lack economic and/or physical access to the quantity and/or quality of food required for a healthy and productive life (Van Keulen, 2006). More than 50% of the global poor live in India, China and sub-Saharan Africa (Pinstrup-Andersen and Pandya-Lorch, 1998; Sanderson, 2005), with the other half scattered throughout South and South-east Asia, the Middle East, South and Central Africa and Latin America.

Although urban poverty is important, poverty in rural areas predominates in these regions, where a large proportion of the population is farming to obtain both food for home consumption and marketable products for the growing urban economies. The continuing poverty and hunger in these regions of the world, as well as their environmental degradation, were central concerns at the World Summit on Sustainable Development in Johannesburg in 2002.

The future

The International Food Policy Research Institute (IFPRI) launched its 2020 vision for food, agriculture and the environment 'to develop and promote a shared vision and consensus for action for meeting food needs while reducing poverty and protecting the environment', as well as 'to generate the information and encourage debate to influence action by national governments, non-governmental organizations, the private sector, international development institutions, and other elements of civil society'. The general aim of the 2020 vision initiative is to support development of a 'world where every person has the access to sufficient food to sustain a healthy and productive life, where malnutrition is absent, and where food originates from efficient, effective, and low-cost food systems that are compatible with sustainable use of natural resources' (IFPRI, 1995; Pinstrup-Andersen and Pandya-Lorch, 1998).

As part of the activities under the Interdisciplinary Research and Education Fund (INREF), initiated at Wageningen University, The Netherlands, in 2001, the Regional Food Security Policies for Natural Resource Management and Sustainable Economies (RESPONSE) programme is a collaborative effort coordinated by Wageningen University and IFPRI in Washington, DC, aimed at identifying: (i) feasible options for agricultural and rural development in less-favoured areas (LFAs) of the world; and (ii) critical incentives needed to enable rural households in LFAs to invest in efficient and sustainable natural resource management.

In this chapter we discuss options for more sustainable NRM in LFAs (i.e. natural resource management in less-favoured areas), by providing increased understanding of the issues at stake. First, basic concepts and approaches are discussed (following section), including the definition of LFAs, their biophysical and socio-economic heterogeneity and the main development pathways and experiences in the search for alternatives for NRM in LFAs. This is followed by an analysis of issues related to productivity and resource use efficiency, with special emphasis on LFAs that are characterized by strong resource limitations at different scales. The penultimate part presents an analytical framework for the evaluation of alternatives at different scales, and the final section presents some final remarks on the design and evaluation of options for more sustainable NRM in LFAs.

In Search of Alternatives in Less-favoured Areas

Less-favoured areas: a relative term in space and time

The term 'less-favoured areas' (LFAs) was first used in the 1970s in the context of the European Common Agricultural Policy (CAP) to designate remote areas with poor soils and/or unfavourable climatic conditions, in danger of depopulation, where financial support for farmers and conservation of the countryside was regarded as necessary (Wathern *et al.*, 1986).

Today, the term LFA is associated mainly with less-developed countries and refers to regions with low agricultural potential because of limited and uncertain rainfall, poor soils, steep slopes and/or other biophysical constraints, as well as to areas that may have higher agricultural potential but with poor infrastructure and limited access to markets, low population density and/or other socio-economic constraints. These areas include most of the semi-arid tropics of Africa and South Asia, mountain areas in South America and Asia, the highlands of East and Central Africa and the hillside areas in Central America and Southeast Asia, as well as large portions of the (sub-)humid tropics of Africa and Latin America (Pender and Hazell, 2000).

It should be emphasized that 'LFA' is, by definition, a relative term, as 'less' implies a comparison. The notion of LFAs may change with time, for example under the influence of technological developments: e.g. before large-scale intensification of agriculture in Europe, sandy soils were considered less favoured, as clayey soils were chemically (more fertile) and physically (higher water-holding capacity) more favourable. Introduction of machinery, fertilizer application and irrigation facilities favoured the sandier soils that could be worked more timely and easily.

In spatial terms, delineation of LFAs depends on the scale and extent of the analysis (i.e. the boundaries and detail of the system under analysis) (Fig. 2.1). For example, from a global perspective and at national scale, Ethiopia as a country is defined as an LFA; however, within Ethiopia, large differences exist between regions in terms of biophysical and socio-economic characteristics. The central and southern regions, with better soils and higher and more reliable rainfall, are of higher agricultural potential (Fig. 2.1, A) than the highlands in the north. From a regional perspective, within the northern highlands, again, areas exist with higher agricultural potential, such as the relatively flat areas with luvisols and cambisols (Fig. 2.1, B), as opposed to the areas

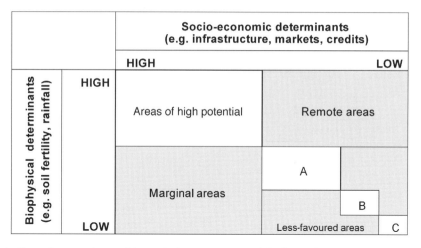

Fig. 2.1. 'Less favoured areas' is a relative concept, mainly dependent on the scale of analysis. Within marginal and remote areas, better regions (A) can be identified and within those, areas of higher (B) and lower (C) potential.

with leptosols on steep slopes (Fig. 2.1, C). Also, at the farm (household) scale, less-favoured areas (fields) in terms of soil fertility are located further away from the homestead, village or roads.

Heterogeneity and diversity in LFAs

One of the main characteristics of LFAs is their diversity and heterogeneity in terms of their biophysical and socio-economic characteristics and the strategies developed by farmers to cope with risks associated with such variability. Intrinsically, LFAs are no more diverse than more-favoured areas; in fact, the latter might present greater opportunities for diversification, but market conditions have led to specialization and homogenization. In LFAs, simultaneous realization of several objectives by NRM (e.g. production of food for home consumption, marketable products, fuelwood, feed for livestock) and the high variability and uncertainty in biophysical and market conditions provide incentives for farmers to diversify.

Underlying aspects of heterogeneity and diversity in LFAs have been discussed in detail by Ruben and Pender (2004) and Van Keulen (2006). Their overall conclusion is that, to design more sustainable alternatives for NRM in LFAs at different scales, such biophysical and socio-economic heterogeneity has to be taken into account and may be exploited.

Heterogeneity and diversity can be analysed, as the term LFA, at different scales. At the field scale, heterogeneity can be observed, either in terms of resource quality such as in hillside fields showing strong within-field fertility gradients, or in terms of management where input application is restricted to part of the fields in a rotational scheme (Kanté, 2001). Digging mini-catchments or 'planting basins' (*zai*) for soil and water conservation, common in (semi-)arid areas, is an example of induced heterogeneity at field scale (Kaboré and Reij, 2004; Sidibé, 2005).

At farm level, heterogeneity is often associated with the fact that single-farm-households manage a number of fields, inherently varying in biophysical resource qualities or as a result of differential management where, for example, distance to the homestead and input use result in strong quality (e.g. nutrients, organic matter, water) gradients (Prudencio, 1993; Tittonell *et al.*, 2005a, b; Giller *et al.*, 2006; Van Keulen, 2006). Also, in socio-economic terms, diversity in on-, off- and non-farm activities contributes to the heterogeneity in LFAs as, dependent on farm-household composition and labour markets, farm-households tend to diversify their activities to cope with the risks associated with agricultural production (Reardon, 1997; Niehof, 2004).

At village, watershed and regional levels, rainwater harvesting is a common example of heterogeneity created to increase the efficiency of water use (Reij *et al.*, 1988); heterogeneity is further increased by differences in farm-household resource endowments, assets, access to markets, credit and institutions. Moreover, in LFAs, livestock activities relying on common pastures are traditionally strong agents in promoting heterogeneity through redistribution of soil nutrients from marginal (pasture) lands to arable fields (Murwira *et al.*, 1995; Williams *et al.*, 2004).

Heterogeneity can thus be considered as both the reason and the consequence of such areas being less favoured, as rural households intentionally diversify the system to cope with risk and benefit from the concavity of the relation between input intensity and output level. Such concavity implies that the average output of two distinct input intensities exceeds the output at the average input intensity (Van Keulen, 2006).

Alternatives for LFAs

Alternatives to stimulate more sustainable NRM in LFAs might include policy measures and technological innovations designed and implemented by different stakeholders at different scales (López-Ridaura, 2005). At national and regional scales, in the agricultural and development policy arena, development pathways for policy formulation have been identified. These development pathways provide broad strategic options for poverty alleviation and sustainable NRM in LFAs in relation to the level of constraints imposed by the biophysical and socio-economic determinants (Ruben and Pender, 2004; Fig. 2.2). At local scales, such as the village, farm-household and field, technological alternatives have been developed for more sustainable NRM in LFAs

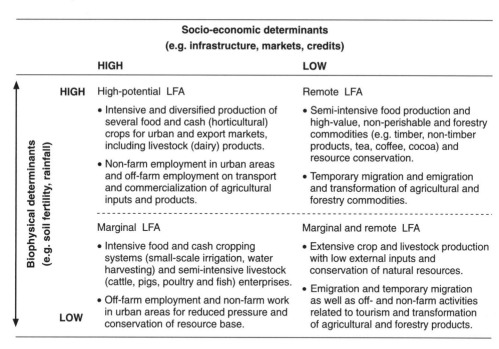

Fig. 2.2. Possible development pathways for policy formulation (after Ruben and Pender, 2004).

(Pretty *et al.*, 2003; IWMI, 2005). Because of their high diversity and heterogeneity in time and space, direct extrapolation is impossible (i.e. transfer of a standard set of activities), and basic principles can be illustrated only by specific experiences.

In areas with favourable agro-ecological conditions, good infrastructure and access to markets, land use intensification for food production, diversification with high-value horticultural crops – as well as commercial dairy production – are feasible options. These options generate important off- and non-farm employment opportunities in the process of transformation, such as transport, storage and commercialization of inputs for agriculture and agricultural commodities. In these areas, however, special attention must be given to the implications of agricultural production, such as the quality of water used for irrigation, the handling of slurry and manure from intensive livestock production and the (mis-)use of inputs for agricultural production that have potentially harmful effects on environmental and human health.

Remote LFAs are commonly associated with relatively high biophysical production potentials, but with weak market access due to their isolation and inadequate infrastructure, as well as fragile environments prone to erosion and resource degradation. Technological innovations and policy measures that encourage production of non-perishable products with high market value such as coffee, cocoa, tea and forestry products, and concurrently effective in the conservation of resources, are appropriate. Transformation of such products for added market value can create important opportunities for increased income generation. Production for subsistence and for markets as well as natural resource management must rely on locally available resources, since external inputs are commonly too costly, as a result of remoteness.

Conservation agriculture (CA) has shown higher and more stable yields, combined with reduced environmental degradation in case studies in the hillsides and mountainous regions of Honduras, Costa Rica, Brazil, Mexico and other Latin American countries, as well as in some areas of Africa and Asia (FAO, 2001). Mimicking natural ecosystems, CA practices include reduction or elimination of tillage operations, maintenance of soil cover and closing of nutrient cycles, practising crop rotation and mixed cropping, growing green manures and cover crops and implementing integrated pest and weed management.

In Costa Rica, for example, the *frijol tapado* (covered beans) system (Kettler, 1997; Meléndez *et al.*, 1999) consists of growing beans (*Phaseolus vulgaris*) using a slash–mulch system, where a semi-determinate bean variety is broadcast into fields covered by selected weeds (e.g. broadleaf and grasses without strong regrowth). The weeds are later cut or chopped and left in the field, so the bean seeds are covered with a green mulch; the beans grow through the mulch and eventually cover it. Increased water availability, reduced soil losses and soil splashing, and avoiding transport of inoculum of bean diseases all result in higher and more stable bean production, as well as in conservation of resources.

In Honduras, the Quesungual agroforestry system (Hellin *et al.*, 1999; Alvarez-Welchez and Cherret, 2002) is a slash–mulch system where secondary forests are partially cleared, leaving selected fodder, fruit and timber trees on steep fields. Direct maize planting into the resulting mulch increases nutrient and water availability for higher maize yields. Yield variation due to erratic rainfall has been reduced, as well as soil losses through erosion. Although CA has also been associated with reduced labour needs for weeding, labour demands for selective tree clearing and weeding and management of mixed stands can represent a constraint for their adoption by labour-constrained households.

Marginal LFAs, characterized by poor-quality resources but by favourable socio-economic conditions – such as access to markets, credit availability and banking services, infrastructure and extension – are located mainly in (semi-)arid regions, such as North Africa and some regions of India, China and Latin America. Alternatives, promoting intensification of agricultural production based on a combination of locally available resources and external inputs, such as fertilizers and irrigation, might increase food production and improve livelihoods in these regions. Perishable cash crops and livestock products for urban markets may represent important income sources for farmers in marginal LFAs, and non- and off-farm employment can generate complementary farm-household income and attractive options to reduce pressure on natural resources (De Ridder *et al.*, 2004).

Integrated soil fertility management (ISFM) and mixed crop–livestock systems, based on the combined use of mineral fertilizers and organic amendments, have been the focus of extensive research for restoring soil fertility and increasing productivity and resource (water) use efficiency in the highly weathered and infertile soils of sub-Saharan Africa (Palm *et al.*, 2001). The synergy between the efficient use of locally available organic amendments (e.g. crop residues and manure) and mineral fertilizers is well documented (Fofana *et al.*, 2004). As the organic matter content of the soil increases, so does its indigenous soil fertility, as well as the recovery of mineral fertilizer (Van der Meer and Van Uum-Van Lohuyzen, 1988), leading to higher nutrient uptake and crop production, which in turn results in larger quantities of organic resources of higher quality, with positive effects on fodder provision for animals.

Breman (2003) has shown that integrated soil fertility management can double, or even triple, yields in the course of 4 years under Sahelian conditions. However, limited access to fertilizers might hamper implementation of ISFM and/or, where the resource base is extremely poor, large quantities of organic inputs are required to restore soil fertility and attain acceptable levels of fertilizer recovery. In semi-arid environments, such as Tigray, northern Ethiopia, biomass production might be insufficient to provide such amounts of manure (Abegaz, 2005).

Remote *and* marginal LFAs are constrained by both biophysical and socio-economic conditions and, therefore, options for agriculture-led development are limited. In these regions, such as much of sub-Saharan Africa and the cold and dry highlands of Eastern Africa and Latin America, extensive low-external input crop and livestock production must be integrated, aiming at the most efficient use of locally available resources for food production for the local population. A limited number of agricultural commodities with high market value are available to serve as a source of cash income for rural households. Temporary migration and emigration can reduce pressure on natural resources, and therefore activities with a low labour demand are needed to sustain the rural population and encourage the conservation of resources.

Key to the success of alternatives for specific settings in LFAs is that they increase the productivity of agricultural activities through the efficient use of the available resources and the conservation of the resource base (i.e. without soil mining, pollution or other degrading processes).

Resource Use Efficiency in LFAs

Efficient use of the available resources is a precondition for attaining sustainable livelihoods, especially in LFAs, where resources are of poor quality and their availability is restricted, highly variable and uncertain.

Resource Use Efficiency (RUE) has historically attracted extensive attention and debate in relation to agricultural production (De Wit, 1992; Zoebl, 1996; Nijland and Schouls, 1997; Kho, 2000). Trenbath (1986), in his treatise of the concept of RU(utilization)E, distinguishes the efficiency of resource capture (in relation to resource availability) and the efficiency of resource conversion (in relation to resource capture):

$$RUE = \frac{Resource}{Efficiency} \quad \frac{Capture}{(R\_Cap\_E)} \times \frac{Resource}{Efficiency} \quad \frac{Conversion}{(R\_Con\_E)} \quad (2.1)$$

Increased RUE can be attained by either increasing capture efficiency, conversion efficiency or both, but RUE (and in fact R_Cap_E and R_Con_E) is commonly associated with the law of diminishing returns, in which increasing resource availability above a certain optimal rate results in decreasing efficiency. A three-quadrant diagram (De Wit, 1953; Van Keulen, 1982) represents the diminishing returns in RUE for plant nutrients (see Fig. 2.3). With the example of nitrogen (N) application and yield, quadrants I and IV illustrate R_Con_E and R_Cap_E, respectively, while quadrant II shows overall resource use efficiency.

While the law of diminishing returns (and its associated reduction in resource use efficiency) is widely accepted, over more than a century the debate has continued on the interaction between different resources and the effects of such interaction on resource use efficiency. Von Liebig's law of the minimum (Von Liebig, 1855) states that the yield of a crop is proportional to the supply of that element (resource) that is essential for the full development of the crop, and that is available in relatively the smallest amount (De Wit, 1992).

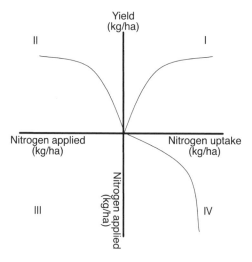

Fig. 2.3. Diminishing returns in Resource Use Efficiency (RUE).

This law is metaphorically illustrated by a barrel with staves of different height in which the shortest stave determines the amount of water the barrel can hold (see Fig. 2.4 (a)). Liebscher's law of the optimum (Liebscher, 1895) states that the closer other production factors (resources) are to their optimum, the higher the use efficiency of the production factor in minimum supply, and thus the higher the yield (Zoebl, 1996). Following the staves metaphor, the law of the optimum can be illustrated with a tilted barrel where the height of the other staves (resources) determines how much the barrel can be tilted and, therefore, the amount of water the barrel can hold (see Fig. 2.4 (b)).

De Wit (1992) showed that, in the agricultural production process, Liebscher's law of the optimum was more common than Liebig's law of the minimum and illustrated that with examples where N uptake by crops and its conversion efficiency was influenced by the availability of both N and other production factors such as water or other nutrients (e.g. phosphorus, P), weed competition and pest and disease incidence. The law of the optimum implies that increasing the availability of another resource, in addition to raising the yield plateau, also positively affects the slope, suggesting a higher resource use efficiency of the most limiting resource (see Fig. 2.5).

Based on the studies of Blackman in 1911, Kho (2000) explored the degree of limitation of different resources and their combined effect on overall RUE in crop growth. By explicitly taking into account the balance of resources necessary for crop growth, he derived a coefficient based on the partial derivatives of the relation between resource

(a)   (b)

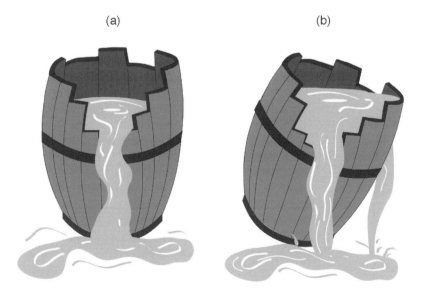

Fig. 2.4. A metaphoric representation of the law of the minimum (a) and the law of the optimum (b) for resource use efficiency.

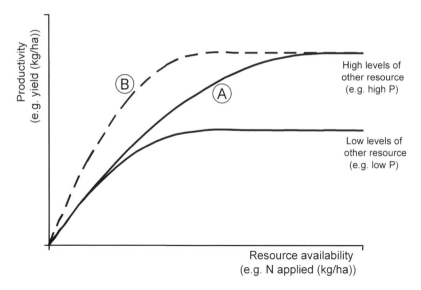

Fig. 2.5. Interaction between resources and resource use efficiency. (A), von Liebig's Law of the Minimum; (B), Liebscher's Law of the Optimum. N, nitrogen; P, phosphorus.

availability and yield that quantifies the degree of limitation of resources in a specific environment. This possibility of explicitly quantifying the combined effect of multiple-resource constraints provides important guidelines for the development of alternative technologies for LFAs where various resources are in limited supply.

One important feature of these analyses is that they describe RUE in general terms and therefore can be applied to different resources at different scales. The examples given here (most common in agricultural research) refer to the interactions between the resources most often yield-limiting under field conditions, water and/or nutrients (often N and P). However, the same laws can be used for resources such as labour, land and capital invested by the farm-household in the production of a specific crop and its yield. Even at regional scale, these analyses can relate resources such as infrastructure, markets and public/private investment to the value of agricultural production.

The different theories on resource use efficiency also apply to both less- and more-favoured areas; however, for LFAs, we may be focused on the wrong end of the response curve, as such areas are characterized by strong limitations of biophysical (e.g. low soil fertility, low and erratic rainfall) and socio-economic (e.g. low labour availability due to migration, poor infrastructure and limited market/credit access) resources.

In LFAs, where yields are far below the potential (the plateau in Figs 2.3 and 2.6), resource use efficiency must be explored at the lower end of the curve. In this segment, we might expect an exponential, rather than proportional (linear), relation between the availability of a given resource and productivity. For cereal production, it is widely accepted that 'below a certain annual rainfall there is no seed yield at all and above that level, there is a practically linear relation until some maximum is reached above which the yield decreases with increasing rainfall, because of the occurrence of periods with severe waterlogging' (De Wit and Van Keulen, 1987; Fig. 2.6).

Such nonlinearity leads to erroneous yield predictions when average climatic data are used to calculate production if the weather variability from year to year is considerable, as yields will be overestimated at the lower end of the curve (A) and underestimated at the higher end (B)[1] (De Wit and Van Keulen, 1987; Nonhebel, 1993; Rabbinge and Van Ittersum, 1998; Fig. 2.6).

Such nonlinear response functions exist for many variables affecting crop growth, such as the relation between radiation and carbon dioxide assimilation, turgidity and stomatal conductance, root density and uptake of water and nutrients (De Wit and Van Keulen, 1987).

In fact, such nonlinearity is also found for resources at higher scales of analysis; for example, at field scale, a minimum degree of weed control is required to produce a marketable yield. Beyond that point, the relation between crop production and degree of weed control is linear, until a point where more intensive weed control will no longer increase yields (Van Heemst, 1985). A similar relation can be elaborated for resources at regional scale, such as infrastructure, where one single kilometre of paved road might not result in a reduction in transaction costs.

For the description of such nonlinear response functions, Yin *et al.* (2003) developed the 'beta growth function' which describes the dynamics of crop growth on the basis of three parameters: (i) the time at

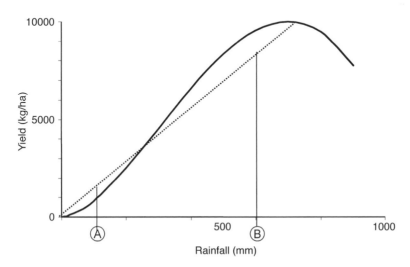

Fig. 2.6. Non-linear relation between rainfall and yield. When averaging data, yield is overestimated at low rainfall levels (A) and underestimated at high rainfall levels (B) (from De Wit and Van Keulen, 1987).

which the maximum growth rate is obtained; (ii) the time at the end of growth; and (iii) the maximum value of biomass (weight). This function is useful in analysing resource use efficiency in general, as time can be substituted by resource availability and the three characteristics of: (i) maximum production; (ii) the resource level needed to attain that production; and (iii) the inflection point at which this relationship changes from exponential to diminishing returns (Eqn 2.2):

$$O = O_{max}\left(1 + \frac{I_{max} - I}{I_{max} - I_{opt}}\right)\left(\frac{I}{I_{max}}\right)^{\frac{I_{max}}{I_{max} - I_{opt}}} \tag{2.2}$$

Where O is the output level at a specific input level, O_{max} the maximum attainable (potential) output level, I the input level, I_{max} the input level at which O_{max} is attained (plateau) and I_{opt} the inflection point in the sigmoid curve. On the basis of this function, RUE can be analysed for different sections of the curve. The absolute value of RUE (output/input), at any point of the function is represented by:

$$RUE = O_{max}\left(\frac{2I_{max} - I_{opt}}{I_{max}(I_{max} - I_{opt})}\right)\left(\frac{I_{max}}{I_{opt}}\right)^{\frac{I_{opt}}{I_{max} - I_{opt}}} \tag{2.3}$$

and the marginal (or Relative) Resource Use Efficiency (R_RUE: ΔOutput/ΔInput) by:

$$R\_RUE = O\left(\frac{(2I_{max} - I_{opt})(I_{max} - I)}{(I_{max} - I_{opt})(2I_{max} - I_{opt} - I)I}\right) \tag{2.4}$$

With this set of equations, productivity and resource use efficiencies can be calculated for different input levels (see Fig. 2.7). Maximum marginal resource use efficiency (point A) is attained at the inflection point; maximum absolute resource use efficiency (point B) is attained at the end of the linear part of the relationship between input and output.

Similarly to the other resource use efficiency analyses, this sigmoid function can be applied to different resources at different scales of analysis, from water use efficiency at the plant or field scale to 'infrastructure use efficiency' at the regional scale.

When in a specific system a given resource is abundantly available, maximum absolute resource use efficiency might be aimed for. However, in situations with limited resources at different scales, as in LFAs (i.e. water and nutrients at the field scale, labour and capital at the farm and/or roads and markets at the regional scale), maximum marginal resource use efficiency might better be targeted. Maximization of a marginal value is widely accepted in development economics, where maximizing marginal utility is considered the objective governing the behaviour of rural households.

It is impossible to identify in general terms the limiting, and non-limiting, resources because of the spatial and temporal heterogeneity of

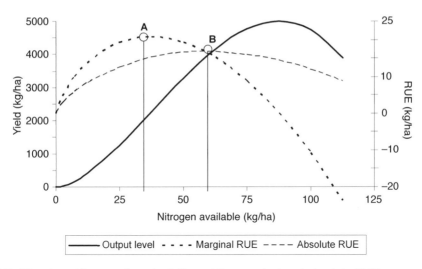

Fig. 2.7. The sigmoid curve of productivity and the marginal and absolute RUEs.

LFAs, in biophysical and socio-economic characteristics, as well as in resource endowments and management. For this reason, general analytical frameworks, flexible and adaptable to different conditions, are needed to assist in the design and evaluation of alternatives for more sustainable NRM.

Analytical Frameworks for Design and Evaluation of Alternatives for LFAs

Analytical frameworks for evaluation of NRM systems (NRM) in LFAs are needed in support of the process of design of specific alternatives for more sustainable NRM; alternatives might include both technological innovations at the field, farm-household and village scales and policy measures implemented at (sub-)regional, national and/or international scales. Thus, such analytical frameworks must be able to generate and integrate accurate biophysical and socio-economic knowledge and information at different scales. Such information enables assessment of the advantages and disadvantages of different alternatives in terms of various indicators, including not only the (physical) production of specific systems but, in addition, their economic viability, their resource-use efficiencies and their ability to cope with the heterogeneity and variability characteristics of LFAs, as a basis for evaluating possible options for their development (López-Ridaura, 2005).

For evaluation at different scales of alternatives for small-scale farming in LFAs, a multi-scale sustainability evaluation framework has been designed to identify and quantify the strengths and weaknesses of different alternatives on the basis of sets of indicators representing the objectives of stakeholders related to the sustainability of NRM. The framework provides a flexible and participatory tool to support stakeholders in: (i) identifying the main issues related to sustainability in specific case studies from a robust, interdisciplinary and theoretical perspective; (ii) selection and assessment of case-specific indicators to evaluate the limitations and potentials of current practices and alternatives; and (iii) integration of the information supplied by the indicators in support of the design of alternatives and the associated decision-making and development processes.

The framework is flexible, in the sense that it permits integration of different sources of information (e.g. models, experiments, surveys, statistics and GIS) and techniques of analysis (e.g. linear programming, multi-agent systems, multi-criteria decision making or fuzzy logic) in the different steps of the evaluation. In the framework, the evaluation process is conceived as a cycle, where stakeholders play the central role (López-Ridaura, 2005; Fig. 2.8). The steps of the framework allow the integration of stakeholders' views in the evaluation process, thus supporting and facilitating their transparent discussion, aiming at joint efforts for the development and promotion of alternatives, taking into account their objectives at different scales. The cyclic structure allows periodic 'updating' of objectives of stakeholders and indicators.

The evaluation cycle is divided into two phases: (i) a *systems analysis phase* (steps 1–3), in which sets of criteria and specific indicators for the different scales are derived; and (ii) a *systems synthesis phase* (steps 4–7), in which quantification and aggregation of indicators are performed and alternatives evaluated by means of scenario analyses.

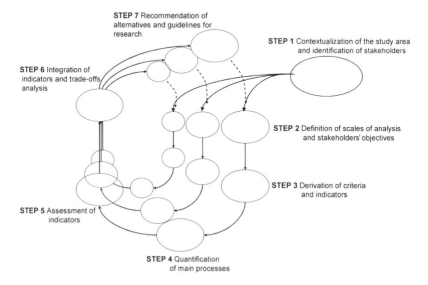

STEP 7 Recommendation of alternatives and guidelines for research

STEP 1 Contextualization of the study area and identification of stakeholders

STEP 6 Integration of indicators and trade-offs analysis

STEP 2 Definition of scales of analysis and stakeholders' objectives

STEP 3 Derivation of criteria and indicators

STEP 5 Assessment of indicators

STEP 4 Quantification of main processes

Fig. 2.8. The multi-scale sustainability evaluation cycle (adapted from López-Ridaura *et al.*, 2005a).

Deriving indicators for multi-scale sustainability evaluation

The *systems analysis phase* of the framework aims at derivation of case-specific criteria and indicators for multi-scale sustainability evaluation. The first step comprises contextualization of the study area, describing the main biophysical and socio-economic determinants and limitations for prevailing NRMs and their most prominent characteristics. Subsequently, the relevant scales of evaluation are defined based on the scales at which the various stakeholders operate and perceive the main challenges for NRM. Finally, strategic indicators are derived that represent their objectives in relation to sustainability of NRM at different scales.

Five scale- and discipline-independent properties or attributes of sustainable NRM were identified as the basis for the derivation of indicators at different scales based on a thorough analysis of different properties used in the field of sustainability evaluation and a systemic perspective to NRM. These properties can be assessed at any scale of analysis and deal, on the one hand, with the performance of the NRM themselves – *productivity, stability* – and, on the other hand, with their ability to cope with changes in their environment or in their own functioning – *reliability, resilience* and *adaptability* (López-Ridaura *et al.*, 2005a).

In operational terms, the degree to which a system is sustainable is characterized by its capabilities of producing, as efficiently as possible, a specific combination of goods and services that satisfies a set of goals (the system is productive) without degrading its resource base (the system is stable) when facing 'normal' (the system is reliable), 'extreme' and/or 'abrupt' (the system is resilient) or 'permanent' (the system is adaptable) variations in its environment and/or co-existing systems.

At farm-household scale, indicators such as income, food self-sufficiency, returns to labour and benefit/cost ratio are commonly used to evaluate the productivity and efficiency of NRM; soil losses and nutrient and soil organic matter balances are common indicators for the stability of NRM in LFAs, while the variation in productivity indicators with rainfall and inputs, their value under extremes of rainfall, temperature and prices of outputs and inputs, as well as the investment costs and the degree of dependence on external inputs are suitable indicators for their reliability, resilience and adaptability. At higher levels of analysis, the economic value of agricultural production, regional food production and employment generation are common indicators for assessing the productivity of the NRM; the degree of deforestation, erosion and overgrazing, as well as the equity of income distribution, are possible indicators for the biophysical and socio-economic stability of the system.

Finally, indicators such as the variation and extreme values of production with variation in rainfall, temperature and input level, as well as access to credit and insurance, are examples of indicators representing the reliability, resilience and adaptability of the NRM. As an example, Table 2.1 presents sets of indicators selected, at regional and farm-household scale, for sustainability evaluation of agricultural activities in the Cercle de Koutiala, an important cotton, grain and livestock production region in southern Mali.

Quantifying indicators for multi-scale sustainability evaluation

In the *systems synthesis phase* of the evaluation framework, step 4 aims at quantification of the main processes and features associated with natural resource management, such as yields (e.g. grain, marketable products and forage) and their variation, with main biophysical and socio-economic factors, labour and inputs needed for production, prices of inputs and outputs, soil losses and nutrient balances. In step 5, information generated in step 4 for the quantification of indicators at different scales is integrated. Step 6 aims at integration and articulation of indicators for an overall assessment of current and alternative NRM at different scales. In step 7, recommendations for alternative, more sustainable NRM are formulated, including actions at different scales of analysis.

Table 2.1. Indicators selected at the regional and farm-household scales for sustainability evaluation in the Cercle de Koutiala, Southern Mali.

	Indicators	
Attribute	Farm-household	Region
Productivity	Gross margin (GM)	Value of agricultural production (VAP)
	Returns to labour	Employment generation
	Benefit–cost ratio	Grain production
	Food (grain) self-sufficiency index (FSF)	Food (grain) self-sufficiency index (FSF)
Stability	Nutrient and carbon balances	Forage self-sufficiency index
	Nutrient and carbon gradients	Fuelwood self-sufficiency index
	Soil loss	Soil loss
		Quantity of biocide usage
Reliability, resilience and adaptability	Variability in GM with rainfall variation	Variability in VAP with rainfall variation
	Variability in GM with price variation	Variability in VAP with price variation
	Diversity of activities	Diversity of activities
	GM in extreme (low) rainfall years	VAP in extreme (low) rainfall years
	FSF in extreme (low) rainfall years	FSF in extreme (low) rainfall years
	GM at low-output and high-input prices	VAP at low-output and high-input prices
	Monetary costs of natural resource management	Number of farmers' organizations
	Dependence on external inputs	Number of farmers belonging to organizations

In the context of quantitative analysis of land use systems, various tools have been developed to generate and quantify indicators for large numbers of alternatives in support of land use policy formulation and the evaluation of technological innovations for NRM (Hengsdijk *et al.*, 1998, 1999; Van Ittersum *et al.*, 1998; Van Keulen *et al.*, 2000). Linear programming, as one of these tools, has been widely used for quantification of indicators and trade-offs, as well as for scenario analyses (Sissoko, 1998; Hengsdijk, 2001; Bos, 2002).

A multi-scale, multiple-goal linear programming (M_MGLP) model has been developed, in which indicators at different scales can be used as objective functions and/or as constraints in scenario formulation and assessment. In addition, constraints such as labour, traction, manure and forage availability can be set to different scales allowing (or not) transfer of such resources across scales of analysis (López-Ridaura *et al.*, 2005b). The M_MGLP model allows explicit quantitative identification of the advantages and disadvantages of alternative NRM in terms of the values of the indicators selected for sustainability evaluation at each of the scales of analysis.

For example, in the Cercle de Koutiala, Mali, alternative activities – including integrated soil fertility management and soil and water conservation practices – are being promoted to stimulate productivity and halt nutrient mining and soil degradation.

Figure 2.9 shows the comparison, at regional and farm-household scales, of two scenarios, current versus alternative agricultural activities, in which regional value of agricultural production was maximized, while maintaining food self-sufficiency at the farm-household scale in dry and normal rainfall years as constraints (not shown in diagram), as well as forage self-sufficiency at regional scale, as pasture land is collectively managed.

Alternative activities may represent important opportunities for development in Koutiala, as they increase productivity and conserve natural resources. However, at the farm-household scale, the increased dependence on external inputs and their high and uncertain prices might represent serious constraints for adoption of such alternatives, as they become economically less attractive and more risk-prone.

With the M_MGLP, trade-off analysis can be performed by running several scenarios, in which an indicator at any scale of analysis, used as constraint, is varied over a range of values, while maximizing another indicator, also at any scale of analysis. López-Ridaura (2005) presents further details on the development of the M_MGLP and the analysis of different scenarios for the Cercle de Koutiala.

The M_MGLP model is of an explorative nature, identifying the opportunities and limitations for NRM rather than predicting behaviour of actors. Results of the M_MGLP model can be dovetailed with predictive studies to establish specific pathways for the implementation of alternatives, including policy measures and technological innovations for more sustainable NRM (Hengsdijk *et al.*, 2005).

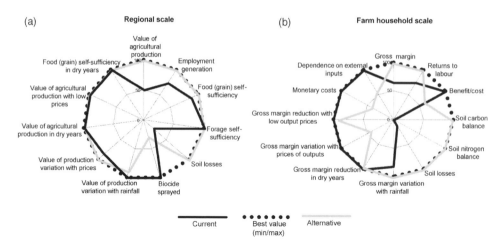

Fig. 2.9. Comparison of current versus alternative activities for regional and farm household scales in the Cercle de Koutiala, southern Mali.

Concluding Remarks

The notion less favoured is relative by definition and its delineation therefore depends on the temporal and spatial scales of analysis adopted for the design and evaluation of alternative resource management systems. For appropriate targeting of policy measures or technological innovations aiming at improved rural livelihoods in LFAs, a clear definition of scales of analysis is needed to identify, for each situation, the specific areas with the strongest limitations of resources, as well as their relations with other more favourable areas.

International, national and regional policies and programmes for poverty alleviation and food security must be based on considerations for sustainable development of those less-favoured nations, regions and/or sub-regions. For example, at the international scale, trade liberalization policies benefit only the poor, if subsidies in rich countries are stopped; at the national scale, subventions on the basis of area cropped or production will yield the greatest benefits for the large and well-equipped farmers. Similarly, in developing technological innovations for increased productivity and conservation of resources at the farm and field scales, identification, increased productivity and resource use efficiency of marginal areas and households are required, and therefore productivity of better-endowed farmers or more fertile fields should not be attained at the expense of degradation of more marginal farms and fields.

Less-favoured areas are heterogeneous and diverse, whether in terms of natural or human systems (Van Keulen, 2006). Such diversity and heterogeneity has important consequences (often positive in the face of variable climates) for the productivity of specific resources and for risk management by farm-households. At the farm scale, in the design of technological innovations, common characteristics of target households and fields must be identified, i.e. relative farm endowments – such as labour and traction to land ratios, age and gender composition of farm-households, remoteness of farms and fields and field characteristics such as slope, soil fertility and water availability. In the design of alternative policies for LFAs, such heterogeneity implies failure of 'one size fits all' approaches. Policy makers must thus identify the specific characteristics of nations, regions, farm-households and fields most likely to be affected.

To identify the most limiting resource(s) for different farm-household types and cope with the temporal variability and spatial heterogeneity, long-term on-farm experimentation on innovative cropping and livestock activities by small-scale farmers is a key activity in the design of alternative, more sustainable natural resource management systems in LFAs. Moreover, exchange of experiences, knowledge and materials among farmers (i.e. farmer field schools or seed exchange gatherings) can contribute substantially to the dissemination and adaptation of such alternative systems. On-station experimentation and modelling research should be limited to understanding of basic processes driving long-term productivity (e.g. micro–macronutrient interactions, soil organic matter decomposition and nutrient availability, crop–weed interactions) and the exploration of future-oriented options not yet practised in a specific setting.

Temporal variability in main biophysical and socio-economic determinants in LFAs is often *buffered* by farm-households through diversified agro-ecosystems and livelihood strategies, as well as through improvements in the quality of the resource base (e.g. improved soil fertility through conservation agriculture or integrated soil fertility management). Several of those efficient and risk-avoiding technologies can be traced back to traditional farming systems as, for example, cultivation of several crops and crop varieties within the same field, traditional soil and water management (e.g. *waru waru* in Peru, *chinampas* in Mexico) or specific combinations of crop and wild species and their role in yield stability in traditional mixed cropping and agroforestry systems. Study of such traditional farming systems and their multifunctional complexity may yield relevant information for the design of alternatives for small-scale farmers in marginal areas (Altieri, 2002).

Increased productivity of agricultural activities is urgently needed in LFAs to meet the growing demand for food to maintain self-sufficiency and increase the marketable surplus to generate more income for increased welfare for the farm-household. Conserving, or preferably improving, the quality of the resource base on which this productivity depends is an equally important condition for the sustainability of NRM.

Resource use efficiency plays a crucial role in determining the production potentials in LFAs that are characterized by strong resource limitations. Identifying the most limiting resources for agricultural production and their maximum absolute and marginal use efficiency for a specific system is essential for design of alternatives for more sustainable NRM in LFAs. At the field scale, limiting resources may include water, and (micro- and macro-)nutrients, while soil properties, such as structure and organic matter content, may play important roles in determining the use efficiency of such resources.

At the farm-household scale, labour may be limiting in, for example, households where part of the labour force is engaged in off-farm activities, has migrated or has been curtailed by diseases such as HIV/AIDS and/or malaria. In other situations, land or capital to invest in agricultural production may be limited. Depending on the main limiting factor, different technologies will be appropriate and, therefore, for each region specific baskets of alternatives for different farm-household types must be available. At higher scales of analysis, infrastructure, access to markets, credit, extension services and/or organization of farmers may be the most limiting resources for agricultural development, and their enhancement might determine the success of alternative activities.

In general terms, in the search for options for more efficient natural resource management systems in LFAs, maximum marginal resource use efficiency should be aimed at for the most limiting resource, and maximum absolute resource use efficiency for resources locally in more abundant supply.

However, it is important to stress that success of specific alternatives is not solely dependent on their advantages in terms of enhanced productivity, resource conservation and/or resource use efficiency, but also on the prevailing socio-economic conditions. For example, Arellanes and Lee (2003) found that, besides plot characteristics, access to irrigation and soil quality, education of the heads of households, regular visits by extension agents and land ownership were major determinants for the adoption of conservation technologies in the hillsides of Honduras. Thus, integral development strategies, including inter-sectorial (agricultural and non-agricultural) actions are needed for adoption of more sustainable NRM in LFAs.

Analytical frameworks for the design and evaluation of alternatives in LFAs should be able to capture the complexity of natural resource management in such marginal and heterogeneous conditions. They should also allow stakeholders to gain insight into the main limitations hampering agricultural productivity and the efficiency of resource use at different scales to evaluate the effect of integrated strategies for agricultural development in a given region, including policy measures and technological innovations.

The multi-scale sustainability evaluation framework presented here offers a structured and coherent set of guidelines, developed from an interdisciplinary and systemic perspective, to select, quantify, assess and integrate case-specific indicators derived from environmental, economic and social concerns (objectives, aspirations) of stakeholders and their desired development pathways. Quantification and integrated assessment of indicators and their trade-offs allow identification of potential tensions between objectives across the scales at which these objectives are pursued, as indicators retain their explicit meaning, allowing stakeholders to discuss them from their own perspective and aspirations.

Independently of the analytical tools employed in the design and evaluation of alternatives for LFAs, a multi-scale sustainability evaluation approach must be adopted, in which the advantages of different alternatives are evaluated in relation to the objectives of stakeholders at different scales of analysis. Such evaluation must involve stakeholders along the whole process of design and evaluation in support of a transparent and open dialogue among stakeholders. This dialogue among actors is an indispensable feature for operationalization of the concept of sustainability and the development of concerted actions for design, evaluation and implementation of more sustainable NRM in LFAs, that will lead to real development and not simply to sustaining the currently unfavourable situation.

Endnote

[1] This also happens when using aggregated soil fertility data, as yields will be over-estimated for very poor soils and underestimated for somewhat more fertile soils.

References

Abegaz, A. (2005) Farm management in mixed crop–livestock systems in the northern highlands of Ethiopia. PhD Thesis, Wageningen University, Wageningen, The Netherlands.

Altieri, M.A. (2002) Agroecology: the science of natural resource management for poor farmers in marginal environments. *Agriculture, Ecosystems and Environment* 93, 1–24.

Alvarez-Welchez, L. and Cherret, I. (2002) The Quesungual systems in Honduras: alternative to slash-and-burn. *Magazine on Low External Input and Sustainable Agriculture* (LEISA) 18, 10–11.

Arellanes, P. and Lee, D.R. (2003) The determinants of adoption of sustainable agriculture technologies: evidence from the hillsides of Honduras. *Proceedings of the 25th International Conference of Agricultural Economists (IAAE)*, Durban, South Africa, 16–22 August, pp. 693–699.

Borlaug, N.E. (2003) *Feeding a World of 10 Billion People. The TVA/IFDC Legacy.* Third Travis P. Hignett Memorial Lecture, IFDC, Muscle Shoals, Alabama.

Bos, J. (2002) Comparing specialised and mixed farming systems in the clay areas of the Netherlands under future policy scenarios. PhD Thesis, Wageningen University, Wageningen, The Netherlands.

Breman, H. (2003) The fertility triangle, motor for rural and economic transformation in the Sahel. *Sahel Agricultural Symposium 2003*, INSAH-IER-Syngenta Foundation for Sustainable Agriculture, 1–3 December, Bamako, Mali.

Conway, G. and Barbier, E. (1999) *After the Green Revolution. Sustainable Agriculture for Development.* Earthscan, London.

Dalgaard, T., Hutchings, N.J. and Porter, J.R. (2003) Agroecology, scaling and interdiscipli-
      narity. *Agriculture, Ecosystems and Environment* 100, 39–51.
De Ridder, N., Breman, H., Van Keulen, H. and Stomph, T.J. (2004) Revisiting a 'cure
      against land hunger': soil fertility management and farming systems dynamics in the
      West African Sahel. Agricultural Systems 80, 109–131.
De Wit, C.T. (1953) *A Physical Theory on the Placement of Fertilizers.* Verslag
      Landbouwkundig Onderzoek 59.4, Staatsdrukkerij, 's Gravenhage, The Netherlands.
De Wit, C.T. (1992) Resource use efficiency in agriculture. *Agricultural Systems* 40,
      125–151.
De Wit, C.T. and Van Keulen, H. (1987) Modelling production of field crops and its
      requirements. *Geoderma* 40, 253–265.
FAO (Food and Agriculture Organization) (2001) *Conservation Agriculture: Case Studies
      in Latin America and Africa.* FAO Soils Bulletin 78, FAO, Rome.
Fofana, B., Breman, H., Carsky, R.J., Van Reuler, H., Tamelokpo, A.F. and Gnakpenou, K.D.
      (2004) Using mucuna and P fertilizer to increase maize grain yield and N fertilizer
      use efficiency in the coastal savanna of Togo. *Nutrient Cycling in Agroecosystems* 68,
      213–222.
Giller, K., Rowe, E., De Ridder, N. and Van Keulen, H. (2006) Resource use dynamics and
      interactions in the tropics: scaling up in space and time. *Agricultural Systems* 88,
      7–28.
Hellin, J., Alvarez-Welchez, L. and Cherrett, I. (1999) The Quezungual system: an indige-
      nous agroforestry system from Western Honduras. *Agroforestry Systems* 46, 229–237.
Hengsdijk, H. (2001) Formalizing agro-ecological knowledge for future-oriented land use
      studies. PhD Thesis, Wageningen University, Wageningen, The Netherlands.
Hengsdijk, H., Van Ittersum, M.K. and Rossing, W.A.H. (1998) Quantitative analysis of
      farming systems for policy formulation: development of new tools. *Agricultural
      Systems* 58, 381–394.
Hengsdijk, H., Bouman, B.A.M., Nieuwenhuysen, A. and Jansen, H.G.P. (1999)
      Quantification of land use systems using technical coefficient generators: a case study
      for the Northern Atlantic zone of Costa Rica. *Agricultural Systems* 61, 109–121.
Hengsdijk, H., Van den Berg, M., Roetter, R., Wang, G., Wolf, J., Lu, C. and Van Keulen, H.
      (2005) Consequences of technologies and production diversification for the economic
      and environmental performance of rice-based farming systems in East and Southeast
      Asia. In: Toriyama, K., Heong, K.L. and Hardy, B. (eds) *Rice is Life: Scientific
      Perspectives for the 21st Century.* International Rice Research Institute, Los Baños
      and Japan International Research Center for Agricultural Sciences, Tsukuba, Japan
      (CD-ROM), pp. 422–425.
IFPRI (International Food Policy Research Institute) (1995) *A 2020 Vision for Food,
      Agriculture, and the Environment: the Vision, Challenge and Recommended Action.*
      IFPRI, Washington, DC.
IWMI (International Water Management Institute) (2005) *Bright Spots. Helping
      Communities to Help Themselves* (http://www.iwmi.cgiar.org/brightspots/accessed
      November 2005).
Kaboré, D. and Reij, C. (2004) *The Emergence and Spreading of an Improved Traditional
      Soil and Water Conservation Practice in Burkina Faso.* EPTD Discussion Paper No.
      114, IFPRI, Washington, DC.
Kanté, S. (2001) Gestion de la fertilité des sols par classe d'exploitation au Mali-Sud. PhD
      Thesis, Wageningen University, Wageningen, The Netherlands.
Kettler, J.S. (1997) Fallow enrichment of a traditional slash/mulch system in southern
      Costa Rica: comparisons of biomass production and crop yield. *Agroforestry Systems*
      35, 165–176.

Kho, R.M. (2000) On crop production and the balance of available resources. *Agriculture, Ecosystems and Environment* 80, 71–85.

Liebscher, G. (1895) Untersuchungen über die Bestimmung des Düngerbedürfnisses der Ackerböden und Kulturpflanzen. *Journal für Landwirtschaft* 43, 49–125.

López-Ridaura, S. (2005). Multi-scale sustainability evaluation. A framework for the derivation and quantification of indicators for natural resource management systems. PhD Thesis, Wageningen University, Wageningen, The Netherlands.

López-Ridaura, S., Van Keulen, H., Van Ittersum, M.K. and Leffelaar, P.A. (2005a) Multiscale methodological framework to derive indicators for sustainability evaluation of peasant natural resource management systems. *Environment, Development and Sustainability* 7, 51–69.

López-Ridaura, S., Van Keulen, H., Van Ittersum, M.K. and Leffelaar, P.A. (2005b) Multiscale sustainability evaluation. Quantifying indicators for different scales of analysis and their trade-offs using Linear Programming. *International Journal of Sustainable Development and World Ecology* 12, 81–97.

Meléndez, G., Vernooy, R. and Briceño, J. (1999) *El Frijol Tapado en Costa Rica: Fortalezas, Opciones y Desafíos.* Universidad de Costa Rica, San José, Costa Rica.

Murwira, K.H., Swift, M.J. and Frost, P.G.H. (1995) Manure as a key resource in sustainable agriculture. In: Powell, J.M., Fernández-Rivera, S., Williams, T.O. and Renard, T.C. (eds) *Livestock and Sustainable Nutrient Cycling in Mixed Farming Systems of sub-Saharan Africa.* Vol. II: Technical papers, ILCA, Addis Ababa, pp. 131–148.

Niehof, A. (2004) The significance of diversification for rural livelihood systems. *Food Policy* 29, 321–338.

Nijland, G.O. and Schouls, J. (1997) *The Relation Between Crop Yield, Nutrient Uptake, Nutrient Surplus and Nutrient Application.* Wageningen Agricultural University Papers 97.3, Wageningen, The Netherlands.

Nonhebel, S. (1993) The importance of weather data in crop growth simulation models and assessment of climatic change effects. PhD Thesis, Wageningen Agricultural University, Wageningen, The Netherlands.

Palm, C.A., Giller, K.E., Mafongoya, P.L. and Swift, M.J. (2001) Management of soil organic matter in the tropics: translating theory into practice. *Nutrient Cycling in Agroecosystems* 61, 63–75.

Pender, J. and Hazell, P. (eds) (2000) *Promoting Sustainable Development in Less Favoured Areas.* Focus 4, 2020 Vision, IFPRI, Washington, DC.

Pinstrup-Andersen, P. and Pandya-Lorch, R. (1998) Food security and sustainable use of natural resources: a 2020 vision. *Ecological Economics* 26, 1–10.

Pretty, J.N., Morison, J.I.L. and Hine, R.E. (2003) Reducing food poverty by increasing agricultural sustainability in developing countries. *Agriculture, Ecosystems and Environment* 95, 217–234.

Prudencio, C.F. (1993) Ring management of soils and crops in the West African semi-arid tropics: the case of the Mossi farming systems in Burkina Faso. *Agriculture, Ecosystems and Environment* 47, 237–264.

Rabbinge, R. and Van Ittersum, M.K. (1998) Tension between aggregation levels. In: Fresco, L.O., Stroosnijder, L., Bouma, J. and Van Keulen, H. (eds) *The Future of the Land. Mobilising and Integrating Knowledge for Land Use Options.* John Wiley & Sons Ltd., Chichester, UK, pp. 31–40.

Reardon, T. (1997) Using evidence of household income diversification to inform study of the rural nonfarm labor market in Africa. *World Development* 25, 735–747.

Reij, C., Mulder, P. and Begemann, L. (1988) *Water Harvesting for Plant Production.* World Bank Technical Paper No. 91, World Bank, Washington, DC.

Ruben, R. and Pender, J. (2004) Rural diversity and heterogeneity in less-favoured areas: the quest for policy targeting. *Food Policy* 29, 303–320.

Sanderson, S. (2005) Poverty and conservation: the new century's 'Peasant question'? *World Development* 33, 323–332.

Sidibé, A. (2005) Farm-level adoption of soil and water conservation techniques in northern Burkina Faso. *Agricultural Water Management* 71, 211–224.

Sissoko, K. (1998) Et demain l'agriculture? Options techniques et mesures politiques pour un développement agricole durable en Afrique subsaharienne. Cas du Cercle de Koutiala en zone sud du Mali. PhD Thesis, Wageningen Agricultural University, Wageningen, The Netherlands.

Tittonell, P., Vanlauwe, B., Leffelaar, P.A., Rowe, E.C. and Giller, K.E. (2005a) Exploring diversity in soil fertility management of smallholder farms in western Kenya: I. Heterogeneity at region and farm scale. *Agriculture, Ecosystems and Environment* 110, 149–165.

Tittonell, P., Vanlauwe, B., Leffelaar, P.A., Shepherd, K.D. and Giller, K.E. (2005b) Exploring diversity in soil fertility management of smallholder farms in western Kenya: II. Within-farm variability in resource allocation, nutrient flows and soil fertility status. *Agriculture, Ecosystems and Environment* 110, 166–184.

Trenbath, B.R. (1986) Resource use by intercrops. In: Francis, C.A. (ed.) *Multiple Cropping Systems.* Macmillan, New York, pp. 57–79.

Van der Meer, H.G. and Van Uum-Van Lohuyzen, M.G. (1988) The relationship between inputs and outputs of nitrogen in intensive grassland systems. In: Van der Meer, H.G., Ryden, J.C. and Ennik, G.C. (eds) *Nitrogen Fluxes in Intensive Grassland Systems.* Martinus Nijhoff Publishers, Dordrecht, The Netherlands, pp. 1–18.

Van Heemst, H.D.J. (1985) The influence of weed competition on crop yield. *Agricultural Systems* 18, 81–93.

Van Ittersum, M.K., Rabbinge, R. and Van Latesteijn, H.C. (1998) Exploratory land use studies and their role in strategic policy making. *Agricultural Systems* 58, 309–330.

Van Keulen, H. (1982) Graphical analysis of annual crop response to fertilizer application. *Agricultural Systems* 9, 113–126.

Van Keulen, H. (2006) Editorial: heterogeneity and diversity in less favoured areas. *Agricultural Systems* 88, 1–7.

Van Keulen, H., Van Ittersum, M.K. and De Ridder, N. (2000) New approaches to land use planning. In: Roetter, R.P., Van Keulen, H., Laborte, A.G., Hoanh, C.T. and Van Laar, H.H. (eds) *Systems Research for Optimizing Future Land Use in South and Southeast Asia.* SysNet Research Paper No. 2, IRRI, Los Baños, Philippines, pp. 3–20.

Von Liebig, J. (1855) *Die Grundsätze der Agricultur-Chemie mit Rücksicht auf die in England Angestellten Untersuchungen*, 2nd edn. Vieweg, Braunschweig, Germany.

Wathern, P., Young, S.N., Brown, I.W. and Roberts, D.A. (1986) The EEC less favoured areas directive. Implementation and impact on the upland use in the UK. *Land Use Policy* 3, 205–212.

Williams, T.O., Tarawali, S.A., Hiernaux, P. and Fernández-Rivera, S. (eds) (2004) *Sustainable Crop-Livestock Production for Improved Livelihoods and Natural Resource Management in West Africa.* ILRI (International Livestock Research Institute), Nairobi and CTA (Technical Centre for Agricultural and Rural Cooperation, ACP-EC), Wageningen, The Netherlands.

Yin, X., Goudriaan, J., Lantinga, E.A., Vos, J. and Spiertz, J.H.J. (2003) A flexible sigmoid function of determinate growth. *Annals of Botany* 91, 361–371.

Zoebl, D. (1996) Controversies around resource use efficiency in agriculture: shadow or substance? Theories of C.T. de Wit (1924–1993). *Agricultural Systems* 50, 415–424.

3   Dimensions of Vulnerability of Livelihoods in Less-favoured Areas: Interplay Between the Individual and the Collective

Johan Brons,[1] Ton Dietz,[2] Anke Niehof[2,*] and Karen Witsenburg[2]

[1]Development Economics Group, Department of Social Sciences, Wageningen University, The Netherlands; [2]Amsterdam Institute for Metropolitan and International Development Studies (AMIDSt), Amsterdam, The Netherlands; *e-mail: anke.niehof@wur.nl; [3]Sociology of Consumers and Households Group, Department of Social Sciences, Wageningen University, The Netherlands

Abstract

The geographical concentration of persistent poverty in so-called less-favoured areas (LFAs) calls for a critical look at the link between poverty and environment. Livelihood studies tend to focus on poverty at the individual level, whereas the concept of LFA implies a problem for the collective. Studies on vulnerability tend to be biased towards external ecological causes at the regional level, while studies on coping and survival usually focus on the household. However, recent insights into the internal and external dimensions of livelihood vulnerability in LFAs provide an argument for linking both dimensions to dynamics at the individual and collective level.

At an aggregate level, individual and household responses to vulnerability lead to intended and unintended effects, while there is also evidence of collective responses to factors originating from the external vulnerability context. These linkages between the external and internal dimensions of vulnerability and responses at the individual, aggregate and collective level should be studied to understand and mitigate current trends of increasing vulnerability of livelihoods in LFAs. Emerging key issues include: (i) analysis of change; (ii) analysis. of livelihood pathways; (iii) aggregate consequences of behaviour; and (iv) cultural dynamics.

Context

The RESPONSE research program is a joint research programme by Wageningen University, The Netherlands and Research Centre (WUR), The Netherlands, and the International Food Policy Research Institute (IFPRI), and was initiated in 2001. RESPONSE is the acronym for Regional Food Security Policies for National Resource Management and Sustainable Economies. The programme has a spatial focus on LFAs.

Though this chapter addresses issues dealt with by the RESPONSE Working Programme 2, namely livelihoods and food security, it has a broader scope because it also reviews recent empirical evidence on the subject other than the findings yielded by this working programme. It includes the results of recent (2000–2005) research on livelihoods in LFAs, in particular work carried out by researchers of the Dutch research schools Mansholt Graduate School (MGS) and the Research School for Resource Studies for Development (CERES). In this chapter, LFAs are not just treated as a given context; instead, the emphasis is on the dynamics of the interfaces between characteristics of LFAs, the livelihoods of the people in those areas and the implications of this for vulnerability.

Introduction

Poverty is considered to be an important constraint for sustainable development at local and global levels (World Bank, 2002; 2004). Much global poverty is geographically concentrated in the so-called LFAs and is related to ecological and social vulnerability. The principal aim of this chapter is to analyse the interaction between livelihoods and the geographical environment in LFAs. Current literature on livelihood draws on a long tradition of socio-economic, geographical and anthropological research. At the same time, the sustainable livelihoods framework, the human ecology literature and agro-economic studies are increasingly integrating questions and findings from the ecological sciences.

Less-favoured areas can be defined in different ways. Place-oriented biophysical features of such areas define such categories like 'drylands', 'highlands', 'uplands' or 'wetlands'. The place-specific, man-made infrastructures and the institutional environment in such areas have to be considered as well because of their intermediary role in the interface between the biophysical environment and household livelihood generation. In addition to these place-derived characteristics, there are also 'space-derived' characteristics that refer to the distance of LFAs to major economic centres, harbours, cities and centres of political power and decision making. Distance is used here in a spatial, political and cultural sense, in the literature often captured by the concept of marginality.

In this chapter, a LFA is seen as an area that combines problematic biophysical characteristics with a poor, man-made physical and institutional environment and a marginal location. The literature on livelihoods in LFAs emphasizes that vulnerability of the environment is an essential part of the vulnerability of livelihoods. In order to assess the importance of environmental conditions in rural areas we focus on vulnerability as the linking concept between environment and livelihoods.

Livelihood studies commonly distinguish an internal and an external side to vulnerability (Chambers, 1990). This chapter relates this distinction to the distinction between micro-level strategies and macro-level outcomes (Krishna, 2004) and to that between the individual and the collective (Rudd, 2003). At the macro- or collective level, changes occur due to aggregate effects, intended and unintended. Typical aggregate effects are market cycles, changes in biodiversity or cultural changes. Intended effects may result from collective responses. These effects modify external vulnerability conditions and, as we shall see, also have a differential impact on internal vulnerability.

Figure 3.1 illustrates the different dimensions of vulnerability and reflects the structure of this chapter. The concepts of internal and external vulnerability and the individual and collective dimensions (left-hand side of Fig. 3.1) will be discussed in the next section. This following section discusses internal and external vulnerability in relation to ecology, institutions, culture and economy. This is followed by an inventory of technical and socio-economic individual responses to vulnerability, after which we turn to the dynamics at the collective level: the unintended aggregate effects and the intended changes through collective behaviour (right-hand side of Fig. 3.1). We conclude that the aggregate effect of responses is a blank spot in livelihood literature and that this has implications for further research and policy-making.

This chapter builds primarily on recent empirical and theoretical studies on vulnerable livelihoods carried out by researchers working at Dutch universities. The cases of LFAs are mainly from three regions: West Africa, the Horn of Africa and Southern Africa, many of these cases focus on semi-arid or sub-humid areas (drylands). Additionally, we use insights

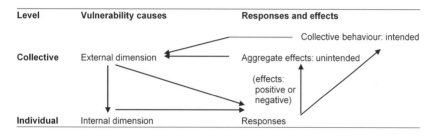

Fig. 3.1. Dimensions of vulnerability.

from studies in areas in Asia to illuminate certain aspects of vulnerability processes or provide evidence of specific effects or responses.

Vulnerability in Livelihood Studies

The concept of livelihood refers to what people do for a living, how they do it and what they gain by doing it. One of the most commonly used definitions is that of Ellis (2000, p. 10): 'A livelihood comprises the assets (natural, physical, human, financial and social capital), the activities, and the access to these (mediated by institutions and social relations) that together determine the living gained by the individual or household.' The focus on vulnerability in current livelihood studies signifies a renewed interest in structural context that could help explain the persistence of poverty in LFAs. In this section we disentangle the ideas that lie behind this concept and the implications these ideas have for further research. We focus on areas where vulnerable livelihoods converge with vulnerable environments.

Blaikie *et al.* (1994, p. 9) define vulnerability as 'the characteristics of a person or group in terms of their capacity to anticipate, cope with, resist and recover from the impacts of natural hazard'. Chambers (1990) distinguishes an external side of vulnerability in the form of risks, shocks and stress to which individuals or households are exposed, and an internal side that refers to the ability to cope without irreversible loss of assets. We will relate this distinction to the dimensions of the individual and the collective. While external vulnerability is basically a collective concern, it is in fact intricately entangled with individual vulnerability and coping behaviour.

Ecological and environmental studies refer to external vulnerability in the description of fragile environmental conditions. Environments are called vulnerable when components of the system reach a certain threshold that renders them unable to bounce back after shocks and disasters (Fraser, 2003). Ecosystems that degrade after deforestation are an example of this sort of vulnerability, but also ecological variability with unpredictable rainfall and high evapotranspiration rates renders an environment ecologically vulnerable (Dietz *et al.*, 2004). Livelihood studies add the human dimension. An environment characterized by unreliable rainfall, drought, flooding and climate change puts livelihoods under stress. The extent to which livelihoods can cope with this has been the focus of interest in much livelihood research.

Current livelihood frameworks most used in the literature draw on the conceptual models from Scoones (1998), Bebbington (1999), Ellis (2000) and Rakodi (2002). The vulnerability context is made operational by reference to trends in climate, population numbers, political change, war, terms of trade, food production and social differentiation, comprising factors that are assumed to be beyond individual control but have a negative impact on livelihoods. While the context constitutes the external side of vulnerability, its internal side refers to the specific characteristics of a person or a group that exacerbate their susceptibility to shocks and stress (Chambers, 1990). The internal side of vulnerability is a result of entitlement failures and lack of access to certain capitals.

Current livelihood frameworks distinguish a pentagon of types of capital from which people can draw for their living (Ellis, 2000; DFID, 2001). These capitals include contextual features such as physical and institutional infrastructures and societal norms, as well as individual assets. Liabilities are the opposite of assets. An individual characteristic like gender becomes a liability in a society based on gender inequality (e.g. Bebbington, 1999; Niehof, 2004). Kevane (2000) provides an example from western Sudan, where gender-discriminatory regulations forced women to close down their roadside shops, thereby blocking their access to income-generating activities.

In LFAs, part of the vulnerability context is located in the biophysical environment and other parts are located in the societal context. In large parts of sub-Saharan Africa, HIV/AIDS forms part of the external vulnerability context of rural households (Blaikie *et al.*, 1994; Barnett *et al.*, 2000), but civil war and violence (e.g. Salih *et al.*, 2001) and the high risks of price volatility on international markets for primary products and services may do so as well. The internal dimension of vulnerability involves the characteristics of individuals (age, sex, education, skills, health status, etc.), households (gender of headship, income, asset ownership, dependency ratio, etc.) and other micro-networks (e.g. Van der Geest and Dietz, 2004).

Parallel to internal and external vulnerability features we distinguish individual and aggregate effects of responses to vulnerability. While for studies on livelihood strategies individuals and households are the focus, for studies that start from the context of LFAs the focus is the collective of people living in such an area. For both types of studies the existence of feedback mechanisms between livelihood strategies and the societal and ecological context needs to be considered (Ruben *et al.*, 2005). Seen from the perspective of households as actors, the vulnerability of poor households forces them to engage in low-return and low-risk activities, which prevents them from fully realizing their economic potential (Brons, 2005).

At the aggregate level, LFAs are characterized by poor natural resources (ecological vulnerability) and poor institutional infrastructures (societal vulnerability). Recent special issues of the journal *World Development* also emphasize the persistent problems of chronic and extreme poverty and of environmental degradation and the consequent need to address these by collective action (Barrett *et al.*, 2005; Green and Hulme, 2005; Sunderlin *et al.*, 2005). Hence, to better understand poverty we need to look at the external and internal side of livelihood vulnerability and the individual and aggregate effects of people's activities in improving their livelihoods.

External and Internal Causes of Vulnerability

In order to look at the causes of vulnerability, we will consider the thematic domains of ecology, institution, culture and economy. This discussion of causes of vulnerability anticipates the next section's analysis of responses to vulnerability.

First, the ecological capacity to cope with shocks and the potential occurrence of shocks are of crucial importance to livelihood systems. Ecological sciences point at the effects of the exploitation of natural resources on the potential incidence of shocks (Fraser, 2003). As a result of poverty, ecosystems are susceptible to shocks because of fragile soil conditions and lack of investments in conservation techniques. The fragility of ecosystems is a function of the disequilibrium conditions of the environment. A disequilibrium environment (a term stemming from the New Range Ecology, see Behnke *et al.*, 1993) is characterized by ecological instability caused by erratic and patchy rainfall. Unpredictability and little surplus generation are indicative of a highly vulnerable environment. Due to limited surplus generation, households do not have the means to invest in natural resource conservation.

However, there are examples of resource conservation as a consequence of increasing population densities (Tiffen and Mortimore, 1994; Zaal and Oostendorp, 2002; Adano and Witsenburg, 2004). Adano and Witsenburg (2004) studied the impact of sedentarization of nomadic people in Kenya and concluded that in some parts of the Marsabit Mountain area, biodiversity and biomass may have increased, with a positive effect on the water retention capacity. This example shows the unintended positive effects on ecological conditions of poor people's activities.

Human (ecological) vulnerability is indicated by high morbidity and mortality (De Bruijn and Van Dijk, 1995; Hoogvorst, 2003). According to Iliffe (1987), there was always a large group of structurally poor people in Africa who had no access to labour, and there have always been poor people whose poverty is a result of stochastic shocks. Vulnerability to poverty is extreme when structural and conjectural poverty converge. 'Arid Ways' is among the few livelihood studies that give an account of how people do *not* cope (De Bruijn and Van Dijk, 1995). Its distressing tale explains the persistence of poverty, showing how individual liabilities like illness, handicaps, ignorance, fear and individual depression are linked to the context of structural poverty, ecological stress and political neglect. Another illustration is the way extremely weak economies engender social problems such as high morbidity, risky behaviour (HIV/AIDS), damaging conflicts (Bryceson and Fonseca, 2005), or alcoholism (Hoogvorst, 2003), thereby increasing societal vulnerability.

Resource scarcity has an impact on social processes and is frequently mentioned as a cause of conflicts, but in many cases the vulnerability of the society catalyses conflict to a much larger extent than the scarcity of resources itself. Political instability seems purposely masked with adverse ecological conditions (Adano and Witsenburg, 2004). Research in northern Kenya revealed that violence causes twice as many deaths in years of abundance than in drought years (Witsenburg and Adano, 2003). This is the opposite of what is usually expected in resource-poor areas.

Secondly, institutional dynamics have differential impacts on people's vulnerability depending on the social organization that lies between the household or the individual and the context. Local institutions as an important societal layer have been the focus of several livelihood studies. Adano and Witsenburg (2004) illustrate inequality in access to resources in a study on the local practices of the land committee in the allocation of land to impoverished pastoralists in settlement areas. These local land committees are organized at community level. The ways in which these committees function are so obscure that villagers seemed to follow different procedures of negotiation, bribes, taxes and form-filling. From survey research it became apparent that those who had few or no animals waited more years to acquire land than those who had large herds. The local land committee, appointed by the council and traditional in its practices, seems a blend of modern and customary institutions.

However, such institutions may reinforce class and gender inequality. In the same book there is a study on institutions of insurance and risk-mitigation. Complex arrangements in livestock holding secure the wealthy layers of pastoral society of resources, but the poorer segments of the society have very little access to these institutions. A similar observation is made in a study on social security systems in a remote village in Indonesia (Nooteboom, 2003). War and political instability may lead to extreme institutional vulnerability, as illustrated in a study on Somali refugees (Horst, 2003).

Informal institutions, like committees of elders in many African villages, are sometimes characterized by high transaction costs relative to formal institutions. For example, agreements that are reached or decisions taken informally among village elders or by local chiefs tend to be unstable, have no formal legal backing, have to be constantly renegotiated, are not transparent and can differ from one day to the next. Poor people cannot bear such transaction costs. In spite of the importance of the workings of informal institutions for livelihoods, the livelihood literature seems to pay scant attention to them.

Thirdly, culture should get a more prominent place in livelihood research. Culture largely defines what people should or should not do, what deviant behaviour is and how this behaviour can finally make or break the structures that keep people poor. Cultural stigmas that discriminate against groups and informal institutions that exclude people can be strong and persistent (Narayan *et al.*, 2000; Negash, 2001). Overcoming poverty requires overcoming culturally underpinned normative structures that keep people poor.

Fraser (2003) acknowledges that culture remains a gap in his framework that connects ecological fragility and social vulnerability. He recommends the use of ethnography for investigating the vulnerability effects of culture. Even though the transformation of structures and institutions has been a theme of research in livelihood studies, culture-based structural (institutional) features are often still ignored in current livelihood research. In an overview of theory and practice in livelihood studies, De Haan and Zoomers (2005) propose to refocus research on access to resources and on the power relations that define entitlements. Gender is probably the most salient and best-studied aspect of culture dynamics (Negash, 2001; Niehof, 2004; De Haan and Zoomers, 2005).

Fourthly, regarding the economic domain, it has been observed that absence of a large enough economic base is a major reinforcing cause of poverty (Campbell *et al.*, 2002; Haggblade *et al.*, 2002; Brons, 2005). Market integration and reduced distance to markets can improve livelihoods considerably. Adano and Witsenburg (2004) describe how livelihoods among settled people on Marsabit Mountain, northern Kenya, improved when households were able to sell products on the market. However, for nomadic people, being integrated into the local meat market is an indicator of adverse conditions, because people sell their animals only when they are in trouble. Dependence on markets can bear serious risks for people whose livelihoods are specialized and non-diversified.

The local economy often influences and is being influenced by migratory processes that may have positive as well as negative implications. People migrate because they intend to earn a living elsewhere, which means that the local situation is not satisfactory. However, migrants come usually from the wealthier layers of society (De Haas, 2004). Migrants (and refugees) are people who exercise their agency to free themselves from restrictive structures. The remittances to the sending areas may have an important welfare-increasing (Horst, 2003), or a multiplier, effect (De Haas, 2004) on the sending area. However, one important negative outcome of out-migration is the paucity of young, healthy, dynamic and risk-taking people in a population. In very marginal areas, where the best part of society has left and investments are absent, the cumulative effects of structural poverty converge to the extent that the people left behind cannot cope (De Bruijn and Van Dijk, 1995; Francis, 2000).

The review of ecological, institutional, cultural and economic processes illustrates a growing and renewed interest in structures and multi-level research. All point to a renewed interest in a context that surpasses the local concern. Besides attention to various geographical levels, the effects of human action on both ecological and societal contexts are also increasingly attracting scholarly attention. Conversely, the vulnerability context appears to have a large differential impact on responses and outcomes due to different levels of internal vulnerability.

Responses and Effects

In response to societal and ecological vulnerability, people seek to secure their livelihoods. Broadly categorized, we distinguish technical and socio-economic responses of households to safeguard subsistence and avert vulnerability. Though both types of responses are closely linked and cannot be analysed independently of one another, for the sake of clarity we start with the responses that are mainly either technological or socio-economic in nature, and then highlight combined responses.

Technological responses

These include: (i) resorting to specific niche production opportunities; (ii) soil and water conservation technologies; and (iii) dissemination of production technologies with higher rewards for labour.

An important study in this context is the research on enset cultivation in Ethiopia (Negash, 2001; Negash and Niehof, 2004). Enset (*Ensete ventricosum* Welw. Cheesman), also called 'false banana', is a major food crop in south-western Ethiopia that can be cultivated in the backyard and requires only low inputs. Enset is fairly drought resistant and provides staple food, medicines and building materials for households. Although the plant is quite resilient, its biodiversity is in jeopardy due to lack of attention from researchers and policy makers. There are other studies that also highlight the role of ecological diversity and home gardens for household food security (Balatibat, 2004; Roa, 2005).

Soil and water conservation techniques have received long-standing attention from development agencies (De Graaff, 1996). Following scepticism over the effects of massive adaptation of micro-level soil and water conservation techniques, some recent studies report successful impacts of these technological interventions on resource conservation, though without explaining the underlying causes for the success (Mazzucato and Niemeijer, 2000; Reij and Thiombano, 2003). While most studies use the livelihood framework to analyse technological impacts, Meinzen-Dick *et al.* (2003) start from an inventory of agricultural production technologies. They show that technological development has a large impact on crop productivity, food prices and informal networks. Although the direct poverty alleviation effect may seem small, enhancement of household and collective social capital improves the prospects for reducing livelihood vulnerability.

Socio-economic responses

These encompass: (i) a change in the portfolio of economic activities; (ii) the use of household and individual networks; (iii) exchange and entrust strategies; and (iv) cultural responses.

A study on income diversification in Burkina Faso (Brons, 2005) points to the inefficacy of economic diversification in overcoming poverty when households diversify their portfolio in a situation of excess capacity. In contrast, other studies emphasize income diversification as a strategic and effective response to vulnerability (Freeman *et al.*, 2004; Niehof, 2004; Van der Geest, 2004). Research in Bangladesh showed that households diversified their sources of income after the devastating floods of 1998 (Ali, 2005). Apparently, diversification may work as a survival strategy, but it has only little effect in local economies that are characterized by excess capacity and little surplus (Haggblade *et al.*, 2002; Brons, 2005).

Migration can be a diversification strategy in response to local vulnerability. However, migration tends to be a structural feature of a society rather than an *ad hoc* response to situations of stress (Henry *et al.*, 2004). Access to migration opportunities appears to be limited to relatively better developed areas, such as some oases in southern Morocco and within the relatively wealthier households in such areas (De Haas, 2004). Remittances to and investment in the sending area may actually increase societal and ecological vulnerability.

In Burkina Faso, access to international migration tends to be limited to a few households, again in the relatively better-endowed villages (Wouterse and Van den Berg, 2004). In the latter study, the effects of remittances on the sending villages were found to be much less important than in the case of Morocco. In other studies, migration brings about opportunities in terms of better income and connection with urban markets (Horst, 2003; Kuiper, 2005), but also threatens household stability and increases exposure to insecurity (Francis, 2000).

A study by Dekker (2004) among rural households in Zimbabwe illustrates the role of social networks in the way that households cope with income shortfalls and the role of bride wealth payments for long-term social security. The study focuses on security mechanisms to control vulnerability. A similar study analyses long-term social security strategies in an upland village in Indonesia, documenting the complexity of the social security system in a village with large income inequalities and economically and ecologically fragile livelihoods (Nooteboom, 2003). Insurance against risk is in this context only partial, not transparent and accessible only to those who are able to contribute to the existing arrangements.

An inventory of responses to drought and famine in Ethiopia yielded a list of mainly socio-economic measures (Teshome, 2003). Households reduce consumption, deplete their assets and minimize the level of consumption, while at the same time they develop activities to regenerate their livelihoods. From this list it can be learned that income shortfalls often trigger fundamental changes in livelihoods.

Responses that are strongly underpinned culturally often receive only marginal attention in livelihood studies, yet culture is important in providing or constraining options and shaping responses, as is illustrated by some recent studies. In Ethiopia, households prefer to have many children to gain social security (Negash, 2001; Negash and Niehof, 2004). In Bangladesh, early marriage strategies aim at safeguarding girls from abuse (Huq, 2000).

Although many studies confirm the structural vulnerability of female-headed households, the study of Mtshali (2002) also shows that, in a patriarchal culture, widowed women can benefit from a situation of having no husband because of lesser control over their mobility. However, in general, a widow's legal or customary lack of entitlements, notably rights to land, puts her in a situation of insecurity (Niehof, 2004). The studies of Balatibat (2004) and Mtshali (2002) provide striking examples of vulnerability of households because of individual risks. Sending households cannot rely on remittances from their migrant members, because migrants fall ill, lose their job or forget their commitments to their household of origin. Many studies show that, as a result of stress, domestic relations may become unstable which, as noted by Francis (2000), frequently leads to gender-related conflicts.

A recent study on famine conditions in Malawi points to serious erosion of social cohesion and commitment among smallholders in a rural peasantry disadvantaged in terms of infrastructure and social services (Bryceson and Fonseca, 2005). The study concerns people who have virtually no means of living (internal vulnerability) and are in a situation of increasing exposure to HIV/AIDS risks (external vulnerability). As coping strategy, women forestall destitution by engaging in prostitution. People's responses to the disastrous and gendered impacts of HIV/AIDS on rural livelihoods in large parts of sub-Saharan Africa, are by now becoming well documented (see, for an overview, Müller, 2004, 2005a, b).

Livelihood studies generally ignore the role of violence and warfare, while it is clear from conflict studies that warfare destroys environmental and physical endowments and obstructs access to otherwise available resources. The destruction of human lives and livelihoods, and the withdrawal of labour from productive to vigilance and defence activities, exacerbate poverty. On the other hand, violence and warfare do result in spoils for some, due to pillage and plunder, and benefits for others such as employment in armies of government agencies and warlords, remittances and indirect benefits. This may also generate institutional and technological change and economic and cultural breakthroughs (see case studies in Salih *et al.*, 2001). The few examples of combined livelihood and conflict analysis (e.g. Adano and Witsenburg, 2004) show so many unexpected results that the need for more such analyses is obvious, and also to counter the many superficial statements on this issue in the literature, often from a political science background (e.g. Homer-Dixon, 1999).

Combined technological and socio-economic responses

These reveal the synergy between societal and ecological processes. An example is provided by the sedentarization of herder families in response to increased vulnerability because of prolonged trends of pastoral decline (De Bruijn and Van Dijk, 1995; Breusers, 2001; Adano and Witsenburg, 2004). Sedentarization often implies a dramatic shift in livelihoods with consequences for those who settle, those who retain a nomadic livelihood and the population in the sending and target areas. In Kenya, settlement combined with a gradual shift towards crop husbandry is the basic response to loss of animals (Adano and Witsenburg, 2004). Herder families prefer to maintain their nomadic livelihoods until adverse conditions force them to settle. Once settled, they encounter all kinds of institutional barriers in rebuilding their livelihoods. In Mali and Burkina Faso, sedentary as well as nomadic households in the pastoral northern zones seek refuge in urban or rural areas (Breusers, 2001; De Bruijn and Van Dijk, 2004). Whether sedentarization is successful appears to depend on access to networks.

Another example of combined responses relates to soil and water conservation at plot level in eastern Burkina Faso (Mazzucato and Niemeijer, 2000). Farm households rely on intensive social networks to gain access to the resources necessary to keep their plots fertile. The study of Mazzucato and Niemeijer documents household strategies to conserve arable fields, but does not go into the causes of chronic poverty in the area. A meta-study on responses in drylands to climate change shows a rich mixture of technological and socio-economic responses at the individual and collective levels (Dietz *et al.*, 2004). The study emphasizes the adaptability of rural households' livelihood strategies. A striking finding is that of the far-reaching processes of change that have occurred at different societal levels (Bryceson, 2002; Van Dijk *et al.*, 2004).

From one generation to another, land use technologies, control over natural resources, economic activities and institutions changed dramatically. These changes can be regarded as responses to increased scarcity of natural resources and demographic change. The responses seem *ad hoc* and unorganized rather than driven by strategy. This can be explained partly by the disequilibrium conditions of vulnerable areas and the absence of surplus generation.

The responses appearing in the above review have in common that they concern mainly household-level responses. The previous section on vulnerability showed that the livelihoods framework incorporates aggregate effects, but does not address the question of how households perceive and make use of the aggregate effect of their activities. The diversification study in Burkina Faso does not analyse why rural households continue to diversify their economic activities or how they perceive the common problem of excess capacity (Brons, 2005).

The social security study in Zimbabwe (Dekker, 2004) analyses social processes in detail but uses the household as the locus of social security and ignores existing visions on the prevailing social security system. In many studies, responses to vulnerability are mostly the result of an inventory of events after a shock in livelihoods (Teshome, 2003). Consequently, the question arises as to what households did in anticipation of shocks that would appear sooner or later. A critical look at the response and outcomes documented by the studies shows that few studies provide information on structural causes of success or failure.

Linkages Between the Household-level and Collective Responses and Constraints

The previous sections show that livelihood studies have focused largely on analysing household practices to avert vulnerability, and emphasize ecological variability and fragility as main causes of livelihood vulnerability. A combined analysis of the internal and external sides of the vulnerability of livelihoods in LFAs shows that institutions cause a differentiating impact on household livelihood vulnerability.

The recognition of the role of institutions is an important step in understanding poverty, but does not resolve the bias in livelihood studies of an overemphasis on the household level. Household practices and outcomes are conceptualized in a pentagon of measurable, mainly material, assets, which are subsequently used in income and security generation. There has been little attention paid to the existence of liabilities and large-scale processes of destitution and marginalization. The conceptual focus on measurable assets has led to a methodological bias of measuring practices such as income generation, wealth accumulation, migration and social security from the perspective of the household. It will also be noted that virtually no livelihood studies pay attention to how actual wealth distribution patterns are perceived by individuals, while the actors' knowledge of and attitude towards wealth distribution can be expected to influence livelihood practices.

It is important to forge this, often missing, link between household livelihood performance and the external vulnerability context shaped by the ecological and institutional environment. As noted in a study of why households in India move in *and* out of poverty: 'Quite different things are happening in different villages and also in different households' (Krishna, 2004, p. 126). The same author also concludes that micro-level motives are complex and varied, and that connecting micro-level strategies with macro-level outcomes is difficult (Krishna, 2004, p. 131). Nevertheless, this is what we believe should be done.

Despite the fact that most studies do refer to contextual processes, an image emerges as if people respond in rather an *ad hoc* way to vulnerability. This is an incomplete, if not false, image. As discussed above, migration studies reveal that, in LFAs, there is a net outflow of relatively well-off and physically strong people. Consequently, such areas are inhabited by a negative selection of those unable to migrate.

Social security systems may lack transparency, and the use of natural resources is often poorly regulated.

An interesting study in this respect is provided by an Overseas Development Institute (ODI) report on the efficacy of community forest management in India (Sarin *et al.*, 2003). The report shows that, where people are dependent on the forest for livelihood needs like fuelwood, fodder and wild plant foods, they develop a collective vision on its value and on the need for preserving it, in which women play a crucial role, and regulate access to and use of forest products. However, once the government steps in by way of laws and regulations, and the forest acquires regional and national commercial significance, the indigenous community management system succumbs under the pressure. The result is more gender inequity and increasing livelihood vulnerability.

We will put forward some conceptual and methodological issues that may help to overcome the one-sided focus on internal vulnerability. The external side of vulnerability and the structural constraints that keep people poor need to be investigated by using appropriate research methodologies. A broader focus such as we propose could shed new light on the dynamic relationship between the individual and the collective.

A first point is that the possibility should be recognized that poverty may seem structural at regional levels but is transient at the household level (e.g. Collier and Gunning, 1999). Adano and Witsenburg (2004) looked at household capital over a number of years, and concluded that those who are poor before and after a crisis are not necessarily the same people. Households react differently to shocks and also, depending on the stage of household formation, are able to climb out of poverty over time.

Numerous case studies yield evidence of transient poverty (Huq, 2000; Mtshali, 2002; Krishna, 2004; Ali, 2005). But also 'change' itself has not been taken seriously. To investigate it, we would need a much more thorough assessment of the past, including archival research. In addition, analysis of panel data, though cumbersome and expensive to collect, is a way of measuring livelihood outcomes before and after a shock. A recent study on rural Bangladesh (Ali, 2005) could use IFPRI panel data of livelihood indicators before and after the floods of 1998, but such data are rarely available. If we want to understand the geographical concentration of poverty, we must also know whether such a region has been inhabited by the same people over time. There is reason to believe that there is much more mobility than the often somewhat static livelihood that studies suggest.

In the analysis of the context, ecological factors seem to have dominated over societal ones. Ecological vulnerability is more tangible and better recorded than societal vulnerability. Strikingly, many livelihood studies refrain from analysing societal processes at the collective level; yet, it is at this level that the causes of persistent poverty and vulnerability are to be found (Barrett and Swallow, 2003).

One exception is formed by the growing body of literature on the consequences of the HIV/AIDS pandemic that shows how externally generated vulnerability impacts negatively on individuals and households through stigmatization and social exclusion. The existence of an extremely poor layer of people in a society suggests that there must be shared perceptions of poverty and inequality that can help explain why people are, and stay, poor. Additionally, a better understanding of culture, norms and informal institutions (like ethnic identity, gender roles and norms, rites of passage) is needed. Culture should not be assumed to be only a repertory that people can draw from. It can also be a liability and a constraint in reaching personal aspirations and achieving societal progress, inducing individuals to engage in deviant behaviour.

Households are exposed to shocks and cope with these by deploying livelihood strategies, which consist of both individual and collective components. Yet, virtually none of the reviewed studies considers the impact of people's activities on the ecological and institutional environment. There is a need for analysis at the aggregate level of communities and regions of the effects of individual (household) behaviour, as well as of the causes and effects of collective action based on collective vision. Pathway analysis provides a promising tool for such analyses (De Haan and Zoomers, 2005).

Pathway effects encompass the externalities of the behaviour of the poor as well as the wealthy. Livelihood behaviour incited by poverty conditions may increase external vulnerability, which may lead to collective awareness and action to redress this. However, the study on communal forestry management cited above (Sarin *et al.*, 2003) shows that this inevitably provokes power struggles that directly impact on people's livelihoods. The socio-economic effects of power structures form a rarely investigated topic in livelihood research. Wealthy individuals in poor regions, for instance, exercise power that is poorly documented in livelihood studies. How people become wealthy, whether their wealth benefits others or is purely exploitative in nature can reveal dynamics that inform us of the interplay between actor and structure, the individual (household) and the group, the people and their leaders.

Cultural changes are another type of pathways effects but, possibly as a consequence of the preoccupation of livelihood studies with tangible assets, culture has hardly featured in livelihood studies. When a mother decides that her daughter should postpone marriage to enable her to finish secondary school first, she deviates from the norm. When more people decide the same, the norm may change. The Bangladesh study cited above (Ali, 2005) shows that culture and customs do change, and that education and labour market participation by young women increase their age at marriage and strengthen their bargaining position *vis-à-vis* their parents and prospective in-laws. Cultural dynamics are frequently reflected in activities that may not directly generate income but yield communal benefits (digging wells, organizing meetings) or protect the environment (building terraces, forest management). Such activities are based on collective vision and reflect collective knowledge and skills. They can shed light on possible institutional or environmental constraints in different situations.

Deviant behaviour at the individual level can inspire institutional change. Individuals in an important position (a chief, elder, trader, priest or teacher) can become role models for change. Also, information on deviant behaviour of groups, rebellion and collective action can be a tool for investigation of institutional change. Collective action challenges hegemonic opinions and existing arrangements – as is shown by the work of scholars like Scott (1976) – which may induce changes in the external vulnerability context. While stressing the importance of collective vision and action for individual livelihoods, it must also be noted that collective perceptions may differ by age, gender, ethnicity and religion, both within as well as across communities. By putting the linkages between the individual and the collective to the fore, we do not want to suggest a naive conception of community. As Schoepf (2001) observes, the study of HIV/AIDS impacts on rural livelihoods has led anthropologists to argue against the reification of the idea of community and acknowledge the differences that may divide local actors. How widely shared 'the collective' is and what the outcomes of collective action are, remain a matter of empirical investigation.

Conclusions

Livelihood studies frequently refer to vulnerability as a main feature of livelihoods in LFAs. Less-favoured areas are defined by unfavourable conditions in terms of ecological potential, infrastructure and institutional environment, and also of distance to main markets and centres of innovation: elements that refer mainly to the external side of vulnerability. On the other hand, many livelihood studies tend to focus on the internal side of vulnerability. Case studies as well as surveys seek to identify how households cope with variability through socio-economic and technological responses. Major attention is given to social security networks, income diversification and soil and water conservation techniques. The analysis of these responses tends to be inward looking: the focus is on household level dynamics rather than on aggregate dynamics and the consequences of responses for the collective.

The way the sustainable livelihoods framework is used may be the reason why the aggregate effects of individual activities are so often ignored in livelihood studies. Livelihood studies focus primarily on income activities of individuals or households. Not only is collective action glossed over, and thus the institutions and power relations involved, but also the activities that people undertake that do not generate an income. Such activities are often communal activities that have effects on and implications for the collective.

Empirical studies on livelihoods have yielded a broad insight into people's responses to vulnerability, yet the precise driving forces and the effects of these responses remain to be further investigated. This chapter has elaborated on two biases in recent livelihood studies, namely first, towards ecological variability for explaining the external side of vulnerability and, secondly, a bias towards household decision making as the internal side of vulnerability. We argue that more attention should be given to the interfaces between the internal and external sides of vulnerability and between individual and collective responses and effects.

Acknowledgements

The authors wish to thank the RESPONSE Program for making this chapter possible and the reviewers of the first version of this chapter, especially John Kerr, for their valuable comments and suggestions.

References

Adano, W.R. and Witsenburg, K. (2004) Surviving pastoral decline. Pastoral sedentarisation, natural resource management and livelihood diversification in Marsabit District, northern Kenya. PhD thesis, University of Amsterdam, Amsterdam.

Ali, A. (2005) Livelihood and food security in rural Bangladesh. PhD thesis, Wageningen University, Wageningen, The Netherlands.

Balatibat, E.M. (2004) The linkage between food and nutrition security in lowland and coastal villages in the Philippines. PhD thesis, Wageningen University, Wageningen, The Netherlands.

Barnett, T., Whiteside, A. and Decosas, J. (2000) The Jaipur paradigm: a conceptual framework for understanding social susceptibility and vulnerability to HIV. *Journal of the South African Medical Association* 90, 1098–1101.

Barrett, C.B. and Swallow, B.M. (2003) *Fractal Poverty Traps.* Cornell University, World Agro Forestry Centre, Ithac, New York.

Barrett, C.B., Lee, D.R. and McPeak, J.G. (2005) Institutional arrangements for rural poverty reduction and resource conservation. *World Development* 33, 193–197.

Bebbington, A. (1999) Capitals and capabilities: a Framework for analyzing peasant viability, rural livelihoods and poverty. *World Development* 27, 2021–2044.

Behnke, R.H., Scoones, I. and Kerven C. (1993) *Range Ecology at Disequilibrium.* Overseas Development Institute, London.

Blaikie, P., Cannon, T., Davis, I. and Wisner, B. (1994) *At Risk: Natural Hazards, People's Vulnerability and Disasters.* Routledge, New York.

Breusers, M. (2001) *Pathways to Negotiate Climate Variability. Land Use and Institutional Change in the Kaya Region, Burkina Faso.* African Study Centre, Leiden, The Netherlands.

Brons, J.E. (2005) Activity diversification in rural livelihoods. The role of farm supplementary income in Burkina Faso. PhD thesis, Wageningen University, Wageningen, The Netherlands.

Bryceson, D.F. (2002) Multiplex livelihoods in rural Africa: recasting the terms and conditions of gainful employment. *Journal of Modern African Studies* 40, 1–28.

Bryceson, D.F. and Fonseca, J. (2005) Risking death for survival: peasant responses to hunger and HIV/AIDS in Malawi. *International Conference on HIV/AIDS, Food and Nutrition Security*, International Food Policy Research Institute, Durban, South Africa.

Campbell, B.M., Jeffrey, S., Kozanayi, W., Luckert, M., Mutamba, M. and Zindi, C. (2002) *Household Livelihoods in Semi-arid Regions. Options and Constraints.* Center for International Policy Research, Jakarta, Indonesia.

Chambers, R. (1990) Editorial introduction: vulnerability, coping, and policy. *IDS Bulletin* 20, 1–7.

Collier, P. and Gunning, J.W. (1999) Explaining African economic performance. *Journal of Economic Literature* 37, 64–111.

De Bruijn, M. and Van Dijk, H. (1995) Arid ways: cultural understandings of insecurity in Fulbe society, Central Mali. PhD thesis, Thela, Amsterdam.

De Bruijn, M. and Van Dijk, H. (2004) The importance of sociocultural differences and of pathway analysis for understanding local actors' responses. In: Dietz, A.J., Verhagen, A. and Ruben, R. (eds) *The Impact of Climate Change on Drylands with a Focus on West Africa.* Vol. 39, Kluwer Academic Publishers, Dordrecht, The Netherlands, Boston, Massachusetts and London, pp. 341–363.

De Graaff, J. (1996) The price of soil erosion. An economic evaluation of soil conservation and watershed development. PhD thesis, Wageningen University, Wageningen, The Netherlands.

De Haan, L. and Zoomers, A. (2005) Exploring the frontier of livelihoods research. *Development and Change* 36, 27–47.

De Haas, H. (2004) Migration and development in Southern Morocco. The disparate socio-economic impacts of out-migration on the Todgha Oasis Valley. PhD thesis, Radboud University Nijmegen, Nijmegen, The Netherlands.

Dekker, M. (2004) Risk, resettlement and relations: social security in rural Zimbabwe. PhD thesis,Vrije Universiteit Amsterdam, Amsterdam.

DFID (2001) *Sustainable Livelihoods Guidance Sheets.* Department for International Development, London (http//www.livelihoods.org).

Dietz, A.J., Verhagen, A. and Ruben, R. (2004) *The Impact of Climate Change on Drylands with a Focus on West Africa.* Kluwer Academic Publishers, Dordrecht, The Netherlands, Boston, Massachusetts and London.

Ellis, F. (2000) *Rural Livelihoods and Diversity in Developing Countries.* Oxford University Press, Oxford, UK and New York.

Francis, E. (2000) *Making a Living. Changing Livelihoods in Rural Africa.* Routledge, London and New York.

Fraser, E.D.G. (2003) Social vulnerability and ecological fragility: building bridges between social and natural sciences using the Irish Potato Famine as a case study. *Conservation Ecology* 7, 9 (online).

Freeman, H.A., Ellis, F. and Allison, E. (2004) Livelihoods and Rural Poverty Reduction in Kenya. *Development Policy Review* 22, 147–171.

Green, M. and Hulme, D. (2005) From correlates and characteristics to causes: thinking about poverty from a chronic poverty perspective. *World Development* 33, 867–879.

Haggblade, S., Hazell, P. and Reardon, T. (2002) *Strategies for Stimulating Poverty-alleviating Growth in the Rural Non-farm Economy in Developing Countries.* Eptd, Washington, DC.

Henry, S., Piché, V., Ouédraogo, D. and Lambin, E.F. (2004) Descriptive analysis of the individual migratory pathways according to environmental typologies. *Population and Environment* 25, 397–422.

Homer-Dixon, T. (1999) *The Environment, Scarcity and Violence.* Princeton University Press. Princeton, New York.

Hoogvorst, A. (2003) Survival strategies of people in a Sri Lankan wetland. Livelihood, health and nature conservation in Muthurajawela. Consumer and Household Studies. PhD thesis, Wageningen University, Wageningen, The Netherlands.

Horst, C. (2003) Transnational nomads. How Somalis cope with refugee life in the Dadaab camps of Kenya. PhD thesis, University of Amsterdam, Amsterdam.

Huq, H. (2000) People's practices. Exploring contestation, counter-development and rural livelihoods. PhD thesis, Wageningen Agricultural University, Wageningen, The Netherlands.

Iliffe, J. (1987) *The African Poor*. Cambridge University Press, Cambridge, UK.

Kevane, M. (2000) Extrahousehold norms and intrahousehold bargaining: gender in Sudan and Burkina Faso. In: Spring, A. (ed.) *Women Farmers and Commercial Ventures: Increasing Food Security in Developing Countries*. Lynne Rienner Publishers, Boulder, Colorado and London, pp. 89–113.

Krishna, A. (2004) Escaping poverty and becoming poor: who gains, who loses, and why? *World Development* 32, 121–136.

Kuiper, M.H. (2005) Village modelling: a Chinese recipe for blending general equilibrium and household modelling. PhD thesis, Wageningen University, Wageningen, Netherlands, p. 261.

Mazzucatto, V. and Niemeijer, D. (2000) Rethinking soil and water conservation in a changing society. A case study in eastern Burkina Faso. PhD thesis, Wageningen University, Wageningen, The Netherlands.

Meinzen-Dick, R.S., Adato, M., Haddad, L. and Hazell, P. (2003) *Impacts of Agricultural Research on Poverty: Findings of an Integrated Economic and Social Analysis*. International Food Policy Research Institute, Washington, DC.

Mtshali, S.M. (2002) Household livelihood security in rural Kwazulu-Natal, South Africa. PhD thesis, Wageningen University, Wageningen, The Netherlands.

Müller, T.R. (2004) *HIV/AIDS and Agriculture in Sub-Saharan Africa*. AWLAE Series No. 1, Wageningen Academic Publishers, Wageningen, The Netherlands.

Müller, T.R. (2005a) *HIV/AIDS, Gender and Rural Livelihoods in Sub-Saharan Africa*. AWLAE Series No. 2. Wageningen Academic Publishers, Wageningen, The Netherlands.

Müller, T.R. (2005b) *HIV/AIDS and Human Development in Sub-Saharan Africa*. AWLAE Series No. 3, Wageningen Academic Publishers, Wageningen, The Netherlands.

Narayan, D., Chambers, R., Sha, M.K. and Petesch, P. (2000) *Crying out for Change. Voice of the Poor*. Oxford University Press, Washington, DC.

Negash, A. (2001) Diversity and conservation of enset (*Eneste ventricosum* Welw. Cheesman) and its relation to household food and livelihood security in South-western Ethiopia. PhD thesis, Wageningen University, Wageningen, Netherlands.

Negash, A. and Niehof, A. (2004) The significance of Enset culture and biodiversity for rural household food and livelihood security in south-western Ethiopia. *Agriculture and Human Values* 21, 61–71.

Niehof, A. (2004) The significance of diversification for rural livelihood systems. *Food Policy* 29, 321–338.

Nooteboom, G. (2003) A matter of style: social security and livelihood in upland East Java. PhD thesis, Radboud University, Nijmegen, The Netherlands.

Rakodi, C. (2002) A livelihoods approach. Conceptual issues and definitions. In: Rakodi, C. and Lloyd-Jones, T. (eds) *Urban Livelihoods. A People-centred Approach to Reducing Poverty*. Earthscan Publications, London.

Reij, C. and Thiombano, T. (2003) *Développement Rural et Environnement au Burkina Faso: la Réhabilitation de la Capacité Productive des Terroirs sur la Partie Nord du Plateau Central entre 1980 et 2001*. Rapport de synthèse, CONEDD, Ouagadougou, Burkina Faso.

Roa, J. (2005) *Food Availability and Access: the Case of Fragile Areas in the Philippines*. Conference paper, Wageningen University, Wageningen, The Netherlands.

Ruben, R., Kuiper, M. and Pender, J. (2005) *Searching Development Strategies for Less-favoured Areas.* Wageningen University and International Food Policy Research Intsitute, Wageningen, The Netherlands.

Rudd, M.A. (2003) Institutional analysis of marine reserves and fisheries governance policy experiments. A case study of Nassau grouper conservation in the Turks and Caicos Islands. PhD thesis, Wageningen University, Wageningen, The Netherlands.

Salih, M.A.M., Dietz, T. and Abdel Ghaffar, M.A. (2001) *African Pastoralism: Conflict, Institutions and Government.* Pluto Press, London, with OSSREA, Addis Ababa.

Sarin, M. with Singh, N.M., Sundar, N. and Bhogal, R.K. (2003) *Devolution as a Threat to Democratic Decision-making in Forestry? Findings from Three States in India.* Working Paper 197, Overseas Development Institute, London.

Schoepf, B.G. (2001) International AIDS research in anthropology: taking a critical perspective on the crisis. *Annual Review of Anthropology* 30, 335–361.

Scoones, I. (1998) *Sustainable Rural Livelihoods: a Framework for Analysis.* Institute of Development Studies, Brighton, UK.

Scott, J. (1976) *The Moral Economy of the Peasant: Rebellion and Subsistence in Southeast Asia.* Yale University Press, New Haven, Connecticut.

Sunderlin, W.D., Angelsen, A., Belchner, B., Nasi, R., Santoso, L. and Wunder, S. (2005) Livelihoods, forests, and conservation in developing countries: an overview. *World Development* 33, 1383–1402.

Teshome, W. (2003) Irrigation practices, state intervention and farmers' life-worlds in drought-prone Tigray, Ethiopia. PhD thesis, Wageningen University, Wageningen, The Netherlands.

Tiffen, M. and Mortimore, M. (1994) Malthus controverted: the role of capital and technology in growth and environment recovery in Kenya. *World Development* 22, 997–1010.

Van der Geest, K. (2004) *'We're Managing!' Climate Change and Livelihood Vulnerability in Northwest Ghana.* African Studies Centre, Leiden.

Van der Geest, K. and Dietz, A.J. (2004) A literature survey about risk and vulnerability in drylands, with a focus on the Sahel. In: Dietz, A.J., Ruben, R. and Verhagen, A. (eds) *The Impact of Climate Change with a Focus on West Africa.* Kluwer Academic Publishers, Dordrecht, The Netherlands, Boston, Massachusetts and London, pp. 117–146.

Van Dijk, H., de Bruijn, M. and van Beek, W. (2004) Pathways to mitigate climate change in Mali: the districts of Douentza and Koutiala compared. In: Dietz, A.J., Verhagen, A. and Ruben, R. (eds) *The Impact of Climate Change on Drylands with a Focus on West Africa.* Kluwer Academic Publishers, Dordrecht, The Netherlands, Boston, Massachusetts and London, Vol. 39, pp. 173–206.

Witsenburg, K. and Adano, W.R. (2003) Ethnic violence, water scarcity and the governance of resources. A case study from Northern Kenya. In: *Faces of Poverty. Capabilities, Mobilization and Institutional Transformation. Proceedings of the International CERES Summerschool,* 2003, pp. 377–396.

World Bank (2002) *Globalization, Growth, and Poverty. Building an Inclusive World Economy.* Oxford University Press, New York.

Worldbank (2004) *World Development Report 2005: a Better Investment Climate for Everyone.* World Development Report, World Bank, Washingthon, DC.

Wouterse, F. and van den Berg, M. (2004) *Migration for Survival or Accumulation: Evidence from Burkina Faso.* Conference paper, Wageningen University, Wageningen, The Netherlands.

Zaal, F. and Oostendorp, R.H. (2002) Explaining a miracle: intensification and the transition towards sustainable small-scale agriculture in dryland Machakos and Kitui districts, Kenya. *World Development* 30, 1271–1287.

4 Market Imperfections

NAZNEEN AHMED,[1] JACK PEERLINGS[2] AND AAD VAN TILBURG[3]

[1]*Bangladesh Institute of Development Studies (BIDS), Dhaka, Bangladesh; e-mail: nahmed@sdnbd.org;* [2]*Agricultural Economics and Rural Policy Group, Wageningen University, The Netherlands; e-mail: jack.peerlings@wur.nl;* [3]*Marketing and Consumer Behaviour Group, Wageningen University, The Netherlands; e-mail: aad.vantilburg@wur.nl*

Abstract

This chapter identifies different types of exchange by actors operating in developing countries. Moreover, it discusses the factors that determine the optimal coordination mechanism of exchange. It then focuses on different types of market imperfections where markets are one of the possible coordination mechanisms. Sub-optimal choice of coordination mechanism leads to economic inefficiencies and, therefore, hampers economic development. Some case studies illustrate the nature of market imperfections and their causes, in a selection of developing countries.

Introduction

Exchange is the main element of economic activities. Different transactions of commodities and factors of production (labour, capital and land) take place between different economic actors (consumers, producers, government). Exchange enables acquisition of commodities and factors that are not in the possession of an individual economic actor. Thus, producers can make use of their comparative advantages and consumers can increase the variety in their consumption bundle. Therefore, exchange leads to welfare improvements for actors involved and calls for coordination of supply and demand of respective commodities and factors.

A number of alternative coordination mechanisms exist to equate supply and demand of commodities and factors. The first extreme is the spot market (Peterson *et al.*, 2001). With spot markets, the intensity of coordination control is low. The invisible hand of the market determines prices and acceptable performance standards. The only control right that actors on each side of a transaction can exercise is to engage in price discovery and make a 'yes or no' decision to enter into the transaction. Supply and demand of commodities and factors meet in a market, and ultimately prices that equate supply and demand are determined. The other extreme is vertical integration, i.e. the creation of one organization that has full control in coordinating activities and exchange; this is also called in-house production. The exact quantity of a produced (and demanded) commodity is administratively determined.

Contracts lie in between the two extreme coordination mechanisms mentioned above. Contracts can be complete: the parties' mutual obligations and provisions for all eventualities are spelled out in advance and for the contract duration; contracts can also be incomplete. In the latter case, it is impossible to reach agreement in advance about all possible events that could affect the exchange. Relational contracts are a form of incomplete contracts. There exists a relationship between parties that can be typified as a relation-based alliance. It is an exchange relationship in which the parties involved share certain risks and benefits. The contracts do not have to be formal, but can also be informal. Knoke and Kuklinski (1991) use the term 'network' for this type of coordination. An actor participates in a social system involving (many) other actors, who are reference points in the actor's decisions. Moreover, the position of an individual actor in the social system is relevant. In this case, coordination arises from mutual control. Concepts such as information and trust are also relevant in this respect.

The best coordination mechanism can be identified for each individual transaction. Best refers here to profit maximization for the actors involved (both seller and buyer) given the constraints they face. Transaction costs are a major determinant of what coordination mechanism is optimal. Besides efficiency (costs), effectiveness (goal oriented) and acceptability are other criteria determining the optimal coordination mechanism. The optimal choice can change over time, e.g. a reduction in transaction costs because of improved communication options can lead to another (more efficient) coordination mechanism.

Sub-optimal choice of coordination mechanism leads to economic inefficiencies, and therefore hampers economic development. A certain coordination mechanism can be optimal given the constraints that actors face, but it can be welfare improving to change these constraints (if possible). For example, setting up a credit system could replace relational contracts as coordination mechanism for credit.

The choice of coordination mechanism in developing countries tends to be sub-optimal, and inefficiencies occur. This is because markets and informal relational contracts are often the only coordination mechanism available in developing countries. The reason behind this is that other coordination mechanisms require a certain institutional environment (e.g. well-defined property rights, contract enforcement, etc.), which is often absent in developing countries. If markets do not function well there is a loss in welfare. The solution lies in making alternative coordination mechanisms possible and/or removing factors that cause market imperfections.

This chapter's aims are as follows: (i) to identify different forms of exchange and actors involved in that exchange; (ii) to discuss the optimal choice of coordination mechanism; and (iii) to explain the sources and nature of various market imperfections and implications of such imperfections for some selected developing countries and regions within these countries. It will concentrate mainly on agricultural commodity and factor markets.

To address the research questions, first the actors and different types of exchange of commodities and factors are identified by using a household modelling framework and general equilibrium framework (following section). The next step is to discuss briefly the factors determining the optimal coordination mechanism. Although markets are one form of the possible coordination mechanisms, in developing countries they are the sole mechanism available. Different types of market imperfections are discussed, then the penultimate section presents empirical evidence on the functioning of market based on the research findings of a number of projects within the RESPONSE (Regional Food Security Policies for Natural Resource Management and Sustainable Economies) programme. This is a joint programme of Wageningen University, The Netherlands, and the International Food Policy Research Institute (IFPRI). The chapter ends with a discussion of the results and the main conclusions.

Identifying Types of Exchange and Actors

Both farm-household and applied general equilibrium modelling frameworks can be used to identify and analyse exchange of commodities and factors of production. In general, there is a need to quantify effects of alternative coordination mechanisms and the removal of imperfections to be able to prioritize actions needed and to better understand mechanisms at work. Both types of models can be a starting point for such analysis. In this section, both frameworks are discussed.

Farm-household model

About one-quarter of the world's population belongs to farm-households, and these households are mainly concentrated in less-developed countries. Farming tends to be organized in family farms in most developed and developing countries (see, e.g. Sadoulet and de Janvry, 1995; and Singh, *et al.*, 1986 for a discussion of family farms and household models). A family farm is an integrated entity deciding on inputs, production, consumption, marketing and factor supply. Such a family farm can be perceived as consisting of a farm and a farm-household. The farm is producing agricultural outputs. The farm-household is supplying factors of production to its own farm as well as to other farms or institutions, and the household is also consuming both farm and non-farm outputs (see Fig. 4.1).

More precisely, the farm uses factor and non-factor inputs (e.g. fertilizers, pesticides, etc.) to produce outputs. Factors may originate from two sources. First, the farm-household itself may supply land, labour and capital. Secondly, it may originate from such other sources as banks and paid workers. The non-household factors have to be paid (e.g. interest to the bank and wages) and these payments can be labelled as paid factor costs. There are (normative) costs linked to the use of factors originating from the farm-household, but these costs are not expenses. The farm-household can be seen as a residual claimant and income from farming can be seen as the reward of supplying factor inputs to the farm.

Apart from factor costs, the farm has to bear the costs of non-factor inputs (non-factor costs). The farm-household may consume farm output directly and may earn revenue from selling farm output. The farm-household may supply factors to other farms (e.g. as paid workers) or to non-farm institutions (e.g. farmer working for a work-for-food programme). Total factor income of the farm-household thus originates from the supply of factor inputs to the farms (farm income) and to non-farm institutions (off-farm income). Total household income equals this total factor income plus other income, which includes the farm-household's income from sources such as remittances or other direct income payments (government transfers or transfers from other households). As a consumer, the farm-household spends total household income on purchasing goods and services, and the remaining is saved.

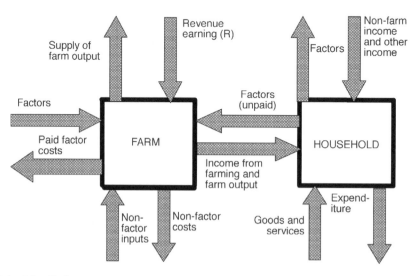

Fig. 4.1. A family farm.

In farm-household models describing the behaviour of family farms it is assumed that: (i) the firm maximizes profit given the market prices of outputs and variable inputs, the quantities of fixed factors (e.g. land in the short-term) and technological constraints; and (ii) the household maximizes utility in terms of consumption of goods, services and leisure (or home time), given income and prices, technology, market wage, total availability of time and household characteristics.

In the farm-household model, the household is assumed to make decisions simultaneously about production, consumption and factor supply. From this farm-household model we can identify several transactions: (i) of farm outputs; (ii) of factors (land, labour and capital) including (formal and informal) credit; (iii) of non–factor inputs; and (iv) of consumer goods and services (other than the particular farm output).

The farm sells its outputs to wholesalers, processors or directly to consumers. Farm-gate prices are the results of interaction between a large number of producers and a (limited) number of buyers. In factor markets, family farms can be both buyers and sellers (e.g. off-farm labour supply). With respect to household consumption, the farm-household buys different goods and services from (retail) markets. Credit markets influence both production and consumption decisions of farm-households. For example, the availability and cost of credit may influence the decision whether to buy a new irrigation tool or not; it may also help to manage farm-household expenses during the lean season. Thus, proper decision making of farm-households depends on the appropriate coordination mechanism for the different transactions to be made.

According to Ellis (1988), peasant farms are dominant in developing countries. He defines peasant farms as family farms characterized by the dominant role the household plays (and the role of women in that household), the partial engagement in markets and imperfections in those markets.

Applied general equilibrium model

An applied or computable general equilibrium (AGE) model can explicitly represent the interactions of different actors in an economy as a whole. This economy can represent the world, a country, region or even a village. During the last three decades, AGE models have been used widely to analyse impacts of policy changes for individual actors in an economy. Construction of AGE models has been discussed by (among others) Robinson (1989) and Ginsburgh and Keyzer (1997). The multi-region AGE modelling approach is described in Hertel (1997). Figure 4.2 presents a simple economy in an AGE framework.

In this economy, agriculture and other industries (including services) purchase factors from the households. In addition, agriculture and other industries purchase non-factor inputs for their production. Total

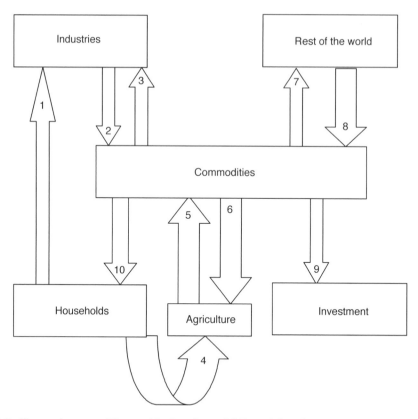

Fig. 4.2. Flows of commodities and factors in an AGE model. 1, factor demand by industries other than agriculture; 2, output supplied by industries other than agriculture; 3, non-factor demand by industries other than agriculture; 4, factor demand by agriculture; 5, output supplied by agriculture; 6, non-factor demand by agriculture; 7, export; 8, import; 9, investment demand; 10, household demand for goods and services.

commodities in this economy come from domestic production of agriculture and other industries and by imports from the rest of the world. Total domestic production and imports are demanded as domestic final and intermediate consumptions, and the remaining is exported to the rest of the world.

An AGE model describes the interaction between different economic actors across markets. Each actor is guided by optimizing behaviour, i.e. producers maximize profit and consumers maximize utility. Optimization behaviour of the producers leads them to make their choice regarding production level and purchase of inputs on the basis of respective prices. Moreover, they have to decide on whether to sell their products in the domestic market or in the international market (export), depending on the relative price. In most AGE models, it is assumed that domestic products and imports are imperfect substitutes to avoid specialization and it is taken into account that most commodities are both imported and exported. The composition of total domestic supply (domestic production plus imports) depends on the relative price of domestic production and imports. Utility maximization by households determines consumption where consumption is a function of income and prices.

In Fig. 4.2, the government (as an actor) is not explicitly mentioned. The government is a producer of public goods and services (administration, health, education, etc.), but sometimes also a producer of private goods (state-owned industries). Moreover, the government intervenes in markets by imposing taxes and providing subsidies (both price subsidies as income transfers). In AGE models different options exist with regard to modelling government behaviour. Assuming the maximization of a public utility function, government behaviour can be modelled similarly to private producer and consumer behaviour.

In an AGE model, equilibrium has to prevail in all the accounts – all commodities, factors and foreign exchange. Usually, the government budget is balanced by allowing its savings or deficit to be residually computed. Equilibrium in other accounts is determined by reconciliation of individual demand and supply forces through the market mechanism. Therefore, the AGE models provide a complete specification of both supply and demand of all markets. The model calculates the relative prices for which all markets are in equilibrium. Thus, according to the AGE framework, the economy is in equilibrium when there is endogenous determination of equilibrium prices, i.e. perfect functioning of different markets in the economy.

An AGE model, therefore, identifies forms of exchange and the actors involved. In the standard model, coordination of exchange takes place in perfectly working markets. However, there are many examples of AGE models that incorporate economic realities as imperfect competition in commodity and factor markets, commodity and factor heterogeneity, subsistence agriculture, etc. (see Ginsburgh and Keyzer, 1997).

Factors Determining the Choice of Coordination Mechanism

This section discusses the main factors affecting the optimal coordination mechanism for exchange of commodities and factors of production (see Williamson, 1987; Douma and Schreuder, 2002). These factors are: transaction costs, information, uncertainty, time inconsistency, transaction-specific investments, bounded rationality and the frequency of transactions.

Transaction costs

Transactions take place within different coordination mechanisms (e.g. markets, contracts or social networks). In practice, there are costs connected with the realization of transactions. Examples of these transaction costs are: (i) costs of searching (for information on prices or quality of goods, etc.); (ii) negotiation costs (time, facilities, etc.); (iii) specification costs (what demands the transferred goods must fulfil); (iv) inspection costs (does the end-product meet the desired specifications?); and (v) arbitration costs (court and representation costs). If transaction costs are high, as is the case in many developing countries, incomplete contracts (for example in networks) seem to have a comparative advantage. If transaction costs are low then the market is the appropriate coordination mechanism.

Information and uncertainty

Uncertainty implies the problem of taking the right decision given limited information. Individuals can increase their knowledge about risk factors by collecting information before taking decisions. Uncertainty and the cost of obtaining information (a form of transaction cost) influence the choice of coordination mechanism.

Lack of information can also take the form of asymmetric information (see, e.g. Douma and Schreuder, 2002). In that case one of the transacting parties (the agent) knows something the other party (the principal) does not know. Information asymmetry can lead to 'hidden information' or 'adverse selection'. Before a transaction one of the parties knows more than the other, and has no reason to let the other know about the additional information he has access to. 'Hidden action' or 'moral hazard' is a form of information asymmetry in which, after a transaction has been concluded, one party undertakes some unobservable action that harms the other party. Elimination or reduction of information asymmetry generates transaction costs and thus influences the choice of a coordination mechanism.

Time inconsistency

Time inconsistency means that the 'rules of the game' are (subsequently) altered by the policy-setting party (e.g. a trader or government). An important aspect to time inconsistency is thus the credibility of a statement, policy, etc. If the actors realize that there is time inconsistency, they will change their behaviour accordingly. Time inconsistency may prevent a transaction or can lead to high transaction cost (e.g. investment in trust).

Transaction-specific investments

Transaction-specific investments are those made in special assets, as well as in specific competences of people. The specificity results from the fact that, for certain production factors there are hardly any alternatives. Transaction-specific investments bind transaction parties strongly to each other. The party who has made such an investment is forced to accept a degradation of transaction conditions because the other party can threaten to break off the relation. This is called the 'hold-up' problem. Protection from the hold-up problem requires extensive and complicated legal contracts (complete contracts). The occurrence of hold-ups leads thus to high transaction costs.

Bounded rationality

Bounded rationality is connected with the human incapability to formulate and solve complex problems. People intend to behave rationally, but in practice that is seldom realized. Bounded rationality is above all important in a complex and uncertain environment that makes it too costly, or even impossible, to get an overview of all the consequences of a decision. Both bounded rationality itself and ways of overcoming it (e.g. research and education) therefore lead to high transaction costs, which favour other coordination mechanisms than the market.

Transaction frequency

The more frequently a transaction is carried out the lower the transaction costs are, in general, because economies of scale reduce the cost per transaction, particularly when special conditions have to be met. A high transaction frequency thus favours the market as a coordinating mechanism.

Market Imperfections

The factors discussed in the previous section lead to the choice of a specific coordination mechanism. Markets are most often the coordination mechanisms selected in developing countries but, at the same time, they are often imperfect. Well-functioning markets are seen as a precondition for economic development. The World Bank (2002) stresses, for example, the importance of well-functioning land and credit markets and formal institutions (e.g. judicial system) that support efficient markets. In this section the nature of different market imperfections is discussed. However, first we look at the characteristics of a perfect market.

Characteristics of a perfect market

The basic characteristics of a perfect market are: (i) perfect competition – no increasing returns, many buyers and sellers (including free entry and exit), all actors are price takers; (ii) perfect information – buyers and sellers have complete information about what they are buying and selling to make the right decisions; and (iii) complete markets – no externalities or public goods, no transactions costs.

Basic conditions of a perfect market lead to economic efficiency. If the conditions of a perfect market do not hold, inefficiencies occur and welfare of the economy is not optimal. The factors discussed in the previous section indicate that the criteria of a perfect market often do not hold. Transaction costs, uncertainty and information asymmetry are often realities, especially in developing countries. Other sources of imperfections are weak institutions, distorting government behaviour (causing and caused by rent-seeking behaviour) and ill-defined property rights. The latter also explain why an alternative coordination mechanism such as complete contracts may not function well in developing countries.

Various market imperfections

Various types and degrees of market imperfections exist. As discussed above, the standard characters of a perfectly competitive market are as follows: (i) homogeneous products; (ii) many buyers and sellers; (iii) market transparency; and (iv) freedom of market entry or market exit. There can be various degrees of failure in attaining these characters, and thus various market failures.

Here we distinguish the following: (i) missing markets; (ii) thin markets; (iii) incomplete markets; (iv) interlocked markets; and (v) well-functioning markets but with imperfect competition.

Missing markets

Sometimes, there does not exist any market for a certain commodity produced or demanded. Janvry *et al.* (1991) identify the non-existence of a market as an extreme case of market failure. This may be because a transaction through a market creates disutility, which is larger than the utility gain of such transactions. According to these authors, market failure is household specific, not commodity or factor specific, which means that markets generally exist but some households fail to participate in markets because of high transaction costs (search costs, transport costs, etc.). As a consequence of non-existence or poor development of markets for outputs, inputs, labour, credit and risk, rural economies become subsistence orientated, relying on family labour and using land and cattle as saving instruments.

Jefferson (2001) identifies missing markets for labour quality as a special kind of missing market under socialist economic systems. As this system does not allow workers to trade both their human capital and intensity of work effort, and there are also restrictions on firm entry, high-quality workers are compelled to exit the state sector, which weakens the average productivity of the economy.

Thus both institutional problems and government regulations or prohibitions contribute to non-existence of markets. Institutional problems include poor infrastructure, which increases uncertainty and risk of involvment in market transactions and may lead to non-transaction.

Thin markets

Under a situation of thin markets it is difficult to attain economies of scale, as there is not enough demand or supply. High transaction costs, weak information (on prices, technologies and potential contracting parties) and a weak institutional framework are considered to be the main contributing factors in creating thin markets. Kydd and Dorward (2004) have noted that economic actors with poor financial and social resources, or political leverage, face high costs in accessing information and enforcing property rights. These inhibit both market development and access to existing markets. As a result, economic and technological development is hampered. Such low level of economic activities leads to thin markets and coordination failure, which again may lead to increase in cost of coping with risk. Moreover, when a market is thin, unit cost of infrastructure development increases (World Bank, 2002).

Incomplete markets

A market is considered to be incomplete if the decisions are taken in an incomplete environment, especially in terms of inadequate or incomplete information. Greenwald and Stiglitz (1986) have noted that, if information is imperfect or markets are otherwise incomplete, competitive markets will not be Pareto-efficient. Inadequate information may take the form of asymmetric information, as discussed above. The World Bank (2002) argues that asymmetric information and missing land markets imply imperfect credit markets and absence of an insurance market in the less-developed economies. As a result, the cost of credit increases and this makes coping with risk more difficult.

Shallow local markets

According to Sadoulet and de Janvry (1995, p. 150), the local market is shallow if there is a high negative covariation between household supply and effective prices. A good harvest in any year will result in oversupply in the market leading to low prices, while a poor harvest will lead to scarcity and high prices. In the latter case, households choose to remain self-sufficient. Poor infrastructure and thus high transport cost may prohibit supplying the commodities to other regions during oversupply or purchasing from other regions during undersupply (lack of arbitrage). Absence of well-integrated markets therefore results in wide variation in prices in response to covariate shocks such as drought.

Interlocked markets

Markets are interlocked if output markets and input markets are linked, e.g. by a credit relationship. A creditor supplies farm inputs, extension, etc. and requires that after harvest the crop is sold to the creditor who pays the market price minus the costs of the inputs. Thus a kind of monopoly is at work. This situation may occur when there are not enough alternatives for obtaining credit. Seasonality in agriculture implies that households have to find ways to balance their budget or income throughout the year. In this case they tend to require credit, and to obtain credit, collateral is needed. If land markets are missing, land cannot serve as collateral.

Distorted markets

A market can be distorted by interventions of government or public authorities. There is widespread use of taxes, tariffs and subsidies by different nations. These market-distorting instruments are used under various justifications such as the protection of local industries (e.g. infant industry argument), encouragement of agricultural production and income support for some specific target groups, etc. Such distortions cause a welfare loss for the economy through an inefficient allocation of resources.

However, discriminatory tariffs or import quota can lead to welfare gains for a single country if the small-country assumption does not apply (Krugman and Obstfeld, 2000). Agricultural policies in developing countries are generally distorted. Almost every country taxes their agricultural sector, often to receive tax revenue or support low-income groups in cities. It has to be realized that, although welfare costs are involved, there is 'value' attached to what one gets back (e.g. more equal income distribution, etc.).

Well-functioning markets but imperfect competition

These include all market structures apart from perfect competition. Market structures can be characterized by the number of suppliers (one, few, many) and number of buyers (one, few, many). There can be monopoly (private or state trading monopolies), oligopoly, monopolistic competition, monopsony, oligopsony, etc. Such imperfect competition in markets may lead to various inefficiencies and welfare loss in an economy. With a monopoly or oligopoly, buyers are negatively affected; with a monoposony or oligopsony, suppliers are negatively affected. The latter is, for example, the case if there are only a few traders to which farms can sell their products. It is important to know who gets the monopoly rents and who bears the burden.

'Workable' competition

The concept of *workable competition* was first developed by Clark (1940). In this market, competition is 'almost' perfect, e.g. almost homogeneous products, sufficient buyers and sellers, reasonable level of market transparency and there are small barriers of entry or exit. Workability under oligopoly depends mostly on the entry and exit possibilities: with moderate entry and exit possibilities, the market behaves quite acceptably. In markets of goods with a public good character (like railways), optimality may be achieved not by deregulation, but by some sort of regulatory mechanism that will ensure workable competition. For public goods (e.g. health care), ensuring workable competition may require extreme government regulations and monitoring.

'Contestable' competition

The theory of contestable market has been developed by Baumol *et al.* (1982). According to these authors, a market is contestable if market equilibrium (price-quantity vector) is *sustainable*. They consider a market equilibrium sustainable if no new firm, with the same technology as existing firms, can offer lower prices. In a perfect contestable market, potential competition controls monopolistic behaviour of the existing firms; entry and exit are considered to be free. However, Gilbert (1989) has argued that, in reality, potential competition is not as powerful as has been considered under the theory of contestable markets. Even with low barriers to entry, the incumbent firms may enjoy preferences (say, in terms of brand recognition) and the market is no longer contestable.

Empirical Evidence

This section presents and discusses some empirical evidences of market imperfections in developing countries. Findings come from a series of PhD research projects under the RESPONSE programme. Several strategic development options for less-developed areas of East Africa, South Asia and South-east Asia were studied under this programme. Primary data were collected in most studies. The market imperfections discussed are at different levels in the economy, from farm level to the world market. Elsewhere in this book a more extensive discussion of the outcomes of the different studies can be found.

Rice marketing in Jiangxi, China (see Chen and Peerlings, Chapter 15, this volume)

Jiangxi Province, located in the south-east of China, is one of the main agricultural provinces in China and rice is its most important cultivated crop. The main topography in Jiangxi Province is hilly and mountainous.

These natural conditions result in high transportation costs and soil degradation. This study focuses on three administrative villages in the north-eastern part of the province with different degrees of market access.[1] Looking at the rice marketing channel in these three villages, it is observed that grain market deregulation and liberalization have strong influences on the market structure.

Liberalization of the rice market has increased competition. Private traders entered the market and started competing with (formerly state-owned) township grain trading companies. However, given the small quantities supplied and traded, the rice market is thin. Moreover, there still exist entry barriers coming mainly from government policies, e.g. requiring registration and paying taxes (market distortions). Both the small quantities supplied (thin markets) and the entry barriers create limited numbers of private traders; this gives market power to the traders (oligopsony).

Trust and informal relationships (networks) among farmers and traders play an important role in determining exchanges. An important reason for this is that private traders do not have the money to pay farmers directly for the product they buy (incomplete market for credit). Search and other transaction costs are relatively high given the small quantities traded. However, there is some evidence that better communication means (mobile telephones) lower search costs. The research further shows that problems with market access have an important effect on how price changes in consumer markets affect prices at the farm gate. Limited market access isolates farmers from changes in the main consumption centres. In other words, there is asymmetric price transformation (incomplete markets). The research also reveals that liberalization is leading rice marketing towards perfect competition.

Vegetable marketing in Benguet, Philippines (see Milagrosa and Slangen, Chapter 10, this volume)

Benguet is one of the six provinces comprising the Cordillera Administrative Region (CAR) in the northern part of the Philippines, a landlocked cluster of 76 towns dominated by mountain ranges. At a plateau 1500 m above sea level, Benguet's uniquely cool climate is useful for vegetable production. Benguet provides for at least 75% of the carrots, potatoes and cabbage demands of the country. The research has looked into the Benguet vegetable marketing system from a transaction costs perspective. The study focuses on the attributes of farmer–buyer relationships of the vegetable trade, the institutional environment in which trade occurs and societal cultural characteristics.

It was found that several man-made and natural resource constraints hinder vegetable trade, and thus economic growth of Benguet. Limited market access due to missing or low-quality farm-to-market roads comprises the bulk of marketing costs and losses for market participants. The Halsema highway is the only artery linking vegetable-producing communities with two major trading markets serving the whole province. The remote communities are always under threat of being cut off from markets due to landslides (missing markets) affecting many roads during the rainy season. The few vegetable trading posts are clustered mainly in certain municipalities and are incapable of handling harvest overflows (shallow markets). Moreover, cold storage warehouses, docking bays for delivery trucks and technology that support timely market information transfer province-wide are lacking (incomplete markets). Agricultural production potential is limited in most municipalities, not only because of depleted soils but also because of dependence on rain-fed agriculture due to lack of irrigation facilities.

The study reveals that social capital in the form of informal networks between farmers and traders is important for vegetable trade in Benguet. The intensity of informal networks in Benguet has not reached a level where it allows the farmers to access resources that are available to members of formal networks. In Benguet, despite strong interpersonal networks, 'trust' within farming communities is rather low. Even at the beginning of the transaction process, farmers and traders are suspicious of each other's tendency for opportunistic behaviour in matters of vegetable weight, grade and price (incomplete market). As a result, transaction (search and negotiation) costs increase. One reason for poor trust levels is market failure due to information asymmetry. On the traders' side, temptation to cheat increases as they are more knowledgeable about market prices than farmers.

Although farmers sell their harvests through commissioners and wholesalers using spot-markets (where price is the most important coordinating mechanism), years of selling their produce in the same market led them to be well acquainted with the traders. For farmers who use the services of a specific trader for a certain cropping season, abruptly breaking off patron–client relations is socially and culturally unacceptable. Moreover, some farmers receive not only production inputs but also gifts and cash advances from traders (interlocked markets). Thus, farmers are caught in (lock-in effect) some sort of relational contract with traders.

Banana marketing in Uganda (see Bagamba *et al.*, Chapter 12, this volume)

Cooking bananas are a key staple food crop in Uganda, being produced mainly for home consumption in low-elevation areas and for both consumption and sale in high-elevation areas. The main market for cooking bananas is in cities, which are largely located within the low-altitude areas. The research has looked into the factors influencing banana production and farm labour allocation decisions in the banana sector. Specifically, the roles played by differences in market access and off-farm employment opportunities are examined using a farm-household model.

In response to poorly functioning and/or missing financial and insurance markets, farms react by product diversification and off-farm employment to spread income risk. Limited access to non-farm income opportunities and imperfections in the labour market can therefore contribute to both an inefficient labour allocation in rural households and a more unequal distribution of income. Options for rural employment are often limited to informal labour exchange amongst households during peak labour demand. As a result, migration and engagement in non-farm activities offer alternatives to rural agricultural employment. Although studies on household income composition have reported substantial shares of income from off-farm and non-farm activities in sub-Saharan Africa, most households are still constrained in accessing the highly remunerative segments of the labour market due to lack of appropriate training, high relocation costs and lack of possibilities for making lumpy investments (e.g. in equipment and machinery). Results showed that labour allocation decisions in Uganda are influenced by production potential, infrastructure (transport costs), exposure to new technologies and household characteristics and structure.

Export performance of Ethiopia (see Jaleta and Gardebroek, Chapter 14, this volume)

Agricultural commodity export is almost the only source of export earnings for Ethiopia. For instance, the share of agriculture in total export value was 98% in 1997/1998, although it declined to 86% in 2001/2002. In addition to dependency on agricultural commodity exports, about 76% of Ethiopia's total export earning comes directly from only three agricultural commodities: coffee, chat[2] and hides and skins.

The decline in the share of agricultural exports in recent years can be explained from both the demand and supply sides. From the demand side, there has been a decline in the world market prices for agricultural export commodities, especially coffee, in recent years. Weather, diseases and other external factors represent supply-side factors that explain the drop in volume of agricultural export. The trade ban on live animals from East Africa by the Middle East Arab countries, due to Rift Valley Fever and the drought of 2002, comprise some of these non-price, external shocks.

Most developing countries, including Ethiopia, depend on the export of primary and traditional agricultural commodities for their foreign currency earnings. However, the world market prices for these commodities are fluctuating, and even declining from time to time. Added to export volume fluctuations, such price fluctuations may exacerbate export income instability. A large amount of variation in Ethiopia's export income is attributed to fluctuations in coffee export income. It has been observed that, to achieve a balanced export portfolio, Ethiopia needs to export more products that fluctuate less in terms of price or volume. In this connection, various market distortions influence the export of agricultural commodities.

Another important aspect is the safety standards of those products. Safety standards or sanitary and phytosanitary measures work as an indirect tariff on the export and thus limit the export up to competitive levels. Other imperfections come from inefficiencies in rural factor and product markets. Improving the efficiency of these markets through reducing transaction costs helps households to buy inputs and sell outputs at effective market prices closer to the actual market prices. In doing so, households can adjust their resource allocation closer to the economically optimal level. The more rural farms that are market oriented and the more farmlands that are allocated to labour-intensive cash crops, the higher will be the levels of rural employment and the income for both landless and land-owning farm-households. One specific inefficiency studied is the asymmetric information about tomato (a potential export crop) prices between farmers and traders. Asymmetric information reduces the bargaining power of the farmers in determining the farm-gate price and therefore leads to low prices (incomplete market). Thus, improved price information reduces the prevailing trade inefficiency at the farm gate.

Apparel export of Bangladesh (see Ahmed and Peerlings, Chapter 16, this volume)

Apparel is the main export industry of Bangladesh, comprising 75% of total export earning and 80% of manufacturing export earning. For the last two decades this industry has been the main source of growth of export and formal employment of unskilled workers. Nearly 1.9 million workers are employed in this industry and 90% of them are female. Around 75% of these female apparel workers are migrants from rural areas, and mainly females from the poorest rural households are coming to work in this industry. The relatively less restrictive Multi-Fibre Arrangement (MFA) import quota on Bangladesh, compared with quotas of traditional apparel exporters like China or India, has induced the growth of the apparel industry. However, the MFA import quotas were fully abolished on 31 December 2004.

Using an applied general equilibrium model for the world (GTAP) shows that abolition of MFA apparel import quotas has had negative effects for Bangladesh. It decreases production, export of apparel, GDP and per capita household utility and welfare. Given that workers in the apparel industry partly originate from rural areas and send remittances, negative effects are transferred to rural areas. Therefore, this is a special case where a market distortion (import quota) led to an (artificial) comparative advantage of Bangladesh apparel, and erosion of the (preferential for Bangladesh) distortion is actually leading to a welfare loss for Bangladesh. In contrast to MFA import quota abolition, attempts by Bangladesh to engage in preferential trade arrangements with its South Asian neighbours, as well as with some East Asian countries, has had positive effects.

The research also looked at the effects of globalization on the level at which workers' rights are addressed. There are two opposite effects: on the one hand, increased competition compels producers to reduce production cost, which often results in violation of workers' rights. On the other, concerns of developed country buyers (which is induced by consumers and trade unions) about workers' rights compel producers to address workers' rights, which increases cost of production. Depending on which pressure mechanism dominates, globalization is either increasing the level of addressing rights or reducing it.

It is pointed out that, if bottlenecks in infrastructure, high cost of investment and corruption are reduced, this will lead to lower cost of production and that will reduce the negative impacts of price pressure caused by MFA import quota abolition.

Table 4.1 summarizes the market imperfections in output markets found in the case study areas. This does not imply that other inefficiencies do not exist in the regions and study areas, but just that they are not discussed here.

Discussion and Conclusions

The aims of this chapter were to: (i) identify different forms of exchange and actors involved in this exchange; (ii) discuss the optimal choice of coordination mechanism; and (iii) discuss the sources and nature of various market imperfections and the implications of such imperfections for some selected developing countries and regions within these countries.

Transaction costs are the main determinant in the choice of which coordination mechanism is optimal. Coordination mechanisms vary from markets to production and consumption within a farm. In developing countries or in particular regions within these countries markets are often, together with informal relational contracts (networks),

Table 4.1. Various market imperfections in the case study areas.

	Jiangxi, China	Benguet, Philippines	Uganda	Ethiopia	Bangladesh
Missing and thin markets	+	+	+		
Incomplete markets	+	+	+	+	
Shallow local markets		+			
Interlocked markets		+			
Distorted markets	+			+	+
Imperfect competition	+		+	+	

the only available coordination mechanism. However, these markets often function poorly. Market imperfections stem from distorting government policies, small number of buyers or sellers, high transport costs (poor infrastructure) and lack of clearly defined property rights (e.g. missing land markets). The chapter gives an overview of different imperfections in markets in developing countries. The empirical evidence describes specific market imperfections in different markets at different levels: from farm-household level to world market level.

Transaction costs are an important determinant of the optimal coordination mechanism. High transaction costs are found to be important in all cases examined – whether it is for rice production in China, tomato production in Ethiopia or vegetable production in the Philippines. Poor infrastructure (especially roads) results in isolation and, consequently, thin commodity and labour markets. The thinner the market, the more the rural households are trapped in self-sufficiency. Labour market imperfections lower the opportunity cost of farm family labour. Lower opportunity costs result from transaction costs, which include the risk involved in job search, transport costs, information gathering costs, time inconsistency, etc. It is also observed that credit and insurance markets are not operating well in the cases discussed, e.g. land is missing as collateral, which constrains production and tends to give an incentive for off-farm employment.

Market imperfections can be solved in different ways depending on the source of the imperfection. The legal market environment should be such that contracts can be enforced. Cooperation or group action as an alternative coordination mechanism could help to lower transport costs, build up countervailing power, lower search costs, etc. However, this requires an institutional reform in several case study countries. Other possibilities include: (i) supplying market information by the government or associations of farmers and/or traders; (ii) providing credit by rural banks or creditors or non-government organizations (NGOs); and (iii) introducing alternative risk-coping mechanisms (e.g. insurance) by farmers' associations. Cooperation along these lines can be set up by private initiatives. However, the government or NGOs could facilitate this. Improving infrastructure and defining and protecting property rights are issues governments must undertake.

Without such measures, actors in the developed countries depend too much on informal relationships with each other, relationships that appear to be unable to solve various inefficiencies. A possible lack of trust in informal relationships makes things even worse.

In this chapter, two different models are discussed that could be used to analyse the effects of using alternative coordination mechanisms or removal of imperfections. In general, there is a need for quantifying effects of alternative coordination mechanisms and removal of imperfections to be able to prioritize actions needed and to better understand mechanisms at work. Some of the studies discussed in this chapter make a contribution in that area.

Acknowledgements

The authors thank Andrew Dorward, Eleni Gabre-Madhin, Ruerd Ruben and Arie Oskam for their comments on an earlier version of this chapter.

Endnotes

[1] In this research, market access is defined as distance for a good to be transported to the consumer market.
[2] Chat is a mildly narcotic plant produced in the eastern, southern and south-western parts of Ethiopia.

References

Baumol, W., Panzar, J. and Willig, R. (1982) *Contestable Markets and the Theory of Industry Structure.* Harcourt Brace Jovanovich, New York.
Clark, J.M. (1940) Toward a concept of workable competition. *American Economic Review* 30, 241–256.
Douma, S.W. and Schreuder, H. (2002) *Economic Approaches to Organisations.* Prentice Hall International, London.
Ellis, F. (1988) *Peasant Economics, Farm Households and Agrarian Development.* Cambridge University Press, Cambridge, UK.
Gilbert, R.J. (1989) The role of potential competition in industrial organization. *The Journal of Economic Perspective* 3, 107–127.
Ginsburgh, V. and Keyzer, M.A. (1997) *The Structure of Applied General Equilibrium Models.* The MIT Press, Cambridge, UK.
Greenwald, B. and Stiglitz, J.E. (1986) Externalities in economies with imperfect information and incomplete markets. *Quarterly Journal of Economics* 101, 229–264.
Hertel, T.W. (1997) *Global Trade Analysis Modeling and Applications.* Cambridge University Press, New York.
Janvry de, A., Fafchamps, M. and Sadoulet, E. (1991) Peasant household behaviour with missing markets: some paradoxes explained. *Economic Journal* 101, 1400–1417.
Jefferson, G.H. (2001) Missing market in labor quality: the role of quality markets in transition. Paper presented at the *4th Annual International Conference on Transition Economics*, 23–25 July 1999, Beijing, China.
Knoke, D. and Kuklinski, J.H. (1991) Network analysis: basic concepts. In: Thompson, G., Frances, J., Levacic, R. and Mitchell, J. (eds) *Markets, Hierarchies and Networks.* Sage Publications, London, pp. 171–182.
Krugman, P.R. and Obstfeld, M. (2000) *International Economics, Theory and Policy.* Addison-Wesley Publishing Company, Reading, UK.
Kydd, J. and Dorward, A. (2004) Implications of market and coordination failures for rural development in least developed countries. *Journal of International Development* 16, 951–970.
Peterson, H.C., Wysocki, A. and Harsh, S.B. (2001) Strategic choice along the vertical coordination mechanism. *International Food and Agribusiness Management Review* 4, 149–166.
Robinson, S. (1989) Multisectoral models. In: Chenery, H. and Srinivasan, T.N. (eds) *Handbook of Development Economics.* Vol. II, Elsvier Science Publishers, Amsterdam.

Sadoulet, E. and de Janvry, A. (1995) *Quantitative Development Policy Analysis.* The John Hopkins University Press, Baltimore, Maryland.

Singh, I., Squire, L. and Strauss, J. (eds) (1986) *Agricultural Household Models, Extensions, Applications and Policy.* The John Hopkins University Press, Baltimore, Maryland.

Williamson, O.E. (1987) *The Economic Institution of Capitalism.* The Free Press, New York.

World Bank (2002) *Building Institutions for Markets.* Oxford University Press, Oxford, UK.

Resource Management Options

5

Soil Nutrient Dynamics in Integrated Crop–Livestock Systems in the Northern Ethiopian Highlands

ASSEFA ABEGAZ[1,2,*] AND HERMAN VAN KEULEN[1,3,**]

[1]Plant Production Systems Group, Wageningen University, Wageningen University and Research Centre, Wageningen, The Netherlands; [2]Mekelle University, Mekelle, Ethiopia; [3]Plant Research International, Wageningen University and Research Centre, Wageningen, The Netherlands; e-mails: *assefa_abegaz@yahoo.com; **herman.vankeulen@wur.nl

Abstract

In the Northern Ethiopian Highlands, one of the least-favoured areas in East Africa, farming systems are characterized by the integrated management of crop and livestock components. The overall objective of this chapter was to increase insight into the functioning of these farming systems, with special attention to the influence of farm management regimes on soil nutrient dynamics, as a basis for formulation of recommendations for technological innovation leading to increased farm productivity, conservation of the natural resources and improved livelihoods for the farming population.

Partial balances of the macronutrients – nitrogen (N), phosphorus (P) and potassium (K) – were studied at farm and field scales and soil organic carbon (C) balance at village scale, and the results indicate that soil nutrient balances were negative at all scales. Nutrient depletion rates differ significantly among farmer wealth groups, with the highest rates recorded for the rich farm group, followed by the medium and poor farm groups. Current levels of organic fertilizer input are much lower than required to maintain a dynamic equilibrium in soil organic matter content. Thus, limited availability of organic inputs is a crucial constraint for attaining sustainability in terms of nutrient elements.

Results of a simulation study on long-term dynamics of soil C, N and P and the consequences of alternative farm management practices for crop-available N and P indicate that, in order to maintain current levels of soil N, organic carbon and P, external inputs in the form of inorganic fertilizers are indispensable. In this chapter, attention focuses on the biophysical aspects of sustainability in terms of soil qualities, with special attention to nutrient elements. Adoption of more sustainable farm management practices, however, is constrained by the (socio-)economic environment.

Introduction

Agriculture is the basis of the Ethiopian economy, accounting for 46% of its GDP and 90% of its export earnings and employing 85% of the country's labour force (UNDP, 2002). About 95% of the agricultural output is produced on integrated crop–livestock subsistence smallholder farms (ADF, 2002). The farm is the primary unit of production, generating a livelihood for the family and serving as an asset for farmers to accumulate wealth (Nega *et al.*, 2003). Moreover, Ethiopia's long-term economic development strategy, 'Agricultural Development-Led-Industrialization' (ADLI), explicitly targets smallholders' private agricultural economy to maintain food security and contribute to economic growth (ADF, 2002).

It is also supposed to initiate development of home industries, stimulate the service sector and create additional purchasing power. These broad objectives underline the importance attached to agriculture in general and to smallholder farms in particular in the Ethiopian economy (Block, 1999). To achieve these targets, appropriate farm management for sustainable agricultural development is vital.

In the highlands of Ethiopia, farm management is characterized by integrated management of crop and livestock components. Farm management comprises a set of decisions and actions by the farmer with respect to their fields, animals and other resources through which they influence the way in which commodities are produced. The appropriateness of such practices should be evaluated on the basis of the degree to which the farm household can realize its objectives, without negatively affecting the quality of the natural resources. This is equivalent to a (partial) evaluation of the degree of sustainability of such farming systems.

A range of sustainability indicators for agricultural production systems has been proposed, including soil chemical and physical characteristics (Hartemink, 1998; Wang and Gong, 1998; Nambiar *et al.*, 2001; Arshad and Martin, 2002), with special emphasis on soil organic matter (SOM) content (Rosell *et al.*, 2001; Freixo *et al.*, 2002; Bessam and Mrabet, 2003). Soil fertility is intimately linked to SOM content, which influences the physical, chemical and biological properties of soil, and therefore indigenous soil nutrient supply (De Ridder and Van Keulen, 1990; Bessam and Mrabet, 2003). Thus, depletion of SOM implies a decline in the reserves of these nutrients and in cation exchange capacity, while it hampers formation of well-aggregated soil structural elements and consequently reduces soil water-holding capacity (De Ridder and Van Keulen, 1990). Hence, the importance of the organic matter content of the soil exceeds that of any other of its properties (Young, 1976).

Limited indigenous supply of N and P, associated with low SOM contents, is a strong constraint for plant growth and dry matter production (Penning de Vries and Van Keulen, 1982). Therefore, appropriate SOM management (De Ridder and Van Keulen, 1990; Ayuk, 2001; Katyal *et al.*, 2001; Quansah *et al.*, 2001), combined with N, P and K management, is a prerequisite for maintenance of soil fertility as a basis for sustainable agricultural development (Breman and Debrah, 2003).

In this chapter we have analysed current soil N, P, K and C balances, and explored long-term dynamics of soil C, N and P under alternative farm management practices and the consequences for crop-available N and P, as a basis for the design of appropriate management practices for Teghane in Tigray State in the northern Ethiopian highlands.

Materials and Methods

Study area

Tigray is situated in the northern highlands of Ethiopia (12° 15′ to 14° 57′ N and 36° 27′ to 39° 59′ E; Fig. 5.1 (a)), with Teghane micro-catchment covering an area of 13.56 km², ranging in altitude from 2710–2899 m above mean sea level, located in its eastern administrative zone, Atsbi-Wonberta district (13° 52′ 53″ to 13° 53′ 37″ N and 39° 42′ 05″ to 39° 43′ 57″ E; Fig. 5.1 (b)), about 70 km north of the capital of Tigray, Mekelle.

The climate is 'Dega' (WBISPPO, 2002), with average annual monomodal rainfall from July to September of 541 mm (coefficient of variation 53% for the period 1901–2002, at a location near Teghane at 14° N and 40° E (Viner, 2003)). In the period 2000–2004, average annual rainfall was 532 mm (Atsbi World Vision, 2004).

Farming system

The dominant farming system in Teghane is a subsistence integrated crop–livestock system. Barley (*Hordeum* spp.) and wheat (*Triticum* spp.) are the major food crops, grown on 66 and 14%, respectively, of the cultivated land in 2002, followed by field pea (*Pisum* spp.) and faba bean (*Vicia* spp.), occupying 10 and 9%, respectively. The main cropping period is the rainy season between July and October. Farm animals

(a)

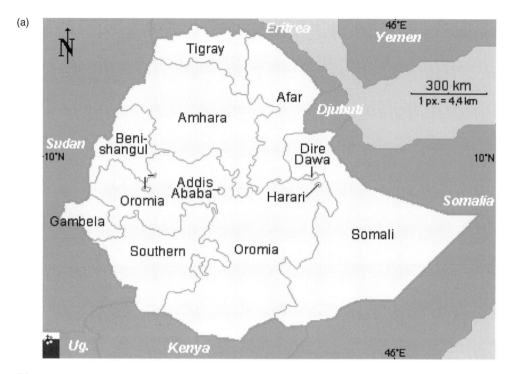

(b)

Fig. 5.1. Geographical map of Ethiopia (a) and Tigray and Teghane (b).

include cattle, sheep and donkeys, providing draught power and manure for cropping. Dried animal manure is used extensively as a source of household fuel. Crop residues are used as feed for livestock.

Field- and farm-scale N, P and K balances

Farm selection

Farmers were classified, during a village farmers' assembly, into three wealth groups (rich, medium and poor) on the basis of the socio-economic conditions in Teghane. The criteria used were: (i) land holding; (ii) herd size (HS); and (iii) the stock of seed/grain for planting and consumption (Assefa, 2005). Through stratified random sampling, five households from the rich, seven from the medium and 12 from the poor group were selected from the list of households in the village.

Framework for analysis of N, P and K flows

In Teghane, nutrient flows between the farm system and the external system/market and internal flows have been monitored for 1 year, following the NUTMON toolbox methodology (Van den Bosch *et al.*, 2001). This resulted in quantification of: (i) stocks of soil nutrients and rates of change; and (ii) the nutrients entering and leaving the system components. Balances for N, P and K were calculated for: (i) the farm scale (aggregation of all farm plots, PPUs); and (ii) individual PPUs. Five PPUs have been distinguished, i.e. barley, wheat, faba bean, pea and natural pasture.

From each FSU (land units with more or less 'stable' soil characteristics), composite soil samples from the 0–20 cm surface layer were collected for laboratory determination of OC (Walkley and Black, 1947), available P (Olsen *et al.*, 1954), total N (Bremner and Mulvaney, 1982) and total K (Knudsen *et al.*, 1982).

Village-scale soil C balance

For computation of the village-scale soil C balance, a model (Assefa, 2005) has been developed that allows analysis of the effect of alternative crop residue management practices on soil C dynamics. In the calculations, the total land area of the micro-catchment was taken into account, except for rock outcrops, water bodies and open woodland. The open woodland was disregarded, as for conservation purposes this land is protected from exploitation. The model accounts for: (i) C inputs in manure (either all manure C or the residual when part of the manure is used as fuel), in crop residues not used as animal feed, in 25% grazing and harvesting losses, recycled in the field and in roots; and (ii) annual soil C losses in the form of CO_2 during decomposition of soil organic matter.

Crop residues should be composted before field application, to avoid negative effects on crop growth of application of material with high C/N ratios, due to immobilization of N during its decomposition (De Ridder and Van Keulen, 1990). Carbon lost via burning of manure as fuel was derived from the proportion of manure deposited in the stable, which was estimated at 58% (Assefa, 2005): livestock are in the stable for 14 h per day and manure excretion is assumed to be evenly distributed over the day (Van den Bosch *et al.*, 2001). Export and/or import of feed resources in the study area were negligible, and were thus not taken into account in the C balance calculation.

Calculation of the soil C balance is based on the top 0.20 m of soil,[1] an annual relative decomposition rate of soil organic matter of 0.06 kg per kg and annual relative humification coefficients of 0.5 for crop residues, roots and feed leftovers and 0.3 for manure (De Ridder and Van Keulen, 1990). Carbon from roots was derived from the harvest index and a shoot:root ratio of 6 (Boons-Prins *et al.*, 1993; Van Keulen, 1995).

Modelling long-term soil nutrient dynamics under alternative farm management regimes

Model description

SOIL N AND ORGANIC C. Soil organic C (OC) and soil organic N (ON) dynamics are closely linked (Parton *et al.*, 1983, 1989; Jenkinson, 1990; Van Keulen, 1995; Struif Bontkes, 1999). Soil organic carbon dynamics are extremely complex, as concurrently processes of *transformation*, *conversion* and *dissimilation* take place, mediated through soil microbes.

In this study, a simplified dynamic model has been applied (described in detail in Assefa, 2005), based on the ON model of Wolf *et al.* (1989), validated by Wolf and Van Keulen (1989) and applied by Hengsdijk and Van Ittersum (2003). Organic carbon dynamics have been linked to the dynamics of organic N via the C:N ratio.

In the model, four organic pools are distinguished in the soil: (i) stable organic N (NSP); (ii) stable organic carbon (OCSP); (iii) labile organic N (NLP); and (iv) labile organic carbon (OCLP). Inputs into the system include four external N sources, i.e. inorganic fertilizer (Ninorf), organic fertilizer (Norgf), rain (Nrain) and biological fixation (Nfix), and one external C source (OCorgf).

Mineral N in the system, either from the external sources or mineralized during decomposition of labile organic material, is either taken up by the vegetation (Ncrop) or lost from the system (Nloss). Part of the N from the labile pool is transferred via microbial action to the stable pool. During decomposition of labile organic material, part of the C is transformed into CO_2 in respiratory processes to provide energy for microbial functioning. Carbon from the labile pool is transferred to the stable pool in association with the N, using the C:N ratio of the labile material.

In the model, 18 transfer coefficients have been defined, describing: (i) partitioning of the external N inputs among losses, crop uptake and the labile pool; (ii) partitioning of N transferred from the labile pool among losses, crop uptake and the stable pool; and (iii) transfer of N from the stable pool to the labile pool. Values for 16 transfer coefficients (except transfers 2 and 6) are presented in Table 5.1 for a situation with moderate risks for losses (Wolf *et al.*, 1989). Losses of organic N from the stable and labile pools via soil loss through water erosion are estimated from the rate of OC loss via erosion by dividing by their respective C:N ratios.

Soil loss through water erosion depends on rainfall, soil erodibility, topography, crop cover and possible anti-erosion measures (Roose, 1977; Renard and Ferreira, 1993), and is calculated in the model using the Soil Loss Equation developed by Roose (1977). Carbon losses through soil erosion are calculated from soil loss and soil bulk density, using an enrichment factor (Knisel, 1980) to account for the higher organic matter content in the topsoil layer.

SOIL P. To explore long-term P dynamics, a module describing P transformations in the soil, commensurate with the N and OC module, has been incorporated (Wolf *et al.*, 1987; Van Keulen, 1995). Unlike N, P is present in the soil in substantial quantities in inorganic forms that can be divided in labile, stable and soil mineral, whereas organic P is divided into stable and labile pools (Wolf *et al.*, 1987; Van Keulen, 1995).

In the model, a stable and a labile pool are distinguished, each containing both organic and inorganic P. Three sources of P inputs are distinguished: weathered soil minerals (Pw), organic fertilizer (Porgf) and inorganic fertilizer (Pinogf). Eight transfer coefficients have been defined, describing: (i) transfer from the stable to the labile P pool; (ii) transfer from the labile to the stable P pool; (iii) transfer from inorganic

Table 5.1. Transfer coefficients (situation, with moderate risks for losses, for nitrogen (N) from inorganic fertilizer (Ninorf), organic fertilizer (Norgf), biological fixation (Nfix), rain (Nrain) and the labile organic pool (NLP) to crop, loss, labile and stable pool and from the stable pool (NSP) to the labile pool (see text for explanation) (from Wolf *et al.*, 1989).

Input	Crop	Loss	Labile pool	Stable pool
Ninorf	0.40	0.40	0.20	–
Norgf	0.15	0.15	0.70	–
Nfix	0.15	0.15	0.70	–
Nrain	0.40	0.40	0.40	–
NLP	0.425	0.425	–	0.15
NSP	–	–	1.00	–

and organic fertilizer/amendments to the labile P pool; (iv) transfer from the labile P pool to crop uptake and losses; (v) transfer from soil mineral P to the stable P pool; and (vi) loss of P through soil erosion.

Model parameterization

SOIL N AND C. Initial pool sizes of SN and OC are computed for the top 0.20 m from total initial organic N and C and the equilibrium ratio of 3 between both pools (i.e. NSP/NLP and OCSP/OCLP). The time constant of conversion of the labile N pool is calculated from its size and its rate of mineralization. The time constant of conversion of stable N is derived from that for the labile N pool, which gave satisfactory results for dryland soils under a range of environmental conditions (Wolf and Van Keulen, 1989). Inputs of Ninorf and Norgf are user defined. Nitrogen supply via biological fixation was derived from results of long-term trials on unfertilized fields (Wolf *et al.*, 1989), whereas N supply via rainfall was estimated from annual rainfall in the area using a transfer function: N deposition = 0.0065* rainfall, mm/year (Van Duivenbooden, 1992).

SOIL P. Initial sizes of the P pools are estimated from P uptake from unfertilized and fertilized soil, which have been derived from one-season P fertilizer trials where crop production and P uptake with and without P fertilizer is established (Janssen *et al.*, 1987; Wolf *et al.*, 1987). The user should specify rate and type of P fertilizer applied. Annual input of mineral P from soil parent material is set to 1 kg/ha (Van der Pol, 1992). The time constants of conversion between the labile and the stable P pool and vice versa were found to be 5 and 30 years, respectively, under a wide range of environmental conditions (Janssen *et al.*, 1987). The enrichment factor for P loss via erosion has been set at 1, assuming that total P distribution in the top 0.20 cm is uniform.

SOIL DATA. In February 2003, the soils of Teghane were surveyed following the topo-sequence survey method, and nine representative soil profiles have been described following the FAO/UNESCO (1990) guidelines. From each profile, samples were taken, bulked for each soil type and analysed for determination of soil physico-chemical properties. Subsequently, a field survey was conducted to delineate the boundaries of soil units and land use/cover (LUC) types, using GPS, supported by topographic maps (EMA, 1997) and aerial photographs from 1994. Following preparation of the soil map, the area of each soil unit was calculated, using the facilities in ArcView (ESRI, 2002).

Soils of the study area were classified as Cambisols, Luvisols and Leptosols (FAO/UNESCO, 1990). Cambisols, covering 26% of the area, located on the colluvial terrace slopes, are intensively cultivated. Most of the Luvisols are located in the valley bottoms or flood plains; they are relatively deep, with favourable physical and chemical characteristics. Traditionally, these soils were under grassland, but in recent years a significant proportion has been transformed to arable land in response to the shortage of land for food crop production. Leptosols, covering 46% of the area, are located on the elevated plateau and on hill slopes, crests and ridges, interspaced with rock outcrops and patches of Luvisols. They are limited in depth to about 25 cm by an underlying continuous hard sedimentary rock, or they contain less than 20% fine particles to a depth of 75 cm (FAO/UNESCO, 1990).

Results and Discussion

Farm- and field-scale nutrient balances

Farm-scale analysis

STOCKS OF SOIL NUTRIENTS AND RATES OF CHANGE. Soil nutrient stocks per farm are defined as the total quantities of the macronutrients present in the top 20 cm of the soil profile, from where crops take up the major part of their nutrients (De Jager *et al.*, 1998; Van den Bosch *et al.*, 1998). These stocks include dissolved ions and nutrients in organic matter, adsorbed to the solid phase or in stable inorganic components. To assess the size of these stocks per farm for the macronutrients N, P and K, between 10 and 25 samples per farm were analysed, depending on farm size and heterogeneity of its soil resources (see Table 5.2).

Total biomass production and nutrient removal (both in absolute and relative terms) were higher from soils with higher total N contents (see Table 5.3). Relative depletion rates were highest for N for all farm groups (but with varying rates, i.e. 1.5% per year for the rich and 0.6% for both the medium and poor groups), followed by K and lowest for P for poor groups whereas, for the rich and the medium farms, rates for P and K were the same.

FARM FIELD BALANCES. Farm field here is defined as the 'aggregated' arable and natural pasture fields. The balances of N, P and K were negative for all three farm groups (see Fig. 5.2), with higher absolute values for the rich than for the medium and the poor farms ($P < 0.05$). The major causes of these high rates of nutrient mining are: (i) very small land holdings; (ii) large families; (iii) large herds (with cattle dung used

Table 5.2. Organic carbon (OC), total nitrogen (TN), total P, available P (P-Olsen) and total K contents of the surface soil (0–20 cm) per farm group.

Farm group	OC (%)	TN (%)	Total P (ppm)	P-Olsen (ppm)	Total K (ppm)
Rich	2.89	0.26	567.5	26.0	5372
Medium	1.95	0.17	500.2	19.3	4036
Poor	1.17	0.12	544.9	9.1	4311

Table 5.3. Nutrient stocks to a depth of 20 cm, calculated with average soil bulk density of 1240 kg/m³, and rates of change in Teghane, northern highlands of Ethiopia, 2002.

	Rich	Medium	Poor
Total N stock (kg/ha)	6448	4216	2976
N flows (kg/ha/year)	−96.0	−24.9	−18.3
N flows (% of stock/year)	−1.5	−0.6	−0.6
Total P stock (kg/ha)	1407	1241	1351
P flows (kg/ha/year)	−11.3	−4.9	−1.3
P flows (% of stock/year)	−0.8	−0.4	−0.1
Total K stock (kg/ha)	13322	10009	10691
K flows (kg/ha/year)	−110.5	−42.1	−25.6
K flows (% of stock/year)	−0.8	−0.4	−0.2

mainly for fuel); and (iv) very limited external inputs. Land is a very scarce resource, with average holdings of 1.6, 0.9 and 0.6 ha per household and household sizes of eight, six and four for the rich, medium and poor types, respectively. Average herd sizes are large in relation to the available grazing resources (Assefa, 2005). Inflows of the inorganic fertilizers, urea and diammonium phosphate (DAP) were very small, i.e. 12.5 kg N and 4.4 kg P/ha for the rich, 10 kg N and 3.5 kg P/ha for the medium and 8.2 kg N and 2.9 kg P/ha for the poor farm group, respectively.

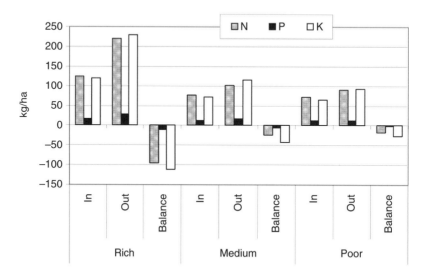

Fig. 5.2. Partial nitrogen (N), phosphorus (P) and potassium (K) balances (kg/ha/year) from farm fields per farm group.

Field-scale/primary production unit balances

The most important primary production units (PPU) were barley, wheat, faba bean and pea and natural pasture. Natural fields, comprising small pasture plots, borders of crop fields and areas around homesteads, were overgrazed and nutrient depletion rates were highest for all farm groups: –115 kg N, –5.8 kg P and –112 kg K/ha for the rich (a), –56.2 kg N and –42.5 kg K/ha for the medium (b) and –56.5 kg N and –34.6 kg K/ha for the poor (c) farms (see Fig. 5.3).

For the latter two groups, P was almost in balance. For *wheat*, all balances were negative for all farm groups (see Fig. 5.3). In *barley* fields, the highest negative balances were recorded for K: –112 for the rich, –50.4 for the medium and –18.2 kg/ha for the poor (see Fig. 5.3). The N balance was positive for the poor group (+5.1) and negative for the rich (–81.9) and the medium group (–8.7 kg/ha). Nitrogen balances for *faba bean* and *pea* were positive for the medium and poor farmers, with values of 4.0 and 20.3 k/ha, respectively. Nitrogen balances for all fields were negative for the rich farm group, with the lowest balance for faba bean and pea (–10.5 kg/ha).

Village-scale soil C dynamics

For the top 0.20 m soil, annual weighted C loss via decomposition was estimated at 2.26 mg/ha, and total C input at 100% feed use at 0.87 mg/ha, resulting in a negative soil C balance of 1.39 mg/ha/year. At 30% best feed use, total C input is 1.3 times higher, comprising 0.06 mg/ha of C from manure (all manure assumed to be applied to the field) and 1.06 mg/ha of C from crop residues and other amendments. To maintain the current soil C content, annual application of 13.0 mg of manure OM or 9.2 mg of crop residue is required. At all levels of feed use, soil C balances are negative, suggesting that in terms of soil carbon, the current system is not sustainable.

Long-term soil nutrient dynamics under alternative farm management regimes

Model calibration

The model was calibrated for Teghane on the basis of OC and N for three soils (Cambisols, Luvisols and Leptosols) and on the basis of P for two soils (Cambisols and Luvisols) (see Table 5.4). Two sources of external organic fertilizer – manure and composted crop residues – and two sources of external inorganic fertilizer – urea and DAP – were considered. Crops in rotation were faba bean/field pea > barley/wheat > wheat/barley. Transfer coefficients were used as given in Table 5.1.

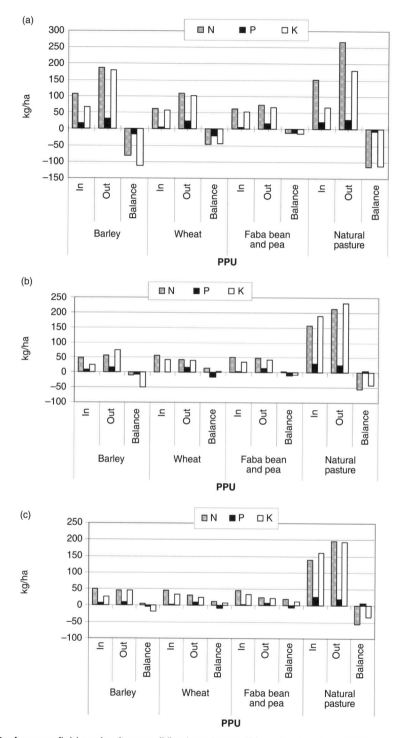

Fig. 5.3. Average field-scale nitrogen (N), phosphorus (P) and potassium (K) flows (kg/ha/year) for rich (a), medium (b) and poor farms (c).

Table 5.4. Soil chemical characteristics and bulk density in the 0–20 cm soil layer of three soil types in Teghane, northern highlands of Ethiopia, 2003.

Soil unit	OC (g/kg)	Total N (g/kg)	P-Olsen (mg/kg)	Exch. K (mmol/kg)	pH (H_2O)	Bulk density (kg/m^3)	Average slope (degree)
Cambisols	11.0	1.4	7.0	15.0	6.4	1285	7
Luvisols	37.0	3.2	6.0	15.4	6.4	1201	3
Leptosols	9.8	1.0	7.4	–	7.4	1121	25

OC AND N MODULE. Initial pool sizes of N and OC (see Table 5.5) have been calculated from total soil N and OC content in the 0–20 cm layer, based on average bulk density (see Table 5.4).

The time constant of conversion of N from the labile pool (see Table 5.5) was estimated on the basis of total mineralized N, derived from average crop N uptake from unfertilized soil (see Table 5.5), taking into account N deposition via rainfall (3.5 kg N/ha/year) and the appropriate transfer coefficients (see Table 5.1).

P MODULE. The system represents a non-steady state situation, as crop P uptake exceeds annual net input of P from weathered parent material (1 kg/ha). The initial size of the labile pool was calculated from crop P uptake from the unfertilized soil by dividing by the uptake fraction of the labile pool, calculated by dividing the recovery of applied P by the fraction of fertilizer P transferred to the labile pool, at 58.9 for Cambisols and 95.2 kg/ha for Luvisols. Initial size of the stable pool equals its time

Table 5.5. Some soil N and P characteristics for model initialization: soil N and P in the 0–20 cm depth, crop-N and -P uptake, initial sizes of stable and labile pools of N and P and time constants of conversion of stable and labile pools for three soils in Teghane, northern highlands of Ethiopia, 2003.

	Soil unit		
	Cambisols	Luvisols	Leptosols
Soil N in 0–20 cm soil depth (kg/ha)	3598.0	7686.4	2331.7
Initial stable N pool (kg/ha)	2698.5	5764.8	1748
Initial labile N pool (kg/ha)	899.5	1921.6	582.9
Initial C:N ratio	7.9	11.6	9.4
Initial stable P pool (kg/ha)	384.0	600.0	na
Initial labile P pool (kg/ha)	58.9	95.2	na
Total crop-N uptake from unfertilized soil (kg/ha)	67.0	129.1	18.2
Total crop-P uptake from unfertilized soil (kg/ha)	10.5	20.0	na
Total crop-P uptake from fertilized soil (75 kg/ha)	21.2	32.6	na
Time constant of conversion of labile N (years)	5.8	6.4	14.7
Time constant of conversion of stable N (years)	116.6	127.9	294.9

na = no data.

constant of conversion multiplied by the P transfer from the stable to the labile pool.

Model validation

The N dynamics module has been validated extensively by Wolf and Van Keulen (1989) for temperate and by Hengsdijk and Van Ittersum (2003) for a tropical region, while the P module has been validated by Janssen *et al.* (1987). For both modules, it was concluded that their performance was satisfactory and that they can be used with confidence in exploring long-term effects of different soil and crop management regimes.

In this chapter, the N and OC modules were validated on the basis of an empirical data set for fields continuously cultivated for 7–53 years in smallholder farms at Lepis in the highlands of Ethiopia (Lemenih *et al.*, 2005a, b). Unfortunately, no data were available for validation of the P module under Ethiopian conditions.

Management options

As the model satisfactorily reproduced the behaviour of the agricultural system under common management in the highlands of Ethiopia (Assefa, 2005), it may be used with confidence for exploration of the effect of alternative farm management practices on long-term dynamics of soil C, N and P and the consequences for crop-available N and P. The model has been applied to three soils in Teghane, varying in chemical characteristics (see Table 5.4), under five alternative farm management regimes:

1. Control: a farming system of continuous cereal crops with complete removal of crop residues and without external nutrient inputs.
2. Alternative 1 (Alt1): cultivation of cereal crops with input of the current level of organic amendments, set to 3.6 mg/ha/year (Assefa, 2005) and recommended chemical N and P fertilizer rates (urea in combination with DAP each at an annual rate of 100 kg/ha (64 kg N and 21 kg P).
3. Alternative 2 (Alt2): composting of all crop residues and the vegetation from pasture land (set at 5.4 mg/ha/year (Assefa, 2005)) and applying this compost to the cultivated fields. Thus, power for traction and other farming practices is assumed to be supplied from sources other than animals and only arable farming is practised, producing crops and/or vegetables or, if livestock production is included, animal feed originates from silage production from cultivated fields and/or concentrates and application of all manure to cultivated fields.
4. Alternative 3 (Alt3): inclusion of leguminous crops in the rotation (legume > cereal > cereal) to control. Under this management regime, annual net N inputs of 60, 40 and 20 kg/ha from biological N fixation have been set for Luvisols, Cambisols and Leptosols (as biomass

production on these soils varies). Amanuel *et al.* (2000) have reported net N inputs of 12–58 kg/ha.

5. Alternative 4 (Alt4): application of the required composted organic amendments and inorganic P fertilizer to maintain the current levels of soil C and P. The C:N ratio of composted organic amendments was set to 17.8, with N content = 0.01687 (Assefa, 2005) and C content of composted crop residues = 0.30 (De Ridder and Van Keulen, 1990).

Soil OC and N dynamics

Organic C dynamics are similar for the control and Alt3 (see Fig. 5.4), because in Alt3 only biological N fixation is added. Initial C content is higher in Luvisols (37.0) than in Cambisols (11.0) and Leptosols (9.8 g/kg), as a consequence of differences in history of cultivation (see soil data section). These differences result in different rates of change in OC at similar input levels. Control management (see Fig. 5.4) results in 44, 42 and 38% depletion of soil OC after 50 years of cultivation in Cambisols, Luvisols and Leptosols, respectively. Despite the higher rate of erosion in Leptosols caused by their steeper slopes (see Table 5.4), the rate of decline in soil OC is lower, due to its lower initial content. Alt1 results in 16% reduction in OC in Cambisols and 32% in Luvisols, whereas in Leptosols it results in a 22% increase. Alt2 results over 50 years in 27% reduction in OC content in Luvisols, whereas in Cambisols and Leptosols it increases by 1 and 57%, respectively.

To maintain a steady state in terms of organic C, required annual input is higher for soils with higher OC content. Hence, under Alt4, composted organic amendments of 5.3, 15.0 and 2.1 mg/ha/year are required for Cambisols, Luvisols and Leptosols, respectively. Inputs above these levels for each soil will result in build-up of OC and vice versa. Under a given management regime, in terms of nutrient recycling and external inputs, soil nutrient depletion is stronger on soils with higher nutrient contents, and the reverse is true for the build-up of soil nutrients.

The rates of reduction in soil N under current management (see Fig. 5.5) are similar to those in OC, as the C:N ratio of the soil organic matter changes only slightly over the 50 years. For both Cambisols and Luvisols, all management options result in N depletion whereas, under Alt1 and Alt2, the N stock in Leptosols gradually increases.

Since under Alt4 the input of organic amendments suffices to maintain soil OC contents, soil N contents after 50 years are 24, 14 and 18% lower than the initial values in Cambisols, Luvisols and Leptosols, respectively. Under Alt1 and Alt2, after 50 years, soil N contents are 26 and 24% (in Cambisols) and 33 and 32% (in Luvisols) lower than the initial values. In Leptosols, over the same period, N contents increase by 7 (Alt1) and 13% (Alt2).

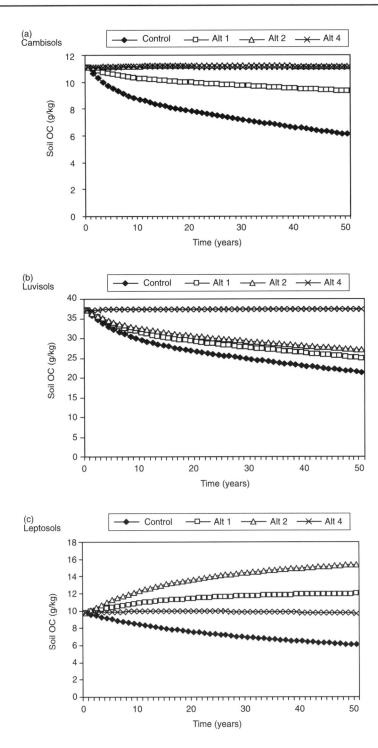

Fig. 5.4. Simulated soil carbon dynamics in three soils of the northern highlands of Ethiopia under alternative farm management practices. Alt 1, etc., see text for explanation.

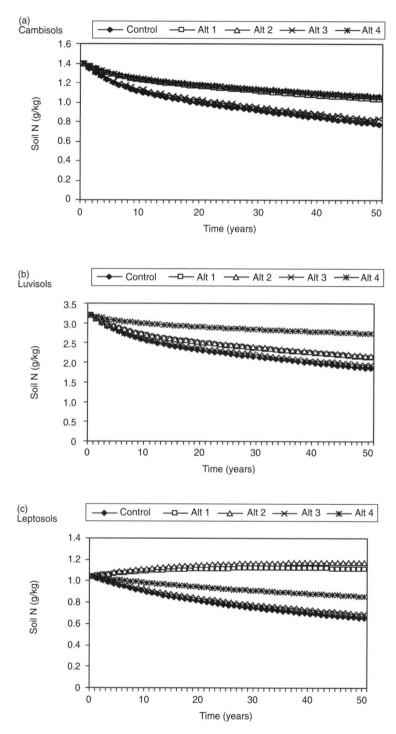

Fig. 5.5. Simulated soil nitrogen dynamics in three soils of the northern highlands of Ethiopia under alternative farm management practices. Alt 1, etc., see text for explanation.

Soil P dynamics

Under the control and Alt2, total P content over 50 years decreases by 46 and 43% in Cambisols and 53 and 52% in Luvisols, respectively. On the other hand, Alt1 results in build-up of total soil P in Cambisols (69% higher after 50 years), but still in depletion (8%) in Luvisols. Under Alt4, maintaining current soil P levels requires, in addition to the annual organic P supply of 5.3 (in Cambisols) and 15.0 (in Luvisols) mg/ha composted organic fertilizer (with P content = 0.00014), annual inorganic P doses of 8 on Cambisols and 23 kg/ha on Luvisols. Inputs above these levels result in build-up of total P and vice versa.

Crop-available N and P

In the control, after 50 years, crop-available N is 87% (in both Cambisols and Luvisols) and 77% (in Leptosols) lower than the initial value. Under Alt1, crop-available N increases in Leptosols (by 14%), whereas in Cambisols and Luvisols it decreases by 31 and 52%, respectively. Under Alt2, crop-available N decreases by 38 and 60% in Cambisols and Luvisols, respectively, whereas in Leptosols it increases by 35%. Under Alt3, crop-available N decreases by 81, 83 and 71% in Cambisols, Luvisols and Leptosols, respectively.

In the control, crop-available P after 50 years is 67% (in Cambisols) and 55% (in Luvisols) *lower* than the initial value. Alt1 results in 55 and 11% *higher* values in Cambisols and Luvisols, respectively. Under Alt2, crop-available P decreases by 54 and 52%, respectively, for these two soils. Under Alt4, crop-available P increases by 7% in Cambisols and 20% in Luvisols.

Yields reflect the changes in crop-available N and P. For example, by applying the calibrated QUEFTS model (Assefa, 2005) to the control (unfertilized soil), 'initial' barley grain yield is estimated at 2.7 and 5.5 mg/ha, respectively on Cambisols and Luvisols. Assuming that supply of K is not limiting, barley yield declines to 0.5 mg/ha on Cambisols and 0.9 mg/ha on Luvisols after 50 years.

Conclusions

To what extent can farm households be food secure on the basis of the current production systems, given the low land productivity and the small size of their holdings? A simple evaluation of food security in relation to cereal production indicates that currently, the rich farmers produce food grain slightly in excess of the minimum annual food requirements, whereas the medium and poor farmers are faced with deficits of 27 and 28%, respectively (see Table 5.6).

Table 5.6. Per capita food grain requirements and annual balances at different production scenarios for three farm groups.[a]

Characteristic	Rich[b]	Medium[b]	Poor[b]
Actual average grain yield household per day (kg) (2003)			
Barley	1891	1016	641
Wheat	300	153	190
Faba bean	160	64	23
Field pea	121	68	8
Total per household	2472	1301	862
Per capita food grain balance (supply/demand, fraction)	1.05	0.73	0.72
Estimated barley grain yield using QUEFTS model at different input levels			
No external inputs			
Barley grain (kg/household)	5641	3086	1182
Barley per capita food grain balance (kg)	402	211	−7.5
Current external input			
Barley grain (kg/year)	5794	3203	1264
Per capita food grain balance (kg)	421	230	13
External N fertilizer (kg/ha/year)	12.5	10.0	8.2
External P fertilizer (kg/ha/year)	4.4	3.5	2.9
Recommended external input			
Barley grain (kg/year	6257	3773	1777
Per capita food grain balance (kg)	479	326	141
External N fertilizer (kg/ha/year)	64	64	64
External P fertilizer (kg/ha/year)	21	21	21
Household characteristics			
Average cultivated land (ha)	0.84	0.62	0.43
Average household size (persons)	8	6	4
Farm household barley grain requirement (kg/yr)[c]	2424	1818	1212
Soil characteristics			
OC (g/kg)	28.9	19.5	11.7
P-Olsen (mg/kg)	26.0	19.3	9.1
Exchangeable K (mg/kg)	10.4	7.7	6.4
Working and non-working days of the year (%)			
Working days of the year		138	
Weekends		104	
Local holidays in weekdays		123	

[a] Of the total population of the study area, 2, 37 and 61% are classified as rich, medium and poor, respectively.
[b] Actual average grain yields (from Assefa, 2005).
[c] Annual per capita barley grain requirement set at 303 kg; daily per capita energy requirement set at 2500 kcal; 100 g raw barley = 301, wheat = 310, faba bean and field pea = 370 kcal (from Garrow *et al.*, 2003).

Current yields at low levels of external inputs are below model-based yield estimates derived from indigenous soil nutrient supply (Assefa, 2005). Moreover, barley yields in unfertilized treatments in on-farm experiments (1.8 mg/ha on Luvisols that have been under cultivation for over 50 years, 2.6 mg/ha on Cambisols and 5.3 mg/ha on

Luvisols, cultivated for 7 years) were much higher than those on farmer-managed fields (maximum grain yield on rich farms was 2.1 mg/ha) (Assefa, 2005). This would imply that other factors than nutrients limit current agricultural productivity. Note however, that our experiments were supplementarily irrigated.

In the study area for almost all crops, weed infestation resulted in up to 30% yield reduction (Assefa, 2005). Moreover, local orthodox Christian holidays take up a substantial proportion of the potential working time, i.e. about 62% of the time consists of non-working days (28% weekends and 34% local holidays during weekdays (see Table 5.6).

These observations could indicate scope for increasing land productivity and thus food security in the short term, because model-estimated yields at the currently recommended chemical fertilizer application rates result in per capita food grain production exceeding the minimum requirements for food security for all farm groups, on the assumption that moisture is not a constraint and crops are free of weeds, pests and diseases (see Table 5.6). However, agriculture can not serve as an engine for economic growth beyond food security, since 61% of the population in the study area are poor, own small parcels of cultivated land and can not afford higher levels of external inputs.

The per capita surplus grain yield of 141 kg/year (for the poor farm group) will barely be enough to cover other expenses such as school fees, medical costs, etc. Moreover, attaining the objective of production increase requires a fundamental change in the current developments, as between 1971 and 2000 the per capita agricultural production index in Ethiopia steadily declined at the rate of 1.15% per year (FAO, 2001) and will further decline with declining soil fertility and increasing population.

Current farm management on smallholder mixed crop–livestock farms in the Ethiopian highlands results in serious soil mining with gradually declining stocks of soil N, P, K and C, leading to decreasing crop-available N and P and lower crop yields. Results of the SOM model indicate that the quantities of external organic amendments required to maintain soil OC at the current level are much higher than the total biomass that can be produced in the area. Model results on introduction of legumes in the rotation and low levels of organic and/or chemical fertilizer application (CSA, 1997; Assefa, 2005) show that these practices cannot compensate for the current nutrient depletion rates so that, in the course of time, grain yields will significantly decrease as a consequence of the reduction in crop-available N and P and soil C. Limited availability of organic inputs is thus a crucial constraint for attaining sustainability in terms of nutrient elements. These results indicate that, to maintain current levels of soil N and organic C and P, external inputs in the form of organic and inorganic fertilizers are indispensable.

These conclusions are in agreement with those of analyses carried out for similar climatic conditions in West Africa (cf. De Ridder *et al.*, 2004). From a soil nutrient point of view, sustainable agricultural development is impossible at the current levels of organic amendments and chemical fertilizers (Alt1), even with the most efficient system of recycling of carbon and nutrients (Alt2).

Moreover, the agricultural development problem in the northern highlands of Ethiopia has not only a soil nutrient dimension but also, and probably equally important, a socio-economic and moisture dimension, and simple solutions are not available. Given the high rate of population increase, the small size of the agricultural holdings and the high demand for organic inputs into the soil to maintain soil nutrients and the associated crop yields in a sustainable way, for long-term poverty reduction and economic growth another strategy is required to attain sustainable increased land productivity. Such a strategy should focus on a substantial expansion of the non-agricultural sectors that can provide gainful employment for a significant proportion of the population from the agricultural sector. These issues warrant further investigation as a basis for new policy formulation.

Endnote

[1] Soil units include: Cambisols (covering 335.5 ha, with an organic carbon (OC) content of 23.54 g/kg and bulk density (BD) of 1.285 g/cm^3), Luvisols (330.2 ha, OC: 13.97 g/kg, BD: 1.300 g/cm^3) and Leptosols (455.5 ha, OC: 9.78 g/kg, BD: 1.121 g/cm^3) (FAO/UNESCO, 1990).

References

ADF (Africa Development Fund) (2002) *The Federal Democratic Republic of Ethiopia: Agricultural Sector Review*. Agricultural and Rural Development Department, Addis Ababa.

Amanuel, G., Kuhne, R.F. and Tanner, D.G. (2000) Biological nitrogen fixation in faba bean (*Vicia faba* L.) in the Ethiopian highlands as affected by P fertilization and inoculation. *Biology and Fertility of Soils* 32, 353–359.

Arshad, M.A. and Martin, S. (2002) Identifying critical limits for soil quality indicators in agro-ecosystems. *Agriculture, Ecosystems and Environment* 88, 153–160.

Assefa, A. (2005) Farm management in mixed crop-livestock systems in the Northern Highlands of Ethiopia. PhD Thesis, Wageningen University, Wageningen, The Netherlands.

Atsbi World Vision (2004) *Atsbi World Vision Climatic Data (2000–2004)*, Atsbi, Ethiopia.

Ayuk, E.T. (2001) Social, economic and policy dimensions of soil organic matter management in sub-Saharan Africa: challenges and opportunities. *Nutrient Cycling in Agro-ecosystems* 61, 183–195.

Bessam, F. and Mrabet, R. (2003) Long-term changes in soil organic matter under conventional tillage and no-tillage systems in semiarid Morocco. *Soil Use and Management* 19, 139–143.

Block, S.A. (1999) Agriculture and economic growth in Ethiopia: growth multipliers from a four-sector simulation model. *Agricultural Econonomics* 20, 241–252.

Boons-Prins, E.R., De Koning, G.H.J. and Van Diepen, C.A. (1993) *Crop-specific Simulation Parameters for Yield Forecasting Across the European Community*.

Simulation Reports CABO-TT No. 32, Group Plant Production Systems, Wageningen University, Wageningen, The Netherlands.

Breman, H. and Debrah, S.K. (2003) Improving African food security. *SAS Review* 23, 153–170.

Bremner, J.M. and Mulvaney, C.S. (1982) *Nitrogen-total*. ANR Analytical Lab., University of California, Riverside, California (http://groups.ucanr.org/danranlab/soil%5Fanalysis/, accessed February 2005).

CSA (Central Statistics Authority) (1997) *Agricultural Sample Survey 1996/1997. Vol. III: Report on Farm Management Practices*. Statistical Bulletin 171, CSA, Addis Ababa.

De Jager, A., Nandwa, S.M. and Okoth, P.F. (1998) Monitoring nutrient flows and economic performance in African farming systems (NUTMON) 1. Concepts and methodology. *Agriculture Ecosystems and Environment* 71, 37–48.

De Ridder, N. and Van Keulen, H. (1990) Some aspects of the role of organic matter in sustainable intensified arable farming systems in the West African semi-arid tropics (SAT). *Fertilizer Research* 26, 299–310.

De Ridder, N., Breman, H., Van Keulen, H. and Stomph, T.J. (2004) Revisiting a 'cure against land hunger': soil fertility management and farming systems dynamics in the West African Sahel. *Agricultural Systems* 80, 109–131.

EMA (Ethiopian Mapping Authority) (1997) *Topographic Maps*. EMA, Addis Ababa.

ESRI (Environmental Systems Research Institute) (2002) *ESRI GIS and Mapping Software*. ESRI, New York.

FAO (2001) *FAO Global Information and Early Warning System on Food and Agriculture. World Food Program Special Report*. FAO/WFP crop and food supply assessment mission to Ethiopia, 9 January 2001, FAO, Rome.

FAO/UNESCO (1990) *Soil Map of the World. Revised Legend*. FAO, Rome.

Freixo, A.A., Machado, P.L., dos Santos, H.P., Silva, C.A. and Fadigas, F.S. (2002) Soil organic carbon and fractions of a Rhodic Ferrasol under the influence of tillage and crop rotation systems in Southern Brazil. *Soil and Tillage Research* 64, 221–230.

Garrow, J.S., James, W.P.T. and Ralph, A. (2003) *Human Nutrition and Dietetics*, 10th edn., Churchill Livingstone, Elsevier Science, London.

Hartemink, A.E. (1998) Soil chemical and physical properties as indicators of sustainable land management under sugar cane in Papua New Guinea. *Geoderma* 85, 283–309.

Hengsdijk, H. and Van Ittersum, M.K. (2003) Dynamics in input and output coefficients for land use studies: a case study for nitrogen in crop rotations. *Nutrient Cycling in Agroecosystems* 66, 209–220.

Janssen, B.H., Lathwell, D.J. and Wolf, J. (1987) Modelling long-term crop response to fertilizer phosphorus. II. Comparison with field results. *Agronomy Journal* 79, 452–458.

Jenkinson, T.S. (1990) The turnover of organic carbon and nitrogen in soil. In: Greenwood, D.J., Nye, P.H. and Walker, A. (eds) *Quantitative Theory in Soil Productivity and Environmental Pollution*. The Royal Society, London.

Katyal, J.C., Rao, N.H. and Reddy, M.N. (2001) Critical aspects of organic matter management in the tropics: the example of India. *Nutrient Cycling in Agroecosystems* 61, 71–88.

Knisel, W.G. (1980) CREAMS: *A Field Scale Model for Chemicals, Runoff and Erosion from Agricultural Management Systems*. Conservation Report No. 26, US Department of Agriculture, Washington, DC.

Knudsen, D., Peterson, G.A. and Pratt, P.F. (1982) *Lithium, Sodium and Potassium*. ANR Analytical Lab., University of California, Riverside, California (http://groups.ucanr. org/danranlab/soil%5Fanalysis/, accessed February 2005).

Lemenih, M., Karltun, E. and Olsson, M. (2005a) Assessing soil chemical and physical property responses to deforestation and subsequent cultivation in smallholders farming system in Ethiopia. *Agriculture, Ecosystems and Environment* 105, 373–386.

Lemenih, M., Karltun, E. and Olsson, M. (2005b) Soil organic matter dynamics after defor-estation along a farm field chronosequence in southern highlands of Ethiopia. *Agriculture, Ecosystems and Environment* 109, 9–19.

Nambiar, K.K.M., Gupta, A.P., Qinglin, Fu and Li, S. (2001) Biophysical, chemical and socio-economic indicators for assessing agricultural sustainability in the Chinese coastal zone. *Agriculture, Ecosystems and Environment* 87, 209–214.

Nega, B., Adenew, B. and Gebre Sellasie, S. (2003) Current land policy issues in Ethiopia. In: Cox, M., Munro-Faure, P., Dey-Abbas, J., Rouse, J. and Bass, S. (eds) *Land Reform: Land Settlement and Cooperatives 2003/3* (special edition), FAO, Rome, pp. 103–124.

Olsen, S.R., Cole, C.V., Watanabe, F.S. and Dean, L.A. (1954) *Estimation of Available Phosphorus in Soils by Extraction with Sodium Bicarbonate.* USDA Circular No. 939, USDA, Washington, DC.

Parton, W.J., Anderson, D.W., Cole, C.V. and Stewart, J.W.B. (1983) Simulation of organic matter formation and mineralization in semi-arid agro-ecosystems. In: Lowrance, R.R., Todd, R.L. and Asmussen, L.E. (eds) *Nutrient Cycling in Agricultural Systems.* Special Publication 23, The University of Georgia Press, Athens, Georgia, pp. 533–550.

Parton, W.J., Sanford, R.L., Sanchez, P.A. and Stewart, J.W.B. (1989) Modeling soil organic matter dynamics in tropical soils. In: Coleman, D.C., Oades, J.H. and Uehara, G. (eds) *Dynamics of Soil Organic Matter in Tropical Ecosystems.* College of Tropical Agriculture and Human Resources, University of Hawaii, Honolulu, Hawaii, pp. 153–171.

Penning de Vries, F.W.T. and Van Keulen, H. (1982) The actual productivity and the role of nitrogen and phosphorus. In: Penning de Vries, F.W.T. and Djitéye, A.M. (eds) *La Productivité des Pâturages Sahéliens. Une Étude des Sols, des Végétations et de l'Exploitation de cette Ressource Naturelle.* Agricultural Research Reports 918, Pudoc, Wageningen, The Netherlands, pp. 196–226 (French, with English summary).

Quansah, C., Drechsel, P., Yirenkyi, B.B. and Asante-Mensah, S. (2001) Farmers' perceptions and management of soil organic matter – a study from West Africa. *Nutrient Cycling in Agroecosystems* 61, 205–213.

Renard, K.G. and Ferreira, V.A. (1993) RUSLE model description and data base sensitivity. *Journal of Environmental Quality* 22, 458–466.

Roose, E. (1977) *Erosion et Ruissellement en Afrique de l'Ouest: Vingt Années de Mesures en Petits Parcelles Expérimentales.* Travaux et documents de L'ORSTOM, Paris.

Rosell, R.A., Gasparoni, S.C. and Galatini, S.A. (2001) Soil organic matter evaluation. In: Lal, R., Kimble, J.M., Follette, R.F. and Stewart, B.A. (eds) *Management of Carbon Sequestration in Soils. Advances in Soil Science.* CRC Press, Boca Raton, Florida, pp. 311–322.

Struif Bontkes, T. (1999) Modelling the dynamics of agricultural development: a process approach. PhD Thesis, Wageningen Agricultural University, Wageningen, The Netherlands.

UNDP (2002) *UNDP Assistance in the 5th Country Program to the Agricultural Sector.* (http://www.sas.upenn.edu/African_Studies/eue_web/undp_p1.htm#top, accessed January 2005).

Van den Bosch, H., De Jager, A. and Vlaming, J. (1998) Monitoring nutrient flows and economic performance in African farming systems (NUTMON). II. Tool development. *Agriculture, Ecosystems and Environment* 71, 49–62.

Van den Bosch, H., Vlaming, J., Van Wijk, M.S., De Jager, A., Bannink, A. and Van Keulen, H. (2001) *Manual to the NUTMON Methodology.* Alterra/LEI, Wageningen University and Research Centre, Wageningen/The Hague, The Netherlands.

Van der Pol, F. (1992) *Soil Mining – an Unseen Contributor to Farm Income in Southern Mali*. Bulletin 325, Royal Tropical Institute, Amsterdam.

Van Duivenbooden, N. (1992) *Sustainability in Terms of Nutrient Elements with Special Reference to West Africa*. Report 160, CABO-DLO, Wageningen, The Netherlands.

Van Keulen, H. (1995) Sustainability and long-term dynamics of soil organic matter and nutrients under alternative management strategies. In: Bouma, J., Kuyvenhoven, A., Bouman, B.A.M., Luyten, J.C. and Zandstra, H.G. (eds) *Eco-regional Approaches for Sustainable Land Use and Food Production*. Kluwer Academic Publishers, Dordrecht, The Netherlands, pp. 353–375.

Viner, D. (2003) *The Climatic Impacts Links Project. High-resolution Girded Datasets* (http://www.cru.uea.ac.uk/cru/data/hrg.htm, accessed February 2005).

Walkley, A. and Black, I.A. (1947) Estimation of soil organic carbon by the chromic acid digestion method. *Soil Science* 63, 251–264.

Wang, X. and Gong, Z. (1998) Assessment of soil quality changes after eleven years of reclamation in subtropical China. *Geoderma* 81, 339–355.

WBISPPO (Woody Biomass Inventory and Strategy Planning Project Office) (2002) *Atlas of a Strategic Plan for the Sustainable Development, Conservation and Management of the Woody Biomass Resources of Tigray. Methodology, Land Use System Analysis*. Ministry of Agriculture, Addis Ababa.

Wolf, J. and Van Keulen, H. (1989) Modeling long-term crop response to fertilizer and soil nitrogen. II. Comparison with field results. *Plant and Soil* 120, 23–38.

Wolf, J., De Wit, C.T., Janssen, B.H. and Lathwell, D.J. (1987) Modelling long-term crop response to fertilizer phosphorus. I. The model. *Agronomy Journal* 79, 445–451.

Wolf, J., De Wit, C.T. and Van Keulen, H. (1989) Modeling long-term crop response to fertilizer and soil nitrogen. I. Model description and application. *Plant and Soil* 120, 11–22.

Young, A. (1976) *Tropical Soils and Soil Survey*. Cambridge University Press, Cambridge, UK.

6 Rural Development and Sustainable Land Use in the Hillsides of Honduras

HANS G.P. JANSEN,[1] JOHN PENDER,[2] AMY DAMON,[3] WILLEM WIELEMAKER[4] AND ROB SCHIPPER[5]

[1] International Food Policy Research Institute (IFPRI), Central America Office, San José, Costa Rica; telephone: +506-255-4011; fax: +506-222-6556; e-mail: h.jansen@cgiar.org; [2] International Food Policy Research Institute (IFPRI), 2033 K St NW, Washington, DC 20006, USA; telephone: +1-202-862-5645; e-mail: j.pender@cgiar.org; [3] Department of Applied Economics, University of Minnesota, 231 ClaOff Building, 1994 Burford Avenue, St Paul, Minnesota, USA; telephone: +1-612-625-1724; e-mail: damo0021@umn.edu; [4] Jagerskamp 31, 6706 EG Wageningen, The Netherlands; e-mail: Willem.Wielemaker@wolmail.nl; [5] Development Economics Group, Wageningen University, Hollandseweg 1, 6706 KN, Wageningen, Netherlands; e-mail: rob.schipper@wur.nl

Abstract

We used data from 376 households, 1066 parcels and 2143 plots located in 95 villages in the hillside areas of Honduras to generate information needed by decision makers to assess the needs and opportunities for public investments, and design policies that stimulate natural resource conservation. We developed a quantitative livelihood approach, using factor and cluster analysis to distinguish between seven household livelihood strategies. Based on a multinomial logit model, we show that livelihood strategies are determined by comparative advantages through a combination of biophysical and socio-economic variables.

While 92% of the sample households earned <US$1.00 per capita/day, households that pursue a livelihood strategy based only on basic grain farming were the poorest. Households following strategies that combine on-farm work with off-farm work attain slightly higher incomes. Household incomes can be increased by the following measures: (i) improving road infrastructure; (ii) widening access to land; (iii) implementing policies that reduce household size and dependency

ratios; and (iv) adoption of sustainable land management technologies that restore soil fertility. Probit models show that the latter can be promoted by agricultural extension programmes and land redistribution. Investments in physical assets should be directed at households that pursue livelihood strategies based on off-farm employment or coffee production, while agricultural training programmes are best focused on livestock producers.

Introduction

Honduras is one of the poorest countries in the Western hemisphere, with poverty being particularly severe in rural areas. Land ownership patterns throughout Honduras have historically meant that large landowners own the majority of the most productive land in the valleys while smaller, poorer agricultural producers work on the hillside land. The hillside areas,[1] which account for 80% of the total area, are generally considered to have lower agricultural potential.

About 60% of the total population of Honduras is considered rural, and 80% of rural people live in the hillside areas (PRONADERS, 2000). Macro-economic reform and structural adjustment strategies alone have proved inadequate to reduce poverty in the rural hillside areas, where 92% of the population live on < US$1.00/day (based on own survey data). Low land productivity and rural poverty are closely associated with insufficient protection of the natural resource base and resulting degradation of soil and water resources (Vosti and Reardon, 1997; Lee and Barrett, 2001). Reductions in agricultural productivity caused by soil erosion (which can can reach 300 t/ha/year of soil loss (Thurow *et al.*, 2002)) not only adversely impact farmers' incomes but also cause negative off-farm impacts such as silting of rivers, resulting in flooding, reduced water quality and diminished reservoir capacity.

Stagnation in the productivity of the basic grains sector[2] in Honduras on which so many rural households depend, disadoption of fertilizer bean in northern Honduras (Neill and Lee, 2001) and increasing demand have resulted in an expansion of the agricultural frontier into primary forest areas and increasing pressure on the natural resource base in hillside areas. Migration out of these areas towards areas with higher potential or to other countries is a selective and costly process.

In Honduras, investments have generally been directed toward more-favoured areas and hillside areas have been left behind. This chapter provides some of the basic information needed to identify opportunities for public investments in hillside areas and to guide the design of promising policies that stimulate natural resource conservation. Using data from a detailed household survey, we: (i) identify and quantify livelihood strategies (LS) through clustering farm households on the basis of the use of their labour and land resources; (ii) isolate the main determinants of the household's LS choice; and (iii) analyse the main determinants of per capita income, including the role of LS and the household's asset portfolio. We also investigate the impact of LS and other factors on land management and soil conservation practices at the parcel level. Finally, we derive a number of policy implications.

The Livelihood Strategy Concept

A livelihood strategy can be defined as a portfolio of activities and choices that people make to achieve their livelihood goals, including productive activities, investment strategies, reproductive choices, etc. (Ellis, 1998; DFID, 1999; Adato and Meinzen-Dick, 2002). These choices are reflected in the way that people use their assets and, as such, are an important part of household behaviour, while also determining well-being.

We implement the LS concept through quantification of the household's asset portfolio and clustering of households. Clustering a sample of households into a limited number of categories that pursue similar livelihood strategies (LS) may be useful to policy makers by enabling them better to target households with certain common characteristics, in this way increasing the efficiency of public investment strategies and other policy measures. In addition, a clear delineation of dominant LS can help in directing technology transfer programmes towards their intended beneficiaries. The alternative to clustering is using continuous measures of asset types, but this does not allow an improvement in the effectiveness of policies through better targeting.

Given the key role of the household's asset portfolio in understanding its livelihood strategy, we based our clustering on the household's use of its main assets, i.e. labour and land. Once we obtained the household clusters (each of which represents a distinct livelihood strategy), we designated each of the clusters through a careful analysis of average time allocation, land use patterns and income shares of households in each of these clusters.

Methods and Data

We used factor and cluster analysis to group the individual household observations into livelihood categories. The factor analysis was based on the correlation matrix of: (i) proportions of time spent on agricultural work on own farm, off-farm agricultural work and off-farm non-agricultural work; (ii) proportions of total agricultural work spent on basic grains, other annual crops, coffee, other permanent crops, livestock activities and off-farm agricultural work; and (iii) proportions of the farm land used for basic grains, other annual crops, coffee, other permanent crops, pastures and forest plus fallow. The rotated factor loadings from the factor analysis served as input to agglomerative hierarchical clustering, followed by k-means clustering to correct for possible misclassification of observations at the boundaries between clusters (Hair *et al.*, 1998; Wishart, 1999).

The household's livelihood choice can be explained based on a set of predetermined asset-based variables that include natural, human and location capital. Livelihood strategies are an important part of a wider set of explanatory, asset-based variables that determine household income and which also include other types of capital (physical, financial and social capital). Finally, land management decisions are influenced by the same set of variables as household income, plus other variables that reflect field-specific characteristics. A household's asset holdings thus has both a direct and indirect (via their impact on the livelihood strategy choice) influence on income and land management strategies.

Our research covered nine provinces (*departamentos*) and 19 counties (*municipios*) selected purposively based on representation of agro-ecological conditions, dominant land use, population density, market access and the degree of presence of projects and programmes. We also included a number of counties in the north-east of Honduras as recipient areas of migrants in the study. The remainder of the sampling process was done in a fully randomized manner. Five communities (*aldeas*) and two hamlets (*caseríos*) were selected in each county and community respectively. Finally, two farm households (*hogares*) were randomly selected in each hamlet.

The household sample contains a total of 376 households, 1066 parcels (defined as a contiguous piece of land based on tenure status) and 2143 plots (sub-parcels defined on the basis of land use). Key socio-economic elements of the household survey (carried out in 2001–2002) included household composition, education, asset ownership, sources of income, sales of crop and livestock products, participation in credit markets, membership of organizations and participation in training and extension programmes. Information collected at the parcel and plot levels included land tenure, cropping patterns, crop production, land management technologies – including use of labour and other inputs – and conservation practices and investments.

The survey also collected detailed biophysical data for a (randomly drawn) sample of two plots on each farm. These included landscape attributes, plot size, soil type, erosion status and presence of physical conservation structures. Soil samples were also taken and analysed for pH, nutrient content, organic matter content and texture, in order to calculate soil moisture availability and soil fertility. Finally, secondary information regarding rainfall, population density, market access and road density was also used (CIAT, 2001).

Results

Cluster analysis

The factor analysis identified six main common factors. The cluster analysis identified seven clear-cut and robust clusters of which the two smallest (12 and eight households respectively) were excluded from the econometric analysis (see below). All variables used in the factor analysis show statistically significant differences between at least one pair of clusters, and provided confidence in the stratification of the household sample into livelihood categories.

Description of livelihood strategies

We named LS based on an investigation of the utilization of land and labour resources in each cluster and corresponding levels and composition of income. A striking common characteristic of all LS is the very high proportions of households that live in extreme poverty (see Table 6.1). Table 6.2 provides data regarding asset ownership for each cluster.

The livelihood of households in cluster 1 is based on extensive livestock farming on relatively large farm holdings. But even these households obtain about 30% of their total household income from working outside their own farm and highly value food security. These households have above-average education and are mostly located in lower-altitude areas with lower population densities. Per capita income is only US$0.58/day, but *average* per capita income is somewhat misleading since, for some households, this livelihood appears to be a poverty-exit livelihood strategy.

Cluster 2 consists of coffee producers who still rely on own production of basic grains for their subsistence needs. Most coffee farms are relatively small and located at higher altitudes. Somewhat surprisingly, market access and education are below average for these households and they farm on relatively poorer soils. On average, these households earned just over 50% of the average per capita income of livestock households. However, the survey was taken during the period when coffee prices had collapsed.

Cluster 3 represents small subsistence farmers who focus on low-profitability basic grains production. This is the poorest group among all households, earning an average of just US$0.15 per capita/day. Most households are located at high elevations and on steep slopes, with little productive assets. These households work little outside their own farms and the probability of a female head is highest for this cluster.

Land use of households in cluster 4 is similar to those in cluster 3. Despite smaller landholdings, these households earn more than double the income of cluster 3 households. Limited access to land 'pushes' these

Table 6.1. Mean household income and percentage composition, by livelihood strategy.

					Livelihood strategy			
Variable	Total sample	Livestock producers	Coffee producers	Basic grains farmers	Basic grains farmers/ farm workers	Mixed basic grains/ livestock/ farm workers	Permanent crops producers	Annual crops/ intensive livestock producers
Households (n)	376	59	28	68	85	116	12	8
Total household income[a]	12,310	20,915	12,536	5,134	13,799	10,798	16,225	9,777
Std. error	1,646	7,531	2,646	1,125	2,491	2,089	3,665	3,499
Per capita income[b]	0.35	0.58	0.33	0.15	0.42	0.29	0.66	0.38
Std. error	0.04	0.18	0.05	0.03	0.08	0.05	0.18	0.24
Composition of household income								
Income from basic grains (%)	34.3	30.4	0.0	79.2	17.5	30.0	22.3	41.4
Std. error	8.0	8.1	20.9	29.4	6.5	13.0	10.2	10.5
Income from annual cash crops (%)[a]	1.3	5.3	0.0	0.0	0.0	1.2	0.0	0.3
Std. error	0.8	4.4	0.0	0.0	0.0	0.8	0.0	0.1
Income from coffee (%)	0.0	4.1	-22.9	0.0	0.6	1.1	0.0	0.0
Std. error	4.2	2.8	51.6	0.0	0.4	1.5	0.0	0.0
Income from other permanent crops (%)	6.9	2.4	4.0	24.4	0.5	4.5	5.2	1.1
Std. error	4.3	1.7	3.6	21.0	0.4	3.9	2.7	1.5
Income from livestock (%)	-3.5	31.0	83.2	-91.2	1.4	0.6	1.3	27.9
Std. error	27.6	8.6	75.4	135.6	1.8	29.6	1.4	7.0
Income from off-farm agricultural work (%)	38.9	13.6	16.7	9.4	76.1	55.5	38.7	4.3
Std. Error	8.9	5.5	9.0	7.1	7.3	25.8	17.3	4.8
Income from off-farm non-agricultural work (%)	20.5	16.9	20.1	72.9	2.5	2.9	24.4	13.7
Std. error	14.6	8.2	9.2	77.1	1.2	1.0	9.3	15.1
Income from transfers (%)	34.7	6.6	1.1	103.3	14.0	33.5	19.7	13.0
Std. error	19.1	5.5	10.8	107.8	1.8	12.7	13.8	1.6

Income from patio production (%)	-1.0	1.3	0.5	-4.1	0.5	-1.8	1.1	0.1
Std. error	1.0	1.4	0.4	2.1	0.3	2.2	0.9	0.3
Net rent income (%)	-18.1	-8.8	0.0	-49.3	-7.0	-19.5	-0.3	-0.3
Std. error	13.0	7.5	0.0	64.0	3.0	13.2	0.2	0.5
Poor households (%)c	92.6	77.1	99.1	97.3	94.4	95.8	95.2	86.2
Std. error	2.4	10.8	1.0	2.8	4.0	1.5	4.7	14.4
Extremely poor households (%)d	92.3	75.5	99.1	97.3	91.8	88.4	94.5	77.0
Std. error	2.4	10.9	0.0	0.0	4.6	1.4	17.3	14.6

a Lps/year.
b US$/day.
c Percentage households with < US$1.50 per capita/day.
d Percentage households with < US$1.00 per capita/day.

Table 6.2. Summary statistics regarding asset-related household characteristics, by livelihood strategy.

| | Total sample (n, 376) | | Livelihood strategy | | | | | | | | | | | | | |
| | | | Cluster 1 (n, 59) | | Cluster 2 (n, 28) | | Cluster 3 (n, 68) | | Cluster 4 (n, 85) | | Cluster 5 (n, 116) | | Cluster 6 (n, 12) | | Cluster 7 (n, 8) | |
	Mean	Std error	Mean	Std. error	Mean	Std. error	Mean	Std. error	Mean	Std. error	Mean	Std. error	Mean	Std. error	Mean	Std. error
Natural capital																
Farm size[a]	14.3	2.4	45.6	9.4	5.0	0.8	3.4	0.4	2.7	0.4	15.3	4.0	3.4	0.4	6.3	1.3
Area of owned land[a]	7.7	1.6	22.7	6.7	3.2	0.8	1.2	0.6	0.6	0.3	9.3	3.4	1.9	0.5	0.5	0.2
Land with formal title (%)	29.4	4.1	36.9	8.9	56.7	12.1	18.5	7.9	14.8	9.2	34.6	8.5	58.4	15.8	6.7	2.5
Rainfall primary season[b]	1,005	17	943	37	917	41	976	41	1,058	45	1,008	25	1,576	74	1,060	53
Altitude[c]	2,231	127	1,220	198	3,845	169	2,009	265	2,412	232	2,569	240	1,661	182	734	111
Rainfall deficit secondary season[d]	14	3	18	7	1	1	33	11	11	4	6	2	9	4	9	10
Soil fertility[e]	2,846	67	2,834	159	2,572	171	2,806	166	2,939	94	2,835	140	2,935	183	3,315	263
Physical capital																
Value of machinery, equipment and transportation[f]	3,698	631	6,023	1,113	6,590	3,612	2,884	947	422	159	4,757	1,527	1,726	930	671	144
Value of livestock[f]	19,703	5,077	87,336	23,146	4,029	848	4,105	1,148	1,994	660	10,394	3,097	892	371	5,547	1,076
Human capital																
Household size	6.1	0.2	6.1	0.5	6.2	0.6	5.7	0.4	5.9	0.4	6.4	0.3	4.6	0.5	7.1	0.9
Dependency ratio[g]	0.9	0.1	0.6	0.1	0.7	0.2	1.1	0.2	1.2	0.2	0.9	0.1	0.9	0.2	0.5	0.2
Female-headed HH dummy (%)[h]	9.4	3.0	14.8	7.1	1.4	1.5	17.7	9.4	12.9	9.7	2.2	1.3	0.0	0.0	3.2	4.1
Female adults in the HH (%)[i]	49.7	1.3	44.7	2.1	51.4	5.1	50.4	3.2	46.2	1.9	52.4	2.4	60.7	3.3	56.3	3.0
Age of household head (years)	47.2	1.4	46.3	2.4	40.7	3.4	43.9	3.6	45.3	3.5	52.3	2.5	49.9	6.7	47.9	1.5
Median years schooling[j]	2.8	0.2	3.4	0.3	1.9	0.3	2.9	0.5	2.3	0.3	3.0	0.2	2.5	0.8	2.5	0.4
Migration index[g]	0.08	0.02	0.21	0.07	0.17	0.13	0.02	0.01	0.03	0.02	0.05	0.01	0.003	0.003		

Participation in programmes/organizations (%)

Conservation training[k]	17.7	4.5	7.4	5.3	29.2	13.7	1.8	1.0	16.1	8.7	32.5	10.7	23.5	15.2	0.0	0.0
Agricultural training[l]	7.6	1.9	5.8	5.2	5.5	3.9	4.1	1.7	4.5	2.5	10.5	4.3	38.7	21.1	12.8	12.5
Conservation extension[m]	7.3	2.1	7.0	3.9	1.0	1.0	12.5	7.5	5.2	3.0	7.2	3.9	18.2	16.6	0.0	0.0
Agricultural extension[n]	5.4	1.6	3.3	3.0	2.5	2.5	1.8	1.2	4.6	2.7	8.3	4.0	37.4	21.2	0.5	0.6
Participation in producer org.[r]	7.2	3.7	7.9	5,5	14.2	7.5	0.3	0.3	1.3	0.9	14.4	10.5	0.6	0.7	0.0	0.0
Participation in rural bank/caja rural[r]	8.2	1.9	0.4	0.4	20.7	13.1	13.9	5.6	5.7	3.1	7.8	2.9	0.0	0.0	10.8	12.0
Participation in NGO programme[r]	9.9	2.9	2.5	1.8	21.2	13.6	2.4	1.4	14.9	8.2	12.7	6.4	18.2	16.6	0.0	0.0

Location capital

Population density[o]	104	12	51	9	81	10	132	28	125	33	99	23	263	103	52	9
Road density[p]	4.0	0.3	2.1	0.3	5.7	0.4	4.0	0.4	4.8	0.6	4.3	0.5	3.3	0.2	2.0	0.2
Market access[q]	73	7	99	13	85	14	60	13	62	18	74	8	88	60	24	18

[a] Manzanas (1 mz = 0.7 ha).
[b] In mm during the *primera* season (May–September); own calculations based on data from the nearest point in the atlas CD of CIAT (CIAT, 2001).
[c] Average altitude from sampled plots on each farm measured by GPS in feet above sea level.
[d] Average from sampled plots during the *postrera* season (October–January) in mm (see endnote 3).
[e] Approximated by potential maize yields (see endnote 5).
[f] Lps.
[g] (HH members < 12 and > 70 yrs (n))/(HH members between 12 and 70 yrs (n)).
[h] 1 = female household head.
[i] Females > 12 yrs as % of total household size.
[j] Median years of schooling of household members > 7 years of age.
[k] Dummy variable (1 = HH has received conservation training).
[l] Dummy variable (1 = HH has received crop technology training).
[m] Dummy variable (1 = HH has received conservation extension visits).
[n] Dummy variable (1 = HH has received crop technology extension visits).
[o] No. of persons/km² in community based on 2001 population census (INE, 2002).
[p] Roads in community (km/km²) (CIAT, 2001).
[q] CIAT (2001); data reflect travel time from centre of community to nearest market outlet, using most common form of transportation. Values based on geographical distance, road quality and slope. Higher values indicate worse market access.
[r] Dummy variable (1 = HH participates).
[s] (Total months lived outside the HH by adult HH members)/(total adult HH members*12 months).

households to be more entrepreneurial and seek out alternative, more profitable employment opportunities, within or outwith agriculture. On the other hand, differences between buying and selling cost of food crops may mean that this group may be worse off than those in cluster 3 in terms of food security.

The livelihood of households in cluster 5 is similar to that of households in cluster 4, but they have larger farms, hire more labour and devote more time to livestock activities. However, by working more on their own farm, these households have lower incomes than households in cluster 4. On the other hand, greater wealth and more diversified income sources may make these households less vulnerable to risks.

Cluster 6 comprises a small group of producers of tree crops (fruits, oil palm) or sugarcane. These households have the highest average incomes in the sample and smaller than average household sizes. They are located in favourable agro-ecological areas with high population densities, high rainfall and good access to paved roads and public transportation, all of which are important for diversification into higher-value permanent crop production.

Finally, the few households in cluster 7 are mostly vegetable growers who work little off-farm due to the labour-intensive character of vegetables. Despite being relatively far from a paved road in areas with relatively low population densities, most households are close to a non-paved road. Average daily income (US$0.38 per capita) is only slightly above average, possibly because of dependence on middlemen.

Determinants of livelihood strategies

Because LS choice is a polychotomous choice variable, we used a multinomial logit model (Greene, 1990) to explain the household's livelihood choice. We assumed that the latter is determined by fixed or slowly changing factors, including the household's natural capital, locational capital and human capital. Natural capital included: (i) the amount of land owned (more land stimulates on-farm activities); (ii) share of total landholdings that have a title (importance of titled land as collateral and stimulates investments); (iii) rainfall during the main growing season (agricultural potential); (iv) rainfall deficit in the secondary growing season[3] (limits double cropping); (v) altitude[4] (proxy for temperature); and (vi) soil fertility.[5]

Location assets are represented by population density (influences both crop choice and production technologies, see Pender *et al.* (2001) and Jansen *et al.* (2006)), market access and road density (both latter assets stimulate cash crop production and off-farm work). Human capital variables included: (i) household size and dependency ratio (both determine labour availability); (ii) gender and age of the household head (female-headed households have specific characteristics and face competing demand on the time of female household heads); (iii) proportion of adults in the household that are female (influences the availability of non-domestic labour); and (iv) the median education of household members (important for off-farm employment opportunities).

The results (see Table 6.3) show that basic grains-only farms are more likely to be found: (i) in areas where secondary season rainfall deficits are greater; (ii) in areas of higher population density; (iii) among households that own less land; and (iv) if the household head is female or younger. In general, subsistence basic grains production is the dominant livelihood strategy in more marginal and land-scarce areas and among poorer households. Livestock production is more likely as the livelihood strategy in areas where: (i) the secondary season rainfall deficit is greater; and (ii) among households with more land, lower dependency ratio or a younger head.

These findings are consistent with the theory of comparative advantage; i.e. crop production is less profitable relative to livestock production in areas of marginal rainfall, provided that households have access to enough land to support their livestock. The diversified basic grains/livestock/farm worker strategy is more common in areas with less rainfall deficit in the secondary season and among households who own more land, have a higher dependency ratio, are male headed but have more female adults, and where the head is older. As households become more mature and acquire more land, female adults and dependants, they seek and are able to diversify into off-farm activities as well as livestock; the opportunities for such diversification are greater in areas of higher agricultural potential.

Few factors are statistically significant predictors of the probability of either the coffee production or basic grains/farm workers LS. Only road density has a weakly significant positive association with coffee production. This association could reflect reverse causality: construction of roads may be greater where coffee is produced because of construction of roads by the Honduran Coffee Institute.

Determinants of household income

Model design

Annual per capita household income was hypothesized to be dependent on the household's livelihood strategy and asset portfolio. In addition to the variables included in the multinomial logit model, variables for physical capital and social capital were also included. Secondly, the targeting issue was addressed, by analysing how some of the programme- and policy-relevant variables interact with the livelihood strategy variables in generating income.

Two different model specifications were analysed. The model specification without interaction variables indicates which of the policy variables are most significant and therefore require better understanding

Table 6.3. Determinants of livelihood strategies (multinomial logit regression).[a]

Explanatory variables[b]	Livestock producers (cluster 1)		Coffee producers (cluster 2)		Basic grains farmers/farm workers (cluster 4)		Basic grains/livestock/farm workers (cluster 5)	
	Coefficient	Std. error	Coefficient	Std. error	Coefficient	Std. error	Coefficient	Std. error
Natural capital								
Rainfall deficit secondary season	−0.00825	0.00681	−0.02432	0.01728	−0.01835***	0.00660	−0.02628***	0.00779
Soil fertility	0.00032	0.00046	−0.00012	0.00049	0.00062*	0.00034	0.00039	0.00043
Summer rainfall	0.00165	0.00148	−0.00475	0.00373	0.00091	0.00104	0.00147	0.00117
Altitude	−0.00095	0.00127	0.00214**	0.00091	−0.00067	0.00086	0.00059	0.00092
Owned land	0.20251***	0.05255	0.02757	0.11059	−0.52985*	0.27374	0.19469***	0.05277
Titled land (%)	−1.00209	1.15818	1.48094	1.04121	0.82708	1.19322	−0.74270	0.91573
Location capital								
Market access	0.13767*	0.07121	0.12030	0.07902	0.09003	0.05954	0.09858	0.06820
Road density	−0.08178	0.25495	0.85183**	0.33211	0.51997**	0.21144	0.31098	0.21955
Population density	−0.00805	0.00630	−0.02907***	0.00858	−0.00429	0.00414	−0.01523***	0.00503
Human capital								
Median schooling	0.07184	0.21005	−0.29940	0.18422	−0.17611	0.17871	0.23879	0.16939
HH size	0.10445	0.12784	0.01362	0.16298	−0.03015	0.10747	−0.00991	0.10817
Dependency ratio	−1.58529**	0.63010	−1.11618*	0.56986	−0.21426	0.46409	0.20291	0.41592
Female HH head	−0.39686	0.78423	−3.96464***	1.50166	−1.29206	0.85613	−3.71278***	0.92736
Female adults (%)	1.57140	1.77866	1.05151	2.97171	−2.53423	1.81363	4.52182**	1.97987
Age of HH head	0.00332	0.02173	0.00336	0.02309	0.02956	0.01983	0.06815***	0.01724
Intercept	−3.45762	2.99668	−0.24639	2.98190	−3.05683	2.40960	−9.31718***	3.08394
Number of observations	59		28		85		116	
Proportion of observations	0.1791		0.0845		0.2142		0.3231	
Mean predicted probability of livelihood	0.1791		0.0845		0.2142		0.3231	

[a] Basic grains farmers ($n = 68$) is excluded category. Strategies 6 and 7 not analysed due to limited numbers of observations. Coefficients and standard errors adjusted for sampling weights and stratification, and robust to heteroskedasticity.
[b] Table 6.2 contains definitions of explanatory variables.
*, **, ***: statistically significant at 10, 5 and 1% level, respectively.

regarding which household types should be targeted when launching public investment programmes that address these policy variables. The model with interaction variables suggests which household types would benefit most from such public investment programmes, and thus helps public policy targeting.

The different specifications of the income model without interaction variables included an ordinary least squares (OLS) model, a median regression (because of concerns about outliers) with bootstrapped standard errors and an instrumental variables (IV) regression (because of potential endogeneity of some of the explanatory variables). Each of these specifications carries its own potential problems: the OLS model is likely to have some endogenous explanatory variables, the median regression model does not correct for sample weights and the IV model may be influenced by weak instrumental variables.

First-stage regressions in the IV procedure confirmed the significance of the instruments for all endogenous explanatory variables. Hansen's J test of over-identifying restrictions was found not to be significant and therefore confirms the validity of our instrumental variables[6]. On the other hand, the Hausman (1978) test indicated that the (more efficient) OLS model is preferred to the IV model and thus supports exogeneity of the potentially endogenous explanatory variables. Table 6.4 reports only the OLS model version model with interaction variables.

Model results

Households that follow a mixed basic grains/off-farm work livelihood strategy earn significantly higher incomes than do pure basic grains farmers (see Table 6.4). The results of the OLS model without interaction variables (not reported) suggest that livestock producers also have higher incomes. Even though the climate variables have insignificant association with income, they still may have indirect impacts, via their effect on LS. Soil fertility has a strong and significant direct positive effect on income, and also an indirect effect through the LS, since better soils are associated with the basic grains/off-farm work livelihood strategy.

Greater land ownership alone does not guarantee higher income. Nevertheless, land ownership indirectly affects income through its effect on LS, though these effects are mixed. More land significantly increases the probability of a household following a livestock-based LS, which is associated with higher income levels in the OLS regression without interaction terms. However, greater land ownership is also associated with lower probability of the household following a basic grains/off-farm work strategy, which obtains higher income. No statistically significant direct or indirect effects of land titling on household income were found.

Unlike population density, market access and road density do not show statistically significant direct associations with income, but there are indirect effects. For example, road density is significantly associated

Table 6.4. Determinants of per capita income, including interaction terms (Lps/year).

Explanatory variable	OLS Regression[a] Coefficient	Std. error
Natural capital		
Altitude	−0.1393	0.1841
Summer rainfall (mm)	−0.4716	0.7180
Rainfall deficit secondary season	2.2529	4.4758
Soil fertility	0.9510***	0.2790
Owned land	5.5853	13.5103
Titled land (%)	−131.6669	437.3759
Physical capital		
Value of machinery/equipment (Lps)	0.0072	0.0243
Value of livestock (Lps)	−0.0007	0.0060
Human capital		
Median years schooling	−25.6262	95.3169
Household size	−232.9582**	96.5350
Dependency ratio	−788.7933**	326.4692
Female HH head	−79.7882	633.1677
Female adults (%)	−448.4123	1230.0750
Age HH head	−10.1592	12.2177
Migration index	1285.3690	1714.7940
Livelihood strategy (cf. basic grains)		
Livestock producer	1186.5730	938.5464
Coffee producer	874.0251	658.4712
Basic grains/farm worker	733.7873**	361.5680
Basic grains/livestock/farm worker	654.8768*	383.3636
Participation in programmes/organizations		
Conservation training	−445.1712	350.1854
Agricultural training	1779.4080*	908.1467
Conservation extension	−333.2486	674.4493
Agricultural extension	−639.9922	614.9435
Producer/*campesino* organization	−779.4486	560.3380
Rural bank/*caja rural*	−46.6406	448.6455
NGO programme	131.4425	423.0769
Location capital		
Market access	62.1511	39.8495
Road density	−58.2827	102.2556
Population density	5.5956***	1.9414
Interaction variables		
Livestock producer * Value of machinery/equipment	−0.0006	0.0481
Livestock producer * Agricultural training	10446.7000***	1687.8050
Coffee producer * Value of machinery/equipment	0.5920**	0.0273
Coffee producer * Agricultural training	−1589.2100	1211.7210
Basic grains/farm worker * Value of machinery/equipment	2.2236***	0.3394
Basic grains/farm worker * Agricultural training	−538.0840	2821.6250
Basic grains/livestock/farm worker * Value of machinery/equipment	0.1094***	0.0352
Basic grains/livestock/farm worker * Agricultural training	−1303.1760	1108.0380
Intercept	567.6548	1498.9580
Number of observations	342	
R^2	0.5339	

[a] Coefficients and standard errors adjusted for sampling weights and stratification, and are robust to heteroskedasticity.
*, **, *** mean statistically significant at 10, 5 and 1% level, respectively.

with higher probability of households pursuing the basic grains/farm-worker strategy which has higher income.

Human capital has less effect on income than expected *a priori*, most likely because of the generally low levels of education in the hillside areas and relatively limited inter-household variation. Dependency ratio and household size have a negative direct effect on income per capita.

Ownership of machinery and equipment has a significant positive association with income in both OLS and IV regressions without interaction variables. The magnitude of the coefficient in the OLS model without interaction variables suggests that 1 additional Lempira (Lps, the currency unit of Honduras) invested in equipment contributes 0.07 of additional annual income per capita, or about 0.42 Lps of total household income on average (during 2001–2002, US$1.00 averaged 16 Lempiras). Livestock ownership, on the other hand, does not have a statistically significant association with household income in any of our models, possibly due to high variance in estimated livestock incomes, which included negative values.

There is no statistical evidence of an impact of short-term agricultural extension or longer-term, conservation-focused training on household income and also no robust statistical evidence that membership of NGO programmes, producer organizations or rural financial institutions has significant impacts on income. However, more general agricultural training is positively associated with household income. In the model with interaction terms, households that have received agricultural training earn about 1800 Lps per capita in extra income. It is hard to believe that agricultural training could have such a large effect on income, though there is a very large and strongly significant positive association of training with incomes of livestock producers. Thus, if there are positive impacts of agricultural training programmes, these positive impacts are greatest for livestock producers.

The positive impact of machinery and equipment is mainly for households pursuing LS involving coffee production (e.g. sprayers) and for households pursuing off-farm employment (which have higher opportunity costs of labour), but much less remunerative for households pursuing basic grains or livestock production only.

Determinants of sustainable land management practices

Adoption of soil conservation measures in the hillside areas is generally low. We analysed the adoption of use of zero burning, zero or minimum tillage and incorporation of crop residues – and used Probit models, since dependent variables are of the dichotomous choice type (Maddala, 1992). Other practices were not sufficiently common to permit reliable econometric estimation. The regressions are estimated using parcel-level data, because this is the level at which data on these land management practices were collected. Compared to the income models, we expanded

Table 6.5. Determinants of land management practices (probit regressions).[a]

Explanatory variable	Zero Burning		Minimum/Zero Tillage		Incorporate crop residues	
	Coefficient	Std. error	Coefficient	Std. error	Coefficient	Std. error
Natural capital						
Altitude	0.00030***	0.00010	−0.00009	0.00008	0.00012	0.00012
Summer rainfall	0.00103***	0.00038	−0.00173***	0.00045	0.00115*	0.00061
Rainfall deficit secondary season	0.00225	0.00292	−0.00364**	0.00184	−0.01211***	0.00462
Soil fertility	−0.00019	0.00014	0.00012	0.00013	−0.00055***	0.00019
Owned land	0.00251	0.00539	−0.00801	0.00566	−0.02065*	0.01055
Titled land (%)	−0.14375	0.36723	0.39666	0.43907	−0.79819	0.58310
Physical capital						
Value of machinery and equipment	0.00000	0.00001	−0.00002**	0.00001	0.00000	0.00001
Value of livestock	−0.00001	0.00000	−0.00001	0.00000	−0.00001	0.00001
Human capital						
Median years schooling	−0.08699*	0.04756	−0.00228	0.04719	0.03489	0.07323
Household size	−0.02897	0.03625	0.03212	0.03475	−0.14574***	0.05228
Dependency ratio	0.16564	0.15702	0.06804	0.15746	−0.51304**	0.24315
Female-headed HH	0.13520	0.41749	−0.53418	0.38125	0.96794**	0.43695
Female adults (%)	−0.14056	0.67252	−0.08206	0.67728	−0.88524	0.92686
Age of HH head	0.00872	0.00693	0.00382	0.00632	−0.00097	0.00886
Migration index	1.66290***	0.48703	0.45864	0.38737	0.73742	0.56510
Livelihood strategy (cf. basic grains farmers)						
Livestock producer	0.15738	0.38872	−0.15527	0.41943	−0.43337	0.49689
Coffee producer	−0.35487	0.38024	−0.73369	0.45014	−1.28124***	0.49666
Basic grains/farm worker	0.78447**	0.30482	−1.12176***	0.34645	−0.53155	0.38804
Basic grains/livestock/farm worker	0.50630*	0.29471	−0.62714*	0.36177	−0.31616	0.37204
Participation in programmes and organizations						
Conservation training	0.24561	0.28844	0.91708***	0.25133	0.27376	0.34648
Agricultural training	0.47078	0.42046	−0.68528**	0.34801	−1.98802***	0.64474
Conservation extension	−0.33692	0.42501	−0.47262	0.31764	1.26840***	0.44800
Agricultural extension	0.86464**	0.38445	0.83674**	0.36259	2.19605***	0.43258
Producers/*campesino* organization	0.09204	0.37328	0.23775	0.35666	−1.59090**	0.63683
Rural bank/*caja rural*	0.31194	0.30474	−0.83068**	0.34971	−1.80687***	0.41727
NGO programme	−0.16339	0.33625	0.30088	0.27397	1.50439***	0.48059
Location capital						
Market access	0.01110	0.02158	−0.04532**	0.02099	0.01773	0.02513
Road density	0.24360***	0.06445	−0.18260**	0.07627	0.36662***	0.08917
Population density	−0.00084	0.00077	−0.00107	0.00111	−0.00244*	0.00126
Parcel characteristics						
Area parcel (mz)	0.00803*	0.00458	0.01253	0.00921	0.02927***	0.00957
Travel time parcel to residence (min)	0.00294*	0.00172	−0.00202	0.00235	−0.00525	0.00352
Travel time parcel to road (min)	−0.00327	0.00481	−0.01164*	0.00700	0.00246	0.00512

Table 6.5. *Continued*

Explanatory variable	Zero Burning		Minimum/Zero Tillage		Incorporate crop residues	
	Coefficient	Std. error	Coefficient	Std. error	Coefficient	Std. error
Position on hill (cf. bottom)						
Top of hill	1.12731***	0.40867	−1.29756**	0.59294	0.19879	0.61334
Hillside	0.14117	0.23112	0.11731	0.23401	0.68288***	0.25952
Slope (cf. flat)						
Moderate	0.19763	0.25252	−0.11885	0.26331	−0.80551***	0.27468
Steep	−0.22478	0.31361	0.46503	0.32946	−1.33237***	0.41959
Land tenure (cf. usufruct ownership)						
Full title	−0.14709	0.27352	−0.16568	0.40627	0.17501	0.52359
Occupied communal land	0.10696	0.38146	0.28144	0.47464	−0.38417	0.52371
Borrowed plot	−1.06753***	0.27888	0.11227	0.27148	−0.29208	0.32117
Rented or sharecropped	−0.87832***	0.32504	−0.05103	0.31612	−0.20013	0.32900
Prior investments on parcel						
Stone wall	1.08242***	0.36211	0.24601	0.41272	1.22839***	0.42266
Live barrier or fence	0.66462**	0.29597	0.81895***	0.28247	−0.55983*	0.28509
Trees planted	−0.77211***	0.27057	0.83221**	0.32330	−0.39188	0.46935
Land use in 1999 (proportion of parcel area; cf. basic grains)						
Other annual crops	−3.32715***	1.05913	−1.61588**	0.76402	0.35532	0.61075
Coffee	−1.33549***	0.34968	−1.80552***	0.40774	−2.11164***	0.58072
Other perennial crops	0.15966	0.35542	−1.61227***	0.43222	−1.69658***	0.63848
Unimproved pasture	−0.58040*	0.32005	−1.82105***	0.48260	−0.64926	0.39474
Improved pasture	0.44644	0.51457	−1.78035***	0.60140	0.39004	0.58035
Fallow	−1.37974***	0.29793	−0.60730**	0.30670	−1.29574***	0.32590
Forest	−0.37226	0.49586	−1.48522***	0.52589	−0.89019**	0.43228
Intercept	−3.09006***	0.94312	2.57981***	0.83330	0.56157	1.28975
Number of observations	776		776		776	
Proportion positive observations	0.3377		0.2321		0.1711	
Mean predicted probability positive obs.	0.3424		0.2419		0.1641	
Hausman test[b]	P = 0.9945		P = 1.0000		NE	
Hansen's J test[c]	P = 0.7624		P = 0.8606		P = 0.6861	

[a] Coefficients and standard errors adjusted for sampling weights and stratification, and robust to heteroskedasticity and non-independence of observations from different parcels from same household (clustering).
[b] Hausman test of exogeneity of livelihood strategies and participation in programmes/organizations (OLS vs. IV linear version of models).
[c] Hansen's J test of overidentifying restrictions in IV linear model.
*, **, *** mean statistically significant at 10, 5 and 1% level, respectively.
NE: the Hausman test could not be computed (negative value of test statistic).

the set of explanatory variables by including parcel characteristics, land tenure, prior investments on the parcel and prior land use.

The results indicate that agro-climatic factors influence a producer's decision to use conservation measures (see Table 6.5). Zero burning is more common at higher altitudes (coffee) and where main season rainfall is higher, perhaps the result of increased risk of run-off and higher intensity of cultivation in areas with better agro-climatic conditions. Zero/minimum tillage is less common where rainfall is higher because weeds are more widespread. Incorporation of crop residues is higher where rainfall is more plentiful but lower on already fertile soils.

Use of zero/minimum tillage is less likely among households that own more machinery and equipment, since some of their equipment is used for tillage. While human capital constraints are not binding for zero/minimum tillage adoption, zero burning is more common among households for whom migration is important, perhaps because this practice can be labour saving (Deugd, 2000). Crop residue incorporation is less common among larger households and households with a higher dependency ratio, reflecting greater demand for such resources and greater poverty among larger households.

Opportunity costs of labour influence the adoption of sustainable land management practices (Neill and Lee, 2001), making households whose livelihood strategy is dominated by off-farm work and located in areas with higher road densities and better market access more likely than basic grains farmers to use zero burning but less likely to use zero/minimum tillage. The result that crop residue incorporation is more likely in areas with greater road density is contrary to the findings with regard to the impacts of road access on minimum/zero tillage, but may be caused by higher returns for labour invested in crop residue incorporation in such areas.

In agreement with Bonnard (1995), lack of land titles is not a major constraint to adoption of land management practices. However, adoption of zero burning is less likely on borrowed and leased plots, probably because this improves soil fertility in the longer term, but perhaps at the expense of short-term fertility due to the release of nutrients by burning.

All three practices are more common among farmers participating in agricultural extension. Zero/minimum tillage is more common among farmers who participated in conservation training programmes but less common among households who participated in longer-term general agricultural training or are a member of a rural bank, or *caja rural*. Financial institution membership also negatively influences incorporation of crop residues. Apparently, training programmes are promoting other technologies or practices to a greater extent. Given the earlier positive association of agricultural training with higher incomes of livestock producers, it may be that these programmes are more oriented to technologies for livestock production than to crop technologies such as conservation tillage. Financial organizations are often associated with and promote rural non-farm activities, which will tend to increase labour opportunity costs, reducing households' interest in labour-intensive farming practices. NGOs, nevertheless, seem to promote the incorporation of crop residues.

Zero burning is more common on parcels that are on top of a hill than at the bottom, (consistent with the earlier result that it is more common at higher altitude), while zero/minimum tillage and incorporation of crop residues are less common, probably because soils tend to be heavier and more difficult to till in valley bottoms. For similar reasons, incorporation of crop residues is less on steep and moderate than on flat slopes. Similar to the findings of Buckles *et al.* (1998) for cover crops, crop residues are more likely to be incorporated on larger plots (possibly because tillage using animal traction is easier on larger plots) but less likely where other land uses besides annual crops are important (tillage practices are used mainly for annual crops).

The slope of the hill has little effect on a farmer's decision to use conservation measures, though the use of crop residues is less on steep hills due to more difficult use of labour and (possibly) equipment.

Finally, zero burning and incorporation of crop residues are more likely on plots where stone walls have been constructed. Use of zero burning and zero/minimum tillage is more likely while incorporation of crop residues is less likely on plots having live barriers or fences. Use of zero/minimum tillage is more likely and zero burning is less likely on plots where trees have been planted. The reasons for all of these associations are not fully clear, though some probably involve complementarity or substitutability between prior investments and current land management practices.

Conclusions and Implications for Policy

Households in the rural hillsides in Honduras hold widely differing asset endowments and follow different livelihood strategies, but the vast majority are extremely poor. Households that follow a livelihood strategy that is exclusively based on basic grain farming are the poorest, because they often live in isolated areas with relatively poor agro-ecological and socio-economic conditions. Opportunities for off-farm work tend to be limited in these areas, but household strategies that combine on-farm with off-farm work are able to give opportunities to earn higher incomes.

Soil fertility has a strong direct positive impact on income, while favourable agro-climatic conditions have an indirect positive income effect because they stimulate more remunerative livelihood strategies. Land ownership and tenure are not the key constraints limiting the potential for higher incomes, but adoption of sustainable land use practices is higher on owner-operated than leased plots and is stimulated by household participation in training programmes and organizations. Agricultural training has a direct positive effect on income, particularly when directed toward livestock producers.

High dependency has a direct negative effect on income and a negative indirect effect through stimulating less remunerative livelihood strategies. Hillside households are not generally recipients of significant amounts of remittances and we found no significant impacts of migration on per capita household income.

While population density has a direct positive impact on per capita income, road density and market access indirectly stimulate higher incomes by promoting livelihood strategies other than basic grains production.

The high reliance of rural hillside households on agricultural and related income means that any strategy targeted at these areas will have to build upon the economic base created by agriculture. Agriculture alone cannot solve the rural poverty problem, but those remaining in the sector need to be more efficient, productive and competitive. Public investment programmes should focus on broadening the physical asset base of poor households and extending the coverage of agricultural training.

Extending households' physical asset bases (particularly machinery and equipment) will increase the returns to land and labour resources, and raise incomes. Such investments should have a primary focus on crop producers, but perhaps with a special focus on households that have relatively high opportunity cost of labour, such as those pursuing off-farm employment or coffee production. Agricultural extension programmes and conservation-oriented training programmes can help in improving income and maintaining soil fertility. With the virtual abolishment of government extension and increasing privatization, the farmer-to-farmer model of extension promoted by NGO-led programmes becomes increasingly important for the economic and environmental sustainability of agricultural production.

Non-agricultural activities are relatively rare in rural Honduras because of the physical distances from urban centres and towns and the lack of good road infrastructure and transport services (Cuellar, 2003). Improving road infrastructure can stimulate LS that emphasize off-farm work with higher returns than working on the own farm. High rates of fertility and dependency are important causes of poverty, and programmes that succeed in lowering both household size and dependency ratios may also help in raising per capita incomes.

Improving access to land (not land titling *per se*) can have an indirect positive impact on income by enabling households to pursue more remunerative LS such as livestock production. Land redistribution programmes seeking to increase smallholders' ownership of land may also be justified on the basis of sustainability considerations, since adoption of zero burning is more prevalent on owned land than on rental land.

Acknowledgements

This chapter is based on the work of H.G.P. Jansen, J. Pender, A. Damon, W. Wielemaker and R. Schipper (2006) Policies for sustainable development in the hillsides of Honduras: a quantitative livelihoods approach. *Agricultural Economics* 34, 141–153. Permission granted by Blackwell Publishers.

Endnotes

[1] 'Hillsides' are defined as areas with slopes of more than 12% (PRONADERS, 2000). 'Hillside areas' include not only hillsides but also flat-floored valleys, 300–900 m in elevation, which are scattered throughout the interior hillsides.

[2] Throughout Central America, the term 'basic grains' refers mainly to maize and beans, but also includes sorghum and rice.

[3] Besides rainfall, moisture availability in the soil is another indicator of agricultural potential. Moisture availability is soil-specific and takes into account not only rainfall but also evapotranspiration, temperature and soil characteristics. Data from soil samples were used to operationalize moisture availability as crop water deficits for annual crops (maize in the main and secondary growing seasons) and permanent crops (coffee). Water deficits were calculated on the basis of monthly temperature, effective rainfall, evapotranspiration and soil characteristics (Wielemaker, 2002, unpublished).

[4] Based on Pender *et al.* (2001), we *a priori* expect altitude to have positive influence on the probability of cluster 2 (coffee farmers).

[5] Soil fertility is yet another indicator of agricultural potential and was approximated by potential yields (nutrient-limited but not water-limited), as calculated by the QUEFTS (Quantitative Evaluation of Soil Fertility and Response To Fertilizers) model (Janssen, 1990).

[6] Details regarding the IV procedure used are available from the authors upon request.

References

Adato, M. and Meinzen-Dick, R. (2002) *Assessing the Impact of Agricultural Research on Poverty using the Sustainable Livelihoods Framework.* EPTD Discussion Paper 89/FCND Discussion Paper 128, International Food Policy Research Institute (IFPRI), Washington, DC.

Bonnard, P. (1995) Land tenure, land titling, and the adoption of improved soil management practices in Honduras. PhD thesis, Michigan State University, East Lansing, Michigan.

Buckles, D., Triomphe, B. and Saín, G. (1998) *Cover Crops in Hillside Agriculture: Farmer Innovation with Mucuna.* International Development Research Centre and International Maize and Wheat Improvement Center, Mexico City.

CIAT (International Center for Tropical Agriculture) (2001) *Atlas de Honduras* (con datos Mitch). CIAT, Cali, Colombia.

Cuellar, J.A. (2003) Empleo e ingreso en las actividades rurales no agropecuarias de Centroamérica y México. In: Serna Hidalgo, B. (ed.) *Desafíos y Oportunidades del*

Desarrollo Agropecuario Sustentable Centroamericano. Economic Commission for Latin America and the Caribbean (ECLAC). Mexico D.F., Mexico, pp. 117–150.

Deugd, M. (2000) *No Quemar. ¿Sostenible y Rentable?* Informe final II, proyecto GCP/HON/021/NET, FAO, Tegucigalpa, Honduras.

DFID (1999) *Sustainable Livelihoods Guidance Sheets.* Department for International Development, UK (http://www.livelihoods.org/info/guidance_sheets_pdfs/section2.pdf).

Ellis, F. (1998) Household strategies and rural livelihood diversification. *Journal of Development Studies* 35, 1–38.

Greene, W.H. (1990) *Econometric Analysis.* Macmillan, New York.

Hair, J., Anderson, R., Tatham, R. and Black, W. (1998) *Multi-Variate Data Analysis,* 5th edn. Prentice Hall, Upper Saddle River, New Jersey.

Hausman, J. (1978) Specification tests in econometrics. *Econometrica* 46, 1251–1272.

Jansen, H.G.P., Damon, A., Rodríguez, A., Pender, J. and Schipper, R. (2006) Determinants of income-earning strategies and sustainable land use practices in hillside communities in Honduras. *Agricultural Systems* 88, 92–110.

Janssen, B. (1990) A system for quantitative evaluation of the fertility of tropical soils (QUEFTS). *Geoderma* 46, 299–318.

Lee, D.R. and Barrett, C.B. (eds) (2001) *Tradeoffs or Synergies? Agricultural Intensification, Economic Development and the Environment.* CAB International, Wallingford, UK.

Maddala, G.S. (1992) *Introduction to Econometrics,* 2nd edn. Macmillan, New York.

Neill, S.P. and Lee, D.R. (2001) Explaining the adoption and disadoption of sustainable agriculture: the case of cover crops in northern Honduras. *Economic Development and Cultural Change* 49, 793–820.

Pender, J., Scherr, S. and Durón, G. (2001) Pathways of development in the hillside areas of Honduras: causes and implications for agricultural production, poverty, and sustainable resource use. In: Lee, D.R. and Barrett, C.B. (eds) *Tradeoffs or Synergies? Agricultural Intensification, Economic Development and the Environment.* CAB International, Wallingford, UK, pp. 171–195.

PRONADERS (2000) *Documento Marco del Programa Nacional de Desarrollo Rural Sostenible.* National Program for Sustainable Rural Development (PRONADERS), Tegucigalpa, Honduras.

Thurow, T.L., Thurow, A.P., Wu, X. and Perotto-Baldivioso, H. (2002). Targeting soil conservation investments in Honduras. *Choices,* Summer 2002, 20–25.

Vosti, S. and Reardon, T. (eds) (1997) *Sustainability, Growth and Poverty Alleviation: a Policy and Agro-ecological Perspective.* Johns Hopkins University Press, Baltimore, Maryland.

Wishart, David (1999) *ClustanGraphics Primer: a Guide to Cluster Analysis.* Clustan Limited, Edinburgh, UK.

7

Resource Use Efficiency on Own and Sharecropped Plots in Northern Ethiopia: Determinants and Implications for Sustainability[1]

GIRMAY TESFAY,[1,*] RUERD RUBEN,[1,**] JOHN PENDER[2,***] AND
ARIE KUYVENHOVEN[1,****]

[1]Development Economics Group, Wageningen University and Research,
PO Box 8130, 6700 EW Wageningen, The Netherlands;
fax: 00-31-317-484037; [2]International Food Policy Research Institute (IFPRI),
2033 K St NW, Washington, DC, USA; e-mails: *tesfay.girmay@wur.nl or
girmay_tesfay@yahoo.com; **R.Ruben@maw.ru.nl; ***j.pender@cgiar.org;
****arie.kuyvenhoven@wur.nl

Abstract

This study analyses the level of resource use efficiency achieved by tenant households on their own and on sharecropped-in plots, and the determinants of the levels of efficiency achieved. Using plot-level and location data from Tigray, northern Ethiopia, it assesses whether tenancy status affects technical efficiency. Stochastic frontier production function analysis results show that a statistically significant level of technical inefficiency exists in the production system, but this was not found significantly associated with the tenancy status of the plot, controlling for other factors. Technical efficiency was found to have significantly positive association with livestock endowments of the tenant household and the population density of the location. As this study is based on cross-sectional data, a comprehensive study – based on a dynamic setting – is critical to assessing the cumulative effect of land contracting on long-term productivity and sustainability of land use.

Introduction

One of the prominent topics in policy discussions on enhancing agricultural productivity and sustainable land management in developing countries is the issue of land tenure security (Feder and Feeny, 1991; Wacher and English, 1992). The issue is raised in relation to its impacts on land transactions, agricultural productivity, conservation and sustainable land use in Tigray region and in Ethiopia at large (Tesfay, 1995; Gebremedhin, 1998, unpublished PhD thesis; Gavian and Ehui, 1999; Pender and Fafchamps, 2001; Deininger *et al.*, 2003; Gebremedhin and Swinton, 2003; Gebremedhin *et al.*, 2003; Pender and Gebremedhin, 2004; Benin *et al.*, 2005).

In an attempt to improve tenure rights, the regional government of Tigray issued a new rural land administration policy in 1997. Accordingly, farmers have rights for an unlimited period of use, inheritance and temporary transfers in the form of contract arrangements, but mortgaging and selling are prohibited (*Negarit Gazeta*, 1997).

Institutional innovations are basic ingredients in sustainable growth of agricultural production (Ruben *et al.*, 2001). Different forms of formal or informal institutional innovations mediate the efficient use of labour, land and draught power resources in farm production (Hayami and Otsuka, 1993). Such contracts evolve as a result of imperfection in resource markets created by institutional gaps and policy restrictions (Bhaumik, 1993; Pender and Fafchamps, 2001). For instance, the restrictions on farmland selling and mortgaging in Ethiopia may be cited (Hagos *et al.*, 1999; Pender and Fafchamps, 2001).

Recently, there has been renewed research interest in Ethiopia on land tenure issues in general and understanding of the nature, development and efficiency of different forms of tenancy (Gavian and Ehui, 1999; Pender and Fafchamps, 2001, 2005; Ahmed *et al.*, 2002; Deininger *et al.*, 2003; Benin *et al.*, 2005). Some studies also assessed the determinants of participation and extent of participation of households in land and labour contract markets (Pender and Fafchamps, 2001, 2005; Deininger *et al.*, 2003; Teklu and Lemi, 2004).

In Ethiopia, farm households use sharecropping, fixed rental, borrowing and exchange for farmland transactions. The findings on efficiency implication of land contracts in Ethiopia are not conclusive and a study on cumulative economic and environmental effects of short-term tenancy is lacking which invites further empirical research on the subject.

This study has three major objectives. First, it aims to measure the level of resource use efficiency on owned and on sharecropped-in plots operated by tenant households focusing on technical efficiency in the context of Tigray. This will contribute to the regional coverage of empirical assessments of the efficiency of tenancy arrangements in Ethiopia. Secondly, it aims to identify the determinants of differentials in technical efficiency achievements. Thirdly, it considers the long-term implications of sharecropping arrangements in the region in light of the available literature and draws policy lessons intended to enhance the contribution that land tenancy arrangements make towards productivity growth and sustainable use of farmland in the region.

The Setting of the Study Area

Tigray is the northernmost region of Ethiopia, bounded by Eritrea to the north, the Sudan to the west and the Ethiopian regions of Amhara and Afar to the south and the east, respectively (Hagos *et al.*, 1999). The population of the region is close to 4 million, and growing at the rate of 3% annually, with an average family size of five persons per household. The region is a dryland area, with annual precipitation ranging from 450–980 mm and high temporal and spatial variability. About 90% of the population are rural and depend mainly on rainfed mixed crop–livestock subsistence agriculture, with the crop sector dominated (90%) by cereals (Gebremedhin, 1998, unpublished PhD thesis). Average landholding size is about 1 ha (Pender and Gebremedhin, 2004). Oxen are the main source of draught power for most agricultural activities. The most critical problems of production systems in Tigray are land degradation, drought, small plot size, crop pests, shortage of feed for livestock and poor marketing systems and infrastructure (Hagos *et al.*, 1999).

Since the 1991 change of government in Ethiopia, the regional government of Tigray has been implementing different institutional reforms and programmes aimed at increasing food security and reducing poverty through sustainable development. On the institutional side, the regional government improved the rural land policy in 1997. Unlike the rural land reform in 1975, the current policy has given recognition to different forms of temporary land transactions, except for mortgaging and selling. In the case of temporary land transactions, the duration is limited to a maximum of 2 years for tenants using traditional technologies, and 10 years when tenants use 'modern technologies'.

Sharecropping is a long-established form of land contract in northern Ethiopia and, unlike the experience of other developing countries, it is not based on dominance and dependency relations between landowners and tenants (Cohen and Weintraub, 1975, p. 50). Nevertheless, sharecropping is dominant in Tigray and it is observed as being practised primarily between households with an excess land:labour or land:draught power ratio and those who are land-deficient relative to their labour and draught power endowment. Temporary land transfer through sharecropping, both from demand and supply sides, is the largest compared with other forms and the trend is increasing in Tigray. For instance, in the 1998 production season, sharecropping accounted for 88% of the total temporary land transfer and 20% of the land operated by tenants. Earlier studies also show the prevalence of sharecropping in high-altitude areas of Tigray, and female household heads sharecrop out more often (Gebremedhin, 1998, unpublished PhD thesis). In the region, the proportion of households participating in sharecropping arrangements increased from 13% in 1991 to 27% in 1998.

Sharecropping contracts in Tigray are informally arranged, without written terms between the parties in the presence of witnesses who are known to both. Contract duration was one season for about 95% and two seasons for the remaining 5% of sample plots in 1998. For the sample of sharecropped-in plots used in this study, tenants have positive expectation of operating these plots for 10 years in 46% of the cases, while 36% do not expect that long a duration and the remaining 18% did not respond.

Conceptual Background and Research Hypotheses

Sharecropping is a widely practised form of tenancy globally, and its economic and social basis and its implications for resource use efficiency and equity have been the subject of a number of studies (Stiglitz, 1974; Bardhan and Rudra, 1980; Otsuka and Hayami, 1988; Hayami and Otsuka, 1993). Understanding the implications of different forms of tenancy for efficiency and equity is a relevant input to land tenure-related debates in Ethiopia (Pender and Fafchamps, 2001, 2005; Benin *et al.*, 2005).

The economics of sharecropping has been modelled, from the tenant perspective since the time of Marshall (1890) and from the landlord side by Cheung (1969), with contrasting conclusions regarding its efficiency (cited in Bhaumik, 1993 and Pender and Fafchamps, 2001). According to Marshall, sharecropping is inefficient as the tenant is likely to undersupply effort. However, Cheung challenged Marshall's view by assuming possibility of monitoring work effort of the tenant without cost. In recent studies, the moral hazard, transaction costs, risk aversion and interlocked markets arguments are used to model the basis for different forms of contracts (Stiglitz, 1974; Hayami and Otsuka, 1993; Agrawal, 1999; Pender and Fafchamps, 2001, 2005).

The risk-aversion behaviour of tenants is a frequently cited reason for the existence of sharecropping (Stiglitz, 1974). Tenants enter into sharecropping arrangements in order to spread the magnitude of risk that they are taking in production and marketing. Both tenant and landlord may also be risk averse and prefer sharecropping arrangements (Ahmed *et al.*, 2002). Thus, sharecropping may be dominant in the dryland areas where risk is an inherent feature. According to Eswaran and Kotwal (1985), sharecropping can also evolve as a result of differences in factor endowments of farm households. A study by Gavian and Ehui (1999) in Ethiopia also indicated economic reasons for the existence of sharecropping, as the value of output share of the landowner was higher than the land rent paid under fixed contracts.

The studies in Ethiopia and elsewhere show mixed results with regard to the efficiency of sharecropping compared with other forms of tenancy. Compared with owner-operated systems, sharecropping is found to be inefficient in some empirical studies, supporting the Marshallian view (Gavian and Ehui, 1999; Ahmed *et al.*, 2002), while others report otherwise or no difference (Nabi, 1986; Kalirajan, 1990; Pender and Fafchamps, 2001, 2005). It is of empirical interest, therefore, to study the share tenancy system in Ethiopia by taking regional contexts.

Another important issue is the implication of share tenancy on the long-term productivity of sharecropped plots. The issue of lack of incentive to maintain future productivity of land capital is raised only in relation to fixed rental or lease arrangements where the tenant manages the land on his own (Agrawal, 1999). Similar concern can be raised in sharecropping where the management aspect is the sole responsibility of the tenant. Under such an arrangement, higher output share of the tenant increases the management effort, although it may have a negative impact on the landlord's motivation (Eswaran and Kotwal, 1985). Braverman and Stiglitz (1986) also found output and cost-sharing arrangements important in stimulating efficient level of input use such as fertilizer and a non-monitorable labour effort. Furthermore, with acceptable terms for share arrangements, crop share contracts may also reduce the tenant's incentive to exploit land attributes (Allen and Lueck, 1992).

Issues of moral hazard in work effort and efficiency are modelled in a static framework. Dubois (2002) suggests that this approach neglects the long-term impact of agricultural tenancy and that the problem should be addressed in a dynamic framework, considering the dynamics of land fertility in formulating the incentives for the tenant. The link between current levels of production efficiency and the long-term productivity of the land is not considered in empirical analysis. Higher exploitation during a production season affects future productivity because of its impact on land fertility (Ray, 2005).

Besides tenancy factors, variations in efficiency achievements are associated with other farmers' socio-economic and environmental circumstances. Some technical factors considered in empirical research include the following: (i) knowledge of farmers regarding agronomic practices and timeliness of farming operations (Kalirajan, 1990); (ii) location factors (Abdulai and Eberlin, 2001); and (iii) farm type as crop or mixed enterprise, farm size and access to irrigation (cited in Battese, 1992).

Socio-economic factors influencing technical efficiency include the following:

- Age of operators (Audibert, 1997; Seyoum *et al.*, 1998; Abdulai and Eberlin, 2001; Ahmed *et al.*, 2002; Gebreegziabher *et al.*, 2005).
- Education level (Kalirajan, 1990; Seyoum *et al.*, 1998; Pender and Fafchamps, 2001; Ahmed *et al.*, 2002).

- Gender (Ahmed *et al.*, 2002; Gebreegziabher *et al.*, 2005).
- Household resource endowment (Abdulai and Eberlin, 2001; Ahmed *et al.*, 2002).
- Family size (Audibert, 1997).
- Family composition in terms of dependency ratio and labour type (Abdulai and Eberlin, 2001; Ahmed *et al.*, 2002).
- Members' health status (Audibert, 1997).
- Primary occupation and wealth status (Ahmed *et al.*, 2002).
- Involvement in off-farm work (Kalirajan, 1990; Gebreegziabher *et al.*, 2005).
- Access to credit (Gebreegziabher *et al.*, 2005).
- Access to extension service (Seyoum *et al.*, 1998).

Ethnic origin of operators is also considered in some studies (Abdulai and Eberlin, 2001; Ahmed *et al.*, 2002). Based on the conceptual review and the results of empirical work elsewhere, we aimed to test the following hypotheses regarding sharecropping (in)efficiency.

Impact of tenancy

The literature is not conclusive about the impact of tenancy status on production efficiency. We expect the possible impact to depend on the context of the area. It is assumed that there is mutual interest among the contracting parties to maintain social relations and a possibility of self-monitoring in the traditional communities of the study area. It is also possible to use social norms by the landowner to enforce contract agreements costlessly in such communities. It is therefore hypothesized that resource use efficiency on sharecropped plots will be comparable to own plots for the optimizing peasant.

Labour and draught-power endowment

Households with higher male labour and oxen endowments have been found to be more efficient in some empirical studies (Abdulai and Eberlin, 2001; Ahmed *et al.*, 2002). A similar effect is expected in this study, as availability of these resources is associated with the timeliness of farming operations, especially during peak periods.

Education and labour market involvement

Better education is usually associated with efficient management of production systems (Kalirajan, 1990; Abdulai and Eberlin, 2001; Pender and Fafchamps, 2001; Ahmed *et al.*, 2002). However, better education may also lead to better off-farm opportunities, reducing the availability of both physical and managerial functions of labour and leading to lower efficiency in farm production. The impact of involvement in non-farm activities on productivity is not conclusive from empirical research results. Abdulai and Eberlin (2001) found lower efficiency of production when households were involved in non-farm activities, while Gebreegziabher *et al.* (2005) found the opposite effect. Thus, the effect of education and off-farm occupation on production efficiency is difficult to determine beforehand.

Access to credit and extension services

Access to credit reduces problems of liquidity and enhances the use of agricultural inputs in production, as often claimed in development theory (Feder *et al.*, 1985). In Tigray, credit is available for purchase of oxen and inputs. Better access to credit and to extension services may increase resource use efficiency.

Other exogenous factors

Two other exogenous factors are considered in this analysis: population pressure and rainfall conditions. In areas with good rainfall conditions there is a positive incentive for tenants to put in extra effort due to better production opportunities and potential for risk reduction. Farmland is scarce in densely populated areas and tenants in these areas may make more efficient use of farmlands, assuming the competition for land and market opportunities for produce are positively influenced by population pressure. Plot-level management may also be better (timely and in the required quality) in areas of high population density because of labour availability, resulting in higher resource use efficiency.

Data and methodology

Data and variable definition

The data come from household- and plot-level surveys conducted in 1999/2000 in the highlands of Tigray by the research project 'Policies for sustainable land management in the highlands of Tigray, Northern Ethiopia', a collaborative work of the International Food Policy Research Institute (IFPRI), the International Livestock Research Institute and Mekelle University, Ethiopia.

The database contains information on basic socio-economic and production activities of 500 households randomly selected, five each from 100 sample villages (see Pender and Gebremedhin, 2004, for sampling). A total of 115 households and the plots they were operating during the 1998 production season are included for the analysis. These households were involved in sharecropping and operated 347 own and 192 sharecropped-in plots during the production season.

Looking at the distribution of sample plots by relative land quality ranking, 21, 40 and 39% of sharecropped-in plots and 47, 31 and 22% of own plots were ranked as good, medium and poor quality, respectively. Nearly 80% of sharecropped plots were medium to poor quality. During the production season, 23, 32, 33 and 12% of sharecropped plots and 28, 30, 33 and 9% of own plots were used for growing *Teff* varieties (a cereal almost exclusive to Ethiopia), small cereals, large cereals and pulses, respectively.

A dataset was created containing basic socio-economic and resource endowment profiles of the sample tenant households:

- Their access to credit and extension services and secondary occupation of the household head.
- The relative quality ranking of both owned and sharecropped-in land by the tenant.
- Plot-level input use and production activities.
- Plot distance from tenant's residence.
- Gross value of grain and straw output, computed based on community-level average prices.
- Indicators of conservation investment (whether the tenant household invested in 1998 and in stock at the end of 1997).
- Manure use practice.
- Terms of sharecropping arrangements.
- Village conditions in terms of population density.
- Rainfall for the 1998 production season.

Crop types are grouped, based on management similarity, as: (i) *Teff*, all varieties; (ii) small cereal, all varieties; (iii) large cereal, all varieties; and (iv) pulses. Small cereals include varieties of wheat and barley, and the large cereals are varieties of sorghum and maize. Based on rainfall data, villages are categorized into relatively low-, medium- and high-rainfall areas. Table 7.1 provides a list of the variables and descriptive information used for the stochastic production frontier analysis and technical efficiency determinants.

Plot tenancy status is defined as a categorical variable as owned plots, plots under equal grain output sharing and plots with two-thirds share of grain output to tenants. The distribution of sharecropped plots by the level of tenant's grain output share is one-third for four plots (2%), one-half for 122 plots (64%), two-thirds for 51 plots (26%), and three-quarters for 15 plots (8%). The 50/50 sharing of grain output is the dominant feature, similar to the findings of Gebremedhin *et al.* (2003) for the highlands of Tigray. Tenants receive 100% of straw output in 90% of the cases and a 50% share in 10%. Input sharing between the tenant and the landowner is low. The landowner's share of inputs is one-third to one-half in 2, 9 and 20% of the cases for fertilizer, labour and seed inputs, respectively. These are also comparable to the 5, 10 and 16% contributions for each input, respectively, found by Gebremedhin *et al.* (2003) from their 1998 community-level survey in the region.

The tenant provides all the draught power and farm equipment and is almost solely responsible for major inputs. This scenario of input

Table 7.1. Descriptive statistics of variables used in stochastic frontier production function estimation and determinants of efficiency (tenant households).

Variables	*n*	Mean	SE	SD
Dependent variable				
Gross value of grain and straw output in Birr[a]	518	441.13	18.52	421.63
Explanatory variables				
Production input variables				
Labour input (person-days)[a]	525	24.87	0.94	21.63
Plot area (m[2])[a]	538	3215.26	124.14	2879.60
Draft-power (ox-days)[a]	527	13.23	0.37	8.51
Seed cost (Birrb)[a]	530	35.52	1.60	36.84
Fertilizer (kg)[a]	534	6.27	0.54	12.56
Agroecological indicator				
Plots in high-rainfall areas (yes = 1)	539	0.232	0.018	0.422
Plots in medium-rainfall areas (yes = 1)	539	0.514	0.021	0.500
Plots in low-rainfall areas (yes = 1)	539	0.254	0.018	0.436
Plot quality ranking				
Plots of good quality (yes = 1)	538	0.372	0.020	0.484
Plots of medium quality (yes = 1)	538	0.342	0.020	0.475
Plots of low quality (yes = 1)	538	0.286	0.019	0.452
Crop factors				
Teff, all varieties (yes = 1)	529	0.27	0.019	0.44
Small cereals/wheat and barley/all varieties (yes = 1)	529	0.31	0.020	0.45
Large cereals/sorghum and maize/all varieties (yes = 1)	529	0.33	0.020	0.46
Pulses, all varieties (yes = 1)	529	0.09	0.012	0.29
Other management practices				
Manure use (yes = 1)	533	0.25	0.018	0.433
Private investment in 1998 (yes = 1)	536	0.080	0.011	0.272
Stock of SWC investment in 1997 (yes = 1)	529	0.416	0.021	0.493
Tenure factor				
Own plots (yes = 1)	539	0.633	0.020	0.483
Half-grain output share of tenant (yes = 1)	539	0.234	0.018	0.424
Two-thirds/three-fourths grain output share of tenant (yes = 1)	539	0.135	0.014	0.343
Tenure security indicator (yes = 1)	486	0.722	0.020	0.448
Number of plots sharecropped-in by tenants (count)[a]	115	2.02	0.048	1.14
Household demographics				
Age of household head (years)	115	46.40	1.12	12.06
Household head education (formal/informal science) (yes = 1)	115	0.37	0.04	0.48
Secondary occupation (non-farming = 1)	115	0.27	0.04	0.44
Institutional credit (received loan = 1)	115	0.68	0.04	0.46
Participation in extension and training (yes = 1)	115	0.35	0.04	0.47
Household resource endowment				
Male labour endowment (adult equivalent)[a]	115	2.33	0.08	0.87
Female labour endowment (adult equivalent)[a]	115	0.34	0.03	0.71
Cultivable land holding size (*Tsimdi*)[c]	115	5.0	0.27	2.90
Potential draught power endowment (TLU)[a]	115	2.60	0.13	1.36
Other livestock – all but oxen (TLU)[a]	115	5.36	0.18	4.2
Location factors				
Population density (persons km[2])[a]	115	118.12	5.63	60.38
Distance of plot from tenant's residence/min/[a]	509	23.2	1.11	25.1

n, Number of observations; SE, standard error of the mean; SD, standard deviation.
[a] included in their natural log forms; dummy variables are defined as yes = 1 or 0 otherwise for the analysis; and, unless and otherwise stated, all variables refer to 1998 production year;
[b] Ethiopian unit of currency;
[c] 1 tsimdi = 0.25 ha.

sharing does not show enough variation as an explanatory factor; grain output sharing shows better variation and this was used as a proxy for the tenancy status of plots. Three dummy variables are thus defined representing a 50% share, a 66–75% share and own plots of the tenant. Other tenure-related indicators considered are the number of plots sharecropped-in by each tenant household and the tenant's expectation of long duration of operation of the plots. The latter is used as an indicator of perceived contract stability or tenure security. Higher numbers of plots sharecropped-in may indicate a higher extent of involvement by the tenant household in the land contract markets.

Analysis

The stochastic frontier production function approach is employed to compute the technical efficiency of resource use at plot level. This methodology is popular in empirical studies of technical efficiency analysis (Kumbhakar and Lovell, 2000; Abdulai and Eberlin, 2001; Ahmed *et al.*, 2002). The frontier production function shows the maximum amount of output obtainable from given quantities of inputs under maximum efficiency. Technical inefficiency is measured from this frontier level, and the composed error specification of the stochastic frontier enables to separate output shortfalls due to technical inefficiency from those caused by random disturbances.

The general model of the stochastic frontier production function (Aigner *et al.*, 1977; Jondrow *et al.*, 1982; Kumbhakar and Lovell, 2000, p. 73; STATA, 2003) is

$$Y_i = f(\beta^* X_i) \exp(v_i - u_i) \quad i = 1, 2, 3 \ldots, N \tag{7.1}$$

Where Y_i is the output for observation i; β represents the vector of parameters to be estimated; X_i represents the vector of input variables for the ith observation; v_i represents the disturbance term with a symmetric distribution ($N(0, \sigma_v^2)$); u_i represents the disturbance term with a half-normal distribution ($N^+(0, \sigma_u^2)$), measuring the technical inefficiency component independently distributed of the v_is; and i is the observation unit, in this case a plot. The technical inefficiency term u_i measures the shortfall of output from its maximum possible value given by the stochastic frontier (Jondrow *et al.*, 1982; Battese and Broca, 1997; Bravo-Ureta and Pinheiro, 1997). The half-normal distributional ($N^+(0, \sigma_u^2)$) assumption for u_i is widely used in empirical work and is adopted here. The technical inefficiency determinants may be expressed as:

$$u_i = \delta^* Z_i + w_i, \quad i = 1, 2, 3 \ldots, n \tag{7.2}$$

Where Z represents the vector of factors that influence the technical inefficiency; δ represents the vector of unknown parameters of the plot-specific inefficiency variables; and w is a random disturbance term obtained by truncations of the normal distribution, with mean zero and variance σ^2. Given the specification of the stochastic frontier production function in equation (7.1), the technical efficiency scores of production for the [ith] plot are predicted as

$$TE_i = \exp(-u_i) = \exp(-Z_i\delta - w_i) \tag{7.3}$$

The Maximum Likelihood estimates of the model parameters are computed using the frontier models routine of the statistical package STATA 8SE, which assumes a Cobb-Douglas production technology (STATA, 2003). The determinants of inefficiency are identified by regressing the predicted scores over the Z variables as:

$$-\ln(TE_i) = \delta * Z_i \tag{7.4}$$

Technical Efficiency Achievement and Determinants

Before proceeding to a stochastic frontier production function analysis, a T-test was conducted to assess the statistical significance of the differences in the mean gross value of output from tenants' own and sharecropped-in plots, controlling for plot area and crop type. The null hypothesis of equality of means was rejected under both equal and unequal variance assumptions, and the results for unequal variance assumption are reported in Table 7.2. Controlling for crop type and area, mean gross value of output from own plots is significantly higher than for sharecropped-in plots.

The stochastic frontier production function is estimated using a Cobb-Douglas function, commonly used in technical efficiency studies elsewhere (Kumbhakar and Lovell, 2000; STATA, 2003) and in Ethiopian studies (e.g. Seyoum *et al.*, 1998; Ahmed *et al.*, 2002; Gebreegziabher *et al.*, 2005). The same assumption is maintained here for reasons of comparability. It is also important to know whether returns to scale are decreasing because, if that is the case, an increase in population growth will have a negative impact on income and sustainability (Pender, 1998).

Table 7.2. Comparison of mean gross values of output from owned and sharecropped-in plots of tenants by crop type per tsimdi (1 tsimdi = 0.25 ha), assuming unequal variance.

Crop type	Own plots		Sharecropped plots		d.f.[a]	T-value	P
	n	Means (SE)	n*	Means (SE)			
Teff, all varieties	96	1315 (185)	45	517 (94.7)	131.7	3.841	0.000
Small cereals	97	616.8 (125)	63	331.7 (52)	126.4	2.103	0.018
Large cereals	110	1184.1 (153)	65	571.5 (78.0)	155.1	3.562	0.000
Pulses	24	495.2 (205)	18	166.4 (33.7)	24.2	1.582	0.063
All crops	327	1003 (86)	191	441.4 (39)	444.6	5.910	0.000

n, Number of observations; SE standard error of the mean.
[a] Satterthwaite's degrees of freedom, the hypothesis tested being that the mean (own) less mean (sharecropped-in) is equal to zero (H_0) against the alternative (H_a) if the difference is greater than zero.

The plot-level gross value of output is considered as a function of the inputs of total labour, plot size, draught power, seed and inorganic fertilizer applied. To handle cases of zero fertilizer input, the method proposed by Battese (1997)[2] is employed. Apart from the conventional inputs dummy variables for rainfall conditions, plot quality indicators, crop type, manure use, the tenant's conservation investments in 1998 and stock of conservation investment at the end of 1997 are also included, as these factors may shift the intercept of the frontier.

The estimated stochastic frontier production function is presented in Table 7.3. A likelihood-ratio (LR) test on the statistical significance of the technical inefficiency within the data confirms that the null hypothesis of technical inefficiency (H_0: sigma $\mu = 0$) is to be rejected ($P = 0.007$). The lambda (λ) value, commonly used as an indicator of the significance of the inefficiency level, is greater than one (1.347). The gamma (γ) value, a measure of the percentage variations in plot output due to technical inefficiency, is 64%. Thus, use of Ordinary Least Squares (OLS) to estimate the frontier function is therefore inappropriate in the presence of inefficiency.[3]

Output response for total labour, draught power and seed and fertilizer inputs use is positive and statistically significant, except for plot area. Output elasticities for total labour, draught power, seed and fertilizer inputs are 0.14, 0.35, 0.24 and 0.15, respectively. A Wald test for the constant returns assumption[4] is confirmed positively. Gross revenue is 26 and 37% lower for small cereals and pulses, respectively, compared with *Teff*. Gross revenue is 29 and 19% lower for plots in low- and medium-rainfall areas, respectively, compared with the plots in high-rainfall areas. Gross revenue from medium- and poor-quality plots is 20 and 28% lower, respectively, compared with that from good-quality plots. Plots with stock of conservation investment in 1997 gave 13% higher gross revenue compared with those without prior conservation.

In order to investigate the determinants of technical efficiency differentials, the technical efficiency scores at plot level were generated using equation (7.3). The distribution of the predicted technical efficiency scores is presented in Table 7.4.

The overall mean of technical efficiency level is 65%, ranging from 18–87%. Technical efficiency scores for a large proportion of sharecropped-in and owned plots are below the overall average. The wide range of variations in technical efficiency achievement reveals the challenge and potential for improving crop production in dryland areas of the highlands of Tigray through better allocation and management of external factors.

Table 7.5 presents a summary of the technical efficiency levels by crop type. The mean technical efficiency on tenants' own plots is marginally higher than from sharecropped-in plots (at 10% level of significance). However, when we disaggregate by crop type, plot quality and rainfall conditions, the statistical significance of the efficiency scores by plot tenancy status is variable (see Tables 7.6 and 7.7). The results in

Table 7.3. Stochastic frontier estimation of plot-level production efficiency in the highlands of Tigray (1998 production season).

Explanatory variables[a]	Coefficient	SE
Production inputs		
Labour input (person-days)[b]	0.146***	0.051
Plot area (m²)[b]	0.089	0.059
Draught power (ox-days)[b]	0.359***	0.077
Seed cost (Birr)[b]	0.244***	0.047
Dummy for fertilizer zero values[c]	0.411***	0.156
Fertilizer (kg)[b]	0.152**	0.060
Agro-ecological conditions (base high rainfall area)		
Plots in medium-rainfall areas (yes = 1)	−0.196***	0.070
Plots in low-rainfall areas (yes = 1)	−0.294***	0.081
Plot quality conditions (base good plot quality)		
Plots of medium quality (yes = 1)	−0.201***	0.067
Plots of low quality (yes = 1)	−0.289***	0.072
Crop factors (base *Teff*, all varieties)		
Small cereals, all varieties (yes = 1)	−0.269***	0.089
Large cereals, all varieties (yes = 1)	−0.067	0.076
Pulses, all varieties (yes = 1)	−0.375***	0.124
Other management practices		
Manure use (yes = 1)	0.007	0.069
Private investment in 1998 (yes = 1)	−0.066	0.100
Stock of SWC investment in 1997 (yes = 1)	0.133**	0.057
Constant	3.471***	0.349
/lnsig2v	−1.572***	0.186
/lnsig2u	−0.976***	0.302
Sigma v	0.455	0.042
Sigma u	0.614	0.092
Sigma square	0.584	0.084
lambda (δ_u/δ_v)	1.347	0.130
Gamma[d] $(\gamma=\lambda^2/(1+\lambda^2))$ or $\gamma=\delta_u^2/(\delta_u^2+\delta_v^2))$	0.644	
Number of observations	484	
Wald chi² (16)	569.39	
Probability > chi²	0.000	
Log likelihood	−426.13	
Likelihood-ratio test of sigma u = 0		
Chibar²(01)	60.6.01	
Probability > = chibar²	0.007	

SE, Standard error of the mean; ***, ** and * indicate 1, 5 and 10% levels of significance, respectively; dummy variables are defined as yes = 1 or 0 otherwise for the analysis.
[a] Dependent variable: gross value of grain and straw outputs (in Birr, the Ethiopian unit of currency);
[b] natural log form;
[c] defined according to the Battese (1997) method for handling zero cases of input use – in our case, fertilizer;
[d] measures the percentage of total output variation due to technical inefficiency.

Table 7.4. Distribution of technical efficiency scores by plot tenancy in the highlands of Tigray (1998 production season).

	Own plots		Sharecropped-in plots		Combined	
Efficiency score range[a]	n	%	n	%	n	%
0–0.2	1	0.32	0	0.00	1	0.21
0.2–0.4	7	2.27	14	8.00	21	4.34
0.4–0.6	89	28.90	43	24.40	132	27.27
0.6–0.8	180	58.44	101	57.39	281	58.06
0.8–1.0	31	10.07	18	10.21	49	10.12
Total	308	100.00	176	100.00	484	100.00
Average score		65.77		64.12		65.17
Cases < the overall mean	135	43.83	79	44.88	214	44.21

[a] Ranges exclude upper boundaries except for the final category.

Table 7.5. Mean technical efficiency levels for owned and sharecropped-in plots by crop type in the highlands of Tigray (1998 production season).

Tenure status and crop type	Plots (n)	Means (SD)	Minimum	Maximum
All plots, aggregate	484	0.651 (0.127)	0.178	0.871
Own plots:				
Teff, all varieties	87	0.665 (0.110)	0.327	0.854
Small cereals (e.g. barley and wheat)	95	0.654 (0.122)	0.282	0.865
Large cereals (e.g. sorghum, maize, millet)	104	0.652 (0.128)	0.178	0.865
Pulses	22	0.670 (0.106)	0.417	0.820
Sharecropped-in plots				
Teff, all varieties	44	0.641 (0.122)	0.314	0.840
Small cereals (e.g. barley and wheat)	58	0.648 (0.132)	0.303	0.871
Large cereals (e.g. sorghum, maize, millet)	58	0.641 (0.144)	0.209	0.855
Pulses (e.g. beans, peas, vetch)	16	0.613 (0.196)	0.233	0.848

SD, Standard deviation.

Table 7.6. Comparison of means for technical efficiency levels for owned and sharecropped-in plots by crop type (T-test assuming unequal variance).

	Own plots		Sharecropped-in plots				
Crop type	n	Means (SE)	n	Means (SE)	d.f.[a]	T-value	P
All crops, aggregated technical efficiency	308	0. 658 (0.006)	176	0.641 (0.009)	320.721	1.308	0.095
Teff, all varieties	87	0.665 (0.011)	44	0.641 (0.018)	79.158	1.096	0.138
Small cereals	95	0.653 (0.012)	58	0.648 (0.017)	113.375	0.223	0.411
Large cereals	104	0.652 (0.011)	58	0.641 (0.017)	107.02	0.487	0.313
Pulses	22	0.670 (0.022)	16	0.613 (0.049)	21.410	1.046	0.153

n, Number of observations; SE, standard error of the mean; [a] Satterthwaite's degrees of freedom; the hypothesis tested is that the mean (own) less mean (sharecropped-in) is equal to zero (H_0) against the alternative (H_a) that the difference is > zero; and T-values and *P* are for H_a: difference > 0.

Table 7.7. Comparison of means of technical efficiency levels for owned and sharecropped-in plots of tenants by rainfall conditions and land quality (T-test assuming unequal variance).

Variable	Own plots		Sharecropped-in plots		d.f.[a]	T-value	*P*
	n	Means (SE)	*n*	Means (SE)			
High-rainfall area	71	0.643 (0.013)	34	0.674 (0.019)	64.497	−1.305	0.901
Medium-rainfall area	150	0.660 (0.01)	97	0.641(0.015)	175.32	1.462	0.072
Low-rainfall area	87	0.664 (0.012)	45	0.632 (0.019)	82.740	1.408	0.081
Good-quality plot	144	0.660 (0.010)	37	0.614 (0.024)	48.819	1.747	0.043
Medium-quality plot	101	0.655 (0.012)	70	0.643 (0.016)	134.107	0.573	0.283
Poor-quality plot	63	0.653 (0.014)	69	0.653 (0.015)	129.49	0.008	0.496

n, Number of observations; SE, standard error of the mean; [a] Satterthwaite's degrees of freedom; the hypothesis tested is that the mean (own) less mean (sharecropped-in) is equal to zero (H_0) against the alternative (H_a) that the difference is greater than zero; and T-values and *P* are for H_a: diff > 0.

Table 7.6 show that the variations in technical efficiency scores on owned and sharecropped-in plots are not significantly different when we compare for each crop type.

A different pattern emerges when comparing mean technical efficiency scores for owned and sharecropped-in plots, controlling for land quality and rainfall conditions (see Table 7.7). In medium- and low-rainfall areas, the technical efficiency achieved by tenants is significantly higher on their own plots than on sharecropped-in plots, but in high-rainfall areas the differences are not statistically significant. The mean technical efficiency scores are higher on tenants' own plots than on sharecropped-in ones when the plots are of good quality, but not significantly so on the medium- and poor-quality plots.

To identify the factors that explain the difference in technical efficiency achievements, regression techniques are used, with equation (7.4) estimated using an Ordinary Least Square model controlling for non-independence of errors within households. The predicted technical inefficiency scores are regressed over the factors hypothesized as influencing the level of efficiency (Z_i). These are drawn from the theoretical and empirical review, and include factors that directly or indirectly affect the management decisions of farmers and the technical efficiency levels that they achieve.

Factors that are believed to have an effect on the managerial skills, timeliness of resource allocation decisions and the implementation of farm operations by farmers are considered. These include:

- The tenure status of the plot.
- The perceived tenure security indicator.
- The number of plots sharecropped-in by the tenant.

- The tenant's age, education status, secondary occupation, access to institutional credit, participation in extension and training and resource endowment (in terms of male and female labour, cultivable land, and draught-power).
- The distance of the plot from the residence.
- The population density of the village.

The variable of interest here is the land tenure status of plots, categorized as either: (i) tenant's own plot; (ii) plots under a half-sharing arrangement; and (iii) plots where a two-thirds or three-quarters share of the grain output goes to tenants. Population density is considered as a proxy for population pressure. The regression results are presented in Table 7.8.

The findings show that tenancy factors were not a statistically significant determinant of technical inefficiency, although the signs were

Table 7.8. Determinants of technical inefficiency.

	OLS	
Explanatory variables	Coefficient	RSE
Tenure status of plot (base, own plot)		
Half of grain output share to tenant (yes = 1)	0.0572	0.0364
Two-thirds/three-fourths grain output share to tenant (yes = 1)	0.0454	0.0401
Number of plots sharecropped-in by tenants (count)[a]	0.0153	0.0375
Tenure security indicator (yes = 1)	−0.0228	0.0288
Household demographics		
Age of household head (years)	0.0002	0.0014
Household head education (formal/informal science) (yes = 1)	−0.0156	0.0304
Secondary occupation (non-farming = 1)	0.0268	0.0442
Institutional credit (received loan = 1)	0.0078	0.0306
Participation in extension and training (yes = 1)	0.0244	0.0327
Household resource endowment		
Male labour endowment (adult equivalent)[a]	0.0443	0.0331
Female labour endowment (adult equivalent)[a]	−0.0472	0.0317
Cultivable landholding size (Tsimdi[b])[a]	−0.0373	0.0360
Potential draught-power endowment (TLU)[a]	−0.0032	0.0580
Other livestock (TLU)	−0.0652**	0.0272
Location factors		
Population density (persons/km^2)[a]	−0.0687*	0.0394
Distance of plot from tenant's residence/min[a]	0.0065	0.0074
Constant	0.8493	0.1938
Number of observations	421	
F (16,112)	1.33	
Probability > F	0.192	
R^2	0.071	

RSE, Robust Standard Errors; ***, ** and * indicate 1, 5 and 10% levels of significance, respectively; and dummy variables are defined as yes = 1 or 0 otherwise;
[a] variables in their natural log form;
[b] 1 tsmidi = 0.25 ha.

positive for all the variables referring to the share tenancy. Technical inefficiency scores were higher for tenants operating large numbers of plots and lower when tenants expected long period of use of plots, although neither were statistically significant.

Technical inefficiency levels are significantly lower in high-population density villages, in line with the Boserupian hypothesis (Pender, 1998). This greater efficiency of tenants in such villages may be due to greater competition for land, which requires tenants to operate at a relatively higher level of efficiency in order to maintain continuity of contracts. The level of technical inefficiency is significantly lower for livestock-endowed households, which could be related to both the wealth status of the household and the dependence of the livestock system on crop residue for feed. Other socio-economic and resource endowment factors of tenants show no significant impact on the level of technical efficiency achievements.

In sum, on both owned and sharecropped-in plots of tenants, there is a significant level of inefficiency in the production systems. However, the regression results do not show significant impact of tenancy status of a plot on technical efficiency. Technical efficiency was significantly positively associated with livestock endowments of households and population density. Previous studies confirmed the existence of other sources of inefficiency, which indicate the need for further empirical research. For example, Gavian and Ehui (1999) identify differences in quality and problems of proper application of inputs (rather than the intensity of application) as possible sources of inefficiency. It is suspected that there may be a problem of timing of input applications on sharecropped-in plots, although it is not possible to support this from the analysis due to lack of data on the specific timing of input use during the production year considered. It is therefore important in further research to control for the quantity, quality and timing aspects of input use to understand how these affect efficiency of production.

Long-term Prospects of Sharecropping Arrangements

The conditions of sharecropping, in terms of output and input sharing, the duration of contracts and obligations regarding land management practices have impact on both the level of efficiency of production and the sustainability of land use. Evaluation of the features of the sharecropping system in Tigray shows that major problems exist in these important aspects. Contract duration is mainly for 1 year and is not likely to give sufficient incentives for tenants to invest in land improvement practices. According to Gebremedhin *et al.* (2003), tenants' investment in soil and water conservation practices and tree planting does not influence the likelihood of contract extensions in Tigray.

Another study, in the Arussi area of Ethiopia, shows one-season land contracts are a major concern for landless tenants, as they give little incentive to invest in long-lasting land improvements (Tolossa, 2003). Under such short contract durations, it is unlikely that landowners threatening to evict tenants will provide much motivation for tenants to invest in and adopt sustainable production systems on sharecropped-in plots (Banerjee and Ghatak, 2004). Thus, the cumulative effects of short-term contracting could be the source of long-term inefficiency of production on continually sharecropped plots, even if the contracts are renewed season after season.

Short duration of contracts may also motivate tenants to focus on short-term objectives that can be achieved through exploitative production technologies (Ray, 2005). Preservation of soil quality and sustainable management could be achieved through other incentives that maintain or enhance plot quality (Dubois, 2002). For instance, land-owners could provide incentives to tenants to manage the land sustainably or to plant less nutrient-mining crops.

Another important aspect of the sharecropping that demands attention is the terms of production input and cost sharing between tenants and landowners. In Tigray, the contribution of the landowner in terms of labour and other inputs is minimal. This minimal contribution of labour may limit the landowner's ability to monitor the work effort of the tenant, as direct participation would lower the cost of supervision. The inability of the landlord to monitor adequately the tenant's input is one of the possible causes of inefficiency of sharecropping, according to the Marshallian view. Likewise, it can be inferred that such a lack of monitoring of the use of fertility-enhancing external inputs, coupled with the difficulty of detecting soil nutrient depletion in one-season arrangements, may lead to a decline in the long-term productivity of farmlands. It is therefore important that the terms of land contracts be designed comprehensively to minimize these problems and that the long-term implication of land contract arrangements gains more attention in empirical research.

Conclusions and Implications for Policy

Efficiency analysis has an important role to play in generating information for policies aiming at enhancing productivity of farm resources. This study employs stochastic frontier production function analysis to evaluate the impact of tenancy arrangements on agricultural productivity. Unlike previous studies in the region (Gebreegziabher *et al.*, 2005), the findings of the current study show that there is a statistically significant level of inefficiency in the production systems of the study area. However, the tenancy status of the plot is not a significant cause of inefficiency when other factors are controlled for.

This study did not find strong evidence of the impact of contract duration on efficiency of production due to the non-significance of this variable in the regression results. About 95% of the plots in the sample are sharecropped for one production season. The regional rural land administration and land use policy also restricts contract duration to 2 years for tenants applying traditional technologies. As the limited information from other studies indicates, such a restriction is not beneficial in terms of sustaining long-term productivity. It is therefore important to consider this in further research and come up with a more empirical evidence of the impact of such restrictions in the region.

The input-sharing arrangements indicate that the tenant is responsible for provision of almost all production inputs. This reduces the possibility for the landowner to monitor and enforce tenants' efforts without a high cost. Other factors also contribute to the differences in technical efficiency achievement. It is therefore important to investigate further empirically the demand and supply side of the contract markets to come up with conclusive results, as this study is based on the demand side information. A comprehensive study will be required to investigate the reasons for the dominance of sharecropping and lack of development in the other forms of land contracts.

Endnotes

[1] The authors are indebted to the IFPRI/ILRI/Mekelle University collaborative research project on 'Policies for Sustainable Land Management in the Highlands of Tigray, Northern Ethiopia', funded by the Swiss Agency for Development and Cooperation, for allowing us to use the project database. The deskwork is supported by the RESPONSE project, Regional Food Security Policies for Natural Resource Management and Sustainable Economies, whom we gratefully acknowledge. The usual disclaimer applies.

[2] A dummy variable is included in the analysis, which takes one for zero-values of the fertilizer input.

[3] The residuals from an OLS estimation show a negative skew (−0.288), which is a further indicator of the presence of technical inefficiency in a dataset (Kumbhakar and Lovell, 2000, p. 73)

[4] The test for the assumption of constant returns to scale was done by imposing a linear restriction as the sum of the coefficients for the labour, land, draught power, seed and fertilizer is equal to unity. The Wald test did not reject the null hypothesis of constant returns to scale.

References

Abdulai, A. and Eberlin, R. (2001) Technical efficiency during economic reform in Nicaragua: evidence from farm household survey data. *Economic Systems* 25, 113–125.

Agrawal, P. (1999) Contractual structure in agriculture. *Journal of Economic Behaviour and Organization* 39, 293–325.

Ahmed, M.M., Gebremedhin, B., Benin, S. and Ehui, S. (2002) Measurement and source of technical efficiency of land tenure contracts in Ethiopia. *Environmental and Development Economics* 7, 507–528.

Aigner, D., Lovell, C.A.K. and Schmidt, P. (1977) Formulation and estimation of stochastic frontier production function models. *Journal of Econometrics* 6, 21–37.

Allen, D. and Lueck, D. (1992) Contract choice in modern agriculture: cash rent *versus* crop share. *Journal of Law and Economics* 35, 397–426.

Audibert, M. (1997) Technical inefficiency effects among paddy farmers in the village of the 'Office du Niger', Mali, West Africa. *Journal of Productivity Analysis* 8, 379–394.

Banerjee, A. and Ghatak, M. (2004) Eviction threats and investment Ghatak. *Journal of Development Economics* 74, 469–488.

Bardhan, P. and Rudra, A. (1980) Terms and conditions of sharecropping contracts: an analysis of village survey data in India. *The Journal of Development Studies* 16, 287–302.

Battese, G.E. (1992) Frontier production functions and technical efficiency: a survey of empirical applications in agricultural economics. *Agricultural Economics* 7, 185–208.

Battese, G.E. (1997) A note on the estimation of Cobb-Douglas Production Functions when some explanatory variables have zero values. *Journal of Agricultural Economics* 48, 250–252.

Battese, G. and Broca, S.S. (1997) Functional form of stochastic frontier production function and models for technical inefficiency effects: a comparative study for wheat farmers in Pakistan. *Journal of Productivity Analysis* 8, 395–414.

Benin, S., Ahmed, M., Pender, J. and Ehui, S. (2005) Development of land rental markets and agricultural productivity growth: the case of Northern Ethiopia. *Journal of African Economies* 14, 21–54.

Bhaumik, S.K. (1993) *Tenancy Relations and Agrarian Development: a Study of West Bengal.* Sage Publications, New Delhi, India and London.

Braverman, A. and Stiglitz, A. (1986) Cost-sharing arrangements under sharecropping, moral hazard, incentive flexibility and risk. *American Journal of Agricultural Economics* 68, 642–652.

Bravo-Ureta, B. and Pinheiro, A.E. (1997) Technical, economic, and allocative efficiency in peasant farming: evidence from the Dominican Republic. *The Developing Economies* 35, 48–67.

Cheung, S.N.S. (1969) *The Theory of Share Tenancy.* University of Chicago Press, Chicago, Illinois.

Cohen, J.M. and Weintraub, D. (1975) *Land and Peasants in Imperial Ethiopia.* Van Gorcum & Co., B.V. Assen, The Netherlands.

Deininger, K., Jin, S., Adnew, B., Gebre-Selassie, S. and Demeke, M. (2003) Market and Non-market Transfers of Land in Ethiopia: Implications for Efficiency, Equity, and Non-farm Development. *World Bank Policy Research Working Paper* 2992, World Bank, Washington, DC.

Dubois, P. (2002) Moral hazard, land fertility and sharecropping in a rural area of the Philippines. *Journal of Development Economics* 68, 35–64.

Eswaran, M. and Kotwal, A. (1985) A theory of contractual structure in agriculture. *American Economic Review* 75, 352–367.

Feder, G. and Feeny, D. (1991) Land tenure and property rights: theory and implication for development policy. *The World Bank Economic Review* 5, 135–153.

Feder, G., Just, R.E. and Zilberman, D. (1985) Adoption of agricultural innovations in developing countries: a survey. *Economic development and Cultural Change* 33, 255–297.

Gavian, S. and Ehui, S. (1999) Measuring the production efficiency of alternative land tenure contracts in a mixed crop-livestock system in Ethiopia. *Agricultural Economics* 20, 37–49.

Gebreegziabher, Z., Oskam, A. and Woldehanna, T. (2005) Technical efficiency of peasant farmers in Northern Ethiopia: a stochastic frontier approach, In: Seyoum, A., Admassie, A., Degefe, B., Nega, B., Demeke, M., Kersemo, T.B. and Amha, W. (eds) *Proceedings of the Second International Conference on the Ethiopian Economy*, June 2004, Vol. II, Ethiopian Economic Association, Addis Ababa, pp. 103–118.

Gebremedhin, B. and Swinton, S.M. (2003) Investment in soil conservation in northern Ethiopia: the role of land tenure security and public programs. *Agricultural Economics* 29, 69–84.

Gebremedhin, B., Pender, J. and Ehui, S. (2003) Land tenure and land management in the highlands of Northern Ethiopia. *Ethiopian Journal of Economics* 8, 46–63.

Hagos, F., Pender, J. and Gebresselassie, N. (1999) *Land Degradation in the Highlands of Tigray and Strategies for Sustainable Land Management.* Socio-economic and Policy Research Working Paper No. 25, International Livestock Research Institute.

Hayami, Y. and Otsuka, K. (1993) *The Economics of Contract Choice: an Agrarian Perspective.* Clarendon Press, Oxford, UK.

Jondrow, J., Lovell, C.A.K., Materov, I.S. and Schmidt, P. (1982) On the estimation of technical inefficiency in the stochastic frontier production function model. *Journal of Econometrics* 19, 233–238.

Kalirajan, K.P. (1990) On measuring economic efficiency. *Journal of Applied Econometrics* 5, 75–85.

Kumbhakar, S.C. and Lovell, C.A.K. (2000) *Stochastic Frontier Analysis.* Cambridge University Press, Cambridge, UK.

Marshall, A. (1890) *Principles of Economics* (1920 edn). Macmillan, London.

Nabi, I. (1986) Contracts, resource use and productivity in sharecropping. *The Journal of Development Studies* 22, 429–442.

Negarit Gazeta (1997) Tigray national state rural land use decree, No. 23/1989 (in Tigrigna and Amharic languages).

Otsuka, K. and Hayami, Y. (1988) Theories of share tenancy: a critical survey. *Economic Development and Cultural Change* 37, 31–68.

Pender, J. (1998) Population growth, agricultural intensification, induced innovation and natural resource sustainability: an application of neoclassical growth theory. *Agricultural Economics* 19, 99–112.

Pender, J. and Fafchamps, M. (2001) Land Lease Markets and Agricultural Efficiency: Theory and Evidence from Ethiopia. Environment and Production Technology Division Discussion Paper No. 81, International Food Policy Research Institute, Washington, DC.

Pender, J. and Fafchamps, M. (2006) Land lease markets and agricultural efficiency in Ethiopia. *Journal of African Economies* 15 (2), 251–284.

Pender, J. and Gebremedhin, B. (2004) Impacts of policies and technologies in dryland agriculture: evidence from northern Ethiopia. In: Roa, S.C. (ed.) *Challenges and Strategies for Dryland Agriculture.* American Society of Agronomy and Crop Science, Society of America, CSSA Special Publication 32, Madison, Wisconsin.

Ray, T. (2005) Sharecropping, land exploitation and land-improving investments. *The Japanese Economic Review* 56, 127–143.

Ruben, R., Kuyvenhoven, A. and Kruseman, G. (2001) Bioeconomic models and eco-regional development: policy instruments for sustainable intensification. In: Lee, D.R. and Barrett, C.B. (eds) *Tradeoffs or Synergies?* CAB International, Wallingford, UK, pp. 115–133.

Seyoum, E.T., Battese, G.E. and Fleming, E.M. (1998) Technical efficiency and productivity of maize producers in eastern Ethiopia: a study of farmers within and outside the Sasakawa-Global 2000 project. *Agricultural Economics* 19, 341–348.

STATA (2003) Stata Reference Manual, Vol. 1 (A–F). StataCorp, College Station, Texas.

Stiglitz, J. (1974) Incentives and risk sharing in sharecropping. *Review of Economic Studies* 41, 219–255.

Teklu, T. and Lemi, A. (2004) Factors affecting entry and intensity in informal rental land markets in Southern Ethiopia highlands. *Agricultural Economics* 30, 117–128.

Tesfay, G. (1995) Methodological review for the economic assessment of soil conservation and water harvesting practices in Tigray, Ethiopia. MSc Thesis, University of North Wales, Bangor, UK.

Tolossa, D. (2003) Issues of land tenure and food security: the case of three communities of Munessa Wereda, South-Central Ethiopia. *Norsk Geografisk Tidsskriff* [*Norwegian Journal of Geography*] 57, 9–19.

Wacher, D. and English, J. (1992) *The World Bank's Experience with Rural Land Titling*. Environment Department Division Working Paper No. 1992-35, Policy Research Division, The World Bank, Washington, DC.

III Livelihoods and Food Security

8 Food Security Through the Livelihoods Lens: an Integrative Approach

JULIETA R. ROA

PhilRootcrops, Leyte State University, 6521-A, ViSCA, Baybay, Leyte, Philippines; fax: 0063-53-335-2616; e-mail: nell_roa@yahoo.com

Introduction

This chapter presents the synthesis of findings of a study that explored an integrative assessment of food security in less-favoured areas in the Philippines. It investigated the linkages between the resource environments (i.e. biophysical, socio-economic), livelihoods and food security of households and individuals using a livelihood systems framework, where the biophysical environment is the entry point of analysis (see Fig. 8.1).

Food security has been defined as 'access by all people at all times to enough food for an active, healthy life' (Maxwell, 1990, p. 3; World Bank, 1996); or, a situation 'when all people, at all times, have physical and economic access to sufficient, safe and nutritious food to meet their dietary needs and food preferences for an active and healthy life' (http://www.fao.org). More explicitly, Maxwell and Frankenberger described 'enough food' as adequate nutrients needed for an active healthy life, and not just simple survival. In order to be food secure, households must have enough resources to earn a living to have access to adequate food. The three composite elements of food security – availability of food, the ability to acquire or be able to afford it, and adequacy of food for a healthy life – are, thus, made explicit (Frankenberger, 1992).

This three-dimensional perspective is the backbone of the framework and methodology used in the study. The study of food security in fragile areas (i.e. easily degraded lands) in one of the most impoverished regions in the Philippines (i.e. Leyte, eastern Visayas region) is purposive because they represent the less-favoured areas. Such types could also be found in the uplands of the even more developed regions in the country.

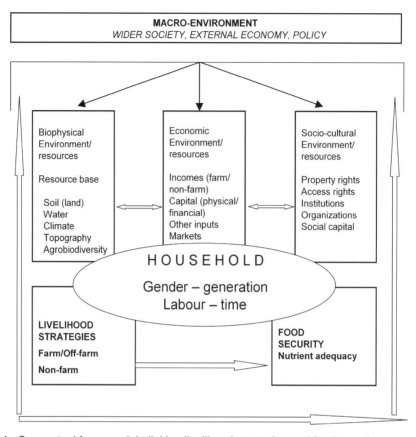

Fig. 8.1. Conceptual framework in linking livelihood strategies and food security.

Degradation of the Philippine uplands (i.e. rainfed areas from sloping to hilly and mountainous terrains) has long been reported as a situation of increasing seriousness and scope (Pelto, 1989; Cramb *et al.*, 2000). In these areas, farming is dominated by permanent or shifting cultivation of annuals (i.e. rainfed rice, maize, rootcrops, few vegetables, tree crops) with or without fallow (Cramb *et al.*, 2000). Here, the poorest of the poor live and till their farms, their labour being the most important resource. Their produce serves limited markets, and institutional failures isolate them from support services such as credit, technology and marketing infrastructure. They face increased nature and market risks, limited livelihood options and are food insecure in some or most parts of the year.

The intertwined challenges of food security, environmental degradation and poverty reduction need to be addressed simultaneously. The study addressed these using the following methods: (i) the livelihood systems perspective; (ii) an experimental attempt at an 'integrated linkage analysis' of the processes within households; and (iii) drawing from various literature in agro-ecology, socio-anthropology, economics, nutrition and food policy. A multidimensional perspective is critical because of the complex situation of the poorest farming households.

Research Area, Hypotheses and Methods

Research location

A number of in-depth, micro-level studies related to food consumption, nutrition and health in the Philippines have been conducted since the 1970s by the University of the Philippines (i.e. Laguna studies), the International Food Policy Research Institute (IFPRI) (e.g. in Ilocos, Southern Tagalog in Luzon; Aklan, western Visayas and Bukidnon, central Mindanao) and the Office of Population Studies, University of San Carlos (i.e. Cebu, central Visayas). Other than the national nutrition surveys of the Food and Nutrition Research Institute (FNRI), no in-depth studies were conducted in the less-favoured and poorest areas such as the Bicol, eastern Visayas, and Muslim Mindanao regions. Thus, this research was purposively carried out in the less favoured areas in eastern Visayas.

Research was conducted in two villages in Leyte, one of the provinces of eastern Visayas in central Philippines (see Fig. 8.2). The fourth largest region in terms of land area (21,432 km²), it comprises 7.1% of the country's total land area. Leyte and Samar are the two main islands, connected by the San Juanico bridge, the latter being part of the Maharlika highway that connects the major islands of Luzon and Mindanao. Leyte is the largest of six provinces (5,713 km², see Fig. 8.2, shaded area), is densely populated and grows at the rate of 1.61% annually – less than the national growth rate of 1.88%, partly due to out-migration. Despite relatively high literacy and labour force participation rates, real per capita income is less than the national average and poverty incidence higher, at 36%. The gross regional domestic product growth, human development index, life expectancy and functional literacy are all below national averages (see Table 8.1).

The research sites are two *barangays* (i.e. barrios) in two towns of Leyte: Alegre in the mid-eastern town of Dulag and Plaridel in the Midwestern town of Baybay, representing two major types of fragile landscape where the households and their farming systems are complex, diverse and risk-prone. Alegre consists of lowland plains prone to river flooding, waterlogging and river instability, and Plaridel is a coastal/upland area subjected to various types of erosion (see Table 8.2). Though officially classified as agricultural, farm incomes have dwindled over the past 10 years or so.

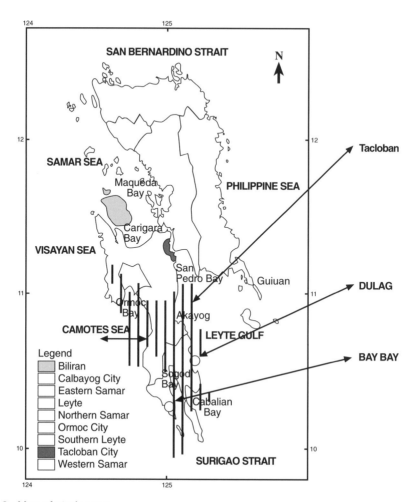

Fig. 8.2. Map of study area.

Hypothesis statements and pattern theory

The series of linked hypotheses specified below follow the mode of pattern theory (see Box 8.1). The latter is an explanatory theory that is less of a chain of cause and effect but more of a web of interrelated concepts and assumptions branching out in several directions. The interrelated hypotheses present a system of ideas that informs; the concepts and relations within them form a mutually reinforcing, closed system. Pattern theory specifies a sequence of phases, or link parts to a whole (Neuman, 1991).

As in any social theory, the hypotheses are systems of interconnected ideas that can be tested against empirical observations. They attempt to explain and provide understanding of the linkages of a set of biophysical and socio-economic endowments to food security. These are stated as

Table 8.1. Selected socio-economic indicators in Leyte, Region 8 and the Philippines, 2000–2004 (from National Statistics and Coordination Board and National Economic Development Authority).

Particulars	Philippines	Region 8	Leyte
Total land area (km^2)	298,170	21,431.7	5,712.8
Population (million, in 2000)	81.1 (2004)	3.61	1.59
Population density/km^2, in 2004	289	173	279
Population growth rate (%)	1.88	1.49	1.61
Literacy rate (simple, %)		90.86	90.6
Male/female	93.9 (1994)	89.2/ 92.7	88.9/ 92.3
Literary rate (functional, %)		79.7	79.5
Male/female	83.8 (1994)	75.7/ 84.2	74.1/ 85.1
Labour force participation rate (%)	66.7	66.5	66.6
Employment rate (%)	88.6	92.8	92.8
Real per capita income (Php[a], in 2000)	21,104	12,262	13,267
Poverty threshold per capita (Php, in 2000)	11,605	9,623	9,790
Poverty incidence (%, in 2003)	28.4	37.8	36.1
Urban (2000)	15.0	19.6	12.6
Rural (2000)	41.4	44.9	47.6
Gross domestic product; 1985 constant prices (million Php)	13,475	20,653.711	na[b]
Agriculture/fishing/forestry (%)		31.0	
Industry (%)		34.5	
Service (%)		34.5	
GRDP growth rate (%, in 2003)	6.4	5.2	na
Human Development Index (2000)	0.629	0.549	0.563
Life Expectancy (2000)	68.9	66.0	67.6
Functional literacy (%)	83.8	80.3	79.4

[a] Philippines unit of currency.
[b] na, data unavailable by province.

multivariate relations that are finally linked by theoretical association. The advantage of pattern theory is that it allows for a simplified, step-wise process of linking various parts of a complex system, a framework for subsets of analysis leading on to the next phase in the relational chain.

In this context, food security can be meaningfully assessed through a series of interlinkages starting from the conditioning factors of the bio-physical environment. The dynamics of the biophysical and socio-economic environments at the micro-level impact on the livelihood decisions of households and individuals, which eventually affect nutritional status. The livelihoods framework is key to this integrative assessment.

Blending of methods and interdisciplinary perspectives

A blend of qualitative and quantitative methods was used for reasons strongly recognized by scientists: the mix is more effective than using either singly. While quantitative analysis provides hard data, the qualitative tools

Table 8.2. Physical description of the village research sites (from Bureau of Soil and Water Management, Region 8 and the National Irrigation Administration. Source: Climatic data: 1961–1997 for rainfall, Tacloban; 1981–1990 for temperature; Baybay, 2000; Dulag, 1998; Saint-Girons, 2003).

Description category	Alegre	Plaridel
Location	Interior village *c*.5 km south-west of Dulag town proper; bounded by Daguitan river to the north and Talisay river to the south	Situated *c*.16 km south of Baybay town proper; bounded by Camotes Sea to the west, Kaipangan mountain range to the east and Inopacan town to the south
Climate	Dry and wet seasons not distinct; lowest precipitation between March and May, highest in December. Summer rainfall is not uncommon. North-east monsoon in November–February when heavy rains occur with probability of typhoons. Highest precipitation in December (329.7 mm), lowest in May (137.5 mm); Temperature is high and even through the year, between 22.6°C (February) and 31.4°C (May–June).	Rainy season, 9 months June–February with south-west monsoon from June–October when 50% of rain falls; north-east monsoon from November–December with high probability of typhoons bringing intense wind and water. Highest precipitation in November (355.2 mm), lowest in May (79.4 mm); dry season not so distinct, March–May Temperature is high and even through the year, between 23.8°C (February) and 31.2°C (April–May)
Topography	5 km from the sea in broad alluvial plain (0–3%), width > 500 m and elevation < 50 m above sea level	Upland sloping to the mountains: 5% plain (0–3%), 15% valley, 15% hilly (3–18%), 65% mountainous (> 18%)
Soils	Alluvial type formed by a series of erosion and sedimentation processes. Soil analysis of three case household farms shows low in Ca and Mg, 70% low in N, 57% low in OM and 30% low in P (2002). BSWM reports slow infiltration rate (0.1–0.7 cm/h), pH 5.1–6.9, low in N, adequate in P and deficient in K (1990s)	Parent material of soil is andesitic of volcanic material from dormant volcano (Mt. Sacripante, 6 km from the centre of Plaridel; still undergoing weathering). Hill soils mostly inceptisols, alfisols and ultisols (USA class); 1st two, still fertile, are earlier stages of transformation of the andesitic parent material; ultisols (aka Guimbalaon clay) the later stage and infertile. Lowlands: soils are of alluvial origin (aka Umingan clay loam); mountains: close to volcano, andisols, black, deep soil with high nutrient content; soft, vulnerable to erosion; commonly soils are sandy, loamy, silty loam, stony and clay loam
Erosion status	Flood waters cause erosion in farms along river banks; town map records severe erosion, especially on river banks	Ranging slight (sheet erosion and rills but < 3/m) to severe (gullied land with exposed subsoil or rock) inland

capture actual behaviour and produce information and insights that give meaning, help interpret and enhance the value of quantitative results from the questionnaire-based formal surveys (Pelto, 1989; Scrimshaw, 1990). The interphase between the local people's 'emic' view (i.e. meanings of culture by the informant) and the scientist's 'etic' view (as observed behaviour) were crucial in confirming the findings in understanding conditions and interrelationships, and in addressing opportunities or constraints with appropriate interventions or policies (Gittelsohn, 1992).

Box 8.1. The hypotheses.

Hypothesis 1
In less-favoured areas, farm productivity affects the decisions of households to diversify their livelihood portfolio.

COROLLARY 1A. Farm productivity is dependent on: (i) land quality (i.e. soil and water characteristics); (ii) techniques used (i.e. variety, capital, fertilizer use/other practices, inputs); (iii) individual and household specific characteristics (i.e. education, years of farming experience, resource base); (iv) seasonality; (v) labour (labour time of male and female workers); and (vi) land tenure.

COROLLARY 1B. Time allocation by gender (whether adult or young) on the farm depends on the time spent by them in activities in other livelihood pursuits and in the domestic sphere (i.e. household chores, reproduction and social maintenance).

Hypothesis 2
Livelihood strategies are determined by: (i) farm productivity; (ii) market opportunities; and (iii) household-specific factors (e.g. age, gender, education level of household head and spouse, dependency ratio, inherited assets, labour availability).

COROLLARY 2A. Livelihoods are said to be secure when households have secure ownership or access to assets and resources needed for: (i) income-earning activities to sustain life; (ii) meeting contingencies; and (iii) managing risks or ease shocks without depleting the resource-base.

COROLLARY 2B. The generation, allocation and use of assets and resources are gender specific.

Hypothesis 3
Nutrient adequacy (i.e. expressed in terms of a health production function) is dependent on: (i) dietary intake; (ii) morbidity; (iii) care; (iv) sanitation; (v) household- and individual-specific characteristics (education level of caregiver, level of nutrition knowledge of caregiver); (vi) livelihood type; and (vii) price of food.

COROLLARY 3A. Individual nutrient intake (as a demand function) is dependent on: (i) the physical activity level; (ii) individual- and household-specific characteristics (age, gender, equivalent household size, dependency ratio); (iii) intra-household distribution of food; (iv) education level of mother; (v) non-bought food; (vi) prices; and (vii) income.

The concepts, framework and methods were drawn from the following sources: (i) the works of social anthropologists and economists on the study of rural livelihoods, households, intra-household resource allocation and food distribution, and food security (Anderson *et al.*, 1994; Reardon, 1995; Bouis and Peña, 1997; Haddad *et al.*, 1997; Pender, 1999; Ellis, 2000; Niehof and Price, 2001; Rola *et al.*, 2002; Quisumbing, 2003); (ii) public health practitioners, nutrition economists and nutritionists (Martorell *et al.*, 1978; James and Schofeld, 1990; Berman *et al.*, 1994; FNRI, 2001; Svedberg, 2000); and (iii) resource economists, agro-ecologists and soil scientists (Scholz, 1986; Turner and Benjamin, 1994; Cramb *et al.*, 2000; Scherr, 2000).

The qualitative tools included focused group discussions (14 FGDs on food security, four on history of land use), case studies (12 case households), key informant interviews, direct observation and informal interviews. Two formal surveys (i.e. socio-economic profile and production; food consumption, nutrition and health) of a sample of 220 sample households (Alegre, $n = 100$; Plaridel, $n = 120$) provided data for quantitative analyses. The second survey on food and nutrient intake was conducted three times over 1 year to capture the normal, peak and lean food months. Multivariate analysis was used to analyse the factors affecting productivity, livelihoods and health. Actual fieldwork and part of the data processing occupied about 2 years (see Table 8.3).

Table 8.3. Summary schedule of methods and activities.

Methods	Activities	Schedule
Preliminaries	Proposal preparation	September–December 2001
	Site selection/local logistical arrangements	January 2002
	Proposal finalized and presented	
	Town and village protocols	January 2002
	Recruitment of field assistants and enumerators	July 2002
Review of literature and secondary data collection	Library and key research institutions data search	July–December 2002/ continuously up to 2004
	Key scientists' consultation	
Non-formal surveys and interviews	Community survey	August–November 2002
	Orientation and concept validation workshops	
	Key informant/other informal interviews	
	Local enterprise/associations interviews	
	Focused group discussions (FGDs):	
	Food security and livelihoods	September–December 2002
	Agro-ecology and land use changes	September–October 2002
Soil analyses	Soil sampling (field)	August–September, 2002 and
	Soil analyses (laboratory)	June–July 2003
Formal survey 1	Pre-testing/training of enumerators	July 2002
	Finalization/production of questionnaires	August 2002
	Conduct of survey (220 × 1)	August–November 2002
Formal survey 2	Pre-testing/training of enumerators	October 2002
	Finalization/production of questionnaires	November 2002
	Conduct of surveys (3 rounds: 220 × 3)	December 2002–November 2003
Focused micro-enterprise survey (Romblon crafts)	Formal survey of local romblon crafts	May–June 2003
	Informal survey of input producers	
Case study	Case studies of 12 households	January–August 2003
Data processing	Qualitative data	Up to December 2003
	Formal survey data	January 2003–August 2005
Research integration	Write-up and consultations	July 2004–June 2005

Synthesis of Results and Conclusions

Food security was assessed through the relationships of the resource environments (i.e. biophysical, socio-economic) with livelihoods. The decisions made by household members are partly influenced by the specific characteristics of households (e.g. size, dependency ratio) and socio-economic circumstances (e.g. wages, value of produce, working children). The dynamics of household decision-making vary (i.e. cooperative, authoritative, mixed), but hinge mostly on the household head and/or the household member who contributes substantially to the coffers. Food security, finally gauged by nutritional status and health, is the outcome of a series of interrelated relationships and decision-making processes along the production–consumption chain.

With biophysical and socio-economic constraints (see Table 8.4), the income of both villages is, on average, < US$1 per capita/day (PCI). With an average PCI among household respondents of only around US$0.90, the relatively 'better-off' village, Plaridel, had more households relying on non-farm income as it had the employment opportunities of the native craft industry, and in a bigger and more active town economy. Alegre's respondent households had an average PCI of around US$0.64 and had more households relying on farm income due to very limited non-farm employment opportunities in a smaller, less active, local economy (see Table 8.5). Most incomes were spent on food supplemented by own production, kin support and gathering from the wilds. Except for protein and niacin, sources of which were found in farms and gardens, all these measures were inadequate in providing households with adequate nutrition.

The discussion and conclusions are grouped into those that address the research questions and those that constitute the lessons learned from the approach. Importantly, these are related since the set of findings has been drawn from an integrative framework of livelihood systems, with the determining factors on crop productivity as the starting point of analysis. This was the logical first step, since farming is a basic activity of farming households, whether or not incomes or profits are gained. Findings showed that the characteristics related to fragile areas (i.e. land quality, erodibility) are basic considerations. Thus, land quality where farms are located and the resulting low crop yields condition livelihood decisions. The other resource environment factors did impact on the diversity of livelihood portfolios; hence the differences in incomes, assets and resources, wealth position, food consumption and, finally, nutritional status and health.

From survey data, the households were classified based on the relative importance of income sources: namely, farm, off-farm and non-farm. This led to the household classification labelled as livelihood types, estimated from the percentage of income by source. These are: livelihood type 1 (LIVETYP1) with about the same percentages from farm, off-farm and non-farm; livelihood type 2 (LIVETYP2), farm income

Table 8.4. Summary of the links between the hypotheses and the research questions.

Subject of hypotheses	Research questions addressed	Significant variables	
		Biophysical/ technology environment	Socio-economic
1 Crop productivity	• How does the situation in ecologically less-favoured areas affect the choice and generation of livelihoods?	Slope (−) Wet season (−) Cropping pattern (+) Fertilizer use (−) Irrigation (+) Variety (+) Loam soil (+)	Non-farm income (+) Sharing tenancy (+) Remittances (−) Livelihood type, non-farm, off-farm incomes (−) Production cost (+) Value of household labour (−) Selling price (−) Dependency ratio (+) Years in farming (+)
2 Livelihood type	• What does the mix of livelihoods households in less-favoured areas engage in, and why? • What are the assets and resources available to these households? What are the conditioning and constraining factors to their use? • How are the livelihoods, budget allocation and resource use differentiated according to age, gender and the household life course?	Idle land (+)	Farm wage worker (+) Household size (−) Size of area cultivated (−) Age of household head (+) Number of adult working children (+) Life cycle (−)
3 Nutrient adequacy Child's health: (0–5 years)	• What are the conditioning and constraining factors that produce differential effects on household food security and intra-household distribution of food and nutrients?	Not applicable	Drinking water (−) Short-term morbidity (+) Total farm full income (+) Primary education of caregiver (+) Livelihood type 3 (off-farm income) (+) Hygiene (+)** Livelihood type 4 (non-farm income) (+)** Sanitation (+)*
(5–12 years)			All (0–5 years variables plus: Elementary education, caregiver (+) Secondary education, caregiver (+) College education (+) Nutrition knowledge (+) Farm income (−)**

*Significant, *P* 0.01–0.50; **significant, *P* 0.50–1.0.

Table 8.5. Socio-demographic profile of households.

Characteristic	Alegre	Plaridel
Households (*n*)/population	180/*c*.900	738/*c*.3700
Sample size (*n* = 200)	91	109
Sex of household head, male (%)	96	89
Average household size	5	5
Average age of household head (years)	*c*.45	*c*. > 50
Age distribution (%)		
< 35 years	25.0	3.8
35–49	39.2	35.7
50 and above	35.8	60.5
Average total household income/year (US$)	1174.90	1638.73
Mean per capita income/day (US$)	0.64	0.90
Livelihood type (%)		
About same distribution	12.8	13.8
> 50% farm income	42.3	31.4
> 50% off-farm income	12.8	8.0
> 50% non-farm income	32.1	46.8

greater; livelihood type 3 (LIVETYP3), off-farm income greater; and livelihood type 4 (LIVETYP4), non-farm income > 50%.

Question-based results in the research

At the outset, the main research question is: 'How do the livelihoods of farming households impact on household food security in a situation of ecologically fragile environments?' Theoretically linked hypotheses were articulated in relation to the specific questions, then identified (see Table 8.4). The corollaries are related qualifiers to further address the research questions.

The research showed that the livelihoods of farming households in less-favoured areas are affected by the interplay of biophysical (e.g. idle land) and socio-economic factors (e.g. non-farm income, share tenancy, remittances, livelihood type, value of household labour). The livelihoods of households are characteristically plural because farming alone could not provide adequately for even their very basic needs. Poor land quality and fewer technology inputs, to start with, already negatively condition productivity resulting in low crop yields. This highlights the critical importance of land quality as the initial conditioning or constraining factor, and is addressed in hypothesis 1 covering the first question.

Different livelihood strategies result in a certain mix of livelihoods or livelihood types (i.e. predominantly farm, off-farm or non-farm) because of differences in resources, opportunities and household characteristics, which help shape the comparative advantages of households. Furthermore, the role of non-farm income and remittances as capital sources, availability of labour (from off-farm work opportunities), security of use rights, production cost, dependency ratio, wages and farming experience all contribute to crop performance, but their impact varies with the crop – whether cash or food – because of their differences in relative importance that affect the way resources (especially labour) are allocated (see hypothesis 1, variables). There were no gender biases found (i.e. in the sense that male or female can perform tasks as the situation permits with no social stigma), while gender assignation of roles and responsibilities exists because of tradition or sociocultural reasons.

Regression analysis identified significant variables linking the first two hypotheses, and clearly reflected the constraints posed by poor land quality of fragile areas and those of socio-economic factors via the allocation and use of resources (i.e. labour, capital, land). For example, the analysis found that idle lands are highly significant and positively related to non-farm, income-based livelihood; and lands tended to be idle when they were degraded (usually on the upper slopes or hillsides), and with inadequate labour (i.e. when used in off-farm or non-farm work) or lack of capital (due to low farm income, other income deficiencies).

The β-values indicated the relative importance of identified biophysical and socio-economic factors, which differed with various crops and livelihood types. These factors helped define the structure of access, use and control of the resources of production, thus determining the benefits or disadvantages to farming households. Those households in the lower socio-economic strata would have little access, use and control, and constitute the impoverished sector. Most households are in this category. This could become a vicious cycle, and bring about poverty and food insecurity.

In the process, the indicators of food security could be gleaned from: (i) the farm and food systems data and crop productivity (i.e. availability); (ii) the livelihood portfolio, income levels and resource generation and control (i.e. access); and (iii) nutritional status of households and individuals (i.e. adequacy).

In hypothesis 1, the regression results showed that low yields of crops arise from poor land quality, inadequate capital resources and lower labour intensity (i.e. due to competing labour use in other crops, other work) resulting in low farm incomes. Thus, households do engage in multi-livelihoods to increase their income. It is precisely because households diversify their labour resources to different uses that there is less intensity of labour use in crop production.

Thus, a vicious cycle of poor land quality, low productivity, low income, low farm labour intensity and low productivity results. To start with, poor land quality is already a constraining factor, and farming becomes a precarious livelihood. Even earnings from the external market-oriented cash crops – such as abaca, rice and coconut – could not generate enough capital for further farm investment. Only the households who received remittances and non-farm incomes could generate financial capital to invest in farms. Yet sustainability depends on regularity and stability of receipts, which often were lacking.

Ethnographic data on poor land quality in the research sites showed degeneration clearly resulted from a mix of man-made and natural causes. There was population pressure starting in the 1960s that led to increasing encroachment into the forest and use of marginal lands. Commercial ventures (e.g. logging, quarrying) of the wealthy and powerful led to the deforestation of the watershed. Crop-shifting cultivation systems by small farmers – such as the continuous growing of annuals in sloping lands with shorter fallow periods without the benefit of sustained conservation measures – resulted in erosion, especially in the hilly lands. Gullying and mass-wasting in Plaridel, and river erosion in Alegre, resulted because of strong wind and water from typhoons that struck the areas from the 1950s to the early 1990s. These latter types of degradation could occur even in the absence of human disturbance (El-Swaify, 1997).

Pender cited several studies on the association of different farm or crops systems with different agro-ecologies, physical and institutional infrastructure and market opportunities (Pender, 2004). Just as in these studies, relatively high-value cash crops such as abaca and rice are grown in areas with better agro-ecologies. Both villages had good market access and infrastructure. Off-farm opportunities in cash crops provided about 10% of income, on average. Non-farm employment is higher in a community linked more to the industrial market and to a more vibrant local economy (i.e. Plaridel versus Alegre), where there are more rural non-farm enterprises (see Table 8.5).

With respect to hypothesis 2, it was found that factors differ in significance and relative importance across livelihood types. The strategies and the resulting livelihood portfolio of households varied because of differences in the biophysical and socio-economic environ-ments, relative importance of crops, non-farm opportunities, wage employment, institutional support and from the resulting dominant livelihood type *per se* (i.e. via resource generation and human capital investment).

Thus, the two villages differed as to predominant livelihood type: Plaridel is more typical of LIVETYP4, where about 47% of households have greater non-farm incomes, and Alegre is more typical of LIVETYP2, where about 42% of households receive more from farm incomes. With low farm productivity, Alegre had lower PCI and was more food insecure (see Table 8.5). For households relying mainly on farming, the regressions showed the significant factors to be farm wage work (–), age of household head (+) and value of farm produce (+). For those living mainly on off-farm income, the significant factors are farm wage work (+), household size (–) and size of area cultivated (–). For those classified as mainly non-farm, idle land (pushed by land degradation) is a highly significant and important factor that motivated people to seek livelihood away from the farm.

Pockets of opportunities existed, such as cash crop production (i.e. abaca, coconut, irrigated rice) or non-farm micro-enterprises with externally oriented markets (e.g. factory- and household-based native crafts). To many, survival means engaging in both cash and food crop farming, whether own farm or wage work, plus engaging in other available work. Non-farm employment was associated with the growth of the indigenous local industry and service sectors, which in turn were facilitated by improved local entrepreneurship, better access to markets and marketing intermediaries, road infrastructure and institutional support systems (e.g. technology and design, financial support, trade promotion). The latter made Plaridel a relatively better off village in terms of socio-economic status of households. In the local mats and bags industry, women gained more access to income-earning opportunities and had money to spend on home, food and children's needs.

Certain livelihoods and the associated major resource, as labour, are differentiated by gender. Cash crop farming and selling, specialized farming (vegetables), heavy farm operations such as land preparation, off-farm employment and non-farm work in carpentry, construction and fishing are male domains; domestic services (paid or unpaid), household-based micro-enterprises such as native craft, retail store, home gardens, light farm work – planting or weeding – and vegetable trading are those of the females. All others may be male or female.

Labour migration for greener pastures, either seasonal or long-term, by male (main in first wave) or female (increasing in current wave) has gained in importance from its starting point in the 1970s, triggered by the insurgency-based decline in the rural economy, and further sustained by the very slow economic recovery. While remittances have helped build up capital resources and assets in some households, labour migration had a negative impact on labour resources both for farm and off-farm work, especially with the cash crops which are major income earners in the villages. Labour is a highly significant and important variable in household livelihood portfolio decisions, both in terms of supply (e.g. number of working children, household size) and as constraining factor (e.g. engagement in off-farm activities, life course).

The age of household head was found to be significant only with respect to mainly farming households (LIVETYP2), since the younger generation tended to seek non-farm employment or migrate, while the older generations more oriented to farming remained in the sector. Generally, the household life course was significant because livelihood decisions are affected by household resources, the dependency ratio and the available workforce.

Assets and resources generation tended to be better among the mainly non-farm household type, and usually from the remittances of members working outside the village, especially those married or working abroad. The mainly farm-based households were observed to have the least welfare and to be more food insecure due to generally low crop yields and to low production from poor land quality and erratic climatic conditions.

In hypothesis 3, the health of children – using weight-for-age as indicator – is used as an indicator of nutrient adequacy. Highly significant factors affecting the health of children, by order of importance, include: (i) the source of drinking water (most important for those 0–5 years); (ii) education level of caregiver (most important for those 5–12 years); (iii) short-term morbidity; (iv) off-farm income (only for those 0–5 years); and (v) nutrition knowledge (only for those 5–12 years).

Non-farm income, hygiene and sanitation were found to be significant factors. Farm income was found to have a positive relation with the health of those 0–5 years, and negative for those 5–12 years, implying that the nutrient needs of children could not be met by reliance on farm incomes alone; hence the significance of off- and non-farm incomes. Although education level and nutrient knowledge of caregiver mattered significantly in children's health, these could not fully impact on nutrient adequacy and better health, basically because of income constraints. Even own food production, such as from gardens and collection from the wilds, and food donations, were inadequate.

For children < 5 years, only protein and niacin were adequately met; and only niacin was met for those 5–12 years old. On average, for all children, energy could only be about 50% adequate; vitamin A, about 40%; vitamin C, 70%; iron, 55%; calcium, 72%; thiamine, 41%; and riboflavin, 64%. On the whole, children < 5 years of age were only slightly better off in nutrient adequacy. There were very few cases of chronic morbidity, yet even short-term morbidity already considerably affected the health of children because they were already weakened due to inadequacy of most nutrients.

In summary, nutrient inadequacy and poor health (the final indicator of food security) of households and individuals factored in low income levels despite multi-livelihoods due to the mixture of biophysical and socio-economic constraints detailed in the discussions above. Even the family support system could not adequately supply nutrient needs. The findings highlighted the diversity of interconnectedness of various factors affecting livelihoods and the initial conditioning impact of the biophysical environment. To address food security in fragile areas, the niggardliness of land is a critical policy concern to start with, especially in the knowledge that the poor in these areas are the most vulnerable. On the whole, food may be available even with less own production because the villages are linked to the food markets. However, the issue of access or entitlement is the more critical in order to achieve adequacy in nutrients and health. The access issue is basically the livelihoods issue: how to create or improve livelihood opportunities and the enabling environments (i.e. policy, institutional support) to capacitate and empower the poor.

In this study, the significant variables identified in the regression analyses involving farm productivity, livelihoods and nutritional status – and the contextual understanding of the complex situation surrounding the resource environments and households – have led to a number of policy insights that are discussed below.

Lessons learned from the approach and methodologies

The application of the pattern theory in the stepwise articulation of relationships in crop productivity, livelihood type and, finally, nutrient status has been made feasible by using the livelihoods framework. The latter made it possible to place the relationships in the context of the biophysical environment, thus necessarily starting with crop productivity relationships. This framework enabled the identification of significant and important factors in the biophysical and socio-economic environments as influencing household livelihood decisions resulting in different livelihood combinations. Using livelihoods as the convergent concept enabled the focus on households as the unit of analysis in making livelihood decisions.

At the same time, the view on livelihoods enabled the backward links to farming productivity; hence, the critical role of the resource environments, commencing with the biophysical environment as conditioning factor. Various livelihood types result in different levels of incomes, resource generation and allocation resulting in some level of socio-economic status of the household. This will then impact on food security in terms of food availability, access and adequacy. In effect, the livelihood approach proved useful in identifying the significant linkage variables among the three sets of relations, as well as their relative importance.

The major use of both quantitative and qualitative tools in data collection and analyses in such a complex set of relationships has been, without doubt, useful: one without the other would have generated less meaningful and less significant findings. Multivariate analyses (MVA) and grounded theory together resulted in outputs where validity and reliability reinforced each other. Qualitative information gathered from the FGDs, case studies and other informal tools provided meaning to the resulting variable coefficients.

Using two sets of interview schedules for the focused formal surveys (i.e. production and socio-economics of households, food consumption and nutrition) provided the minimum data set required for the comprehensive analyses of food security as exemplified here. They correspond to the Family Income and Expenditure Survey (FIES) and the National Nutrition Survey (NNS) that were already undertaken in the Philippines, except that the instruments used here differ in sampling design and period of administration from both the FIES and NNS. Thus, the linkage analyses in this thesis could not be performed with the current set-up. This implies that the analyses performed here could be applied with some minimal modification of the FIES instrument (e.g. stratify for area agroecology, add farming component) and the matching of the period of administration with the NNS. This means that, to apply the integrated approach to food security assessment in the Philippine context does not need another set of national surveys but utilization of the existing system with the suggested modifications. Location-specific programmes or projects can therefore use these two data sets (instead of another survey usually carried out with projects) and enrich these with information from qualitative tools in relevant areas. Thus, it is recommended that the conduct of the FIES and the NNS be rationalized.

Implications for policy, research and development

It was not the intention that this study produce a blueprint for policy action, but it was intended to inform it. Importantly, it is hoped that the results will contribute to some rethinking of 'strategic notions' of what makes a difference in the development process that can transform the farming households in less-favoured areas into a viable and vibrant sector composed of healthy people. Often, it is difficult to strike a balance between government responsibilities, necessary public action and scarce resources. Ideas and actions of opportunities to be seized abound, and are frequently missed for a host of reasons. The diversity and the complexity of circumstances among resource-poor farming households in less-favoured environments are simply too wieldy to deal with. Too many location-specific or household-unique cases mean that such that a blanket policy, frequently ineffective, could not apply.

To policy advocates, policy-makers and development workers alike, the orientation and direction could be that the challenges posed by the three components of food security (i.e. availability, access and adequacy) be pursued interdependently, with matching policies and programmes that address technologies or innovations, infrastructures, institutions and governance. Without these, improving rural livelihoods, alleviating poverty and food insecurity will be less likely (Lacy *et al.*, 2003).

Recognizing this critical role of livelihoods to reduce the incidence of poverty and food insecurity, Pender reiterated the argument echoed by a number of observers that 'The key to the development of sustainable rural livelihoods is investment in an appropriate and socially profitable mix of physical, human, natural, financial and social capital, taking into account the diversity of contexts in developing countries' (Pender, 2004, p. 340).

Policies and programmes that impact on the land quality of fragile areas and related technologies

These would require the following basics:

1. Intensifying the integration of environmental conservation and ecologically friendly land-use strategies in agriculture and land-use programmes, with priority given especially to fragile areas. This, however, should be grounded on a very basic requirement that, to date, is still inadequate: an updated comprehensive characterization of agro-ecosystems nationwide, including in particular an assessment of land quality properties versus land use. The first step could be the collation and synthesis of the fragmented available primary documentation of land use *vis-à-vis* the different agro-ecosystems from the different commodity-based research and extension institutions, based on or affiliated to state universities or ministries, in addition to those of the Bureau of Soils and Water Management (BSWM). These will need the collaborative efforts of the Department of Agriculture through the leadership of the BSWM, together with the regional/local research and extension units of strategic or zonal state universities and the local government units, and the Department of Environment and Natural Resources (DENR). This collaboration is geared towards a more cost-effective implementation of already existing programmes by these agencies.

2. Farm-specific land quality enhancement technologies should be appropriate, available and affordable, and could use indigenous materials and be sensitive to the socio-economic situation of resource-poor farmers in less-favoured areas. This is easier said than done, as experience has shown the non-adoption, for example, of bio-organic fertilizers in areas where they were tested or introduced. The economic incentive that encourages adoption is critical, which is related to the benefits or profitability of the crop or farm systems. Hence, land quality enhancement technologies should be developed (i.e. better with farmers themselves) and viewed in the context of the crop or farm systems to be applied, as well as matching with market opportunities. The triple criteria of appropriateness, availability and affordability of the technology or innovation require a mechanism that may be location specific because of the diversity of farm and farmer circumstances (i.e. especially that of mixed livelihood systems that intensively use household labour and zero capital). Scientists and researchers will need this type of sensitivity in technology generation and dissemination. The demands are both the 'hard' technology (e.g. soil fertility replenishment) and 'soft' technology (i.e. the 'how' of innovations, their delivery and use-markets matching).

3. In the case of degeneration of land quality in a watershed, this may call for a long-term, partly or fully subsidized programme of regeneration of the watershed, either community-based or of wider watershed involving various communities. Sourcing of funds will depend on the nature and interests of stakeholders, i.e. partly when a large, private company or interests are involved and mainly by government with external donors, for poor communities. This would involve the DENR, the BSWM and/or the NIA, as the case may be, the LGUs of influence communities and the relevant research institute or university.

Policies and programmes that impact on livelihoods and food security

1. A nationwide data management system that allows for the interlinked analyses of resource environments, livelihoods and food security to generate programme interventions and policies related to:

- Resource environments (i.e. the biophysical and socio-economic).
- Livelihoods (i.e. farm, off-farm, non-farm or other categories) in relation to their respective challenges, constraints, opportunities and growth areas by livelihood type.
- Food availability, consumption, nutritional status and health.

2. The current Family Income and Expenditure Survey (FIES) conducted by the National Statistics and Coordination Board (NSCB) and the National Nutrition Survey (NNS) by the Food and Nutrition Research Institute (FNRI) already include the variables (as applied in this thesis) needed for the comprehensive analyses of food security, including the linkage relationships from production to consumption and nutrient intake. However, under the current set-up, the sampling designs and periods of administration of the two surveys are different and, thus, could not be used properly for such type of analyses. To do so requires the rationalization and coordination of their sampling designs and implementation by both national agencies. A review of the current design, implementing rules and guidelines in the conduct of these surveys, including some modification of the rural/farming sector (e.g. the inclusion of a variable that allows for agro-ecology stratification and land quality specification), may be in order. Coordinating with the Department of Agriculture is another possibility, especially in relation to agro-ecology and land-use characterization.

3. One serious review of the agrarian reform programme based on an overall cost–benefit evaluation that includes the social aspects; and findings on incentives, resource use and livelihood opportunities of agrarian reform beneficiaries. It has been suggested that an agrarian reform based on the ideology of 'land to the tiller' or land redistribution needs serious rethinking for the obvious reason that, despite some reasonably good performance in land distribution, the performance in terms of agricultural productivity has not been bright. Philippine farms are the least productive in Asia, especially so in less-favoured areas. This study found that land ownership *per se* is not necessarily a critical factor in productivity, and tenancy was not found to be a deterrent to crop productivity. Instead, security of land use rights, commonly family land use rights and handed-down, family-based tenancies, were found to provide enough motivation for productivity. In addition, respect for the indigenous system of property rights was found to be more facilitative of productivity or rural development programmes rather than was the introduction of another legal order. The agrarian reform programme needs some return to the basics, and some rightsizing of ministries for cost-effectiveness, better coordination and better integration with related components such as land suitability and farm systems, utilization and markets. Instead of having a separate department, the agrarian reform programme could be integrated with the Department of Agriculture.

4. Optimize the implementation of Republic Act No. 8435, or the Agriculture Fisheries and Modernization Act (AFMA). This legislation already defines the concept of food security (i.e. a novel departure from the long history of food self-sufficiency focus) and the goals and strategies identified to achieve the development of the fisheries and agriculture sectors for poverty reduction and improved quality of life. The law stressed that the food security goal be clearly understood, and not mistake it for food self-sufficiency. This definitely has implications for research, development and extension (RDE) in terms of prioritization and budget allocation.

One important research issue relevant to farming in fragile areas is that of cash and food crops (e.g. abaca or coconut + rootcrops, banana), and may well be expressed in the RDE budget prioritization. As this study has shown, cash and food crops are not necessarily competitive by themselves, or with other farming household activities. Two crops are technically complementary in a farming system, i.e. rootcrop or banana is intercropped or mixed with abaca in order to maximize labour use and the return to land. In RDE budget prioritization, the purely cash crop argument (i.e. comparative advantage, investible surplus) to generate incomes may not be realistic with farming households in less-favoured areas, who must also partly or fully produce for their own food in addition to other income-generating activities. This implies that RDE agriculture development programmes will need to adopt at least a systems perspective that recognizes interdependence, complementarity and synergy of different parts.

5. The need to coordinate the social services for education and health and population management education (especially nutrition education) by the Department of Health, Food and Nutrition Research Institute, the National Nutrition Council and the Departments of Agriculture, Education and Culture and Sports. There is also the need to utilize the role of public media and collaborate with non-government organizations (NGOs) for extensive and intensive health and nutrition education in rural areas, social marketing for indigenous nutrient-rich foods and intensive home gardening, agro-industry skills development, market promotion and primary health care. The potentials of tapping the overseas foreign workers (OFW) remittances to fund rural development programmes – particularly for women and out-of-school youth skills training, and certain deficiencies in existing health services – particularly water and sanitation, especially in rural areas – will need to be explored. These are critical factors that affect the health situation of farming households, especially the children.

6. Non-farm livelihoods are crucially important in the overall livelihood system of farming households. Non-farm income remittances were used as capital (e.g. seed, animal power, fishing boat, labour wage), as insurance for risk, for human capital investment (e.g. education of siblings) and for household maintenance (i.e. medical expenses, food sustenance). These were usually achieved through: (i) the exploitation of indigenous resources, capacities and market niches; (ii) appropriate technology and innovations; and (iii) support systems, such as market promotion and expansion, financial sources, standards and regulations, capacity for organization strengthening, entrepreneurship development, leadership and management of non-farm livelihoods (Haggblade *et al.*, 2002; Ruben and Pender, 2004). Interventions for the rural non-farm sector are in response to identified constraints in either the demand (e.g. absence or limited markets, low incomes, unattractive price) and/or the supply (e.g. unstable raw materials, product quality, technology, lack of capital) sides. Demand-side policies could include pro-agriculture policies, government purchase, trade policies, linkages and marketing assistance; supply-side policies include macro-economic policies that affect input cost and output prices, standards and regulation, zoning, rural infrastructure, credit supply and availability, technology, entrepreneurship and management assistance, extension, raw materials and skills training. These interventions will need to be tailored to the specific circumstances of different settings because their prospects, opportunities and constraints differ, as the research in these two villages has shown. Thus, the requirement is for an overall rural development that stresses the growth of farm, off-farm and non-farm components within a village or clusters of villages depending on the effective catchment or micro-region, as the case may be, and that considers the diverse influence of the biophysical and socio-economic factors. The diversity and complexity of the conditions of poor farming households in less-favoured areas may render this a formidable task. Models and approaches which worked in other areas of more or less similar circumstances could be tried, noting the principle of comparative advantage, systems-orientation, the role of institutions and relevant support systems, leadership and management capacities among stakeholders and sensitivity to the perspective of various actors.

Food security, in the tri-dimensional context elaborated here, should be an integral part of poverty assessment. This allows for a systems-oriented analysis of livelihoods and food security issues. There is a mixed coverage of food security issues in real practice. All those involved in the development process, both government and non-government, need to intensify capacity building in addressing food security concerns in poverty reduction programmes at local and national levels.

References

Anderson, M., Bechhofer, F. and Kendrick, S. (1994) Individual and household strategies. In: Anderson, M. *et al.* (eds) *The Social and Political Economy of the Household.* Oxford University Press, New York, pp. 19–68.

Berman, P., Kendall, C. and Bhattacharyya, K. (1994) The household production of health: integrating social science perspectives on micro-level health determinants. *Social Science Medicine* 38, 205–215.

Bouis, H.E. and Peña, C. (1997) Inequality in the intrafamily distribution of food: the dilemma of defining an individual's 'fair share'. In: Haddad, L., Hoddinott, J. and

Alderman, H. (eds) *Intrahousehold Resource Allocation in Developing Countries. Models, Methods and Policy*. IFPRI, The Johns Hopkins University Press, Baltimore, Maryland and London, pp. 179–193.

Cramb, R.A., Garcia, J.N.M., Gerrits, R.V. and Saguiguital, G.C. (2000) Conservation farming projects in the Philippine uplands: rhetoric and reality. *World Development* 28, 911–927.

Ellis, F. (2000) *Rural Livelihood Strategies in Developing Countries*. Oxford University Press, London.

El-Swaify, S. (1997) Factors affecting soil erosion hazards and conservation needs for tropical steeplands. *Soil Technology Journal* 11, 3–16.

FNRI (2001) *Philippine Nutrition. Facts and Figures*. Department of Science and Technology, Food and Nutrition Research Institute, Philippines, 109 pp.

Frankenberger, T.R. (1992) Indicators and Data Collection Methods for Assessing Household Food Security. College of Agriculture, University of Arizona, Tucson, Arizona.

Gittelsohn, J. (1992) Applying anthropological methods to intrahousehold resource allocation. In: *Proceedings of the Conference on Understanding how Resources are Allocated within Households*, IFPRI, Washington, DC.

Haddad, L., Hoddinott, J. and Alderman, H. (1997) Introduction: the scope of intrahousehold resource allocation issues. In: Haddad, L., Hoddinott, J. and Alderman, H. (eds) *Intrahousehold Resource Allocation in Developing Countries. Models, Methods and Policy*. IFPRI, The Johns Hopkins University Press, Baltimore and London, pp. 1–16.

Haggblade, S., Hazell, P. and Reardon, T. (2002) *Strategies for Stimulating Poverty-alleviating Growth in the Rural Nonfarm Economy in Developing Countries*. EPTD Discussion Paper No. 92, Environment and Production Technology Division, International Food Policy Research Institute (IFPRI) and Rural Development Department, The World Bank, Washington, DC, 111 pp.

James, W. and Schofeld, E. (1990) *Human Energy Requirements: a Manual for Planners and Nutritionists*. Oxford University Press, New York.

Lacy, W.B., Lacy, L.R. and Hansen, D.O. (2003) Global food security, environmental sustainability and poverty alleviation: complementary or contradictory goals? In: Lal, R., Hansen, D., Uphoff, N. and Slack, S. (eds) *Food Security and Environmental Quality in the Developing World*. Lewis Publishers, London.

Martorell, R., Jean-Pierre, H. and Klein, R. (1978) Anthropometric indicators of change in nutritional status in malnourished populations. In: Underwood, B. (ed.) *Methodologies for Human Population Studies in Nutrition Related to Health*. National Institute of Health, Maryland, 195 pp.

Maxwell, S. (1990) Food security in developing countries: issues and options for the 1990s. *IDS Bulletin* 21, 2–13.

Neuman, L. (1991) *Social Research Methods. Qualitative and Quantitative Approaches*. Allyn and Bacon, Needham Heights, Massachusetts.

Niehof, A. and Price, L. (2001) *Rural Livelihood Systems: a Conceptual Framework*. Wageningen-UPWARD Series on rural Livelihoods No. 1, WU-UPWARD, Wageningen, The Netherlands.

Pelto, G.H. (1989) Introduction: methodological directions in nutritional anthropology. In: Pelto, G.H. *et al.* (eds) *Research Methods in Nutritional Anthropology*. United Nations University Press, Tokyo, 8 pp.

Pender, J. (1999) *Rural Population Growth, Agricultural Change and Natural Resource Management in Developing Countries: a Review of Hypotheses and some Evidence from Honduras*. EPTD Discussion Paper No. 86, Environment and Production Technology Division, Institute of Food Policy Research Institute (IFPRI), Washington, DC, 86 pp.

Pender, J. (2004) Development pathways for hillsides and highlands: some lessons from Central America and East Africa. *Food Policy* 29, 339–368.

Quisumbing, A.R. (ed.) (2003) *Household Decisions, Gender, and Development: a Synthesis of Recent Research*. International Food Policy Research Institute (IFPRI), Washington, DC, 274 pp.

Reardon, T. (1995) Using evidence of household income diversification to inform study of rural non-farm labour market in Africa. *World Development* 23, 1495–1506.

Rola, A., Paunlagui, M. and Elazegui, D. (2002) *Household Food Security: Concept and Evidence from Laguna, Philippines Villages*. Institute of Strategic Planning and Policy Studies, College of Public Affairs, University of the Philippines, Los Baños, Laguna, Phillipines, pp. 1–22.

Ruben, R. and Pender, J. (2004) Rural diversity and heterogeneity in less-favored areas: the quest for policy targeting. *Food Policy* 29, 303–320.

Scherr, S.J. (2000) A downward spiral? Research evidence on the relationship between poverty and natural resource degradation. *Food Policy* 25, 479–498.

Scholz, U. (1986) Regional overview of environmental and socio-economic aspects of tropical deforestation in the Asian and Pacific region. In: *Environmental and Socio-economic Aspects of Tropical Deforestation in Asia and the Pacific*. Economic and Social Commission for Asia and the Pacific, Bangkok, Thailand, pp. 5–24.

Scrimshaw, S. (1990) Combining quantitative and qualitative methods in the study of intra-household resource allocation. In: Rogers, B.L. and Schlossman, N.P. (eds) *Intra-household Resource Allocation: Issues and Methods for Development Policy and Planning (papers prepared for the Workshop on Methods of Measuring Intra-house-hold Resource Allocation*, Gloucester, Massachusetts, October 1983. United Nations University Press, Tokyo, pp. 86–98.

Svedberg, P. (2000) *Poverty and Undernutrition: Theory, Measurement and Policy*. Oxford University Press, New York, 348 pp.

Turner, B. and Benjamin, P. (1994) Fragile lands: identification and use for agriculture. In: Ruttan, V. (ed.) *Agriculture, Environment, and Health: Sustainable Development in the 21st Century*. University of Minnesota Press, Minneapolis, Minnesota and London, pp. 104–145.

World Bank (1996) *A Strategy to Fight Poverty*. ESCAP Region Operations Division, World Bank, Washington, DC.

9 Changing Gender Roles in Household Food Security and Rural Livelihoods in Bangladesh

AHMED ALI[1] AND ANKE NIEHOF[2,*]

[1]Bangladesh Rural Advancement Commission (BRAC), Dhaka, Bangladesh;
[2]Sociology of Consumers and Households Group, Department of Social Sciences, Wageningen University, The Netherlands and Research Centre, PO Box 8060, 6700 DA Wageningen, The Netherlands; tel: +31-317-482622; fax: +31-317-482593; *e-mail: anke.niehof@wur.nl

Abstract

This chapter addresses the changing gender roles in household food and livelihood security in a flood-prone area in Bangladesh. It looks at the contribution of women to household food security and women's use of social capital for sustaining their livelihood. IFPRI panel data collected before and after the devastating floods of 1998 were used, as well as the results of fieldwork carried out in the same area during 2001–2003. Taken together, the data show the significant and changing role of gender in rural livelihood and food security.

It was found that, at household level, gender inequality still results in the relatively poor nutrition of women in spite of their crucial role in the food system. In particular, young women and widowed or divorced women without sons are now taking more responsibility for sustaining their livelihoods by engaging in economic activities. Female literacy and educational attainment are on the rise. Kinship and marital relations are changing, though not always to the advantage of women, and dowry is increasingly becoming a social problem. It is concluded that, though social change has not obliterated gender disparities, the gender gap seems to be narrowing.

Introduction

This chapter presents part of the results of a PhD research project on livelihood and food security in a fragile area in Bangladesh (cf. Ali,

2005). It focuses on gender roles in household livelihood and food security in a situation of change. To capture the changes, an analysis was performed of quantitative panel data that compare household livelihood and food security indicators before and after the devastating floods of 1998, and fieldwork was carried out during 2001–2003 that included the use of several qualitative research methods and retrospective questioning. Altogether the different types of data, collected at different times and referring to different periods, reveal changes and continuities in gender roles in the constraints women are facing and the social capital they use to overcome those.

There is a fundamental link between gender and food security (IFAD, 2001; Niehof, 2003). This applies especially to societies like Bangladesh where considerable gender disparities, a culture of male domination and an unequal intra-household distribution of food prevail, and where men have greater control over material resources, knowledge and ideology than women (Cain *et al.*, 1979; Palriwala, 1990). Their control is rooted in social, economic and cultural systems (Sen, 1994). Women play a crucial role in both food production and consumption and are the gatekeepers to family nutrition. At the same time, due to gender discrimination, they tend to take less and less of the nutritious foods than would be sufficient to satisfy their calorie and nutritional requirements (HKI, 1994).

Bangladesh is a patriarchal society, based on patrilineal kinship and virilocal residence. Male supremacy is fostered through a strict division of labour (Jahan, 1995), and men mediate women's access to resources and social networks (Kabeer, 1994). Since the 1980s, women have been increasingly involved in livelihood activities as a result of the changing social context and NGO initiatives, and studies conducted on gender status in Bangladesh are now revealing women's role in livelihood generation. Though women's economic activities still largely remain unrecognized (Paris, 2004), their contribution to household food and livelihood security is becoming more visible. At the same time, the sociocultural context is changing; the educational gap between girls and boy is narrowing. While the percentage of illiterate males (> 15 years old) is 50%, and that of females of the same ages 69%, primary school enrolment of girls now surpasses that of boys, though still more boys than girls are enrolled in secondary education (UNFPA, 2005, p. 108).

Against the backdrop of the changing macro-context, this chapter focuses on gender roles in household food security and vulnerability. It looks specifically at the change and continuity of gender roles in the micro-context of rural livelihoods and family relations, and at how women use social capital to feed their families and sustain their livelihoods. The study area is Manikganj district, which is located at a distance of about 65 km north-east of Dhaka. It is a fragile area because it is prone to flooding.

Conceptual Framework

The conceptual framework draws on existing theoretical frameworks for livelihood analysis (Chambers, 1989; Ellis, 2000; Niehof, 2004). It captures the relationships between the core concepts in this chapter, while gender and time are cross-cutting themes. Household food security is seen as part of household livelihood security.

Definition of concepts

First, the major concepts involved will be briefly discussed. These are: livelihood, household food security, vulnerability, social capital, gender and the temporal perspective.

Livelihood

Livelihood is about what people do for a living and what they gain by it. The concept refers to activities, the resources and assets needed for carrying them out and the outcomes resulting from these activities. Resources and assets can be material or tangible in nature, as well as intangible. Ellis (2000, p. 10) defines livelihood as follows: 'A livelihood comprises the assets (natural, physical, human, financial and social capital), the activities, and the access to these (mediated by institutions and social relations) that together determine the living gained by the individual or household.' The key elements of livelihood – activities, assets and access – are affected by gender relations and gender roles, which determine access and control by women.

Household food security

Food security is seen as access by all people at all times to the food needed for an active and healthy life (Maxwell and Frankenberger, 1992). At the household level, food security refers to the ability of households to procure adequate food for meeting the dietary needs of their members. This research conceptualizes household food security as the availability of and access to food of the household and adequate food consumption by all household members to maintain their nutritional adequacy. In the analysis of the quantitative panel data, non-staple calorie intake was taken as indicator of household and individual food security, since this best measures actual food security status (Hoddinott and Yohannes, 2002).

The research used the household as the unit of analysis. Rudie (1995, p. 228) sees the household as: 'a co-residential unit, usually family based in some way, which takes care of resource management and the primary needs of its members.' This definition is considered appropriate because of its emphasis on resource management and provision for primary needs. Although the household was used as the unit of analysis, intra-household food distribution and individual food security were investigated as well.

Food security is dependent upon livelihood status. Well-off households with secure livelihoods will also be food secure. Poor households with vulnerable livelihoods will struggle to be food secure and are unable to create a surplus and strengthen their asset base. Most households in the research area have problems in achieving food security and protecting their asset base, especially in a situation of adversity (ill health) and disaster (floods).

Vulnerability

Vulnerability is the opposite of security. When a livelihood is insecure we call it vulnerable. Vulnerability has been defined in different ways in the literature (Chambers, 1989; Ellis, 2000). It has an external and internal dimension: the first refers to the fragility of the external environment of households and shocks that may originate from it (such as floods and war), the second to individual adversity and lack of means (death of a spouse, poverty). The degree of vulnerability reflects the extent to which households are capable of dealing with shock and adversity while maintaining their asset base (Chambers, 1989; Niehof, 2004). This research focuses on food vulnerability as part of the vulnerability context of households.

Social capital

Social capital has different dimensions and comprises both horizontal and vertical linkages (Putnam, 1995; Bastelaer, 2000). To Woolcock and Narayan (2000): 'It's not what you know, it's who you know.' In addition, social capital refers to the norms and networks that enable people to act collectively. 'Social capital refers to features of social organization such as networks, norms, and social trust that facilitate coordination and cooperation for mutual benefit' (Putnam, 1995, p. 67).

A crucial issue in livelihood is people's ability to cope with vulnerability and stress. One way of coping is the use of social capital: 'Social strategies, i.e. the sharing of resources within the community through family or clan, is also a common practice of coping with food deficit' (Campbell, 1991, p. 145). Social capital is important for accessing other types of capital (Niehof, 2004). Though not available in cash, social capital is sometimes more influential than other forms of capital, especially for a household at risk.

In this chapter we look at the role of the different forms of social capital in sustaining livelihood and achieving food security. In the quantitative analysis of the IFPRI panel data, credit taken without collateral was taken as the proxy indicator for social capital (Ali, 2005).

Gender

Gender refers to the socially defined roles of and relationships between men and women in society. Actual gender definitions, norms and notions vary according to the sociocultural context and change through time. At the household level, the division of labour is gendered. The fact that Bangladesh is a male-dominated society influences decision making at the household level and women's control over and access to resources, including access to social capital (Molyneux, 2002). This ultimately affects women's food security and nutritional adequacy. Women and girls, more often than men, suffer from food insecurity and nutrition deficiency. In this chapter, we will look at issues relating to livelihood, food security and nutrition from a gender perspective.

Conceptual model

Figure 9.1 pictures the relationships between the concepts discussed above. Although, strictly speaking, social capital is part of the asset base of individuals and households, to emphasize its importance we put it in a separate box. Social capital is also gendered; the forms of social capital of which one can avail oneself and the ways in which they are used, differ between men and women.

Temporal perspective

Time is always a crucial factor in livelihoods, because livelihood strategies are developed in anticipation of future developments and because the livelihood portfolio of activities and assets is developed within a certain historical context. In this chapter we use a temporal perspective by looking at gender roles in relation to household food, livelihood security and nutrition and the changes therein during the past 20 years.

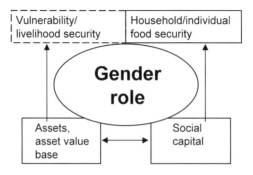

Fig. 9.1. Conceptual model of the key concepts of livelihood analysis.

Research Methodology

In the research project on which this chapter is based, both quantitative and qualitative data were used. The quantitative data set was derived from survey data collected by IFPRI in the sub-district of Saturia, district of Manikganj. Several rounds of surveys were conducted before the floods of 1998 (during 1996–1997) and after the 1998 floods (during 1999–2000) among the same households. For more information on the IFPRI sampling frame see Bouis (2000). For the quantitative analysis the sample consisted of the 307 households, of which data from all rounds were available (Ali, 2005).

In the same area, complementary fieldwork was carried out during 2001–2003. A survey was conducted to fill some gaps in the data set. Eight focus group discussions were held, six with women and two with men, in which the following topics were covered: (i) kinship; (ii) food and livelihood security; (iii) nutrition and diet; (iv) gender and women's empowerment; and (v) health and family planning. Six life histories were collected, three from men and three from women. The case study method was applied to 17 cases that represented different livelihood situations in terms of livelihood activities and vulnerability (cf. Ali, 2005). The findings presented in this chapter are based mainly on this fieldwork. Before presenting them, we will first briefly review some of the results of the panel data analysis.

Selected Results of the Panel Data Analysis

According to the panel data, female household members had both before and after the floods significantly less non-staple calorie intake ($P < 0.01$) than male household members. Whether the household head is male or female does not appear to have an influence on household food security (Ali, 2005, pp. 173–174). Food vulnerability proved to be significantly influenced by social capital both before and after the floods (Ali, 2005, p. 99).

The 1998 floods affected the livelihoods in the area in several ways. The data show shifts in livelihood portfolios when comparing the situation before and after the floods. After the floods, households diversified their livelihood portfolios to include various minor sources of income. Fish production declined because fish ponds had been destroyed by the floods. The supply of credit without collateral (social capital) increased significantly during the post-flood period, due to the NGOs' policies of supplying more credit to the flood-affected people and increased informal borrowing from moneylenders, relatives, neighbours and friends. Collateral-based credit also increased after the floods. From the data it appears as though the households tried to retain their asset base and diversified their sources of income. Poor nutrient intake among adults and schoolchildren was observed during the post-flood period, with the nutrient intake of women and girls being the worst (Ali, 2005, pp. 100, 176–179).

Results of the 2001–2003 Fieldwork

Changes in women's livelihood

Though particularly women from landless and female-headed households are economically active and richer women less so, Table 9.1 shows a general trend of a broadening range of economic activities of women during the past 20 years.

Paddy husking, boiling paddy and selling the processed product at the local market were once an important source of income for women, but the introduction of rice mills and mobile paddy husking units has put a stop to this.

Table 9.1. Women's livelihood activities (from Ali, 2005, p. 126).

Livelihood activity	In the past	At present
Harvesting commercial vegetables	Absent	Common
Embroidering and stitching garments	Absent	Common
Factory work (garments, tobacco, textiles)	Absent	Common
Working in NGOs	Absent	Common
Regular private jobs	Absent	Common
Full-/part-time jobs in NGOs	Absent	Common
Small-scale commercial poultry rearing	Absent	Common
Cocoon rearing	Absent	Common
Vending foods at the markets	Absent	Common
Traditional cattle rearing	Less common	Common
Earthwork labour sale	Rare	Common
Regular government service	Rare	Common
Sowing seedlings and weeding (rice fields)	Absent	Increasing
Vending women's and children's garments	Absent	Less common
Cultivating commercial vegetable plots	Absent	Less common
Running a small grocery store	Absent	Less common
Sewing quilts and crafts (rope, mats, fans, etc.)	Common	Common
Cooking for agricultural labourers	Common	Common
Harvesting root crops and postharvest processing	Common	Common
Traditional poultry and goat rearing	Common	Decreasing
Pottery	Common	Less common
Household work as maid	Common	Less common
Gleaning	Common	Less common
Paddy husking	Common	Rare
Boiling paddy and selling it after processing	Common	Rare

Gender roles in food security

This section looks at gender roles in food procurement and household food security.

Food production and preparation

Except for helping their husbands in the harvesting of potatoes, in former times women were rarely found working in the fields. Because labour has become expensive, women now work in the rice fields, to plant seedlings and to do weeding and harvesting. Women's involvement in vegetable cultivation is a new trend. Vegetable cultivation is quite labour intensive. Women do the planting, weeding, irrigation and harvesting and are in charge of pest control. Girls sometimes sell the vegetables at the weekly market. Gleaning of cereals and potatoes and gathering wild vegetables, very common in the past, are traditionally performed mostly by women and children.

Poultry rearing is also women's work: men sell the chicks, but women sell the eggs. Caring for cattle is mainly women's work, men being responsible only for cutting grass and showering the animals. Income from selling eggs and vegetables belongs to women, but money resulting from the selling of poultry and other livestock goes to men. Among poor families, postharvest processing of paddy is also women's work. When hired labour is used, women have to cook the meals for the labourers. Food preparation is exclusively done by women.

Food gifts

Though only a small portion of total food consumption, food gifts are part of rural culture. When special foods are prepared, people give some of it to kin or neighbours. Food giving is women's business, but sometimes men intervene, especially when family is involved. Food giving occurs between people of the same status and from rich to poor. The exchange of food gifts could be observed during the fieldwork, but families who cook special foods do not always offer some of it to their neighbours. Wives may ask neighbours or relatives for nice vegetables for their husband if they don't have any themselves.

Food borrowing

As the panel and survey data show, food borrowing is common among kin and neighbours. Among kin, the brother or brother's wife are the first to turn to. Even well-off families sometimes borrow food, especially non-staple foods. The poor often have to borrow food prior to weekly market, which they have to repay immediately after shopping. Sequential borrowing is not allowed. Borrowing food is mainly done by women. When a woman does not manage to borrow food in the vicinity of the household, the husband tries to borrow food elsewhere or purchases food at the market on credit. Long-term food shortages are taken care of by men through food loans and credit.

Food loans

Food loans are also common in the study area. They are more pronounced in a situation of crisis like floods. The poor usually take food loans during preharvest or flood periods. The loan is in the form of rice and has to be paid back with interest (also rice), but no collateral is required. This type of borrowing relies on personal relationships, client–patron relationships and social networking. The loans, often provided by the big landowners, have to be repaid after harvesting and a high rate of interest is charged. In the event of severe floods, such as in 1998, it was difficult to procure a food loan because there was such a high demand and the rate of interest increased. Food loans are the responsibility of men.

Money lending

When people are unable to borrow food or get a food loan, they have to borrow money or sell their assets. Borrowing arrangements vary from short to long term. Borrowing money is usually done by men, but women also borrow money, from their relatives, neighbours, local village organizations and NGOs. Borrowing from NGOs is solely women's business. Though NGOs generally provide loans only for production purposes, the money is often used for consumption, especially for buying food. This practice of using loans for consumption may put households into a cycle of debt. As was revealed in the case studies, many household have membership of more than one NGO. They borrow from one NGO to pay the instalments of another NGO loan or to pay back moneylenders.

Food relief

The study area is classified as a food-assisted area by the World Food Program and the government of Bangladesh. Food assistance is provided to vulnerable groups, such as widows, deserted and separated women and households with a sick husband. The Food-for-Education programme aims at increasing girls' education as well as food security among poor households. In the event of severe floods and during post-flood periods, people are assisted through Vulnerable-Group-Feeding (VGF) programmes, in which women are prioritized.

Gender roles in food security strategies at different times

Households with food security problems may change their course of action or decide on new strategies. To do this successfully, access to information and social networks is crucial. While in the past this posed a problem for women, at present women have better access due to their increased mobility and involvement with NGOs. Women in female-headed households in theory have more manoeuvring space to choose their own strategy and use new information, but sociocultural constraints often prevent them from doing so. The options available to women to enhance their household's food security depend also on the circumstances, as is discussed below.

Poor households in normal times

In normal times, households try to consume staples from own production and purchase from the market. Women and children of poor households additionally engage in postharvest gleaning. For vegetables, households rely on own production from home gardens and purchase from the market. Women from poor households also gather wild vegetables or acquire vegetables as wages in kind for picking vegetables. A study on nutrition in the Philippines (Garcia, 1990) found that the diet of the poorest-income groups was surprisingly high in fish, fruits and vegetables, because they receive a relatively high proportion of their non-staple foods as wages in kind. Seasonal migration on a contract basis for wages in rice is common among men from poor households.

Short-term emergency

Households rely on their own stocks of food or purchase food at the market to meet short-term emergencies. If this is not possible, they borrow food or money to buy food from kin or neighbours. Food borrowing or borrowing small amounts of money (up to 100 taka) is done by women, while borrowing larger amounts is done by men. In the survey it was shown that it is the husband's brother's wife who lends food and small amounts of money. When men borrow money they tend to go to their brothers or parents. For short-term emergencies, households rarely borrow money from NGOs or moneylenders.

Normal flood and preharvest periods

Preharvest floods are a yearly phenomenon for which households have to take precautions. They try to store foods and use their own stocks or save money and buy food. Households cut down their food consumption and reduce the frequency of meals during such periods to make their stocks last. If the household's resources are not enough, food is borrowed from kin or neighbours, followed by selling poultry. When these strategies fail, households try to borrow money from kin or neighbours without interest. When this fails, they try to get a collateral-free loan from neighbours or NGOs on an interest-only basis. Only when these options are exhausted will households try to procure a food loan from a rich person or a moneylender, for which high interest will be charged. It is the women who are always concerned about food storage during (normal) flood periods. In anticipation, they often save a handful of rice from their daily budget to store for flood or crisis periods.

Severe floods

In the case of severe floods, households will first try to implement the same strategies that are applied when there are normal floods. When this fails, they will sell assets like utensils, poultry, goats or even cattle, followed by renting out or mortgaging of land. Vulnerable women are targeted for food relief on a priority basis and provided with VGF cards.

Adverse conditions

In the case of divorce or desertion, women usually return to their parental houses, after which the parents try to marry them off again. Upon the death of a husband, a woman either returns to the parental house or stays at her husband's house, but this depends on the age of the woman, the number and ages of children and the attitude of the husband's family. When the husband is sick, women will rely on stored food, savings, food borrowing and food loans and, finally, borrowing money from kin, neighbours, friends, NGOs and moneylenders.

Gender and nutrition

Women and girls suffer most during the preharvest period when households experience food rationing. Evidence from the panel data shows that, while there are no significant differences in nutrient adequacy between boys and girls of pre-school age, schoolgirls and women seem to sacrifice themselves for boys and men. This may cause malnutrition among women later in life and low birth weight of their offspring (Ali, 2005, pp. 176–179; Bosch, 2005).

Women usually eat last of all household members, which causes them to have a smaller share of rich foods or even none at all. In the past, daughters had to take their food after sons, but now they are served together. Young children are served first. Formerly, the mother-in-law was in control of the kitchen and wives were not allowed to serve themselves and eat before their mother-in-law. They were given food by their mother-in-law only twice a day: at the end of the morning and at night. However, the roles are now reversed: the wife is in control of the kitchen and has three meals a day, while the mother-in-law sometimes only eats twice a day.

Food beliefs and taboos also affect the nutrition of girls and women, though nowadays these are less observed than in the past. Girls and women, particularly pregnant and lactating women, suffer from nutrient deficiencies due to social customs and food beliefs and taboos (cf. Bosch, 2005). A pregnant woman will reduce her food intake later in pregnancy for fear that the baby will become big and labour will be difficult, as was also observed by Niehof (1988) in Madura, Indonesia. These traditional practices contribute to the about 30% of babies born with low birth weight. In 1990, this percentage was as high as 50% (World Bank, 2003). At present, due to mass campaigns and better antenatal care, pregnant women are eating better, but insufficient food intake by pregnant women is still widely prevalent. While formerly, women had to observe all kinds of dietary restrictions after childbirth, such practices are now rare.

Gender in kinship and marriage

Kinship occupies a key position in support provision, in which the husband's family is more important than the wife's family, though the role of the latter cannot be ignored. Survey data revealed that father, mother, brothers and brothers' wives are the most important sources of support (Ali, 2005, p. 36). Support from the wife's family is especially important during the initial years of marriage. The economic status of the wife's family and the distance at which they live play a crucial role in determining the amount of support. It is obligatory for a woman's family of origin to help their daughter settle after marriage. Support of divorced or deserted women and widows by parents, brothers and uncles is a cultural obligation. It is also quite common for orphaned children to be raised by their maternal grandparents.

Culture allows brothers to enjoy their sisters' share of the parental land and entitles women to protection from their brother(s). In the past, women did not inherit land from their parents but were compensated with a share of the produce or a token amount of money by their brothers, or even not compensated at all. While women formerly rarely claimed their share from their brothers, nowadays they are increasingly demanding it. Women are now inheriting land from their parents or get more money from their brothers in return for their share, but the money ultimately belongs to the husband or sons. If a woman inherits land from her parents she has ownership but no authority. While formerly, brothers rarely handed land over to their sisters, at present, because of the increased value of land, women may force their maternal uncles to give them their share. However, women often fear that insisting on their share will cause bad relations with their brother(s). The case studies and life histories testify to the importance of the claims women have on their parents' property, whether these are cashed in or not.

Dowry

Despite the Dowry Prohibition Act of 1980, dowry is a serious social problem in the area. The amount of dowry is increasing. Nowadays, even for the marriage of a poor girl, a minimum dowry of 5000 *taka* (US$90) is required. Many husbands press their wife to get more dowry money, but payment of additional dowry for one daughter causes higher demands from the other daughters' in-laws. Sometimes the wife's family has to provide the husband with a job. Failure to comply causes family problems that may end in divorce. Access to credit at the village level is also responsible for dowry increase, as people now use NGO credits for this purpose. The income opportunities that young girls from poor households now have may result in further increase of dowry as well, because young girls are saving money for their marriage.

Violence

Domestic violence against women has always been prevalent in Bangladesh. In this research we recorded cases of violence against wives because of dowry disputes. In dowry disputes it is often the man's relatives who are the source of domestic violence, either by forcing the man to commit violence against his wife or by committing it themselves. Political and legal awareness have lessened the abuse of women, but wife beating is generally still not seen as violence but rather as the husband legitimately exercising his authority. During our field work we came across cases of harassment by NGO workers of women who had failed to pay the instalments. A husband may also become violent towards his wife when he wants her to borrow from a NGO and she refuses.

Divorce

Often, divorce is not a unilateral decision made by the husband but one imposed by his family and forced upon the husband to implement. Violence against women, whether mental or physical, is a prime indicator of impending divorce or desertion. When a woman has to live with her in-laws, mental abuse is part of her life. Physical violence is also not rare. It was very common in the past and continues to be a social problem. Many women think that it is part of life. Upon divorce, desertion or widowhood, young women mostly return to their parental homes and remarry. A problem arises when the brothers' wives see their sister-in-law as a burden for their own family. An older widowed woman may remain at her husband's house with her children, but her in-laws can make her life miserable.

When women are married they rarely get the bride price that was agreed upon: it is paid to them in the event of divorce. Similarly, the dowry has to be returned upon divorce. Together, this money provides some security for the divorced woman. However, sometimes women are cheated by their husband and do not receive it. A woman's decision to remarry also depends on the gender of her children. Young widows or divorced women without children or with only daughters tend to remarry. A woman with a son may prefer to remain single and raise her son herself, hoping that he will take care of her in the future. At present, employment opportunities for daughters of divorced, widowed or deserted mothers provide new livelihood options.

Gender and vulnerability

Asset ownership, access to land, labour and income opportunities are crucial for a secure livelihood. The assets women bring with them in the form of gifts or dowry at the time of their marriage are appropriated by their husband's family. The only asset a woman can claim her own is jewellery, though her husband still may try to sell it. In the past, women could not even think of holding money of their own, but presently they do. But when they do have money they will hide it from their husbands. Borrowing money among villagers is common. However, when women lend money to men and in-laws, they risk losing it. This is why women who lend money usually do so in somebody else's name.

Female household heads

Divorce, desertion, widowhood or a sick husband force women to take charge of the household. A female-headed household lacks entitlements and support from the community. Female household heads face social stigma and are barred from livelihood activities that males have access to. They depend on their in-laws, parents or brothers for livelihood support. If they are not supported, they are forced to find ways to support themselves. When female household heads are barred from working by their guardians, they will send their (young) children to work. When they do enter the labour market they are faced with gender discrimination in payment, earning a maximum of 60% of their male counterparts. In addition, older women rarely get work in the factories or industries: they can become only manual labourers, and the lack of resources and assets adds to their vulnerability. In times of hardship, women may get support from the Vulnerable Group Development programme or NGOs. If not, they will try to borrow money that they will be unable to repay. In the end, they are caught in a downward spiral of increasing vulnerability.

If female household heads have land, they cannot cultivate it themselves. An option is to hand it over to sharecroppers. Because they are often exploited by sharecroppers, they may prefer to hire labour to cultivate the land. We also found that many women rented their land out instead of cultivating it in a sharecropping arrangement. The money they gain can be lent to farmers with interest paid in the form of paddy. However, farmers tend to exploit women as well. Some women are lucky to get the land of their deceased husband, but this is rare. We came across one case where the husband had put the land in his wife's name to protect her rights from the claims of her in-laws. This problematic relationship of women to land is revealed in many of the cases and life histories.

Labour is also a crucial factor in livelihood generation. The absence of healthy adult men in the household makes a livelihood more vulnerable, as is illustrated by the cases of women with an old or sick husband. In such cases, women have to take up farming or enter the labour market. The sudden death of a husband may force a woman to look for paid work, even if she has small children. This constitutes a double burden to women, involving both reproductive and economic responsibilities.

Gender and social capital

Many forms and uses of social capital are gender related. Women play an important role in the acquirement and handling of food, using their social capital, particularly their social network. Increasing labour market participation of young women and girls has given them a better bargaining position within their family. Such girls are no longer a burden to their family but, instead, are becoming assets. They contribute to the household income and are able to share in the costs of their marriage. This enables parents to find better husbands for their daughters and decreases their worries about their daughter's marriage.

NGOs constitute a new source of social capital from which women can benefit. NGOs' policies of targeting women have given women access to credit for income-generating activities, which strengthens their position. However, in fact this is often only pseudo-empowerment, because though their husband or sons have to rely on them to get the loan, it is women who have to bear the burden of repaying it and suffer the consequences of default (see also Goetz and Sen Gupta, 1996; Kabeer, 2001). NGOs have recently introduced the system of requiring the husband or son to co-sign the loan paper as witness and guarantor (cf. Rozario, 2002), but this is of little help to women.

Conclusions and Discussion

The position of women in the research area is changing. Women became more empowered, both socially and economically, as a result of the changing social structure, improvements in education and NGO initiatives aiming at empowering women. Since the early 1980s, NGOs have been working in the research area to empower women through social development, health, education and credit programmes. Opportunities for girls' education were created by the establishment of non-formal schools by NGOs in the early 1980s and the government's Food-for-Education initiative. These efforts have narrowed the gender gap in education. Ahmed and Del Ninno (2002) found that Food-for-Education programmes have had a greater impact on the education of girls than that of boys.

Women are now taking more responsibility in sustaining their livelihoods by engaging in economic activities, particularly young women and widowed or divorced women without sons. But, in doing so, the older women in particular still face many constraints. Female household heads who are not supported by their in-laws, parents or brothers have to sustain their family but they are still bound to gender-specific livelihood activities, and when they do enter the labour market they have to endure gender discrimination in payment. Still, social change, though not obliterating gender disparities, brings more opportunities, especially for the younger generation of women.

However, cultural patterns and gender roles prevailing in the context of household and family change more slowly and still hamper women's health, nutrition and well-being. Women play a crucial role in household food security and family nutrition. However, our data on individual food security suggest that women's own nutrition is still negatively affected by gender inequality at the household level. We found that traditional and cultural norms still cause women to eat less food and less nutritious foods than needed to satisfy their calorie and nutritional requirements. These alarming findings are corroborated by the literature (HKI, 1994; Rozario, 2002). Gender-specific food taboos and beliefs that used to aggravate women's nutritional vulnerability are nowadays – fortunately – rarely observed.

Another persistent and even growing problem that has significant gender effects is dowry. Dowry is a serious social problem that Rozario (2002) attributes to the monetarization of the village economy and the increase in waged labour. Naved (1994) sees people's helplessness regarding dowry as being responsible for its proliferation. Amin and Cain (1997) explain the problem by a demographically induced shift in the marriage market. However, young women's participation in the labour market increases their age at marriage, which may ease the squeeze of the marriage market. The government, together with civil society, is working to eradicate dowry through social awareness programmes, but so far these efforts have been unsuccessful. Dowry disputes still cause a lot of domestic violence.

While violence against women is a more general social problem, Kabeer (2001) found that incidents of violence against women in poor households declined with women's access to credit. Schuler *et al.* (1996) and Kabeer (2001) also found that older women, women with sons and educated women were beaten less by their husbands. Women's access to credit, training for employment and income-generation activities has enhanced their social mobility (Rozario, 2002) and political and legal awareness (Hashemi *et al.*, 1996), and provided them with the ability to make decisions concerning their own fertility (Zaman, 1999). Pitt and Khandker (1994) found that increased income among borrowers resulted in increased household consumption and better child education.

However, our results paint a less rosy picture. In our research, poor women reported how harassment by NGO workers upon delayed or non-payment of instalments forced them to give up their NGO membership. Karim (2001) also noted the intimidation of borrowers and the increase in violence against women due to repayment pressure. The underlying problem is the lack of control by women over the loans obtained in their name. Goetz and Sen Gupta (1996) reported that 63% of loans were controlled by men and that the lending process victimized the female borrowers, thus contributing to disempowerment.

Female-headed households are particularly vulnerable for a variety of reasons, including gender discrimination in the labour market. Adult women are still barred from *permanent* industrial jobs and must rely mostly on jobs in underpaid manual labour. Female-headed households lack male labour and sometimes land as well. Henderson *et al.* (1997, p. 132) reported that the well-being of female-headed households may be constrained by 'labour shortage together with an overall marginalization as well as cultural and religious proscriptions'. Most households without a man are unable to plough the lands they possess. These factors add to the vulnerability of female-headed households.

Livelihood activities are still largely defined by gender, although less so than in the past. We found that women are now taking up new income-generating activities (vegetable cultivation, self-employment), though – at the same time – some traditional sources of income for women are in decline (e.g. paddy husking). Women have become secondary income earners by their involvement in income-generating activities. They are taking on part of the burden of men in the production sphere, but men do not share in the reproductive work of women. Thus, poor women's involvement in income generation adds to their burden, which may result in them compromising child care or leisure time (Moser, 1996). We also observed that women often have no access to their earnings except for the money they earn from poultry rearing.

Social capital plays a crucial factor in averting vulnerability because it provides access to resources. Women's main sources of social capital appeared to be their kinship network, especially their brothers, and NGO membership. It was clearly shown in this research that women use NGO membership to obtain credit and diversify the household's livelihood portfolio, though they also use NGO credit for feeding the family. However, using a loan for consumption purposes makes payback difficult, which may result in the household being caught in a vicious cycle of debt, leading to increasing vulnerability (cf. Glewwe and Hall, 1998; Huq, 2000; Rozario, 2002).

Acknowledgements

The RESPONSE programme provided the PhD scholarship of the first author in the framework of which the research for this chapter was carried out. We would like to thank the International Food Policy Research Institute (IFPRI) in Washington, DC, USA, for providing access to their data and the Neys-van Hoogstraten Foundation (NHF) in the Netherlands for funding the fieldwork. Last, but not least, we are most grateful to the reviewers for their constructive comments on the earlier version of this chapter.

References

Ahmed, A.U. and Del Ninno, C. (2002) *Food for Education Program in Bangladesh: an Evaluation of its Impact on Educational Attainment and Food Security.* FCND Discussion Paper No. 138, International Food Policy Research Institute, Washington, DC.

Ali, A. (2005) Livelihood and food security in rural Bangladesh: the role of social capital. PhD thesis, Wageningen University, Wageningen, The Netherlands.

Amin, S. and Cain, M. (1997) The rise of dowry in Bangladesh. In: Jones, G.W., Douglas, R.M., Caldwell, J.C. and D'Souza, R.M. (eds) *The Continuing Demographic Transition.* Oxford University Press, Oxford, UK, pp. 290–306.

Bosch, A.M. (2005) Adolescents' reproductive health in rural Bangladesh: the impact of early childhood nutritional anthropometry. PhD thesis, Groningen University, Groningen, The Netherlands.

Bouis, H.E. (2000) Commercial vegetable and polyculture fish production in Bangladesh: their impacts on household income and dietary quality. *Food and Nutrition Bulletin* 24, 482–487.

Cain, M., Khanam, S.R. and Nahar, S. (1979) Class patriarchy and women's work in Bangladesh. *Population and Development Review* 5, 405–438.

Campbell, D.J. (1991) Strategies for coping with severe food deficits in rural Africa: review of the literature. *Food and Foodways* 4, 143–162.

Chambers, R. (1989) Editorial introduction: vulnerability, coping and policy. *IDS Bulletin* 20, 1–7.

Ellis, F. (2000) *Rural Livelihoods and Diversity in Developing Countries.* Oxford University Press, London.

Garcia, M. (1990) Resource allocation and household welfare: a study of the impact of personal sources of income on food consumption and health in The Philippines. PhD thesis, Institute of Social Studies, The Hague, The Netherlands.

Glewwe, P. and Hall, G. (1998) Are some groups more vulnerable to macroeconomic shocks than others? Hypothesis tests based on panel data from Peru. *Journal of Development Economics* 56, 181–206.

Goetz, A. and Sen Gupta, R. (1996) Who takes credit? Gender, power, control over loan use in rural credit programs in Bangladesh. *World Development* 24, 45–63.

Hashemi, S., Schuler, S.R. and Riley, A.P. (1996) Rural credit programs and women's empowerment in Bangladesh. *World Development* 24, 635–653.

Henderson, H.K., Finan, T.J. and Langworthy, M. (1997) Household and hearthold: gendered-resource allocations in Chad. *Culture and Agriculture* 19, 130–137.

HKI (1994) *Summary Report on the Nutritional Impact of Sex-based Behaviour.* Helen Keller International, Institute of Public Health Nutrition, Dhaka.

Hoddinott, J. and Yohannes, Y. (2002) *Dietary Diversity as a Food Security Indicator.* FCND Discussion Paper No. 136, International Food Policy Research Institute, Washington, DC.

Huq, H. (2000) People's perception: exploring contestation, counter-development, and rural livelihoods. Cases from Muktinagar, Bangladesh. PhD thesis, Wageningen University, Wageningen, The Netherlands.

IFAD (2001) *Gender and Households* (http:/www.ifad.org/gender/approach/hfs/wfs/wfs_3.htm, accessed 12 March 2002).

Jahan, R. (1995) Men in seclusion, women in public: Rokey's dream and women's struggle in Bangladesh. In: Basu, A. (ed.) *Challenges in Local Feminism.* Westview Press, London.

Kabeer, N. (1994) *Reserved Realities: Gender Hierarchies in Development Thought.* Verso, London.

Kabeer, N. (2001) Conflicts over credit: re-evaluating the empowerment potential of loans to women in rural Bangladesh. *World Development* 29, 63–84.

Karim, L. (2001) Politics of the poor? NGOs and grass-roots political mobilization in Bangladesh. *PoLAR* 24, 92–107.

Maxwell, S. and Frankenberger, T.R. (1992) Household Food Security: Concepts, Indicators, Measurements. IFAD and UNICEF, Rome and New York.

Molyneux, M. (2002) Gender and the silences of social capital: lessons from Latin America. Development and Change 33, 167–188.

Moser, C. (1996) *Confronting Crisis: a Comparative Study of Household Responses to Poverty and Vulnerability in Four Poor Urban Communities.* Environmentally Sustainable Development Studies and Monographs Series No. 8, The World Bank, Washington, DC.

Naved, R.T. (1994) Empowerment of women: listening to the voices of women. *Bangladesh Development Studies* 22, 121–155.

Niehof, A. (1988) Traditional medication at pregnancy and childbirth in Madura, Indonesia. In: Van der Geest, S. and Whyte, S.R. (eds) *The Context of Medicines in Developing Countries: Studies in Pharmaceutical Anthropology.* Kluwer Academic Publishers, Dordrecht, The Netherlands, Boston, Massachusetts and London, pp. 235–353.

Niehof, A. (2003) Gendered dynamics of food security. In: Ulmer, K. (ed.) *No Security without Food Security. No Security without Gender Equality.* APRODEV, Brussels, pp. 59–66.

Niehof, A. (2004) The significance of diversification for rural livelihood systems. *Food Policy* 29, 321 338.

Palriwala, R. (1990) Introduction. In: Dube, L. and Palriwala, R. (eds) *Structures and Strategies: Women, Work and Family.* Sage, New Delhi, pp. 15–55.

Paris, T. (2004) Changing women's role in homestead management: mainstreaming women in rural development. Keynote paper presented at *Dialogue on Women's Contribution to Rural Economic Activities: Making the Invisible Visible,* 22 April, Dhaka.

Pitt, M. and Khandker, S. (1994) *Household and Intrahousehold Impacts on the Grameen Bank and Similar Targeted Credit Programs in Bangladesh.* The World Bank, Mimeo, Washington, DC.

Putnam, R.D. (1995) Bowling alone: America's declining social capital. *Journal of Democracy* 6, 65–78.

Rozario, S. (2002) Gender dimension in rural change. In: Toufique, K.A. and Turton, C. (eds) *Hands not Land: How Livelihoods are Changing in Rural Bangladesh. Part 2: Original Research Papers*, pp. 121–130 (http//www.livelihood.org/lessons/docs/ handsland.pdf).

Rudie, I. (1995) The significance of eating: cooperation, support and reputation in Kelantan Malay households. In: Karim, W.J. (ed.) *'Male' and 'Female' in Developing Southeast Asia*. Berg Publishers, Oxford, UK, pp. 227–247.

Schuler, S.R., Hashemi, S., Riley, A.P. and Akhter, A. (1996) Credit programs, patriarchy and men's violence against women in rural Bangladesh. *Social Science and Medicine* 43, 1729–1743.

Sen, A. (1994) *Population: Delusion and Reality*. Review of Books, New York.

UNFPA (2005) *State of the World Population 2005. The Promise of Equality*. United Nations Population Fund, New York.

van Bastelaer, T. (2000) *Does Social Capital Facilitate the Poor's Access to Credit? A Review of the Microeconomic Literature*. Social Capital Initiative Working Paper No. 8, The World Bank, Washington, DC.

Woolcock, M. and Narayan, D. (2000) Social capital: implications for development theory, research, and policy. *The World Bank Research Observer* 15, 225–249.

World Bank (2003) *World Development Report 2002*. The World Bank, Washington, DC.

Zaman, H. (1999) *Assessing the Impact of Micro-credit on Poverty and Vulnerability in Bangladesh*. Policy Research Working Paper No. 2145, The World Bank, Washington, DC.

10 Does Social Capital Matter in Vegetable Markets? The Social Capital of Indigenous Agricultural Communities in the Philippines: Socio-cultural Implications and Consequences for Local Vegetable Trade

AIMEE MILAGROSA[1],[*] AND LOUIS SLANGEN[2],[**]

[1]Centre for Development Research (ZEF), University of Bonn, Germany;
[2]Agricultural Economics and Rural Policy Group, Social Sciences Department,
Wageningen University and Research Centre, 6708 KN Hollandsweg 1,
Wageningen, The Netherlands; e-mails: *Aimee.milagrosa@wur.nl;
**Louis.Slangen@wur.nl

Introduction

One of the earliest studies of Philippine social capital was by Buenavista (1998), who observed the fishing practices and hierarchical patron–client relationships of dynamite fishermen. Buenavista indicated that illegal dynamite fishing was widely practised because patrons (fishermen) and clients evolved a gift-giving culture that encouraged this environmentally unsustainable practice. Benigno (2002) discussed a study conducted by the Nomura Research Institute (NRI), a Japanese firm involved in corporate strategy. He stated that foreign investors shied away from the Philippines, not only because of the country's socioeconomic or infrastructure problems but also because of certain 'undesirable' Filipino traits.

The study alluded that Filipinos associate many aspects of their lives with self-interests. The prevailing attitude of 'family first' shows that the Filipinos are more family- than nation-oriented, unlike the Japanese or Koreans. Because of its archipelagic structure and geographical/language barriers, technology and knowledge transfer is difficult. These factors

© CAB International 2007. *Sustainable Poverty Reduction in Less-favoured Areas*
(eds R. Ruben, J. Pender and A. Kuyvenhoven)

largely contribute to the slow development of a national-trust culture. Considering that the Philippines is predominantly agricultural, the issue now is whether the Nomura research findings can sustain themselves in agricultural settings.

The neglect of policy makers, together with several agro-ecologic and socio-economic constraints, has made the province of Benguet in northern Philippines a less-favoured area. However, the authors of this chapter wished to investigate whether indigenous agricultural communities of the province, who cultivate vegetables in a long-standing tradition, are also less endowed in terms of social capital. This chapter attempts to answer the following research questions: (i) does social capital exist in indigenous agricultural communities of Benguet and in what form?; (ii) what factors influence social capital?; and (iii) how does social capital affect vegetable marketing in the province?

In order to answer these research questions, the chapter's specific objectives were to: (i) determine which factors motivate social capital formation among farmers and traders and in the farming communities as a whole; (ii) analyse social capital levels for farmers and traders and per municipality; (iii) evaluate whether respondents' attributes such as education, gender, religion, age and ethnicity influence community social capital; and (iv) examine how social capital levels affect the present and shape the future of the vegetable industry.

The Economy and People of Benguet

The economy

Benguet is one of the six provinces comprising the Cordillera Administrative Region (CAR), a landlocked cluster of 76 towns dominated by mountain ranges. A plateau 1500 m above sea level, Benguet has a climate useful for temperate vegetable production. Benguet provides at least 75% of the carrot, potato and cabbage demands of the country (Pekas *et al.*, 1998). The importance of agriculture to the economy was strongly felt in 2000, when the share of agriculture region's Gross Regional Domestic Product (GRDP) plunged from 16 to 2% annual growth when crops failed due to pests (Cabreza and Caluza, 2002).

Since vegetable production and marketing provide income and employment for the majority of the people, good supporting infra-structures are expected to be in place. Unfortunately, the bulk of costs and losses for vegetable market participants stems from poor or limited market access due to missing or low-quality farm-to-market roads. The Halsema highway is the only highway linking vegetable-producing communities to La Trinidad and Baguio, the two major trading posts that serve the whole province. The remoteness of communities is emphasized on the occasions when they are permanently under threat of being cut off from markets due to landslides, which affect many roads in the rainy season. Cold storage warehouses, docking bays for delivery trucks and technology that support timely market information transfer province-wide are lacking (Dalmo *et al.*, 1994).

Agroecologic constraints in the form of steep slopes and poor soils have challenged provincial vegetable growers for years. Terracing, hillside planting and intensive agriculture have defined vegetable production in the region. Benguet is also located along the typhoon belt, the common course taken by storms when they pass through the Philippines. This makes it prone to unusually heavy rains during the rainy season that result in leaching, erosion and environmental damage. Agricultural potential is limited in most municipalities, not only because of depleted soils but because of dependence on rain-fed agriculture due to lack of irrigation facilities.

Despite its contributions to the regional economy, vegetable production is becoming less profitable. Agricultural wage rates in Benguet farm households were running at only 137 PhP/day in 1999 (€1.96) compared with 184 PhP/day (€2.63) for non-agriculture rates.[1] Declining land fertility, high costs of inputs, low vegetable prices and the removal of import quotas have slowly led to the centuries-old vegetable farming tradition becoming financially disadvantageous (Aquino, 2003).

The people

Benguet is the most populous province in the CAR, with 330,129 residents in 2000 (NSO, 2002). In terms of culture, Benguet is a melting pot of inhabitants from various ethnic backgrounds. A total of 13.4% of the population are Ilocano, 29.2% are Ibaloi and 43% are Kankanaey. Tribal affiliation represents the major dialects spoken, thus most of the population speaks Kankanaey, followed by Ibaloi and Ilocano. Kankanaeys from the northern part of the province originate from South China, while Ibalois from the southern part of the province are of north Malaysian origin. More recently however, interaction between the two tribes resulted in a mixing of traditions and beliefs, such that the difference at present is mostly linguistic and not cultural.

Fierce tribal wars that previously separated groups have given way to free trade and harmonious inter-ethnic relations. Within agricultural communities, farmers and traders interact freely with each other: they are parts of the same informal social networks within the municipalities. Farmers and traders may be involved in different lines of work, but there is no delineation between them in terms of intra-community social participation. Since farmers' and traders' inclination is primarily agricultural, they interact repeatedly within communities with similar backgrounds.

Farming has been the way of life for most of the people in the province, with the Kankanaeys having relatively more involvement in agriculture and agricultural trade (Russel, 1989). According to the National Economic Development Authority (NEDA) in 2002, the agricultural sector of CAR employs 60% of the economically active population. However, the region is one of the country's poorest, with 31.1% of the regional population living below the poverty threshold (NSCB, 2001).

Applying Social Capital Theory to Benguet Vegetable Markets

Davis (1973) conceived the term 'economic personalism' to describe the kinship and *suki* system (favoured buyer) of social organization in Baguio markets. The suki system is defined by 'highly personal relationships where individuals interrelate in ... more than purely economic dimensions'. Farmers and traders try to include reciprocity and relationship building into vegetable trade in order to increase benefits from transaction agreements. Beugelsdijk and Schaik (2001) stated that social capital works by increasing communication, interaction, information transfer and cooperation between transacting partners without the influence of power and market. Trust can make people go beyond the requirements of the contract through early delivery, higher quality or some other means to support their good intentions and sustain trust. Increasing the level of connectedness by participating in informal networks leads to collective action in the pursuit of common goals. Temptations to achieve short-term personal wealth through the transaction are superseded (Beugelsdijk and Schaik, 2001).

Repeated interactions lead to increased trust levels. According to Beugelsdijk and Schaik (2001), trust is present when you expect your partner not to exploit your vulnerability based on the expectation that your partner will perform the duties that are expected of them. In Benguet however, despite highly personal relations, information asymmetry and opportunistic behaviour is not only entrenched but also used as a strategy by farmers and traders for profit gains (Milagrosa, 2001). A study by Batt (1999) on Benguet potato farmers and their seed suppliers showed that, despite repeated interactions, no connection between trust and the length of farmer–trader relationship could be established. As farmers buy potato seed supplies from preferred sellers, they simultaneously consider offers from others. Farmers' commitment to a relationship is related to their satisfaction with trader performance rather than to trust.

Data and Methodology

Data

Seven of the 13 municipalities of Benguet[2] were selected as survey areas. These municipalities were selected because they are representative of the province in terms of demographics, vegetable production capabilities and range of geographic variations. The municipalities of Buguias, Atok and Kibungan are the top vegetable producers by volume, while Bakun, Itogon and Bokod are the lowest producers. La Trinidad was chosen because it has the La Trinidad Trading Post, where most of the vegetables harvested in the region are traded.

Within municipalities, *barangays* (local districts) were chosen randomly. Within barangays, farmers were chosen using purposive sampling. Farmer and trader sampling was a problematic issue because of the expectation of the availability of complete lists of farmers and traders at the provincial level during the research design stage. The planned random sampling could not be implemented.[3] Thus, it was decided that the research would follow the 'snowballing' procedure used by previous researches in the province (e.g. CHARM and VLIR[4]).

The interviewed farmers totalled 450. They were selected primarily because they are representative of farmers who grow vegetables with the intention of selling them in La Trinidad and Baguio City markets for profit. Sampling for traders was also conducted purposely. The interviewed traders totalled 195. Overall, there were fewer traders than farmers per municipality and, for this reason, fewer traders were interviewed. Although selected purposively, the farmers and traders interviewed represent a wide diversity of conditions in terms of crops planted, preferred distribution channels, production costs, access to markets and profits. The results of the studies are therefore representative of a broad range of circumstances that the farmers and traders face within the municipalities of Benguet province.

Methodology

The World Bank (WB) asserts that no standard measure of social capital can be achieved, since social capital measurements are dependent on the definition rendered by researchers. However, the WB suggested three approaches to social capital measurement: (i) quantitative studies, e.g. by Knack and Keefer (1997) or Narayan and Pritchett (1999); (ii) comparative analysis, e.g. by Putnam (1993) or Light and Karageorgis (1994); and (iii) qualitative studies, e.g. by Portes and Sensenbrenner (1993), Gold (1995) and Heller (1996). No approach is superior to any other in measuring social capital. Grootaert and Van Bastelaer (2002) argued that empirical social capital data analyses could utilize any approach since no standard calculation procedures existed.

Because of a lack of previous work on social capital in the province, the study preferred to test the simpler quantitative–additive approach (assuming equal weights). Grootaert and Van Bastelaer (2002) point out that what is important is to attempt to capture social capital in its cognitive and structural dimensions. For this reason, first, a Likert scale measured farmer and trader perceptions on social capital statements.

Secondly, an equation was used to model cognitive and structural social capital. Cognitive social capital is the intangible aspect of social capital in the form of trust, local ethics, traditions and morals. Cognitive social capital was measured by 22 statements on optimism and satisfaction, common goals and perceptions, trust and civic associations. Structural social capital is tangible and deals with membership in formal networks, government organizations, church and clubs. Structural social capital was measured by active membership in religious, political, cooperative and recreational groups.

The social capital index

Except for membership in local organizations, all items were obtained using a five-point scale. To normalize the five-point scale, the individual value for cognitive social capital indicators $\dfrac{SCIndicator_{ij} - 1}{4}$ was used, where *ij* refers to the cognitive social capital dimension of farmer *i* in municipality *j*. The product was multiplied by a factor representing the within-group weight (*wgw*) of the variables being analysed. The within-group weight depends on the number of items measuring the indicator. Thus, for the cognitive social capital CSC_{ij}:

$$CSC_{ij} = \left(\frac{(\sum_{j=1}^{J} associatedness_{ij}) - 1}{4} * wgw \right)$$

$$+ \left(\frac{(\sum_{j=1}^{J} trust_{ij}) - 1}{4} * wgw \right)$$

$$+ \left(\frac{(\sum_{j=1}^{J} goalsperceptions_{ij}) - 1}{4} * wgw \right)$$

$$+ \left(\frac{(\sum_{j=1}^{J} optimismsat_{ij}) - 1}{4} * wgw \right)$$

(10.1)

i=1..., I and j=1..., J

where CSC_{ij} is the cognitive social capital of farmer *i* in municipality *j*; $associatedness_{ij}$ is the associatedness levels of farmer *i* in municipality *j*; $trust_{ij}$ is the trust levels of farmer *i* in municipality *j*; and so on. The values for structural social capital were obtained as actual memberships of respondents in formal organizations. Thus, structural social capital equals:

$$SSC_{ij} = \sum_{j=1}^{J} membership_{ij} \qquad\qquad (10.2)$$

i=1..., I and j=1..., J

where SSC_{ij} is the structural social capital of respondent i in municipality j and $membership_{ij}$ is the membership of farmer i in municipality j to the various formal organizations presented in the questionnaire.

The resulting indicator cognitive values were weighted equally and standardized to 50. Thus, 0 represents no cognitive social capital and 50 represents full cognitive social capital. Since active membership in local organizations was provided using forthright answers, structural social capital values were calculated by obtaining the percentage equivalent and then standardizing responses to 50.

The outcomes reflect actual memberships into specific formal institutions. This would mean for each farmer (and therefore, each municipality) a value of 0 for no membership at all and a value of 50 for membership in all formal organizations enumerated. To achieve a social capital index, structural and cognitive values were simply added. Following Grootaert and Van Bastelaer (2002), a preference for separate presentation of structural and cognitive social capital *first*, before aggregating a single social capital index *second*, was conducted. This is because the two indicators capture different dimensions of social capital that are both significant in their own right.

Deconstructing Social Capital of Indigenous Agricultural Communities in Benguet

Overall social capital: the big picture

Principal component analysis

As shown in Table 10.1, six components were loaded from the initial factor analysis. These components explain 66% of the variance before and after Varimax rotation. To determine which factors are relevant, the Kaiser criterion, where initial Eigenvalues < 1 are excluded, was used. Coefficients in the final rotated component matrix results were sorted by size. The Kaiser-Meyer-Olkin Measure of Sampling Adequacy (KMO) is equal to 0.834, while the Bartlett's test for Sphericity significance is at 0.000, both indicating that Factor Analysis would be useful for the data since it contains significant inter-variable relationships.

FACTOR 1.   The statements that load highly on Factor 1 all seem to relate to the quality of casual peer-to-peer associations. This factor is labelled

Table 10.1. Rotated component matrix results[a] for aggregated farmer and trader statements (from own survey).

Statements	Factors					
	Factor 1 Informal networks	Factor 2 Core trust	Factor 3 Institutional trust	Factor 4 Poverty perception	Factor 5 Common goals	Factor 6 Life satisfaction
I get along well with people in my community	0.876	0.084	0.127	−0.034	0.139	0.077
I get along well with other farmers	0.850	0.075	0.108	0.012	0.114	0.018
I get along well with family and friends	0.823	0.101	0.050	−0.029	0.157	0.056
I get along well with other traders	0.708	0.133	0.178	0.068	0.070	0.045
I participate actively in the community and volunteer for community work	0.310	0.097	0.187	−0.192	0.211	0.131
I trust family and friends	0.158	0.779	0.080	0.131	0.037	0.061
I trust the church and its people	0.158	0.740	0.280	0.073	0.052	0.176
I trust other farmers	0.130	0.737	0.231	0.089	0.057	0.096
I feel safe in my neighbourhood	0.111	0.734	0.262	0.027	0.131	0.070
I can safely say I am trustworthy	−0.070	0.702	−0.049	0.328	0.136	−0.062
I trust municipal police	0.164	0.212	0.844	0.103	0.011	0.126
I trust the legal system	0.177	0.219	0.840	0.098	0.030	0.149
I trust municipal government and their policies towards agriculture	0.224	0.256	0.689	−0.240	0.169	0.155
People are poor because they are lazy and have no willpower	0.004	0.230	0.004	0.881	0.056	−0.006
People are poor because they are not given the same chances as others	0.002	0.205	0.078	0.880	0.105	−0.036
Local government should concentrate on fighting rising input prices	0.134	0.070	0.088	0.086	0.817	0.074
The country must create more job opportunities	0.252	0.010	0.096	0.003	0.760	0.058
Community members should get more involved in policy making	0.075	0.189	−0.047	0.077	0.732	0.004
The local government treats everyone equally	0.074	0.012	0.269	−0.148	0.135	0.691
I am satisfied and happy with my life and it will get even better in the future	0.388 / −0.096	−0.088 / 0.298	0.045 / −0.154	−0.296 / 0.338	0.027 / 0.077	0.649 / 0.564
Do you agree that most people could be trusted?	0.017	0.171	0.271	0.103	−0.032	0.481

[a] Extraction method: Principal Component Analysis; rotation method: Varimax with Kaiser Normalization; rotation converged in six iterations.

'informal networks', a component showing the strong positive correlation of being on good terms with people in the community (Eigenvalue of 0.876) to Factor 1. Farmers who value informal networks find it highly important to be on good terms with community members.

FACTOR 2. The second set of variables relates highly towards trust within the immediate environment, particularly trust of his family, neighbours, farmers, the church and respondents' own feelings of trustworthiness. This factor is labelled 'core trust', showing that respondents assign highest importance to familiars and to religion. Putnam (1993) refers to this as the bonding element of social capital.

FACTOR 3. Factor 3 shows attitudes related to trust in the formal institutional environment, with emphasis on the legal system, police and municipal government, and is labelled 'institutional trust'. A positive significant relationship between trusting the municipal police and institutional trust shows that respondents who have high scores in trusting institutions also tend to trust the municipal police highly. This is referred to as the bridging element of social capital (Putnam, 1993).

FACTOR 4. The fourth factor is labelled 'poverty perceptions'. Two factors loaded heavily for this component: poverty because of laziness and poverty because of lack of life opportunities.

FACTOR 5. The fifth factor is a measure of community aspirations and is termed 'common goals'. It relates to objectives shared by farmers and traders in terms of community goals and what the local government should focus on. Fighting rising input prices loaded heavily for this factor (Eigenvalue 0.817). It means that farmers and traders who loaded highly on the 'common goals' attribute find the issue of rising input prices important.

FACTOR 6. This last factor loads heavily on statements related to 'life satisfaction'. Equal treatment, life satisfaction and optimism largely contribute to this component. Respondents who find equal treatment from the government important would load highly on the life satisfaction component.

Structural and cognitive social capital

A social capital index shown in Table 10.2 for each municipality was computed by adding cognitive and structural social capital values. Social capital index is therefore:

$$SCI_j = CSC_{ij} + SSC_{ij} \tag{10.3}$$

where SCI_j = social capital index for municipality j.

Table 10.2. Aggregated social capital of farmers and traders in Benguet vegetable markets, by municipality and mean values (from own survey, 2003).

Social capital	Atok	Bakun	Bokod	Buguias	Itogon	Kibungan	La Triniad	Mean
Cognitive	30.3	30.2	28.5	30.6	30.6	29.3	29.2	29.8
Structural	8.3	7.0	4.7	7.8	6.5	7.0	8.6	7.1
Total SC	38.6	37.2	33.2	38.4	37.1	36.3	37.8	36.9

Municipalities were well below the 50 mid-point on which all calculations were benchmarked. Four municipalities had cognitive scores higher than the mean; only three out of seven municipalities had structural scores higher than the average. A one-sample t-test comparing cognitive, structural and total social capital indices proved that municipal means differed significantly from the assigned median of 50. Initially, high cognitive and structural social capital scores for municipalities were expected in accordance with social capital theory and existing local anthropological literature. In particular, ethnicity, the remoteness of the research area and common agriculture-related goals were predicted to bind the societies together. During the interviews, a surprising trend of low trust and low membership in formal organizations began to emerge, irrespective of tribal affiliation and municipal location. Only statements regarding common goals were the ones that held out.

Social capital between groups: farmer : trader comparisons

Principal component analysis

To ascertain whether social capital formation was distinct among farmers and traders, factor analysis was conducted separately. Table 10.3 shows the ranked components extracted from farmers. Rotated component matrices show that different factors drive social capital formation among farmers and traders. Among farmers, the Eigenvalue of the first factor extracted – informal networks – explains 28.7% of the total variance. Other components explaining social capital were core trust, institutional trust, common goals, poverty perception and life satisfaction. It appears that farmers are defined more by bonding social capital in the form of informal associations. Social capital is high for those that are in his immediate environment.

For traders, the Eigenvalue of the first factor extracted – outer core trust – explains 21.6% of the total variance. Compared with farmers, traders seem to distinguish between different levels of trust. Whereas farmers assign highest importance to informal networks, traders consider trust towards neighbours, the church and farmers as being most valuable. It appears that traders are defined more by bridging social capital in the form of outer core trust. Social capital is relatively higher

Table 10.3. Rotated component matrix results[a] for farmer statements on social capital formation (from own survey, Benguet, 2003).

	Factors					
Statements	Factor 1 Informal networks	Factor 2 Core trust	Factor 3 Institutional trust	Factor 4 Common goals	Factor 5 Poverty perception	Factor 6 Life satisfaction
My life will get even better in the future	−0.163	0.416	−0.146	0.061	0.236	0.459
The local government treats everyone equally	0.089	0.008	0.321	0.104	−0.156	0.676
I am satisfied and happy with my life	0.376	−0.048	0.035	0.036	−0.303	0.648
Do you agree that most people could be trusted?	0.048	0.165	0.171	−0.029	0.131	0.565
I can safely say I am trustworthy	−0.002	0.749	−0.075	0.090	0.313	−0.068
I trust family and friends	0.173	0.807	0.097	0.048	0.119	0.063
I feel safe in my neighbourhood	0.192	0.691	0.295	0.187	0.040	0.145
I trust the church and its people	0.183	0.711	0.294	0.062	0.081	0.187
I trust other farmers	0.110	0.745	0.254	0.074	0.077	0.055
I trust municipal government and their policies towards agriculture	0.241	0.225	0.716	0.163	−0.243	0.176
I trust municipal police	0.176	0.195	0.861	0.062	0.120	0.108
I trust the legal system	0.216	0.210	0.827	0.023	0.127	0.222
I participate actively in community and volunteer for community work	0.377	0.096	0.129	0.242	−0.061	0.269
I get along well with family and friends	0.829	0.099	0.095	0.196	−0.027	0.086
I get along well with people in my community	0.881	0.068	0.161	0.163	−0.046	0.082
I get along well with other farmers	0.860	0.110	0.114	0.139	−0.012	0.003
I get along well with other traders	0.719	0.190	0.196	−0.010	0.132	0.080
The country must create more job opportunities	0.222	0.050	0.082	0.753	−0.003	0.073
Community members should get more involved in policy making	0.076	0.108	−0.014	0.744	0.135	0.039
The local government should concentrate on fighting rising input prices	0.135	0.118	0.104	0.805	0.013	−0.009
People are poor because they are lazy and have no willpower	0.018	0.256	−0.004	0.032	0.885	−0.036
People are poor because they are not given the same chances as others	0.003	0.211	0.079	0.122	0.891	−0.029

[a] Extraction method: Principal Component Analysis; rotation method: Varimax with Kaiser Normalization; rotation converged in seven iterations.

for those that are in his external environment. This is perhaps an allusion to the skills needed because of the nature of their job. The rotated component matrix for traders is shown in Table 10.4.

What is interesting to note in the separate factor analyses is that 'core trust' loaded heavier among farmers but loaded less important and farther for traders. It could be that, because more live in remote municipalities as compared with traders, they are less exposed to opportunism in business relations. Therefore, their core trust ranks heavier than for traders. For traders, 'core trust' loaded negatively. The negative sign (−0.540) on factor 6 indicates that there is a significant negative relationship between traders' own trustworthiness and 'core trust' perceptions. For farmers who have a positive attitude towards core trust, the issue of self trustworthiness is irrelevant.

Structural and cognitive social capital

Shown in Table 10.5 are the social capital scores comparing farmers with traders. From mean values, farmers shared fewer common goals and perceptions and had lower optimism and satisfaction than traders in general. However, farmers had better community relations than traders. This however, is not statistically different. The total social capital of traders was higher than that of farmers, at 33.3 versus 37.1; this difference is statistically significant.

A closer look however, reveals that although in both social capital types traders scored higher than farmers, the real difference lies in their respective membership of formal associations. A paired samples t-test proves that cognitive scores for farmers and traders are not statistically different from one another, but that structural social capital scores are. Traders are more active in formal organizations than farmers, leading to higher structural social capital values. Overall, traders have higher social capital scores because they are more active in formal organizations than farmers.

Although farmers and traders are part of the same social networks and organizations, it appears that the real cause of their social capital originates from different elements. Farmer social capital is affected by the bonding aspect of social capital: he trusts persons within the family circle and those within his immediate environment. Trader social capital is more the bridging type, the type of capital that comes from knowing people outside the immediate social network (Woolcock, 1999; Grootaert and Van Bastelaer, 2002).

Socio-cultural Highlights

Social capital and gender

Does gender influence social capital? Table 10.6 shows significant gender correlation with some social capital indicators.

Table 10.4. Rotated component matrix results[a] for trader statements on social capital formation, Benguet, 2003.

	Factors						
Statements	Factor 1 Outer core trust	Factor 2 Informal networks	Factor 3 Institutional trust	Factor 4 Common goals	Factor 5 Poverty perceptions	Factor 6 Core trust	Factor 7 Life satisfaction
My life will get even better in the future	−0.033	0.041	0.104	0.119	0.385	−0.403	0.594
The local government treats everyone equally	0.148	−0.036	0.066	0.171	−0.102	0.134	0.759
I am satisfied and happy with my life	−0.068	0.366	0.077	0.008	−0.226	0.269	0.614
Do you agree that most people could be trusted?	0.216	−0.047	0.352	−0.073	0.320	0.380	0.305
I can safely say I am trustworthy	0.353	−0.240	0.246	0.312	0.229	−0.540	−0.116
I trust family and friends	0.458	0.126	0.336	0.070	0.061	−0.514	0.002
I feel safe in my neighbourhood	0.812	−0.090	0.087	0.018	0.145	−0.057	−0.022
I trust the church and its people	0.827	0.101	0.241	0.032	0.163	−0.001	0.098
I trust other farmers	0.786	0.190	0.239	0.030	0.109	−0.052	0.065
I trust municipal government and their policies towards agriculture	0.404	0.146	0.573	0.187	−0.120	0.304	0.040
I trust municipal police	0.217	0.101	0.847	−0.089	0.003	−0.036	0.153
I trust the legal system	0.168	0.025	0.854	0.107	0.001	−0.024	0.014
I participate actively in community and volunteer for community work	−0.004	0.059	0.128	0.117	−0.090	0.570	0.081
I get along well with family and friends	0.052	0.825	−0.004	0.064	−0.061	−0.115	0.036
I get along well with people in my community	0.073	0.853	0.097	0.134	−0.020	−0.057	0.099
I get along well with other farmers	0.040	0.792	0.065	0.079	0.084	0.157	0.037
I get along well with other traders	0.057	0.573	0.045	0.364	−0.135	0.251	−0.044
The country must create more job opportunities	−0.081	0.300	0.129	0.785	0.020	0.096	0.024
Community members should get more involved in policy-making	0.377	0.066	−0.247	0.667	0.049	−0.060	0.104
The local government should concentrate on fighting rising input prices	−0.024	0.120	0.119	0.817	0.178	−0.032	0.164
People are poor because they are lazy and have no willpower	0.193	−0.060	−0.060	0.129	0.894	−0.049	−0.023
People are poor because they are not given the same chances as others	0.168	−0.012	0.011	0.066	0.869	−0.157	−0.093

[a] Extraction method: Principal Component Analysis; rotation method: Varimax with Kaiser Normalization; rotation converged in seven iterations.

Table 10.5. Social capital comparison between farmers and traders (from own survey, Benguet, 2003).

Indicator	Farmers	Traders
Associatedness	7.81	7.69
Trust	6.57	6.57
Common goals and perceptions	9.21	9.46
Optimism and satisfaction	6.06	6.36
Total cognitive	29.70	29.96
Total structural	3.57	7.13
Social capital	33.26	37.10

Table 10.6. Correlation coefficients between social capital indicators and gender (from own survey).

Social capital indicator	Correlation coefficient (2-tailed test of significance)
Optimism about life getting better	0.079*
Satisfaction and happiness about life	0.121**
Volunteering for community work	0.084*
Good relationship with people in community	0.085*

*, ** and *** indicate significance at the 0.10, 0.05 and 0.01 levels (2-tailed), respectively.

To confirm whether there is a difference between males and females in terms of the gender-correlated social capital indicators, the Kruskal–Wallis test was conducted. We applied the non-parametric ANOVA test of Kruskal and Wallis because one of the variables is nominal (e.g. gender) and the other is ordinal (five-point scale). The hypothesis and results are shown in Table 10.7.

The test confirms that women are more satisfied and happy with their lives. However, there is no difference between men and women in terms of optimism, community participation in terms of voluntary work and strength of informal community networks.

Social capital and educational attainment

Does education influence social capital? Table 10.8 shows that education is significantly correlated with some social capital indicators.

To test whether there is a difference between highly educated (those with a university degree) and poorly educated people (those with primary school and secondary education only), the Kruskal–Wallis test was conducted. The hypotheses and their corresponding results are presented in Table 10.9.

Except for trust towards the church, the tests confirm the alternative hypothesis that respondents with higher education generally exhibit more trust than poorly educated people.

Table 10.7. Kruskal–Wallis tests for gender and social capital variables (from own survey).

Hypotheses	Asymptotic significance (1-tailed)	Decision rule
H0: There is no difference between men and women in terms of optimism H1: Women are more optimistic	0.362	Accept HO
H0: There is no difference between men and women in terms of satisfaction and happiness H1: Women are more satisfied and happier with their lives	0.030	Reject HO
H0: There is no difference between men and women in terms of voluntary work H1: Men do more voluntary work.	0.389	Accept HO
H0: There is no difference between men and women in terms of informal community networks H1: Women have better relationships in the community	0.245	Accept HO

Table 10.8. Correlation coefficients between social capital indicators and educational attainment (from own survey).

Social capital indicator	Correlation coefficient (2-tailed test of significance)
Satisfaction and happiness about life	0.135**
General trust	0.105**
Trust of family and friends	0.093*
Trust of church	0.081*
Trust of municipal government	0.085*

*, ** and *** indicate significance at the 0.10, 0.05 and 0.01 levels (2-tailed), respectively.

Table 10.9. Kruskal–Wallis tests for education and social capital variables (from own survey).

Hypotheses	Asymptotic significance (1-tailed)	Decision rule
H0: There is no difference between the well educated and poorly educated in terms of life satisfaction and happiness H1: Well-educated people are more satisfied and happier with their lives	0.000	Reject HO
H0: There is no difference between the well educated and poorly educated in terms of general trust H1: Well-educated people have higher levels of general trust	0.003	Reject HO
H0: There is no difference between the well educated and poorly educated in terms of trust of family and friends H1: Well-educated people have more trust towards family and friends	0.015	Reject HO
H0: There is no difference between the well educated and poorly educated in terms of trust of church H1: Well-educated people have more trust towards the church.	0.700	Accept HO
H0: There is no difference between the well educated and the poorly educated in terms of trust of municipal government H1: Well-educated people show more trust towards the municipal government.	0.060	Reject HO

Social capital and religion

Does religion influence social capital? Table 10.10 shows that religion is significantly correlated with some social capital indicators.

To test whether there was a difference between actively religious and non-religious people, the Kruskal–Wallis test was conducted. The hypotheses and their corresponding test results are shown in Table 10.11.

Test results lead to the conclusion that the religious and non-religious are not similar in terms of life satisfaction and happiness, volunteerism, trust towards the government and quality of relations with other community members.

Social capital and age

Does age influence social capital? Table 10.12 shows that age is significantly correlated with some social capital indicators.

Table 10.10. Correlation coefficients between social capital indicators and religion (from own survey).

Social capital indicator	Correlation coefficient (2-tailed test of significance)
Satisfaction and happiness about life	0.436**
Volunteering for community work	0.325**
Trust in municipal government	0.305**
Good relationship with people in community	0.313**

*, ** and *** indicate significance at the 0.10, 0.05 and 0.01 levels (2-tailed), respectively.

Table 10.11. Kruskal–Wallis tests for religion and social capital variables (from own survey).

Hypotheses	Asymptotic significance (1-tailed)	Decision rule
H0: There is no difference between religious and non-religious people in terms of satisfaction and happiness		
H1: Religious people are more satisfied and happier with their lives	0.000	Reject HO
H0: There is no difference between religious and non-religious people in terms of volunteering for community work		
H1: Religious people volunteer more for work	0.000	Reject HO
H0: There is no difference between religious and non-religious people in terms of trust of municipal government		
H1: Religious people trust the municipal government more	0.000	Reject HO
H0: There is no difference between religious and non-religious people in terms of good relationships with people in the community		
H1: Religious people have better relationships with people in the community.	0.000	Reject HO

Table 10.12. Correlation coefficients between social capital indicators and age (from own survey).

Social capital indicator	Correlation coefficient (2-tailed test of significance)
Getting along with other farmers	0.081*
Membership of religious group	0.132**
Membership of farmer cooperative	0.133**
Membership of local government	0.900*
Membership of neighbourhood group	0.144**

*, ** and *** indicate significance at the 0.10, 0.05 and 0.01 levels (2-tailed), respectively.

To test whether there was a difference between young and old respondents, the binomial test was conducted (because age is scalar). The hypotheses and their corresponding test results are in Table 10.13.

Test results lead to the conclusion that older people tend to be more active in organizations such as religious groups, cooperatives and local neighbourhood gatherings. Young and old are similar in commitment to politically inclined groups. In general, older respondents have better-quality relations with other farmers than do the younger respondents.

Table 10.13. Binomial tests for age and social capital variables (from own survey).

Hypotheses	Asymptotic significance (1-tailed)	Decision rule
H0: There is no difference between old and young people in terms of getting along with other farmers		
H1: Older respondents are better at getting along with other farmers	0.000	Reject H0
H0: There is no difference between old and young people in terms of membership of religious groups		
H1: Older respondents are more active members of religious groups	0.000	Reject H0
H0: There is no difference between old and young people in terms of membership of farmer cooperatives		
H1: Older respondents are more active members of farmer cooperatives	0.002	Reject H0
H0: There is no difference between old and young people in terms of membership of the local government		
H1: Older respondents have more active memberships of the local government	0.411	Accept H0
H0: There is no difference between old and young people in terms of membership of the local neighbourhood groups		
H1: Older respondents have more active memberships of local neighbourhood groups	0.011	Reject H0

Social capital and ethnicity

Ethnicity impacts how we behave and act. Dialect spoken is a major attribute that bonds members of an ethnic group. Does ethnicity, measured through mother tongue, influence social capital? Five dialects were spoken most by growers and traders. These are Ilocano (4% of respondents), Ibaloi (31%), Kankanaey (58%), Tagalog (1%) and Ikarao (6%). Table 10.14 shows that ethnicity through mother tongue is correlated with several social capital indicators.

There is a significant negative correlation between ethnicity and trust in the municipal government, volunteerism and active membership in the local government. There is significant positive correlation between ethnicity and relationship with other traders and membership in a trader cooperative. The next step is to test whether there is a significant association between ethnicity and social capital elements. Test hypothesis and test results using Kruskal–Wallis are shown on Table 10.15.

Kruskal–Wallis test results show that ethnicity is independent of participation in local government politics but plays a large role in terms of trust towards the government, informal community participation/ volunteerism and relationship/membership with traders and their organizations.

Does Social Capital Matter? Consequences for Local Vegetable Trade

Low social capital exists amongst indigenous agricultural communities in Benguet, northern Philippines. Our tests have shown that consistently low scores on membership in formal associations and trust override high scores on common goals and informal networks. Considering that vegetable exchange in the province is characterized by interpersonal trade between farmers and traders, low social capital could be one important limiting ingredient towards efficient market transactions. Perhaps one of the reasons why a favoured buyer system has sustained itself is because of low social capital. On the one hand, the system

Table 10.14. Correlation coefficients between social capital indicators and ethnicity (from own survey).

Social capital indicator	Correlation coefficient (2-tailed test of significance)
Trust municipal government	−0.074*
Participate actively in community and volunteer work	−0.100**
Get along well with other traders	0.074*
Active membership in a trader cooperative	0.095**
Active membership in the local government	−0.080*

*, ** and *** indicate significance at the 0.10, 0.05 and 0.01 levels (2-tailed), respectively; a Kendall's tau_b was used.

Table 10.15. Kruskal–Wallis tests for ethnicity and social capital variables.

Hypotheses	Asymptotic significance (1-tailed)	Decision rule
H0: Trust in the municipal government is independent of ethnicity	0.023	Reject HO
H1: Trust in the municipal government is dependent on ethnicity		
H0: Active participation in the community and volunteerism are independent of ethnicity	0.010	Reject HO
H1: Active participation in the community and volunteerism are dependent on ethnicity		
H0: Getting along well with traders is independent of ethnicity	0.054	Reject HO
H1: Getting along well with traders is dependent on ethnicity		
H0: Active membership of traders' cooperative is independent of ethnicity	0.005	Reject HO
H1: Active membership of traders' cooperative is dependent on ethnicity		
H0: Active membership of local government is independent of ethnicity	0.197	Accept HO
H1: Active membership of local government is dependent on ethnicity		

ensures that those who are favoured can easily dispose of their harvests in the market. On the other, those who are not within this system are excluded. This means that key players on both sides are not able to fully exploit market possibilities. Favoured farmers are compromised to sell their crops to selected traders, who may not have the highest price offers, while favoured sellers are compromised to buy crops from selected farmers who may not have the best vegetable grades.

Aside from low membership rates, perhaps one of the reasons for the failure of farmer cooperatives in the province to evolve as a bargaining force to reckon with is because of low solidarity among farmer-members. Low solidarity can derive from low social capital when farmers look into cooperatives for pursuing personal interests. This results in low collective bargaining powers among farmers.

The importance of formal networks must not be overlooked, because active membership in organizations could serve as a conduit between farmers and traders to other market institutions. Stone and Hughes (2001) showed that personal ties could emerge from involvement in formal associations because they foster repeated interaction between people with common interests. Formal institutions can also provide access to resources that would not have been possible in an informal relationship.

Low trust within Benguet markets spells higher transaction costs for parties. In the province, contracts are normally unwritten and incomplete because of unstable vegetable prices. Trust should be called upon in order to maintain the contract and oversee transaction completion. When trust is low, negotiation and enforcement costs increase because neither party is convinced that the other is honest in their transactions. When both parties can rely on each other to meet their ends of the deal, risk and uncertainty are reduced. Explicit cooperation can be expected, even without explicit contracting.

High social capital facilitates information exchange about prices and markets (Chloupkova and Bjornskov, 2002). In Benguet, where market information is scarce and unreliable, social capital is needed in order to disseminate critical market news in the quickest manner. Farmers and traders can rely on dense informal networks at the micro-level as a cheap but effective means for spreading information.

Conclusions

A total of 450 farmers and 195 traders from seven municipalities of Benguet were asked for opinions on 22 social capital statements and membership on seven community associations. The quantitative–additive method was used to calculate cognitive and structural social capital scores and create social capital index. Six components that underlay social capital were extracted from pooled Principal Component analysis. These were: (i) informal networks; (ii) core trust; (iii) institutional trust; (iv) poverty perception; (v) common goals; and (vi) life satisfaction.

Independent factor analyses for farmers and traders showed that informal networks and outer core trust, respectively, loaded heavily in terms of social capital motivations in the province. Social capital scores for farmers showed that they had significantly better community relations than traders. Traders scored higher membership in formal organizations for this reason: their overall social capital index was higher. Collectively, social capital is in its strongest in the form of common goals and informal networks. Membership in formal associations and low trust reduced social capital. In sum, all municipalities scored below the assigned 50 median point for social capital.

Social capital in Benguet is influenced, in varying degrees, by gender, education, religion, age and ethnicity. In particular, women represent more prolific resources that can be tapped into for increasing social capital levels through better community relations. Investing in education for the population will lead to increases in trust, not only within the community but especially towards the government, where it is much needed. The influence of religion to move people towards volunteerism, cooperation and government trust should not be underestimated.

Religious groups can be called upon for their manpower and support in times of need, and therefore should receive proper recognition from local government and society. In terms of age, it appears that the young are overlooking the positive effects of formal associations. Providing them with activities that promote collective action, venues for exchange as well as recognizing and rewarding their efforts will encourage more interaction in informal and formal settings. In this manner, community participation and a sense of responsibility is instilled in them during their formative years. Ethnicity seems to affect community relations in a more profound manner than was initially assumed.

However, the exclusionary nature of ethnicity-based social capital works against the foundation of community participation and development of government trust. Measures to override the negative effects of ethnicity, not only in intra-community relations but also in livelihood-related decisions, should be explored. Social capital affects vegetable production and marketing in the province in more profound ways than were expected. Low social capital resulted in the encouragement of the favoured buyer system that limits marketing possibilities for farmers and traders. Low solidarity resulted in the failure of farmer cooperatives to provide bargaining leverage to farmers in marketing crops. Because contracts are incomplete, market participants incur higher negotiation and monitoring costs as they cannot rely on trust alone to oversee transaction completion. Social networks are not sufficient to facilitate valuable information exchange about prices and markets.

This study highlights the important role of carefully directed policies from the local governments of the agricultural communities in fostering social capital. Because infrastructure and resource are at their control, the local governments can initiate efforts to increase intra- and inter-community social interaction. Tests showed that farmers' and traders' social capital stem from different sources. Therefore, the focus should be on increasing the bridging element of social capital: initially, by increasing positive interaction between farmers and traders. Later, steps to build links across different networks and organizations in the external environment could be taken. By providing opportunities for local cohesion, local citizens can be mobilized to think and act collectively.

Endnotes

1 Philippine Peso:EURO exchange rate = 70:1 as of January 2004.
2 The total population for the seven municipalities eligible for the survey is 26,329 farmers.
3 There was actually an incomplete list of registered farmers' and traders' cooperatives available. However, the author decided not to use them because: (i) the lists themselves were incomplete; (ii) they showed only registered cooperatives (as opposed to the non-registered); and (iii) a bias towards cooperative members would arise.
4 CHARM is the US$41M Cordillera Highland Agriculture Resource Management Project funded by the Asian Development Bank. VLIR is the Vlaamse Interuniversitaire Raad, having cooperation projects with two Philippine Universities.

References

Aquino, C. (2003) *The Philippine Vegetable Industry: Almost Comatose* (http://www.ppi. org.ph/index.htm, accessed 18 June 2005).

Batt, P.J. (1999) *Modelling Buyer–Seller Relationships in Agribusiness in Southeast Asia* (http://www.bath.ac.uk/imp/pdf/17_Batt.pdf, accessed Nonember 2003).

Benigno, T.C. (2002) Why investors avoid us: the trust factor. *Philippine Star*, Manila.

Beugelsdijk, S. and Van Schaik, A.B.T.M (2001) *Social Capital and Regional Economic Growth.* Tilburg University, The Netherlands.

Buenavista, G. (1998) Social Capital in Community-based Resource Management (http://poverty.worldbank.org/library/view/5862/).

Cabreza, V. and Caluza, D. (2002) Pattern shows Cordillera region dependent on veggie industry. *Philippine Daily Inquirer*, A-17.

Chloupkova, J. and Bjornskov, C. (2002) Could social capital help Czech agriculture? *Agricultural Economics* 48, 245–249.

Dalmo, D.B., Francisco, C.S.J. *et al.* (1994) *Marketing and Information Needs Assessment Report.* Bureau of Agricultural Statistics, Department of Agriculture, Philippines.

Davis, W.G. (1973) *Social Relations in a Philippine Market,* University of California Press, Los Angeles, California and London.

Gold, S. (1995) *From the Workers' State to the Golden State: Jews from the Former Soviet Union in California,* Allyn and Bacon, Boston, Massachusetts.

Grootaert, C. and Van Bastelaer, T (2002) *Understanding and Measuring Social Capital: a Multidisciplinary Tool for Practitioners,* World Bank, Washington, DC.

Heller, P. (1996) Social capital as a product of class mobilization and state intervention: industrial workers in Kerala, India. *World Development* 24, 1055–1067.

Knack, S. and Keefer, P. (1997) Does social capital have an economic pay-off? *Quarterly Journal of Economics* 52, 1251–1287.

Light, I. and Karageorgis, S. (1994) The ethnic economy. In: Smelser N. and Swedberg, R. (eds) *The Handbook of Economic Sociology.* Princeton University Press, Princeton, New Jersey.

Milagrosa, A. (2001) Marketing of vegetables in the Cordillera region Philippines: a transaction cost analysis. MSc thesis, Department of Agricultural Economics, University of Gent, Gent, Belgium.

Narayan, D. and Pritchett, L. (1999) Cents and sociability: household income and social capital in rural Tanzania. *Economic Development and Cultural Change* 47, 871–897.

NSCB (2001) *Primer for the Cordillera Admnistrative Region (CAR).* National Statistics Coordination Board, CAR, Philippines.

NSO (2002) *Philippine 2000 Census.* Philippine National Statistics Office (PNSO), Philippines.

Pekas, B.T., Badival, H.B., Fang-asan, V.F., Andiso, E.C. and Compay, D.F. (1998) *Inflow and Outflow of Selected Highland Vegetables in Baguio City.* Department of Agriculture (CARFU), Philippines.

Portes, A. and Sensebrenner, J. (1993) Embeddedness and immigration: notes on the social determinants of economic action. *American Journal of Sociology* 98, 1320–1350.

Putnam, R. (1993) *Making Democracy Work: Civic Traditions in Modern Italy,* Princeton University Press, Princeton, New Jersey.

Russel, S.D. (1989) *Informal Credit and Commodity Trade in Benguet Upland Luzon.* Cordillera Studies Centre, University of the Philippines, Baguio, Philippines.

Stone, W. and Hughes, J. (2001) The nature and distribution of social capital: initial findings of the Families, Social Capital and Citizenship survey. In: *Competing Visions National Social Policy Conference*, Australian Institute of Family Studies, Melbourne, Australia.

Woolcock, M. (1999) Managing Risks, Shocks, and Opportunity in Developing Economies: the Role of Social Capital. World Bank, Washington, DC.

IV Markets and Institutional Development

11 Making Markets Work for the Poor: the Challenge in the Age of Globalization[1]

ELENI Z. GABRE-MADHIN

Ethiopia Strategy Support Program, International Food Policy Research Institute, PO Box 5689, Addis Ababa, Ethiopia; telephone: 251 (0)11-646-3215; fax: 251 (0)11-646-2927; e-mail: e.gabre-madhin@cgiar.org

Objectives and Organization of the Chapter

Even as the forces of market liberalization and globalization sweep across the world, the jury is still out on whether market reforms have benefited the poor. It is yet unclear how the poor can interact with the market in a way that improves, rather than undermines, their welfare. What do we mean by making markets work for the poor? If markets cannot be intrinsically pro-poor, in our view, the relevant question is then: how can the poor interact with the market in a way that improves, rather than undermines, their welfare? It can be argued that, at present, the poor bear the brunt of market volatility (World Bank, 2000) while benefiting little from its opportunities for positive gain. This implies that, even when markets work, they do not necessarily work for the poor.

In this chapter, we aim to explore and refine what the implications of being poor are for dealing with the market. We start by reviewing the experience of market reform and its impact on the poor, particularly in Africa in the following section. This is followed by analysis of how to get markets right, then we lay out the dimensions of poverty that influence how the poor interact with the market. The dimensions of poverty that we consider are assets, space, exclusion, power and vulnerability. In the penultimate section, we review various donor-driven approaches and best practices for integrating smallholders into the market and raise the concept of a Commodity Exchange as a promising new direction, followed by conclusions in the final section.

The Stakes for Smallholder Agriculture in the Age of Globalization

The role of small-scale agriculture has witnessed a resurgence of interest in the recent poverty reduction debate, as it is presented as a 'growth-equity win-win' (Vorley and Fox, 2004). Nevertheless, the evolving global agrofood system and the advances of market liberalization have raised unique challenges for smallholder integration into the global market and have polarized the debate between staples versus high-value agriculture, and between domestic versus export-led market integration. Thus, it is argued that smallholders need not be marginalized in the changing global agrofood system, as they may be involved in export horticulture as employees on large plantations, commercial farms or packing plants, or as independent farmers sometimes working under contracts with exporters. There have been many concerns about the impact of contract farming on poor households, but some recent studies suggest that, under certain circumstances, there are rewards for smallholder contract farmers (Stringfellow and McKone, 1996; McCulloch and Ota, 2002).

In the broader debate on whether smallholders have benefited from globalization, the winners have been those that are: (i) vertically integrated with agribusinesses or are organized into farmer organizations for collective strength; (ii) have access to better infrastructure and credit; and (iii) have benefited from the role played by the public sector and others in capacity building (Narayanan and Gulati, 2002).

Others have argued that increased domestic demand for staples, coupled with investments in productivity-enhancing technology and measures to reduce marketing costs, will have significant potential for poverty reduction and growth (Diao and Hazell, 2004). This perspective is in sharp contrast to the view that the growth potential of agriculture lies largely in non-staples production (Maxwell, 2004). At yet another level, the debate has centred on whether interventions in the post-reform era should focus on building market linkages for smallholders through supply chain development or whether broader interventions to build institutions for markets such as warehouse receipts systems and market information are more appropriate.

Even as the forces of market liberalization and globalization sweep across the world, it seems the jury is still out on whether market reforms have benefited the poor. To begin, few studies have explicitly studied this question. Thus, while we know with increasingly sophisticated methods that market reforms have been good for markets themselves, it is not entirely clear whether markets are good for the poor (Jayne and Jones, 1997; Kherallah *et al.*, 2002). The few studies that have directly addressed the question of markets and poverty have indicated at best mixed or even negative outcomes (Peters, 1996; Sahn *et al.*, 1997; Deininger and Okidi, 2001; Dercon, 2001; Christiaensen and Pontara, 2002).

Most of this literature finds that, among the poor, those that have benefited from markets are those with better endowments or better access to markets. These studies have viewed poverty impacts as changes in producers' terms of trade, changes in price volatility facing producers or direct income effects. With few exceptions, the poor are treated as net sellers of agricultural goods and the analysis of impact has been partial, focusing on output price or income derived from production, rather than on the notion of returns to assets or to production inputs. Most studies have also taken a static perspective, focusing on the poverty impacts of those already in the market rather than addressing the question of changes in access to markets of the poor over time resulting from market reforms.

Yet, the net market position of the poor does matter a great deal in their interaction with the market, i.e. whether they are net buyers or net sellers of agricultural goods. Recent studies have revealed that, even in highly subsistence-oriented agricultural systems, a significant share of rural households are net buyers: 46% in Ethiopia (Clay *et al.*, 1999) and 42% in Mali (Dione, 1989). Physical access to markets is also a highly relevant indicator. The fact that the poorest countries in the world are also those with the lowest densities of roads and telecommunications suggests that markets can have only limited impact on the poor in these contexts, at least if viewed statically.

Another set of studies has attempted to contribute indirectly to this question through evaluation of the impact of market reforms on supply response, which is viewed as a proxy for the incomes or well-being of the poor. These studies point to the overwhelmingly disappointing impact of market reforms on production.

More recently, as commitment to the concept of 'pro-poor growth' has grown, a literature has emerged around the notion of 'pro-poor markets' (DFID, 2000; Christiaensen and Pontara, 2002). Embedded in the idea of pro-poor markets is the concept that market outcomes should disproportionately benefit the poor. This concept seems problematic in that it is not at all clear how market processes, in and of themselves, would accrue gains in a disproportionate manner to a relatively less-endowed segment of the population. In fact, the opposite may be argued if one considers that market outcomes are, in fact, path dependent in that markets reward those with greater initial assets (Christiaensen *et al.*, 2002). Thus, without *ex ante* or *ex post* asset redistribution, markets will at best be poor-neutral rather than pro-poor.

How have Smallholders Responded to Market Liberalization?

Earlier empirical analyses on the impact of market reforms on smallholder agriculture suggest mixed outcomes. Von Oppen *et al.* (1998), in a comparison of aggregate productivity of smallholder farms in India, Kenya and the Sudan, found that improved market access results in increased on-farm productivity. They found that, in the Nakuru district of Kenya, a 10% improvement in market access resulted in an increase of 1.5% in aggregate productivity, of which 0.4% was achieved through specialization and 1.1% through intensification. However, they also found that gains from market access increased with farm size (Kamara and Von Oppen, 1999).

In contrast, Jayne and Jones (1996) argue that market reforms have led to the demise of what was an emerging smallholder maize revolution in the 1980s in parts of eastern and southern Africa, where the pre-reform policy regime featured state-led investments in inputs, credit and purchasing centres.

Prior to reforms, a large proportion of smallholders benefited from implicit transport subsidies in pan-territorial pricing alongside input subsidies and concessional credit. Per capita smallholder grain production in Zimbabwe and Zambia increased by 51 and 47%, respectively, between the late 1970s and late 1980s. In Kenya and Tanzania it rose by 30 and 69%, respectively, between the 1970–1974 and 1980–1984 periods. At the same time, production growth in this region was achieved at a cost greater than the value of the output, and state-led provision of services to smallholders proved both politically and economically unsustainable (Jayne and Jones, 1996). With the partial or complete removal of explicit subsidies to smallholders, hybrid maize seed purchases and fertilizer use declined in the early 1990s in this region, and population growth has outpaced grain production growth in most of eastern and southern Africa. While part of the food output decline in the early 1990s was due to the 1992 drought, the downward trend in production growth since the 1980s remains.

Sahn and Arulpragasam (1994) argue that production in Malawi has not risen following market reforms, primarily because real producer prices have not risen. Moreover, smallholders exhibit price responsiveness by reallocating resources among crops. Thus, existing data suggest that an inverse relationship exists between maize production and the relative price of cash crops. In this view, this effect explains the lack of intensification of maize production above population growth in Malawi. In Zambia, Chiwele *et al.* (1998) note that the results of a survey of smallholders in the post-reform era suggest that, while the majority of rural households had access to agricultural extension and credit services, smallholders faced problems in marketing their output under the liberalized system. Smallholders became more vulnerable *vis-à-vis* private agents because of the cash liquidity constraints that forced them to sell at harvest time rather than store output on-farm. Similarly, credit constraints led to barter transactions at disadvantageous terms for smallholders. Elsewhere, in Tanzania, Ghana and Mali, liberalization has increased the smallholders' role in storing maize, although storage losses are considerable (Coulter, 1994).

The effect of market reforms on smallholders depends in large part on the extent to which the pre-reform policies taxed or subsidized smallholder production (Kherallah *et al.*, 2002). In the eastern and southern African context, market reforms have resulted in the removal of input and credit subsidies, the positive effects of which have not been offset by the gains from lower-cost, private distribution systems in the short term.

In contrast, reform in other contexts, such as in West Africa and elsewhere where smallholders have been taxed heavily, has had an initially positive, although limited, impact on production. Generally, it appears that market reforms have improved the distribution of inputs and outputs but have not led to increased demand by farmers for modern inputs. The increased costs of inputs after reforms are partly offset by increased output prices and an improved input distribution system. However, due mainly to lack of access to credit, the use of modern inputs consequent gains in productivity remain low (Badiane *et al.*, 1997).

The process of state disengagement in food markets raises two major concerns. In the case of farmers, the key issue is whether and how private sector agents will fill the gap left by the parastatal agency. The two main concerns are: (i) whether farmers benefit or are penalized by their reliance on parastatals; and (ii) whether private food markets function well. In the case of consumers, the key issue is whether consumers had access to food at subsidized prices prior to reform and whether liberalization will harm the poor as subsidies are removed.

In sum, reforms have had limited impact on grain production and agricultural productivity; and have increased price instability for both producers and consumers (Jones, 1998). Remaining issues are: (i) the low level of investment and specialization by private traders; (ii) the failure of market development to progress to more sophisticated arrangements such as forward trading and quality premiums; and (iii) high transport costs (Beynon *et al.*, 1992). Almost universally, private sector agents are constrained by limited access to credit and storage facilities, as well as by problems in securing transport (Beynon *et al.*, 1992; Badiane *et al.*, 1997). As a result, turnover of stocks is rapid and seasonal storage is rare, which serves to exacerbate the volatility of prices. The challenge is twofold: first to get markets right in and of themselves; secondly to make markets work for the poor. To do so implies going beyond improving markets *per se* to addressing the dimensions of poverty that lead to weak outcomes for smallholders, even when markets function effectively.

Getting Markets Right: the New Agenda

Beyond the structural adjustment-led era of the last two decades of the 20th century, the fundamental market problem in the 21st century is *not* whether to free or restrict markets: it is to understand how markets function, what roles different institutions play in supporting market exchange and how to design, transfer and maintain these institutions. This implies a shift in policy thinking from the earlier dominant perspective of 'getting prices right' to that of 'getting markets right'.

Getting prices right implies that, once policy incentives are aligned, market order would emerge spontaneously or endogenously and that markets would take care of themselves. Getting markets right implies that market order depends on an underlying set of institutions and supporting infrastructure, requiring guidance from a 'visible hand' rather than from Adam Smith's invisible hand.

Thus, in the new agenda, getting markets right also implies a concerted need and challenge for the public sector to facilitate the role and performance of the private sector. Thus, in the post-reform era, the relation between the state and private actors must be initially defined and redefined as the market itself evolves.

Privatization, institution building and infrastructure development are complex tasks that need long-term investment and commitment. These types of reforms are not easy to implement given the short-term nature of policy-making. In addition, these changes are more difficult to incorporate in policy-adjustment lending programmes of international donor organizations. In the case of sub-Saharan Africa, in particular, this means that the steps ahead for further reform in Africa will be more difficult to achieve and will require readjustment of both government and donor behaviour.

In practical terms, getting markets right suggests the following: building markets in which buyers and sellers are well coordinated, transaction costs are low, contracts are enforceable, risks are manageable, exchange can be impersonal, price volatility is dampened and transactions are liquid and highly responsive to shifts in supply and demand. To achieve the above, efforts to transform the market must occur over a sustained period of time in which market development is progressively achieved.

Moreover, market adjustment requires a gradual alignment of incentives and behaviours within the context of institutions and even social norms. Moreover, these efforts require a balance between policy incentives, the broader infrastructural environment and the development of appropriate market institutions. These can be considered the '3 I's of market development': incentives, infrastructure and institutions.

Looking more closely at the elements within the framework of the '3 I's', incentives involve the overall policy environment and the stability therein, the general investment climate, the macro-economic framework as well as tax and trade policies. Infrastructure for markets is comprised of telecommunications, transport, storage and logistics in terms of physical capacity as well as research, skills and extension in terms of technical capacity.

While roads are often given the bulk of attention in discussions of market failure, the modern wave of globalization suggests that we are in the midst of not only an information revolution but also a logistics revolution, in which success in the market is ultimately determined by processes such as, among others, just-in-time delivery.

Finally, market institutions, which have, perhaps, been most obviously neglected and whose role has been least understood in the post-reform era, concern market information, grades and standards, contract enforcement, the coordination of market actors, trade and producer associations, market regulation, industry-wide forums for dialogue and trade finance. While each of these dimensions implies a significant role for the state, the private sector – defined as the producers, traders, processors and service providers such as those found in transport and storage – plays a pivotal role (see Fig. 11.1).

It should also be noted that there are significant interactions between these three dimensions. For example, in the case of transport, while it is common to perceive transportation as being a function of access to good roads, a large part of transport costs is related to the coordination of supply and demand in the transport market. Thus, in contexts where there is weak information regarding demand for transport and frequent delays in the system, costs tend to rise considerably as the costs are covered mainly by the 'fronthaul' or first leg of the trip, as the 'backhaul' or return trip is frequently under-utilized because of information gaps. Moreover, even with good roads and appropriate coordination, bad policies such as restrictive import policies or licensing or tax disincentives can still result in high transport costs and market failure if these policies result in collusion or thin transport service provision. Thus, the key challenge in market development is not to view these three elements in isolation but, rather, to approach them in an integrated or holistic manner.

The three dimensions of market development are significantly interrelated and jointly affect market outcomes. This integrated approach also clearly delineates what are the public and private roles in the system and what the relations are between them. The few successes of market reforms in Africa suggest that success depends on precisely adopting this integrated approach in which the public sector creates a space for private actors.

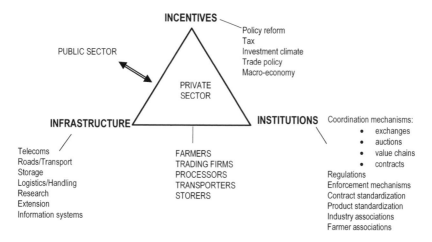

Fig. 11.1. Integrated perspective on market development.

Markets and Poverty

To understand how markets can be made to work for the poor, we must first determine the relevant dimensions of poverty. The understanding of poverty itself has evolved considerably in recent years. Traditional notions of poverty primarily encompassed material deprivation or the command over market-purchased goods. Over the years, it has broadened to include attainments in health and education, as well as vulnerability and exposure to risk, powerlessness and lack of voice.

Lack of command of material possessions

This is a key dimension of poverty. The conventional measure to capture material deprivation entails analysis of income or consumption levels of the household *vis-à-vis* an established poverty line. Rowntree, an early analyst of poverty, estimated the minimum budget required for a family of six to obtain food, shelter, clothing, fuel and sundries, and thereby estimated the poverty line. This approach more or less characterizes present-day methods. The preferred measure is to use current consumption over all marketed commodities consumed by the house-hold from all sources (Lipton and Ravallion, 1995).

Assets

In order to better understand the determinants of poverty in all its dimensions, it is useful to think about the assets owned by people and the returns they obtain from these assets (World Bank, 2000; DFID, 2002). These assets include natural, physical, financial, human and social elements.

For example, studies from sub-Saharan Africa show the importance of initial private endowments in education and land as key conditioning factors that impact the extent to which households benefit from reforms (Christiaensen *et al.*, 2002). Similarly, in India, Datt and Ravallion (2002) found that initial literacy affects the extent of pro-poor growth in Indian states or the capacity of the poor to participate in opportunities created. Asset ownership impacts the gains from market reform because it affects both transactions costs and productivity.

Access to physical assets such as infrastructure (roads, electricity) or storage and transport facilities, along with the spatial dimension of poverty, also has an impact on transaction costs. Empirical studies evaluating the characteristics of groups who benefit from food market reform show that access to roads and land were significant, for example in Ethiopia (Dercon, 1995). Inadequate infrastructure and distance from markets imply that transaction costs rise due not only to higher transport costs but also to the increased costs of screening, bargaining with and monitoring distant trading partners (Staal *et al.*, 1997).

A case study of dairy transactions in East Africa illustrates some of these issues (see Box 11.1). It represents a case where liberalization of the sector provided producers with more options with respect to the marketing outlets they could use, but the ability to use these alternative outlets – some of which offered higher prices and more stable payments – was determined by their level of transactions costs. Communities further from markets restrict the parties with whom they make transactions.

Box 11.1. Dairy transactions in East Africa (from Staal *et al.*, 1997).

A study of transactions in the East African dairy market shows how access to infrastructure, storage and transport systems and spatial factors impact transactions costs, and thereby the marketing behaviour of small-scale producers.

Until the liberalization of the dairy industry in 1992, the Kenya Cooperative Creameries (KCC) had a monopoly on fluid milk sales in urban areas. Dairy cooperatives sold their milk to private buyers or collection centres belonging to KCC, which was effectively a parastatal situation. Prices that KCC paid to the cooperatives were set by the government, and the pricing policies adopted were pan-seasonal and pan-territorial. Nevertheless, these pricing schemes severely compromised KCC's financial position over time. The squeeze on its balance sheets meant that KCC was unable to increase milk prices to keep pace with increases in input prices and, furthermore, payments by KCC to the cooperatives were delayed. As a result, producers began to shift sales to the informal raw milk market, and supplies to KCC fell substantially.

In 1992, the Kenyan government liberalized the dairy industry and revoked KCC's monopoly on urban milk sales. There was a rapid development in milk market innovations, including 'self-help' groups, which collect and market raw milk. In addition, cooperatives began themselves to market a greater portion of the milk to urban markets. However, in order for cooperatives to sell to other marketing outlets, where they typically received higher prices and more stable payments, they had to be willing to incur the costs of searching for alternative market outlets, screening trading partners and bargaining with them, and monitoring contractual agreements with new trading partners. In addition, the cooperatives, many of whom did not have adequate transportation or cooling equipment, incurred losses due to spoilage.

A survey of dairy cooperatives in three districts provides evidence of the impact of these transaction costs on dairy marketing. It shows that all three districts enjoyed growth in alternative marketing outlets, but the district closest to Nairobi had a dramatic decline in the share of output going to KCC. In contrast, the districts further away showed only a marginal decline in milk sales to KCC, despite the fact that producers received lower prices and unstable payments than they would have obtained through alternative channels in Nairobi.

Hence, for districts further away, which are likely to have higher transactions costs due to the transport costs, spoilage and screening/monitoring, the transaction costs outweighed the losses imposed by the uncertainty of payments by KCC and the lower prices. Hence, while liberalization permitted the growth of new marketing channels and outlets, the level of transaction costs determined the extent to which different producers could actually utilize different marketing channels.

Asset ownership affects the gains from reforms, not only by increasing transactions costs but also from the important implications for productivity. Asset ownership, in particular access to credit, notably impacts productivity of existing assets, which would allow the poor to expand their scale of current activities or to enter into value-added activities. For example, the importance of financial capital is particularly evident for the underlying competitiveness of the poor in land markets. Imperfect credit markets, subject to information and moral hazard problems, lead to credit rationing for small and near-landless farms. Hence, the rural poor may not be able to compete for land and, moreover, face a severe disadvantage in improving the productivity of their land and labour without access working capital.

In a theoretical model, Carter and Zegarra (2000) show how constraints in access to credit (and multiple market imperfections) can interact to reduce the underlying competitiveness of producers in the land market and result in different land market outcomes. In addition, weak credit supply for smallholders constrains adoption of more productive agricultural technology, despite extensive liberalization.

Limited access to finance

Thus, a major cause of material deprivation is limited access to finance. Credit constraints stemming from problems of information asymmetry and moral hazard are comprehensively analysed in the literature (Stiglitz, 1989; Hoff and Stiglitz, 1990). These factors, combined with inadequate access to collateral, restrict access to finance, particularly for the poor. In Kenya, for example, only 4% of the poor have access to credit through banks and another 3% through cooperatives. Even informal mechanisms can exclude the poor from obtaining credit. For example, a study of moneylenders in Chambar, Pakistan, shows that the average rate charged on interest for credit was 79% per annum. These rates reflect a combination of high screening and administrative costs, and the size of loans (Aleem, 1990). Access to physical assets such as technology, production inputs and infrastructure also contribute to material deprivation (World Bank, 2002).

The notion of poverty has broadened to include access to human capital, gaining intellectual support from Sen's definition of poverty in terms of 'capabilities'. Capabilities are not only intrinsically important, but are also instrumentally significant for achieving incomes (Glick and Sahn, 1997; Sen, 1997). The relationships observed between human capital and health and between education and income poverty are stark. In many countries, indicators such as child mortality rates, malnutrition, incidence of illness, access to schooling and educational achievement are significantly worse for the income-poor than for the non-poor. For example, in a sample of Asian and African countries, the proportion of children born in the previous 5 years but no longer living is between three and six times larger for the poorest decile than for the richest.

Steady improvement in this variable is observed as wealth increases (Kanbur and Squire, 2000).

With regard to illness, the prevalence of tuberculosis in India is four times higher in the poorest 20% of the population than in the richest, and the prevalence of malaria more than three times higher (World Bank, 2000). Similarly, with respect to education, a study of poorest households by income level in five Latin American countries shows that low-income households attain fewer years of education than richer households. In the poorest households, the number of years of education for 25-year-olds range from 20–50% of that attained by counterparts in wealthier households. In each country, the number of years of education increases as one moves up the income scale (Kanbur and Squire, 2000).

Access to information

Market information systems, grades and standards are amongst the key assets that impact transactions costs. In many countries, market information systems perform poorly or are non-existent due to both inadequate funding and the ability of government agencies to collect reliable market information. Following liberalization, new information systems to replace those previously administered by the state marketing channels are very underdeveloped, particularly since the private sector has not been able to assume the institutional role previously fulfilled by the state (Jones, 1998; Chaudhury and Banerji, 2001).

Poor producers and consumers, who lack the scale to collect their own information, are particularly adversely affected. Transactions costs, such as costs of acquiring information or search for marketing and trading partners, increase. Similarly, inadequate access to standardized systems of grades and standards, which can provide a greater level of certainty about the quality of produce, also increases search and screening costs. It implies that traders visually inspect each product. More generally, inadequate access to market information implies that the poor are unable to plan their production, harvesting and sales according to market demand, or to sell their products in the most lucrative markets.

Limited access to information has several implications for the livelihood strategy that the poor adopt. First, reliance on trader information and limited access to reliable information implies that smallholders are often forced to sell in a buyers' market, with consumers buying in a sellers' market. Secondly, access to information, particularly for communities far from markets, impacts their decision to commercialize. For example, studies show how poor women in a region called Banaskantha, in the northern part of Gujarat, India, did not commercialize their traditional craftwork because they had no understanding of the market value or demand for their products. As a result, they simply used to barter their craftwork with occasional traders for plastic buckets, which were significantly below the value of their own handicrafts. When programmes designed to educate these women about market prices and commercial value for their products were initiated, they began to commercialize and generate a livelihood from this source. Finally, information asymmetry also impacts the power relations between poor producers and traders, and lowers the bargaining power of the former.

Risk and vulnerability

In addition to asset ownership, risk and vulnerability are also key dimensions of poverty, and have recently been brought to the fore (World Bank, 2000). Nearly all poverty reduction strategies and programmes (PRSPs) conducted by 21 countries and participatory poverty assessments in countries such as Zambia, Djibouti and Tajikistan show that the poor associate poverty with insecurity, uncertainty, exposure to risk and vulnerability. While all individuals face risk from a variety of sources, the importance of the agricultural sector to the poor and their vulnerability in labour markets[2] exacerbate the risks they face. As many poor earn their income from agriculture, they are particularly exposed to natural disasters, seasonality, year-to-year variability and commodity price volatility. Given their limited ability to cope with risk due to resource constraints and absence of formal risk insurance markets, the poor are left vulnerable. With limited options to manage risk through formal market mechanisms, they experience significant fluctuations in income, which makes consumption smoothing difficult.

One study of South Indian villages shows that the coefficient of variation of annual income from main crops ranges from 0.37–1.01 (World Bank, 2000). In rural China, where those in the poorest wealth decile are also the least well insured, 40% of an income shock is passed on to current consumption, in contrast to the richest third of households whose consumption is protected almost 90% from income shock (Jalan and Ravallion, 1997).

Vulnerability and risk also have implications for livelihood strategies adopted by households. Liquidity-constrained households may choose risk management strategies to lower the variability of household income, such as choosing crops with a lower yield or price variability, diversifying crops, etc; these strategies can have costly negative effects on productive efficiency (Walker and Jodha, 1986; Alderman and Paxson, 1994; Morduch, 1995).

These strategies may be observed in a number of countries. For example, poor Indian farmers devote a larger share of land to traditional varieties of rice and castor than to high-return varieties. Similarly, Tanzanian farmers without livestock grow more sweet potatoes – a low-risk, low-return crop – than do farmers who own livestock. As a result, returns to farming per adult household member are 25% higher for the wealthiest group than for the poorest (World Bank, 2000). In a study of cattle-rearing in Western Tanzania, Dercon analysed the link between wealth, risk and activity choice. In the presence of credit constraints, risk and fixed set-up costs to obtain cattle, poor households are less likely to invest in cattle-rearing activities, though it is a more profitable enterprise (Dercon, 1998).

Finally, vulnerability and risk also impact accumulation and depletion of productive assets which, as discussed earlier, impact benefits from market participation. Households facing high borrowing costs or those constrained in their ability to borrow may attempt to smooth consumption through accumulation and depletion of productive assets, which can increase costs and detrimentally affect productive efficiency (Rosenzweig and Wolpin, 1993).

Lack of power and voice

Poverty is also associated with the lack of power and voice, both in the economic sphere and with respect to state institutions. While empirical evidence concerning income poverty and powerlessness is scarce, participatory poverty assessments reveal the nature of concerns felt by the poor. For example, in the economic sphere, concerns about exploitation by middlemen and traders are consistently voiced by the poor. With respect to state institutions, the poor report being harassed, ignored and unable to make their voices heard. In Kenya, for example, the poor complained that district officials tended not to go through the villages or would pass through quickly without talking to the poor about their problems. Interviews with officials indicated that officials were not aware that health clinics regularly charged fees to the poor (Narayan and Nyamwaya, 1996). In Mexico, leaders provide services in exchanges for votes. This trading is the only way that the poor can acquire access to land, housing and urban infrastructure. The vertical hierarchy of this patronage system led many of the poor to perceive a lack of opportunity to act individually or as a collective group in their own interests.

Social marginalization

Poverty, as defined in economic terms, is often accompanied by social marginalization, which can take the forms of exclusion and discrimination. These can be based on rigid sociopolitical hierarchies, or on discrimination based on ethnicity and caste, or gender. For example, evidence from India shows that scheduled castes and tribes face a higher risk of poverty. Among rural scheduled caste women in India, the literacy rate was 19% in 1991, as compared with 46% for men, and compared with a 64% average for the country as a whole (World Bank, 2000). Social stratification or discrimination can also be exacerbated by geographic isolation. For example, the disproportionately high poverty rates among indigenous groups in Latin America reflect their greater distance from markets, schools, hospitals and post offices. Evidence shows that indigenous groups also receive less education on average than non-indigenous groups.

Similarly, gender-based inequality, which varies by country, depends on a number of factors, among them local kinship rules. Systems of kinship, which determine rules of inheritance and rights under marriage, can perpetuate political, legal, economic and educational inequalities. In Botswana, Lesotho, Namibia and Swaziland, rules of marriage determine a woman's autonomy and control. In these countries, married women have no independent right to manage property, except unless otherwise specified in a prenuptial contract. Similarly, in a study of six countries in Latin America, women's access to and capacity to utilize land productively was hindered by customary rights (World Bank, 2000).

A growing literature on social capital explains the role of personalized relationships, trust and social norms in facilitating market interactions in a world characterized by imperfect information and enforcement. Personalized relationships can facilitate the following benefits: (i) circulation of information about prices and market conditions; (ii) the provision of trade credit; (iii) the prevention and handling of contractual difficulties; (iv) the regularity of trade flows; and (v) the mitigation of risk. In the presence of information asymmetries and cost of monitoring and enforcement, transactions may take place only between groups for whom people have information or are known to be reputable. In this case, the inability of the poor to engage in reciprocity may imply that they are effectively excluded from transactions (World Bank, 2000).

Social capital can also impact knowledge diffusion, and thereby the productivity improvements that producers can make. A study of how local social structures affect fertilizer adoption among rural households in Tanzania shows that households with ethnically based and participatory social affiliations may be more likely to diffuse new information successfully, and to adopt new technologies. In addition, to the extent that the poor live far from urban markets and do not have the social networks in these markets, knowledge diffusion is affected. Research on diffusion of knowledge and innovation highlights the importance of proximity, both cognitive and spatial, and face-to-face contact between agents (Maskell and Malmberg, 1999).

Geographic space

Finally, poverty is also defined in terms of geographic space. Across and within countries, poverty is concentrated in less-favoured rural areas and tends to be associated with distance from cities and the coast. Most countries that undertake PRSPs prioritize rural and agriculture-based development, based on the preponderance of the poor who live in generally less-favoured, rural areas. In Thailand, the incidence of poverty in the remote rural north east was almost twice the national average in 1992, and although only one-third of the population lives there, it accounted for 56% of all poor. In Peru, 20–30% of the rural households in the poorest quintile are in the less-favoured mountain regions, while fewer than one-tenth are in the better-endowed coastal area (World Bank, 2000).

The literature on spatial poverty has attempted to identify whether location can make the difference between growth and contraction in living standards for otherwise identical households. Empirical work in rural China shows that indicators for geographical capital do indeed have an impact on living standards. One explanation is that neighbourhood endowments of physical and human capital influence the productivity of a household's own capital (Jalan and Ravallion, 1997).

In sum, market participation by smallholders can be viewed as a function of the extent to which these various dimensions of poverty constrain their engagement: asset ownership, access to credit, access to information, social connectedness, power and geographic location.

Empirical evidence emerging in various African countries confirms the importance of assets, measured directly – as well as indirectly by land size – for maize sales. The results shown for Zambia in Table 11.1 reveal that sales, as well as the concentration of market surplus, are driven by assets (Jayne *et al.*, 2005).

Making Markets Work for the Poor

Market development efforts in the post-liberalization era have focused on two types of interventions: fostering reliable market linkages for smallholders – particularly to high-value export markets – and strengthening the institutional arrangements that govern markets. Both interventions are reviewed briefly below.

Building market linkages for smallholders: the value chain approach

The premise for interventions has been that market forces alone will not ensure the integration of smallholders into the global market, because of the high transaction costs associated with involving numerous, small-

Table 11.1. Assets and commercialization in Zambia (from Jayne *et al.*, 2005).

	n	Farm size (ha)	Asset values (US$)	Gross revenue, maize sales (US$)	Gross revenue, crop sales (US$)	Total household income (US$)
Top 50% of maize sales	23,680	6.0	1,558	690	823	2,282
Rest of maize sellers	234,988	3.9	541	74	135	514
Households not selling maize	762,566	2.8	373	0	36	291

scale and geographically dispersed producers. A review of interventions by Joffe and Jones (2003) considers that efforts have focused on two areas: establishing rural retail networks for inputs and creating farmer-based enterprises linked to global markets. In these efforts, either non-governmental organizations (NGOs) or donors have played a very active sponsoring role. Activities included in this effort include the following:

- Identifying and training rural retailers.
- Facilitating supply contracts between input suppliers and retailers.
- Providing partial credit guarantees to suppliers.
- Providing demonstrations to farmers on technologies.
- Facilitating the formation or strengthening of farmer marketing groups (associations, clubs, cooperatives).
- Undertaking commodity market studies and providing information services.
- Facilitating contractual agreements with buyers.

Non-governmental organizations have been pioneers in these efforts. What has come to be known as the 'Rockefeller model' has focused on establishing rural input retailer networks in eastern and southern Africa. Similarly, what might be considered the 'USAID model', through partners such as CLUSA, ACDI/VOCA and Technoserve, has been heavily engaged in the creation of producer market-oriented organizations, operating as business enterprises in western, eastern and southern Africa. These approaches have demonstrated early successes in linking smallholders to the global value chains and in developing a business orientation in collective action groups. However, in considering scaling-up of these efforts, it remains unclear to what extent programme costs outweigh the benefits or whether the initiatives will survive beyond the lifetime of the projects (Joffe and Jones, 2003).

Building institutions for markets: the market development approach

With regard to traditional commodity or staples markets in particular, the requirements that have emerged following the dismantling of state marketing enterprises are:

- Mechanisms for transparently grading and standardizing products for market, from the production level onward throughout the market chain.
- Market information that is accessible to all market actors.
- Fostering competitive practices among all market actors, across all levels of the chain.
- Financial markets to respond to market needs for trade finance, for inventory finance, and for alternative financial products.
- Evolution of dispute settlement and regulatory systems according to market needs, and in a way that relies also on the private incentives

for self-regulation, notably through the potential role of trade associations.

- Risk transfer through mechanisms such as forward contracts and transferable warehouse receipts.
- Concerted efforts to build capacity throughout the marketing system, including cooperatives, small and medium private traders, and public actors.

Up to the present time, interventions concerning the above have tended to involve the creation of long-term institutions and have thus involved national governments to a greater extent. However, the experience of sustained efforts is limited and the impact has generally been mixed. Efforts have been focused on three of the above areas: market information systems, grades and standards and warehouse receipt systems.

Market information

In many countries, market information is collected, analysed and disseminated by a number of organizations – federal and regional government organizations, cooperatives, donors, international organizations and NGOs. The data collection methodologies and procedures vary considerably from organization to organization and must be standardized in order to make such data comparable and commercially valuable. A clear conceptual framework regarding the levels of the market and the quality standards for which price data are quoted by the different organizations needs to be devised and implemented in collaboration with the different organizations engaged in data collection.

Grades and standards

With regard to a viable system of grades and standards, which is vital to market development, one key issue is how to translate standards to the very basic level of production in the commodity chain. The biggest challenge in standards implementation is translating standards to the farm level. Currently, there is a wide gap in the implementation and enforcement of standards on various products, and many of the prepared standards have been shelved across various countries.

Inventory finance

Banks have generally been reluctant to engage in inventory finance linked to a warehouse receipts system, because of the high risks in agriculture and an insufficiently secure receipts system.

In many countries at present, the recent market development agenda still remains fraught with internal tensions and critical concerns. At the heart of these concerns is the need to consider market development as an integrated whole rather than the sum of piecemeal interventions targeting different sets of actors. This is as much a matter of perspective as design, and can be viewed as the 'fallacy of composition' argument that considers that the sum of the parts equals the whole.

An illustration of this fallacy is the promotion of contractual arrangements between farmer groups and industrial buyers without consideration of the broader whole that is the market mechanism, in which buyers and sellers must arrive at an appropriate market-clearing price, determined through an accepted and transparent system of measurement of quantity and quality, and within a system that ensures that contracts are enforced and property rights are secure. A second example might be the tensions inherent in the promotion of a system of inventory credit, a financial instrument designed to meet price stabilization objectives in the absence of accompanying measures to provide transparent information on product prices, qualities, stocks and warehouse performance and a viable dispute settlement mechanism, all of which are essential in providing incentives for participation of the financial system.

The promise of the new commodity exchanges: toward an integrated approach

Achieving a holistic perspective on market development with significant impacts on smallholders remains an important challenge. A successful commodity exchange facilitates transactions between market participants – farmers, processors, traders, consumers, food aid agencies, parastatal agencies and others – in a low-cost environment. The lowering of costs is passed on to market actors, who can then directly benefit from a higher share of the final price. This in turn generates incentives for increased market volume, and provides an incentive for increased participation in the market.

As an institution, a commodity exchange itself depends on a number of linked institutions, which are critical to its functioning. These core institutions are: (i) a market information system; (ii) a system of product grading and certification; (iii) a regulatory framework and appropriate legislation; (iv) an arbitration mechanism; and (v) producer and trade associations. In addition, a warehouse receipts system is a very important related institution. A commodity exchange also depends on the functioning of 'allied' sectors: banking, insurance, transport, IT services and even inspection services. Thus, while these sectors are not strictly part of an integrated institutional development plan, they must, none the less, be engaged and involved and taken on board as the exchange development proceeds (see Fig. 11.2).

When linked to a negotiable warehouse receipts system, the increased liquidity as market transactions increase over time evolving to futures trading implies that the thinness of markets lessens, and the market can be expected to enable the transfer of risk from market actors such as farmers to those who are keen to absorb risk, such as speculators.

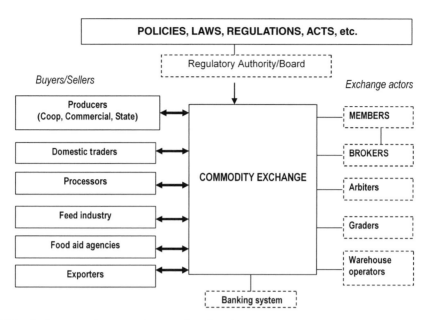

Fig. 11.2. An integrated approach to market development.

A new, emerging breed of technology-intensive commodity exchanges presents a promising avenue. While commodity exchanges have, historically, had tremendous power to transform markets when appropriately designed and implemented, what distinguishes the new, emerging exchanges is their reliance on information technology, which has the effect of addressing the powerful constraint of spatial exclusion from markets for the poor.

The world has witnessed over the past decade a dramatic pace in the rapid creation of new exchanges and the expansion of existing ones across the world. Commodity exchanges are no longer the domain of the highly industrialized countries with world-reference price-setting exchanges that dominate the commodities market, such as those in New York, Chicago and London. Over the past decade, a large number of new exchanges have been established in developing countries and, while many have not survived, others have come to occupy significant positions in the market such as those in Malaysia, South Africa and Brazil. These exchanges include those focusing on both spot and forward trading and the futures trade. Thus, the share of the US exchanges in total exchange trading in the world has dropped by nearly one-half, from a dominant share of 58% in 1991 to as low as 27% in 2003 (see Fig. 11.3). This trend is explained by the rapidly growing exchange trading in the Asia–Pacific region, as well as the resurgence of exchanges in Latin America in the 1990s.

The Asian region, in particular, has seen dramatic expansion in its exchanges and it is likely that Asia will account for the bulk of global

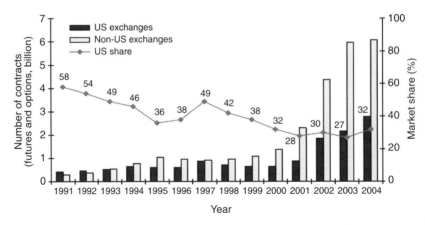

Fig. 11.3. Global future and options trading volumes (billions of contracts), 1991–2004 (from UNCTAD, 2006).

futures trade in the years to come. The dominant countries have traditionally been South Korea, Japan and Malaysia. While Japan and South Korea are strong traditional players, Malaysia's Kuala Lumpur Commodity Exchange, recently merged into the Bursa Malaysia Derivatives, has been growing significantly and has become the reference price for palm oil. More recently, however, the three Chinese exchanges created in the early 1990s and the three Indian exchanges established in 2002–2003 have experienced phenomenal growth in the last few years and are now in the top 20 world exchanges.

China has had an interesting history of commodity exchanges. In the early part of the 20th century, China had dozens of exchanges, but these had disappeared by the 1930s. Following reforms and market opening, the first exchange was re-established in 1990. By 1993 more than 40 exchanges had appeared, trading staples such as wheat, maize and soybeans. Following a period of uncontrolled growth, at least ten major scandals during the early 1990s rocked the markets, resulting in large losses by traders and leading to two major 'rectifications' by the China Securities Regulatory Commission (CSRC). These corrections reduced the number of exchanges from 40 to 15, and further to three: the Dalian Commodity Exchange (DCE), the Zhengzhou Commodity Exchange (ZCE) and the Shanghai Futures Exchange (SHFE).

Following these reforms, China has adopted a much more regulated and cautious approach, continuing its ban on financial future and exchange-traded options and on foreign investors in the Chinese market, and limiting the number of futures contracts traded on the three exchanges. Since 2000, the three Chinese exchanges have steadily grown into the world's largest exchanges, with the Dalian Commodity Exchange overtaking the Chicago Board of Trade in 2004 to become the world's second largest commodity futures exchange in the world, following NYMEX (New York Mercantile Exchange).

This cautious approach adopted by the Chinese is in sharp contrast to the recent Indian experience with national, multi-commodity exchanges, which have also experienced explosive growth. Like China, India also had an earlier experience with commodity exchanges, with the first organized futures market for cotton in 1921. In a similar fashion to that in China, trading in forward and futures contracts disappeared through a price control system and/or overt bans. Forward contracting was re-established in the 1950s, following which a large number of largely ineffective, local, single-commodity exchanges opened, although commodity futures trading was not fully legalized until early in 2003. Thus, the older, traditional, local, single-commodity exchanges still in operation are largely confined to the Bombay Oilseeds and Oils Exchange and the International Pepper Futures Exchange, amongst others.

The new move toward establishing national, multi-commodity exchanges in 2002/2003 has lead to a dramatic turnaround in India. Three national exchanges – all demutualized, implying that the management of the exchange is separate from the ownership, with permanent authorization from the Government of India to trade any permitted commodity – have opened up and experienced phenomenal growth over a short period of time. These exchanges, the National Commodities and Derivatives Exchange Mumbai (NCDEX), Multi-Commodity Exchange Mumbai (MCX) and National Multi-Commodity Exchange (NMCE) in Ahmedabad, are characterized by high-technology, web-based trading, with a focus on finding solutions to inclusion of dispersed rural populations (through innovative VSAT[3] solutions) and domestically oriented commodities, many of which are unique to India.

These new Indian commodity exchanges have been growing at a rate of 277% annually (see Fig. 11.4), from roughly US$15 million in 2002–2003 to US$478,000 million in 2005–2006 (NCDEX, 2006), with exponential growth in the volume of contracts. If the current growth rates are maintained, these exchanges are positioned to rank among the world's largest exchanges.

The Indian model gives particular insights into the potential for a dramatic impact on smallholder livelihoods. The commodities market in India is vast, with over 30 major markets in operation alongside 7500 small, localized markets (known as *mandies*). As a result, the Indian Gross Domestic Product (GDP) is hugely dependent on smallholder-produced agrarian commodities. To meet the challenge of reaching a large number of rural markets, the new exchanges have been driven by a technology-intensive approach, and have thus rolled out an electronic trading network infrastructure highly accessible in a low-cost fashion to rural communities through diverse mechanisms, such as VSAT, PSTN[4] lines, leased lines, the Internet and VPN[5] connectivity, to ensure a level playing field for all participants across the country. The VSAT equipment is relatively simple and low-cost and can be operated in difficult environments. For smaller users, access to the exchange can be channelled through a broker in a nearby town, such as a rural bank branch, a rural cooperative or the *panchayat*, or local government office.

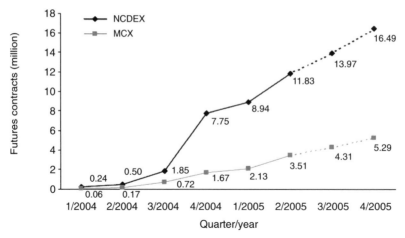

Fig. 11.4. Growth of India's commodity exchanges, 2004–2005 (from UNCTAD, 2006).

In under 3 years, the three national exchanges have succeeded in establishing links with over 4800 such trading terminals in the rural market towns, enabling wide access to the national exchange, which explains the phenomenal growth. In addition, the exchanges are rapidly expanding the presence of public price ticker boards across the country, with national spot and future prices being provided in real time in rural bus and railway stations, mandi market yards and rural warehouses, with the effect of transforming the information asymmetry of smallholders.

Essentially, the new exchanges embody a new approach to market development in that they reverse the traditional model of smallholder access to the market, because they are designed to enable the market to reach the smallest user. The impact of this market in reaching rural smallholders has multiple dimensions. First, the rural smallholder overcomes the spatial constraint, identified as one of the key dimensions of poverty, in that he or she is able to trade on the national commodity market instead of at the nearest local market, at a relatively low level of transaction costs. Secondly, through the establishment of price tickers displaying prices publicly in real time around the country, the rural smallholder is able to overcome the problem of access to market information. Thirdly, by using the commodity market to sell forward and otherwise hedge risk, the smallholder producer has a powerful mechanism for managing market risk. Lastly, through the linked function of banks and warehouse receipts, the smallholder has access to inventory finance and thus overcomes the problem of lack of credit.

The financial risks and costs to smallholder producers can be distinguished according to pre- and post-production, as shown in Fig. 11.5. Thus, pre-production risks include both credit needs as well as insurance needs; post-production risks include market price risk and credit flow, as well as proper storage.

As shown in Fig. 11.6, these new exchanges offer integrated solutions to farmers at both the pre- (a) and post-production (b) stages. Thus, in the pre-production phase, the exchange enables the futures prices to be known, which has an impact on cropping decisions, on enhancing liquidity through forward contracting, as well as on the increased ability to obtain bank lending and/or crop insurance products.

In the post-production phase, farmers can sell forward to hedge price risk and can also benefit from the reduced transaction costs of grading and storing products in exchange-linked warehouses, which enables their access to inventory credit. Early analysis of the impacts of the new exchanges in India on producer prices suggests that producer prices may be both less volatile and higher in the post-exchange environment (NCDEX, 2006).

Another important function of the national exchanges is their facilitation of government procurement at a lower cost through the exchange, thus reducing welfare losses. Thus, in 2005, the exchange enabled the purchase of 50,000 tons of paddy rice from 35,000 farmers on behalf of the national government agency. Thus, the exchanges

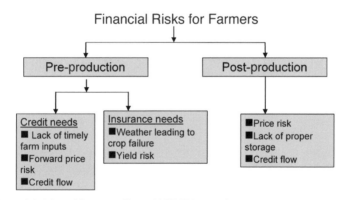

Fig. 11.5. Financial risks of farmers (from NCDEX, 2006).

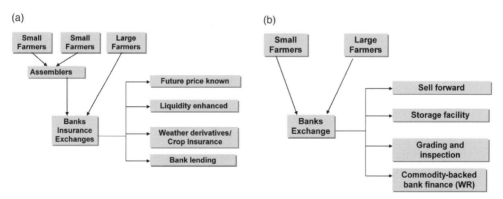

Fig. 11.6. Exchange integrated solutions (from NCDEX, 2006). (a) Pre-production solutions through exchange; (b) post-production solutions through exchange.

provide an alternative to the national support price and enable hedging for the Government of India.

Unlike the Asian exchanges that have quickly evolved into futures trading, some of the newly established exchanges in the transition economies of Eastern Europe have focused mainly on spot and forward contracts for commodities, such as those in the Czech Republic, Slovakia, Yugoslavia, Bulgaria and Uzbekistan. Others, such as in Poland, Hungary and Slovenia, have focused on domestically oriented futures trading. These exchanges have not witnessed significant evolution over the 1990s and early 2000s, and the Budapest Commodity Exchange is still probably the strongest.

Africa remains the weakest region in terms of exchange development. With the exception of South Africa's SAFEX, which is the commodity futures arm of the Johannesburg Stock Exchange, other efforts to inaugurate exchanges have experienced a start-and-stall pattern, largely due to policy reversals, excessive government intervention or to inappropriate business models initially.

In sum, there has been a remarkable surge in the establishment of exchanges in the 1990s, virtually in all the developing regions. The new exchange model described in the recent Indian experience offers a promising avenue of integrated market development in a way that enhances the livelihoods of the poor. It is important to recognize, however, that there is no blueprint or silver bullet in commodity exchange development. Further scrutiny of this emerging model and its potential for replicability in other contexts presents an exciting research agenda.

Conclusions

This chapter has highlighted the debate and tensions concerning the fate of smallholder agriculture in the context of globalization and, in particular, market liberalization. The review of the impact of market reforms on the poor has revealed the limited and mixed impact of liberalization on smallholder productivity and on the welfare of the poor. The lack of a clearly positive impact of market reforms is attributed to the fact that post-liberalization, private markets remain weak, segmented and characterized by high transaction costs and risk. Thus, the new agenda that has been highlighted in this chapter is that of getting markets themselves right, moving away from the earlier, narrow reform agenda of getting prices right.

The new agenda of market development is a long-term agenda, requiring a progressive and integrative perspective that addresses the '3 I's' of market development in a holistic fashion. These 3 I's are incentives (policies), institutions (to reduce transaction costs and risk) and infrastructure (physical and human capacity). Within the new agenda, there is an important role for the state *vis-à-vis* all of these dimensions of market development and a central role for the private sector.

However, this chapter has noted that, even so, getting prices right, while necessary, would not suffice for making markets work for the poor. To do so, first, the dimensions of poverty that influence the relationship between the market and poverty must be understood. Thus the focus on assets, vulnerability and risk, exclusion and voice, power and space as key dimension of the markets–poverty nexus.

In reviewing the market development experiences to date and in what ways the above challenge has or has not been met in current best practices, the chapter has highlighted the dichotomy in approaches between highly donor-driven, short-term, value chain development efforts in contrast to longer-term, state-led, market institution building. To date, the latter has demonstrated less impact, though more sustainability in the long term, than the former. With regard to market institutions, it has been emphasized that piecemeal interventions do not have the intended results.

Thus, the chapter's focus has been on a third approach, an integrated market development, as currently embodied in the promising direction of recent emerging commodity exchanges. Emphasis on new directions in commodity exchange development – notably the technology-driven, emerging exchanges with tremendous potential to reach the rural poor and directly address some of the key dimensions of poverty – have been discussed.

Looking forward, we conclude broadly by stating the key requirements facing policy makers and development practitioners in addressing the urgent issue of market development, both for global, high-value chains as well as for domestic or traditional bulk commodity markets as being the following:

- Determination of the appropriate role of the state *vis-à-vis* the private sector.
- Specific efforts to address smallholder engagement in both the global and domestic markets.
- Going beyond policy reforms to investments in institution building and infrastructure.
- An integrated rather than a piecemeal approach to market development.

Endnotes

1. Paper submitted for the Plenary Session of the International Association of Agricultural Economics 2006 Conference, 15 August 2006.
2. Studies show that, typically, the first workers to be laid off during public sector cutbacks are those with low skills, who then join the ranks of the urban poor – a phenomenon observed in both Africa and Latin America during the structural adjustment reforms of the 1980s and 1990s (World Bank, 2000).
3. Very small aperture terminal satellite.
4. Public switched telephone network.
5. Virtual private network.

References

Alderman, H. and Paxson, C. (1994) Do the poor insure? A synthesis of the literature on risk sharing institutions in developing countries. In: Bacha, E.L. (ed.) *Economics in a Changing World. Vol. 1 Development, Trade and the Environment.* Macmillan, London, pp. 48–78.

Aleem, I. (1990) Imperfect information, screening and the cost of informal lending: a study of rural credit markets in Pakistan. *World Bank Economic Review* 4, 329–349.

Badiane, O., Goletti, F., Kherallah, M., Berry, P., Govindan, K., Gruhn, P. and Mendoza, M. (1997) *Agricultural Input and Output Marketing Reforms in African Countries.* Final Donor Report, International Food Policy Research Institute (IFPRI), Washington, DC.

Beynon, J., Jones, S. and Yao, S. (1992) Market reform and private trade in Eastern and Southern Africa. *Food Policy* 17, 399–408.

Carter, M.R. and Zegarra, E. (2000) Land markets and the persistence of rural poverty: post-liberalization policy options. In: López, R. and Valdés, A. (eds) *Rural Poverty in Latin America.* St Martin's Press, New York.

Chaudhury, N. and Banerji, A. (2001) *Agricultural Marketing Institutions.* Background paper for the World Development Report 2001, Institutions for Markets, World Bank, Washington, DC.

Chiwele, D., Muyatwa-Sipula, P. and Kalinda, H. (1998) *Private Sector Response to Agricultural Marketing Liberalisation in Zambia.* Nordic Africa Institute Research Report No. 107, Uppsala, Sweden.

Christiaensen, L. and Pontara, N. (2002) *Poor Markets or Markets for the Poor? The Role of Agricultural Marketing Arrangements* (mimeo). Poverty Reduction Group, World Bank, Washington, DC.

Christiaensen, L., Demery, L. and Paternostro, S. (2002) *Economic Growth and Poverty Reduction in Africa: Messages from the 1990s* (mimeo). World Bank, Washington, DC.

Clay, D.C., Molla, D. and Debebe, H. (1999) Food aid targeting in Ethiopia: a study of who needs it and who gets it. *Food Policy* 24, 391–409.

Coulter, J. (1994) *Liberalization of Cereals Marketing in Sub-Saharan Africa: Lessons from Experience.* National Resources Institute (NRI) Marketing Series 9, Chatham, UK.

Datt, G. and Ravallion, M. (2002) *Is India's Economic Growth Leaving the Poor Behind?* Policy Research Working Paper, World Bank, Washington, DC.

Deininger, K. and Okidi, J. (2001) *Growth and Poverty Reduction in Uganda, 1992–2000: Panel Data Evidence* (mimeo). World Bank, Washington, DC and Economic Research Council, Kampala, Uganda.

Dercon, S. (1995) On market integration and liberalization: method and application to Ethiopia. *Journal of Development Studies* 32, 112–143.

Dercon, S. (1998) Wealth, risk and activity choice: cattle in Western Tanzania, 1998. *Journal of Development Economics* 55, 1–42.

Dercon, S. (2001) *Economic Reform, Growth and the Poor: Evidence from Rural Ethiopia.* Working Paper Series 8, 1–41, Centre for the Study of African Economies, UK.

DFID (2000) *Making Markets Work Better for the Poor: a Framework Paper.* Department for International Development, UK (www.livelihoods.org).

Diao, X. and Hazell, P. (2004) *Exploring Market Opportunities for African Smallholders.* 2020 Africa Conference Brief No. 6, International Food Policy Research Institute, Washington, DC.

Dione, J. (1989) Informing food security policy in Mali: interactions between technology, institutions and market reforms. PhD dissertation, Michigan State University, East Lansing, Michigan.

Glick, P. and Sahn, D. (1997) Gender and education impacts on employment and earnings from Conakry, Guinea. *Economic Development and Cultural Change* 45, 793–824.

Hoff, K. and Stiglitz, J. (1990) Introduction: imperfect information and rural credit markets – puzzles and policy perspectives. *The World Bank Economic Review* 4, 235–250.

Jalan, J. and Ravallion, M. (1997) *Are the Poor Less Well-insured? Evidence on Vulnerability to Income Risk in Rural China.* World Bank Policy Research Working Paper, World Bank, Washington, DC.

Jayne, T.S. and Jones, S. (1996) *Food Marketing and Pricing Policy in Eastern and Southern Africa: Lessons for Increasing Agricultural Productivity and Access to Food.* International Development Working Papers 56, Department of Agricultural Economics, Michigan State University, East Lansing, Michigan.

Jayne, T.S. and Jones, S. (1997) Food marketing and pricing policy in Eastern and Southern Africa: a survey. *World Development* 25, 1505–1527.

Jayne, T.S., Villarreal, M., Pingali, P. and Hemrich, G. (2005) HIV/AIDS and the agricultural sector: implications for policy in Eastern and Southern Africa, Food and Agriculture Organization of the United Nations. *The Electronic Journal of Agricultural and Development Economics* 2, 158–181.

Joffe, S. and Jones, S. (eds) (2003) *Stimulating Private Investment and Market Development for Agriculture: New Approaches and Experience.* Report to the World Bank under the DFID programme of advisory services, Washington, DC.

Jones, S. (1998) *Liberalized Food Marketing in Developing Countries: Key Policy Problems.* Oxford Policy Management, Oxford, UK.

Kamara, A.B. and Von Oppen, M. (1999) Efficiency and equity effects of market access on agricultural productivity: the case of small farmers in Machakos district, Kenya. *Zeitschrift für Ausländische Landwirtschaft* 38, 65–77.

Kanbur, R. and Squire, L. (2000) The evolution of thinking about poverty: exploring the interactions. In: Meier, G. and Stiglitz, J. (eds) *Frontiers of Development Economics: the Future in Perspective.* Oxford University Press, New York.

Kherallah, M., Delgado, C., Gabre-Madhin, E., Minot, N. and Johnson, M. (2002) *Reforming Agricultural Markets in Africa.* International Food Policy Research Institute (IFPRI), The Johns Hopkins University Press, Baltimore, Maryland.

Lipton, M. and Ravallion, M. (1995) Poverty and policy. In: Nehrman, J. and Sriniovasan, T.N. (eds) *Handbook of Development Economics.* North-Holland, Amsterdam, vol. 3, pp. 2551–2657.

Maskell, P. and Malmberg, A. (1999) Localized learning and industrial competetiveness. *Cambridge Journal of Economics* 23, 167–185.

Maxwell, S. (2004) *New Directions for Agriculture in Reducing Poverty.* Launch of DFID Consultation, Overseas Development Institute (ODI), London.

McCulloch, N. and Ota, M. (2002) *Export Horticulture and Poverty in Kenya.* Working paper 174, Institute for Development Studies, Sussex, UK.

Morduch, J. (1995) Income smoothing and consumption smoothing. *Journal of Economic Perspectives* 9, 103–114.

Narayan, D. and Nyamwaya, D. (1996) *Learning from the Poor: a Participatory Poverty Assessment in Kenya.* World Bank Social Policy and Resettlement Division, Washington, DC.

Narayanan, S. and Gulati, A. (2002) *Globalization and the Smallholders: a Review of Issues, Approaches, and Implications.* MTD Discussion Paper 50, International Food Policy Research Institute (IFPRI), Washington, DC.

NCDEX (2006) *National Commodity and Derivatives Exchange Limited.* Circular NCDEX Trading (http://www.ncdex.com).

Peters, P.E. (1996) *Failed Magic or Social Context? Market Liberalization and the Rural Poor in Malawi.* Development Discussion Paper 562, Harvard Institute of International Development, Harvard, Massachusetts.

Rosenzweig, M. and Wolpin, K. (1993) Credit market constraints, consumption smoothing and the accumulation of durable production assets in low income countries: investments in bullocks in India. *Journal of Political Economy* 10, 223–244.

Sahn, D.E. and Arulpragasam, J. (1994) The stagnation of smallholder agriculture in Malawi: a decade of structural adjustment, *Food Policy* 16, 219–234.

Sahn, D., Dorosh, P. and Younger, S. (1997) *Structural Adjustment Reconsidered: Economic Policy and Poverty in Africa*. Cambridge University Press, Cambridge, UK.

Sen, A.K. (1997) *On Economic Inequality*. Clarendon Press, Oxford, UK.

Staal, S., Delgado, C. and Nicholson, C. (1997) Smallholder dairying under transactions costs in East Africa. *World Development* 25, 779–794.

Stiglitz, J. (1989) Markets, market failures, and development. *The American Economic Review* 79, 197–203.

Stringfellow, R. and McKone, C. (1996) *The Provision of Agricultural Services through Self-help in sub-Saharan Africa*. Report to Overseas Development Institute (ODI), London, Natural Resources Institute (NRI), Chatham, UK.

UNCTAD (2006) *Overview of the Worlds's Commodity Exchanges*. Study prepared by the UNCTAD Secretariat, UNCTAD/DITC/COM/2005/8, Geneva, Switzerland.

Von Oppen, M., Munyemana, A., Bernard, M., Hundt, A., Agbo, B., Lose, S., Floquet, A., Bothe, M., Weller, U. and Van den Akker, E. (1998) Assessing the impact of technology options over time and at different levels from production to markets. In: Lawrence, P., Renard, G. and Von Oppen, M. (eds) *The Evaluation of Technical and Institutional Options for Smallfarmers in West Africa*. Margraf Verlag, Weikersheim, Germany, pp. 111–137.

Vorley, B. and Fox, T. (2004) Global food chains – constraints and opportunities for small-holders. Paper prepared for the *OECD POVNET Agriculture and Pro-poor Growth TaskForce Workshop*, Helsinki, 17–18 June 2004.

Walker, T.S. and Jodha, N.S. (1986) How small farm households adapt to risk. In: Hazell, P., Pomareda, C. and Valdés, A. (eds) *Crop Insurance for Agricultural Development*. John Hopkins University Press, Baltimore, Maryland.

World Bank (2000) *World Development Report 2000/2001: Attacking Poverty*. Oxford University Press, New York.

World Bank (2002) *World Development Report 2002: Building Institutions for Markets*. Oxford University Press, New York.

12 Market Access, Agricultural Productivity and Allocative Efficiency in the Banana Sector of Uganda

FRED BAGAMBA,[1] KEES BURGER,[2] RUERD RUBEN[3*] AND
ARIE KUYVENHOVEN[4**]

[1]Banana Research Institute, Uganda; [2]Development Economics Group,
Department of Social Sciences, Wageningen University, The Netherlands;
[3]Centre for International Development Issues Nijmegen (CIDIN), Radboud
University, Nijmegen, PO Box 9104, 6500 HE Nijmegen, The Netherlands;
[4]Agricultural Economics and Rural Policy Group, Wageningen University,
Hollandseweg 1, 6706 KN Wageningen, The Netherlands;
e-mails: [*]r.ruben@maw.ru.nl; [**]arie.kuyvenhoven@wur.nl

Abstract

This chapter analyses the factors influencing productivity and farm labour allocative efficiency in a banana-based production system in Uganda. A production function was estimated to analyse factors contributing to differences in banana production in three regions. Following a household economics theoretical framework, farm household behaviour regarding resource allocation to crop production is analysed, specifically elucidating the effect of imperfections in the labour and food markets.

Where organic amendments (specifically animal manure) are applied in significant quantities, productivity is significantly increased. Land and labour allocation decisions are influenced by land characteristics, household characteristics, distance from motorable roads, access to off-farm opportunities and the need to satisfy subsistence needs. The separability condition is rejected in favour of non-separability between production and consumption decisions. Improvement in labour and food market conditions favours the production of sweet potato in the central region, matooke (a banana cultivar) and coffee in Masaka and matooke in the south-west.

Introduction

Earlier studies on farmer behaviour and resource allocation efficiency in traditional agricultural systems were sparked off by Schultz's (1964) assertion that farmer operators in less-developed areas cannot significantly increase their farm production either by reallocating their farm resources at their disposal or by making further traditional investments. He defined traditional agriculture as being in economic equilibrium, this state having been achieved after a considerable period of time during which the state of technology, preferences and motives had remained constant. The rate of return to increased investment under existing technology was thus considered too low to induce further investment. Agricultural development at this stage, therefore, would depend more on breaking the existing equilibrium and adopting new technology involving the introduction of new modern inputs. Thus arose the view that only dramatic shifts in farm technology (seed, fertilizers and insecticides coupled with provision of credit) manifested themselves in many rural development programmes of the 1960s and 1970s (Ellis, 1993).

On the other hand, because farmers were deemed allocatively efficient, engineering price changes would cause them to change their production methods and to innovate. Hence, policies such as fertilizer price subsidization and credit schemes were promoted in the 1980s to stimulate adoption of improved technologies. Farmer education and extension work were considered low-cost methods of achieving increases in productive efficiency under the hypothesis of 'allocatively efficient but technically inefficient' (Ellis, 1993).

Earlier studies on resource allocation efficiency in agriculture in developing countries supported the hypothesis that farmers were allocatively efficient (Hopper, 1965; Chennareddy, 1967; Sahota, 1968). These studies described farmers involved in a technologically stagnant agriculture as being aware of resource substitution possibilities, and only resources such as fertilizers that were not within easy reach of individual farmers showed high marginal returns, in which case fertilizer use would be less than optimal.

A number of criticisms, however, were voiced against Schultz' propositions on agricultural transformation: The model was criticized for being based only on the farm firm and profit maximization criteria, disregarding other economic factors such as risk, uncertainty and the associated differences in marginal utilities that farm operators attach to prospective gains and losses (Adams, 1967). Adams noted that acceptance of the claim that farm operators were economically rational and efficiently allocated resources at their disposal does not necessarily entail belief in efficient resource allocation at the sector level. Feder (1967) shared that view of economic inefficiency at sectoral level, which he attributed to estate owners (absentee landlords) whose incomes were supplemented and often exceeded by non-farm incomes, a factor that forms a disincentive to farm their land well.

Results by Randhawa and Heady (1964) were in agreement with the above view where additional production was realized from re-planning available resources without improvement in technology. Mubarik and Flinn's (1989) results also showed substantial improvement in the profitability of Basmati rice in Gujaranwala district, India, through better use of existing technology. A number of studies provide evidence of agricultural inefficiency in Africa, and show heterogeneity across households in terms of their access to the best available technology (Aguilar and Bigsten, 1993; Adesina and Djato, 1996; Heshmati and Mulugetya, 1996; Croppenstedt and Demeke, 1997; Seyoum *et al.*, 1998; Mbowa, *et al.*, 1999; Olowofeso, 1999).

Transaction costs in different markets in developing countries determine whether a particular household does or does not participate in the given market. Households facing different market opportunities may make different decisions related to production, which affects efficiency. In the absence of credit and insurance markets, liquidity-constrained farmers might limit their investments in purchased inputs and hired labour. Imperfections in output markets could force farmers into subsistence production, leaving no or limited surplus for market sales.

In a survey to document production dynamics in Uganda's highland cooking banana production, farmers attributed the decline of banana production in central Uganda to soil exhaustion, pest and disease pressure and socio-economic constraints and changing opportunities (Gold *et al.*, 1999). A number of biophysical studies have since been carried out to verify and quantify the effect of biophysical constraints on banana production (e.g. Smithson *et al.*, 2001; Ssali *et al.*, 2003).

However, the socio-economic constraints and opportunities associated with differences in productivity among smallholder farmers have not been adequately investigated. Specifically, understanding the interactions between non-farm employment opportunities, market access and farm production is essential in formulating appropriate policies for improving the banana subsector and the overall agricultural sector. This study was carried out to elucidate the factors influencing banana productivity and efficiency among smallholder farmers in Uganda. The remainder of this chapter is organized as follows: the theory is discussed in the second section, while the following sections cover the empirical specification of the model and a description of the data used in the study. Results are then presented and discussed, with the final section providing concluding remarks.

Agricultural Household Model

Traditionally, economists have used a profit function approach to explain household behaviour. This is possible when markets are functioning well, with farmers facing low transaction costs, thus rendering consumption and production decisions separable (Benjamin,

1992). When farmers face missing or incomplete markets (e.g. labour, credit and insurance markets), production decisions cannot be separated from consumption decisions and the proposition that profit is the principle driving factor for production decisions is not plausible. Where development of reliable markets for consumer goods (crop inputs and foodstuffs) is lacking, households allocate their time preferentially to essential non-market household activities, including the provision of secure food supply (Low, 1986).

Analysis of farm household behaviour in a situation where markets are not functioning well requires a household approach, which involves simultaneous estimation of production and consumption decisions. Estimation of the complete structural system of equations for consumption and production behaviour and reduced form approaches are often used. One of these approaches focuses on the time allocation of farm households under labour market imperfections, in which one estimates households' shadow wages and incomes, based on first-order conditions for utility maximization in the context of a non-separable household model, which are then used as regressors in subsequent labour supply estimations (Jacoby, 1993; Newman and Gertler, 1994; Skoufias, 1994; Mishra and Goodwin, 1997; Abdulai and Regmi, 2000; Barrett *et al.*, 2005). The separability condition is rejected if the shadow wage rates are significantly different from the market wage rate.

Household behaviour under functioning labour markets

The basic household model postulates the existence of perfect markets for goods produced and consumed by farm households, which enables the households to separate production and consumption decisions by first maximizing profit from food production and using income from profit to maximize utility from consumption.

Following a household economics framework, the farm household's utility function is

$$u = u(c, l; z)$$

where c is a vector of home produced goods and l is time spent in leisure and social activities. The vector z parameterizes the utility function and summarizes household characteristics, such as the number of people in each age and sex category, education level and distance to market. The c is produced using a vector of purchased goods, x, and a vector of quantities of own time. A significant part of c is also obtained from own production, the rest being sold to the market at competitive prices. Limiting the consumption of goods, c, to a staple crop (e.g. bananas), the farmer's problem is:

$$Max\ u(c, l; z)\ \text{w.r.t.} \qquad c, l, l_o, l_f \text{ and } l_h \tag{12.1}$$

and $l \geq 0,\ l_o \geq 0,\ l_f \geq 0,\ l_h \geq 0$ (non-negativity constraints) s.t.

$$c = pq(l_F;x) - wl_h + wl_o + y \text{ (cash constraint)} \tag{12.2}$$

$$1 + l_f + l_0 = E \text{ (labour constraint)} \tag{12.3}$$

$$\text{and } l_f + l_h = l_F \tag{12.4}$$

where $q = f(l; x)$ is a twice differentiable, convex production function. Land area, x, is allocated to production of q and is assumed to be fixed and exogenous. Labour l_F is the sum of family and hired labour, $l_f + l_h$, and w is the normalized price for hired labour and off-farm labour l_o. The household is endowed with resources; time E and exogenous income y. Hired labour and family labour are perfect substitutes and have the same wage rate w.

From Equation (12.3), we have:

$$l_o = E - 1 - l_f \tag{12.5}$$

Substituting (12.5) for l_o in Equation (12.2), we obtain,

$$c = pf(l_F;x) - wl_h + w(E - 1 - lf) + y \tag{12.6}$$

From Equation (12.6) we obtain the full income constraint:

$$c + wl = y + pf(l_F;x) - wl_h - wl_f + wE = M \tag{12.7}$$

Maximizing M leads to an indirect utility function: $u = \phi(y + \pi(w;x) + wE(\beta), w, \beta)$. Utility is maximized through maximizing the full income M, which itself is maximized by maximizing profits: $\pi = \pi(w; x)$. This is the recursive property, where the household first maximizes profits and then maximizes utility from the income obtained from the profits.

The first order condition with respect to l_F from profit maximization is:

$$f_{l_F}(l_F;x) = w^* \tag{12.8}$$

where $w^* = w/p$. $f_{l_F}(l_F; x)$ is the marginal product of labour and w is the market farm wage rate for labour. We can derive the reduced form for farm labour demand as:

$$l_F = l_F(w^*,x) \tag{12.9}$$

Equation (12.9) shows that labour use is determined by the real farm wage rate and the fixed land resource. Demand for labour is independent of consumption decisions taken by the household. The first order condition in Equation (12.8) indicates that an increase in wage rate reduces labour demand while an increase in output price would result to an increase in labour use.

The Lagrange for utility maximization is

$$z = u(c,l;z) + \lambda(M - pf(l_F;x) + wl_h + wl_f - wE - y) \tag{12.10}$$

This leads to the following first order conditions:

$$\frac{\partial z}{\partial c} = u_c - \lambda p = 0 \tag{12.11}$$

$$\frac{\partial z}{\partial l} = u_l - \lambda w = 0 \tag{12.12}$$

From Equations (12.11) and (12.12), we have the following first order condition for utility maximization:

$$\frac{u_c}{u_l} = \frac{p}{w} \tag{12.13}$$

From Equation (12.13), we can derive the reduced form for farm labour supply as

$$l_f = l_f(w, p, M, z) \tag{12.14}$$

The separation property provides a representation of the dual nature of the farm household both as a producer and worker. The household is able to attain maximum utility through the market either through hiring more labour – in the case of labour deficit, or by selling labour to the market – in the case of labour surplus. The condition in Equation (12.13) shows that consumption decisions include the choice of home time, l, which is traded off with the consumption of goods, c, that would need more income and hence more work. Unless the commodity c is a given good (in which case its consumption increases when the price rises), the relationship between the consumption of c and price, p, should be negative. Likewise, leisure, l, is a normal good and the household should reduce its consumption when there is a rise in wage rate. A rise in price of the farm product, c, raises its output and full income, reduces family time committed to its production, increases use of hired labour, increases market surplus and reduces consumption of the farm output by the household. A rise in market wage results in reduced farm output and full income, increase in family time in farm production, reduction in use of hired labour, an increase in consumption of home produced goods and a reduction in consumption of purchased goods.

However, there are several substitution and income effects involved that affect the net outcome of the price and market wage rate changes (Ellis, 1993). The outcomes depend on household consumption preferences between the three consumption choices: own farm produce, non-farm time and consumer good, which cannot be anticipated by theory but by empirical estimations. The general impact of income (profit) effect is to give the household greater scope to pursue its preferences. Ellis (1993) observes that peasant farmers strive to obtain economic efficiency, although this might not be attained in the strict neoclassical sense due to the nature of the peasant economy in which they operate. Profit maximization conditional on multiple goals is pursued, and resource constraints and markets confronted by the farmers may exist even if the strict efficiency is not observed. Farmers take into account risk and uncertainty, have household goals other than profit maximization (e.g. food security, social status and income sustainability) and face imperfections in different factor markets (land, labour and capital).

Household behaviour under missing or incomplete labour markets

The above model assumes the following: (i) identical work preferences for farm and non-farm production; (ii) perfect substitution of family labour and hired labour; and (iii) complete markets (Benjamin, 1992). Neither does the model consider different prices facing farm households, where the retail price is higher than the farm-gate price. High retail product prices lower the opportunity cost of family labour, while low prices increase the opportunity cost of labour, leading to greater chances for off-farm employment.

In the presence of labour market imperfections, the market wage does not reflect the opportunity cost of family labour, and different household members will have different opportunity costs for their time. Lower opportunity costs result from transaction costs confronting the different household members, which include the risk involved in job search, transport costs and information-gathering costs. These must be subtracted from the market wage on offer to express the opportunity return to labour at the farm gate. If the individual's real opportunity cost is higher than the marginal farm productivity, the household member engages in off-farm employment, and vice versa.

Under the household economics assumption of profit maximization, farm households seek to minimize the costs of subsistence production in order to maximize returns to family labour. The farm household allocates the time of different household members in such a way that the goods are produced at the lowest possible costs (Gronau, 1973; Evenson, 1978). Households minimize costs in the production of the subsistence good by employing members with the lowest potential wage rates, unless the higher-wage earners are more efficient in production of the subsistence good by an amount that offsets the difference in wages (Low, 1986). On the contrary, higher returns to family labour can induce high demands for leisure and home time, leading to a substitution away from the relatively scarce resource of family time towards time-saving production technology that involves the use of hired labour (Goldman and Squire, 1982).

A market fails when the cost of transaction through the market creates disutility exceeding the utility obtained from the transaction, which results in the market not being used. Even if the market exists, the gains for a particular household may be below or above the cost, which results in some households participating in the market while others do not (De Janvry *et al.*, 1991). There is a wedge between selling and buying price when markets are incomplete, which depends on: (i) transportation costs; (ii) mark-ups from traders; (iii) the opportunity cost of time involved in selling (search costs) and in buying (recruitment and supervision costs); and (iv) the risk associated with price uncertainties due to the seasonal nature of agriculture, which results in certainty-equivalent prices that are lower for sales and higher for purchases. Poor infrastructure (especially roads) results in isolated, shallow food and labour markets and, the shallower the market is, the more the prices can be expected to be correlated with movements in shadow prices, which traps particular households within the range of self-sufficiency (De Janvry *et al.*, 1991).

Imperfections in the hired labour market

Small farmers confront wage rates that are different from those faced by large farmers due to imperfections in the labour market. Specifically, small farmers confront a low opportunity cost of labour, which is lower than the social wage, while by contrast larger farmers confront higher prices for labour that are above the social wage (Ellis, 1993). Transaction costs have the effect of raising the wage rate above the level that would occur in the absence of the transaction costs. Such costs include monitoring and supervision for hired labour, incentive, labour-retaining, efficiency and moral hazard costs.

The full income for a household facing transaction costs in the hired labour market is expressed as follows:

$$M = pf(l_F;x) - (w + T_w)l_h - w_f l_f + w_f E + y \qquad (12.15)$$

where w = social cost of labour, w_f is the opportunity cost of labour for the household and T_w is cost incurred by the household above the social cost as a result of transaction costs.

The household maximizes utility by maximizing the full income and first order condition for profit maximization is:

$$pf'(l_F;x) = w + T_w \qquad (12.16)$$

From Equation (12.16), we get:

$$f'(.) = w / p + T_w / p = w^* \qquad (12.17)$$

Labour will be hired in by the household if $f'(.) > w^*$ and hired out if $f'(.) < w_f$. Households are self sufficient in labour if $w_f \le f'(.) \le w^*$.

Imperfections in the off-farm labour market

Consider the case of imperfections in the non-farm sector in which some household members are segregated or the labour market is rationed such that the household can obtain work only up to a certain number of hours per week:

$$l_0 = l_{max} \qquad (12.18)$$

The full income, M, can no longer be determined by the profits from production alone but also by the conditions in the non-farm labour market. The farmers' utility problem is solved through maximizing income and leisure concurrently. The farmer's utility maximization problem is:

$$\text{Max } u(c, E - l_f - l_0; z) \qquad (12.19)$$

st:

$$c = pf(l_F;x) - w_h l_h + w_f l_o + y \text{ (budget constraint)} \qquad (12.20)$$

$$l = E - l_f - l_0 \text{ (time constraint)} \qquad (12.21)$$

$$l_o = l_{max} = E - 1 - l_f \text{ (off-farm labour constraint)} \qquad (12.22)$$

Utility is maximized through maximization of full income subject to the off-farm labour constraint. The Lagrange function for the problem is:

$$z = \lambda(pf(l_F;x) - w_h l_h - w_f l_f + w_f E + y) + \psi(l_{max} - l_o) \qquad (12.23)$$

The first order condition with respect to family labour working on farm is

$$\frac{\partial z}{\partial l_f} = \lambda pf(l_F;x) - \lambda w_f + \psi = 0 \qquad (12.24)$$

From Equation (12.19), we get:

$$pf'(l_F;x) = w_f - \frac{\psi}{\lambda} \qquad (12.25)$$

$$f'(.) = 1 / p(w_f - \frac{\psi}{\lambda}) = w^* \qquad (12.26)$$

where $\lambda \geq 0$ is the Lagrange multiplier associated with the budget constraint, $\psi \geq 0$ is the Lagrange multiplier associated with the time constraint, p is farm gate price of farm output while w_h and w_f are wage rates for hired and off-farm labour respectively. The rest of the variables are as already defined. Labour will be hired out if $f'(.) < w^*$ and hired out if $f'(.) > w_h$. Households are self-sufficient in labour if $w^* \leq f'(.) \leq w_h$.

The first order condition in Equation (12.26) shows that the shadow wage rate for the household facing market imperfections is not the market wage rate but is a function of exogenous prices and other factors, z, that affect household consumption decisions. In particular, factors that affect access to off-farm opportunities (proximity to urban areas, good infrastructure and education and gender of individual household members) would influence farm production and consumption decisions.

Production function estimation

The production function is comprised of farm inputs that are representative of the production system, including labour and other variable inputs and fixed inputs (land and organic amendments). Included in the production function are factors that are hypothesized to affect the production potential.

In the absence of fertilizer and other chemical inputs, the production function can be specified as:

$$y_i = f(l_{ij}, x_{ik}; \xi) + \varepsilon_i; i = 1, ...n, j = 1, ...,2 \text{ and } k = 1, ..., m. \qquad (12.27)$$

where y_i is output realized from farm i; l_{ij} refers to the different types of labour input (family and hired labour) used by the farmer; x_{ik} refers to fixed factors: land (x) and organic inputs (animal manure, crop residues and grass mulch) and ξ refers to farm and plot characteristics (pest and disease incidence, soil characteristics and access to technical information as proxy for technology). ε_i is the error term. Family and hired labour is further categorized as male, female and child labour.

Total labour input, l, can be expressed as follows:

$$f_m + a_1 f_f + a_2 f_c + b(h_m + a_1 h_f + a_2 h_c) = l \tag{12.28}$$

where f and h refer to family and hired labour respectively. The subscripts m, f and c refer to male, female and child labour, respectively. a_1 and a_2 measure the efficiency with which female and child labour substitute for male labour, while b measures the substitutability between family and hired labour.

The relationship between output, land and organic amendments can be expressed in the form:

$$y = \alpha[x(1+\mu_1 M)(1+\mu_2 G)(1+\mu_3 C)]^\theta \, l^\beta e^\varepsilon \tag{12.29}$$

where x and l, respectively, refer to crop area and total labour input while α, θ and β are the parameters that are estimated which refer to, respectively, the constant, and elasticities of crop area and total labour input. M, G and C refer to quantities of manure, grass mulch and crop residues with μ_1, μ_2 and μ_3 as their respective coefficients, which measure their contribution to land productivity. The variable ε refers to the error term. With small values of $\mu_1 M$, $\mu_2 G$ and $\mu_3 C$, Equation (12.29) can be Taylor approximated into the following equation:

$$y = \alpha[xe^{\mu_1 M} e^{\mu_2 G} e^{\mu_3 C}]^\theta \, l^\beta e^\varepsilon \tag{12.30}$$

If w is wage rate and p is price of output, the producer's restricted profit is $p'y - w'l$. The producer is assumed to choose the combination of labour and output that maximizes profit subject to the technology constraint:

Max $p'y - w'l$, st. $g(y, l, x) = 0$.

The profit function takes the form of the production function. Assuming a Cobb-Douglas function, $y = \alpha l^\beta x^\theta$, maximizing $\pi = p\alpha l^\beta x^\theta - wl$ leads to the first order condition:

$$\frac{\partial \pi}{\partial l} = p\alpha\beta l^{\beta-1} x^\theta - w = 0 \tag{12.31}$$

From Equation (12.31), we derive the labour demand:

$$l = [\frac{w}{p}(\alpha\beta x^\theta)^{-1}]^{\frac{1}{\beta-1}} \tag{12.32}$$

We can obtain a linear function from Equation (12.32) by log transformation:

$$\ln l = \frac{1}{\beta-1}\log(\frac{w}{p}) - \frac{1}{\beta-1}\ln\alpha - \frac{1}{\beta-1}\ln\beta - \frac{\theta}{\beta-1}\ln x \tag{12.33}$$

The log transformation of the production function gives:

$$\ln y = \ln\alpha + \beta\ln l + \theta\ln x \tag{12.34}$$

Equations (12.33) and (12.34) are estimated simultaneously to obtain the production function estimates α, β and θ.

Productivity and Allocative Efficiency Estimation

Model specification

The production function

A Cobb-Douglas production function was estimated with the output and input variables (labour and crop area) transformed into logarithm form. For female labour, a_1 was fixed at 0.8 while a_2 for child labour was fixed at 0.5. The coefficient for hired labour, b, was initially varied between 0.8 and 1.2 and finally fixed at 1.0 because there was no significant impact on the parameter estimates. The production functions were estimated under the assumption of constant returns to scale. Parameters estimated gave an insight into the sources of productivity differences between regions and groups of farmers.

The 3SLS procedure was used to estimate the production function and labour equations simultaneously. Household characteristics (age, gender, education and household size and structure, distance from tarmac road and credit access) were included in the labour equation to capture any effects from market imperfections in the labour, commodity and credit markets. Farm size and number of extension visits in the previous 6 months were included in the production function as proxy for farm characteristics and production technology. Regional dummies were included to capture the diverse soil and agroclimatic characteristics of the different regions (central, Masaka and south-west).

Allocative efficiency scores

The marginal products of labour estimated from the production functions were used to test for allocative inefficiencies within different production regions and groups of farmers. For the proposition of allocative efficiency to hold, the marginal product of labour should be equal to the real or normalized wage rate.

$$\text{MPL} = w^* \tag{12.35}$$

where $w^* = w/p$.

To test for allocative inefficiency, the following function was estimated:

$$\text{Ln(MPL)} = a + b\text{Ln}(w^*) + e.$$

The null hypothesis of allocative efficiency holds if the joint F-test for parameters a and b, being equal to 0 and 1, respectively, is not rejected.

Allocative inefficiency scores were computed as $Ln(MP_L/w^*)$ = Ln(ai). The score equal to zero implies the household employs labour efficiently; above zero the household uses less labour than is optimal and a negative score implies that the household uses more labour than is optimal.

Data sources and description

Primary data were generated from a stratified random sample that included 660 households, of which 540 were usable. The sample was drawn from 33 villages located in major banana production zones in Uganda, stratified by altitude and previous exposure to new banana varieties and improved practices. Differences in farm and biophysical characteristics such as soil productivity and pest and disease incidence are highly correlated with elevation. The final sampling frame consisted of 27 sub-counties. One village was randomly selected from each sub-county in 24 of the sub-counties. Three villages were selected from each of the three other sub-counties (Bamunanika, Kisekka and Ntungamo). The three sub-counties were purposively included in the sample for the objective of collecting soil data (Rufino, 2003; Bagamba *et al.*, 2004).

The three sub-counties represent three production levels: (i) Ntungamo represents areas with high production; (ii) Kisekka represents areas with medium production; and (iii) Bamunanika represents areas with low production. From each village, 20 households were randomly selected from a list of all households within the village.

The units of observation were the village and the household. Village-level data included prices and distance to tarmac road (highway). Household-level data included demographic characteristics, production, income and inputs. The variables are defined and data are summarized in Table 12.1.

Results and Discussion

Production function estimates were obtained for the major crops grown. The results from the first-stage estimation of the production functions are included in Tables 12.2 to 12.5. Final estimates of the production functions are presented in Tables 12.6 to 12.9.

The estimated coefficients have the expected positive sign for land and labour inputs. For the overall sample, high elasticities of labour are obtained for beans, coffee and cassava, with the lowest for maize and sweet potato (see Table 12.6). Manure has a positive and significant effect (significant at 10%) on productivity. Farm size has a negative effect on output, especially for cassava and maize, which is consistent with much of the literature on farm size and productivity effects (Carter, 1984;

Table 12.1. Descriptive for exogenous variables used in production analysis.

Variable	Definition	Overall ($n = 563$)	Central ($n = 297$)	Masaka ($n = 129$)	South-west ($n = 137$)
Farm size	Farm size (acres)	4.024	4.408	4.380	2.860
		(8.407)[a]	(5.448)	(14.350)	(5.640)
Ext	Number of extension visits	0.682	0.610	0.790	0.740
		(1.859)	(2.050)	(1.760)	(1.490)
W	Casual wage rate	344.5	432.7	267.8	227.7
		(156.6)	(157.4)	(106.1)	(27.1)
Hhsz	Family size (adult equivalent)	4.707	4.850	4.200	4.860
		(2.199)	(2.220)	(2.050)	(2.220)
Plab	Proportion of adults in household	0.506	0.500	0.510	0.510
		(0.239)	(0.240)	(0.270)	(0.200)
Gender	Male = 1; female = 0	0.805	0.808	0.767	0.830
		(0.397)	(0.394)	(0.420)	(0.380)
Age	Age of household head (years)	45.28	46.50	44.10	43.80
		(15.95)	(16.40)	(16.00)	(14.60)
Age_2	Age squared (years)	2303.9	2432.6	2196.2	2126.5
		(1599.0)	(1676.5)	(1581.2)	(1419.8)
Edhh	Education of household head (years)	5.364	5.810	4.870	4.870
		(4.175)	(4.560)	(3.240)	(4.000)
D	Distance from tarmac road (km)	13.66	12.31	20.29	10.30
		(18.09)	(9.04)	(31.75)	(12.99)
Dd	D = 1 if ⩽ 10 km; otherwise = zero	0.520	0.470	0.581	0.577
		(0.5)	(0.5)	(0.5)	(0.5)
FWY	Mean village income from farm wage × 1000	45.63	57.47	15.32	48.50
		(75.40)	(95.75)	(18.26)	(45.10)
CY	Mean village income from casual wage × 1000	65.20	79.40	17.58	79.30
		(99.80)	(123.30)	(17.02)	(71.00)
RY	Mean village income from regular wage × 1000	106.10	111.80	33.52	162.10
		(171.6)	(78.4)	(35.4)	(314.3)
SY	Mean village income from self employment ×1000	269.3	363.6	136.1	190.1
		(280.1)	(308.2)	(143.1)	(235.8)

[a] Values in parentheses are standard deviations.

Benjamin, 1995; Barrett, 1996; Byiringiro and Reardon, 1996; Heltberg, 1998; Kanwar, 1998; Pender *et al.*, 2004).

Smaller farmers face constraints in the off-farm labour market, especially for female and child labour (Bardhan, 1973). Faced with the off-farm labour constraints, the implicit cost of labour proved to be lower on smaller farms and, therefore, smaller farmers tend to apply more labour per acre than the larger farmers (Kanwar, 1998). The high labour use intensity on small farms is responsible for the inverse relationship between farm size and productivity. The extension variable is positively related to output, with the exception of maize, being statistically significant (5%) for coffee, beans and sweet potatoes.

Table 12.2. Determinants of labour use intensity for different crops, overall sample.

Variable[a]	All crops (n = 560)	Banana (n = 509)	Coffee (n = 174)	Bean (n = 347)	Maize (n = 237)	Sweet potato (n = 207)	Cassava (n = 212)
Constant	17.079	6.250	3.144	4.480	4.373	5.456	5.524
Ln (A)	1	1	1	1	1	1	1
Ln (w/p)	−2.058	−1.427	−1.499	−1.791	−1.127	−1.129	−1.451
Hhsz	−0.039*	−0.038*	−0.026	−0.077	−0.37*	−0.093***	−0.091*
	(−1.70)[b]	(−1.72)	(−0.19)	(−1.01)	(−1.82)	(−2.18)	(−1.89)
Plab	0.216	0.175	0.464	0.487		0.186	0.146
	(1.14)	(0.97)	(0.38)	(0.75)		(0.51)	(0.37)
Gender	−0.020	0.100	0.154	0.012	−0.107	0.305	0.024
	(−0.17)	(0.92)	(0.21)	(0.03)	(−0.45)	(1.47)	(0.10)
Age	0.053***	0.073***	0.085	0.101***	0.123***	0.077***	0.085***
	(6.04)	(8.19)	(1.49)	(3.55)	(6.63)	(3.71)	(4.52)
Age_2	−0.0005***	−0.0007***	−0.0007	−0.0010***	−0.0010***	−0.0010***	−0.0010***
	(−4.74)	(−6.90)	(−1.04)	(−2.77)	(−5.75)	(−3.05)	(−3.75)
Edhh	0.048***	0.022*	0.029	0.027	0.043*	0.072***	0.051**
	(4.09)	(1.95)	(0.36)	(0.68)	(2.01)	(3.39)	(2.06)
D	−0.003	−0.008***	0.082**	−0.010	−0.370*	−0.005	−0.004
	(−1.30)	(−3.26)	(2.00)	(−1.25)	(−1.82)	(−0.39)	(−0.94)
FWY	0.0070***	0.0030***	−0.0003	0.0020	0.0030*	−0.0005	0.0020
	(10.89)	(4.89)	(−0.06)	(0.65)	(1.74)	(−0.23)	(1.40)
CY	0.002***	0.003***	0.006	0.001	0.004***	0.001	0.003***
	(3.670)	(7.350)	(0.980)	(0.490)	(5.410)	(1.220)	(3.270)
RY	−0.00020	−0.00006	−0.00200	−0.00100	0.00300***	−0.00100*	−0.00200***
	(−0.61)	(−0.25)	(−0.98)	(−0.97)	(2.73)	(−1.69)	(−2.61)
SY	0.0003*	−0.0003*	−0.0010	−0.0010	−	−0.0001	0.0010*
	(1.71)	(−1.84)	(−0.85)	(−0.80)		(−0.39)	(1.96)
R^2	−1.331	0.37	0.199	−0.441	0.213	−0.014	−0.086
Adjusted R^2	−1.382	0.355	0.140	−0.493	0.179	−0.077	−0.152
Durbin-Watson stat	0.810	1.357	1.743	1.057	1.181	1.43	1.105

*, ** and *** represent 10, 5 and 1% levels of significance, respectively. [a] See Table 12.1 for explanation of variables. [b] Values in parentheses are standard deviations.

However, Pender *et al.* (2004) observe that training and extension programmes tend to work with farmers who are more productive. Regional dummies (south-west and Masaka), included to capture regional differences in biophysical and environmental conditions, are both positive, except for sweet potato production. The results indicate that crop productivity is higher in the south-west, followed by Masaka and lowest in central. However, the central region has a comparative advantage in sweet potato production.

Production function estimates for the different regions are presented in Tables 12.7 (central), 12.8 (Masaka) and 12.9 (south west). In the central region, labour elasticities for individual crops are close to 0.3, although when aggregated the elasticity is close to 0.5. Manure has a positive effect on matooke productivity, the relationship being significant

Table 12.3. Determinants of labour use for different crops, central Uganda.

Variable[a]	All crops (n = 294)	Banana (n = 246)	Coffee (n = 105)	Bean (n = 183)	Maize (n = 177)	Sweet potato (n = 141)	Cassava (n = 170)
Constant	16.927	6.423	3.283	4.561	4.983	6.596	5.889
Ln (A)	1	1	1	1	1	1	1
Ln (w/p)	−1.990	−1.497	−1.443	−1.542	−1.259	−1.385	−1.580
Hhsz	−0.035	−0.095	0.065	−0.069*	−0.062	−0.036	−0.059
	(−1.35)[b]	(−0.56)	(1.17)	(−1.83)	(−1.32)	(−0.82)	(−1.19)
Plab	−0.371	0.353	−0.545	0.384	0.117	−0.235	−0.161***
	(−1.61)	(1.26)	(−1.01)	(1.13)	(0.28)	(−0.58)	(−3.90)
Gender	−0.080	−0.095	−0.070	0.456**	0.203	0.159	0.276
	(−0.57)	(−0.56)	(−0.23)	(2.25)	(0.82)	(0.74)	(1.15)
Age	0.060***	0.062***	0.027	0.143***	0.083***	0.056***	0.099***
	(5.19)	(4.29)	(0.94)	(7.92)	(4.04)	(2.94)	(4.42)
Age_2	−0.0006***	−0.0006**	−0.017	−0.001	−0.001	−0.0006***	−0.0009***
	(−4.68)	(−3.60)	(−0.49)	(−6.86)	(−3.46)	(−2.81)	(−4.08)
Edhh	0.044***	0.028*	−0.017	0.004	0.008	0.045**	0.030
	(3.37)	(1.74)	(−0.49)	(0.18)	(0.37)	(2.02)	(1.22)
D	−0.004	−0.033***	0.095**	−0.043***		−0.028*	−0.031
	(−0.49)	(−3.24)	(5.35)	(−2.95)		(−1.91)	(−2.19)
Dd					0.521**		
					(2.14)		
FWY	0.0070	0.0030***	0.0070**	−0.0008	–	0.0020	0.0050
	(11.36)	(4.34)	(2.94)	(−0.63)		(0.63)	(0.31)
CY	0.0020***	0.0020***	0.0030	0.0005	0.0030	0.0020**	0.0030***
	(4.46)	(3.45)	(1.60)	(0.59)	(3.75)	(2.42)	(3.40)
RY	0.0040***	0.0006	0.0100***	−0.0030**	0.0040	0.0040***	−0.0010
	(5.58)	(0.64)	(4.04)	(−2.53)	(2.93)	(2.84)	(−0.8)
SY	0.0002	0.0003	0.0010	−0.0010**	–	−0.0006	0.0006
	(1.21)	(1.28)	(1.57)	(−2.29)		(−1.59)	(1.25)
R^2	−0.132	0.304	0.571	0.275	0.223	0.184	−0.281
Adjusted R^2	−0.180	0.268	0.515	0.224	0.176	0.107	−0.389
Durbin-Watson stat	1.126	1.515	1.991	1.602	1.191	1.589	1.098

*, ** and *** represent 10, 5 and 1% levels of significance, respectively. [a] See Table 12.1 for explanation of variables. [b] Values in parentheses are standard deviations.

at 5%. Farm size has the expected negative effect on productivity. The extinction variable has a positive effect on cassava (significant at 10%). This could be explained by the delivery of cassava mosaic-resistant planting material to farmers. The effect is also positive for coffee, beans and sweet potato, although not significant. In Masaka region, high labour elasticities are obtained for matooke and coffee but low for beans, maize, sweet potato and cassava.

The effect of farm size on productivity is significant (and negative) only for cassava production. Positive and significant (5%) effects of extension

Table 12.4. Determinants of labour use for different crops, Masaka.

Variable[a]	All crops (n = 129)	Banana (n = 126)	Coffee (n = 69)	Bean (n = 65)	Maize (n = 60)	Sweet potato (n = 66)	Cassava (n = 42)
Constant	18.751	8.905	5.023	5.348	5.177	5.566	4.69
Ln (A)	1	1	1	1	1	1	1
Ln (w/p)	−2.246	−2.289	−1.762	−1.339	−1.314	−1.368	−1.111
Hhsz	−0.113***	−0.073	−0.045	0.008	0.028	−1.640	
	(−2.69)[b]	(−1.45)	(−0.66)	(0.10)	(0.27)	(−1.38)	−0.298***
							(−3.83)
Plab	0.395	0.038	0.657	0.623	0.182	0.427	0.085
	(1.38)	(0.51)	(1.45)	(1.20)	(0.215)	(0.480)	(0.120)
Gender	0.274*	0.041	0.717**	−0.384	−0.330	0.528	−0.227
	(1.74)	(0.23)	(2.48)	(−1.18)	(−0.70)	(0.98)	(−0.52)
Age	0.102***	−0.002	−0.013	−0.038	0.054	0.074*	−0.001**
	(6.50)	(−0.09)	(−0.45)	(−0.72)	(0.89)	(1.69)	(−2.29)
Age_2	−0.00100***	0.00003	0.00010	0.00030	−0.00100	−0.0006	0.09800**
	(−5.64)	(0.13)	(0.43)	(0.62)	(−1.12)	(−1.29)	(2.01)
Edhh	−0.023	−0.016	−0.015	−0.031	0.016	0.052	0.098**
	(−1.077)	(−0.650)	(−0.430)	(−0.730)	(0.300)	(0.810)	(2.010)
D		−0.023***	0.145***	0.120***	0.019***	0.065	0.004
		(−4.16)	(4.81)	(3.80)	(2.87)	(1.36)	(0.56)
Dd	−1.508***						
	(−5.07)						
FWY	0.024***	0.053***	−0.015***	−0.145	−0.018*	−0.013*	−0.012
	(3.14)	(4.92)	(−3.56)	(−0.29)	(−1.70)	(−1.81)	(−1.30)
CY	−0.005	−0.016	0.030***	1.002	0.104***	0.010	−0.027
	(−1.030)	(−3.710)	(4.140)	(0.768)	(2.930)	(0.850)	(−0.850)
RY	−0.020***	−0.021	−0.007***		−0.023***	−0.002	0.005
RY				−0.144***			
	(−4.95)	(−3.73)	(−4.23)		(−3.02)	(−0.70)	(0.71)
				(−5.56)			
SY	−0.0020***	−0.0002	−0.0030***	−0.0340	−0.0080***	−0.0040	−0.0020
	(−3.01)	(−0.30)	(−3.99)	(−0.73)	(−3.98)	(−2.46)	(−1.19)
R^2	−0.685	0.287	0.342	0.482	−0.071	−0.073	0.396
Adjusted R^2	−0.859	0.211	0.200	0.362	0.345	−0.316	0.146
Durbin-Watson stat	2.141	1.776	2.053	(2.389)	1.074	1.947	1.719

*, ** and *** represent 10, 5 and 1% levels of significance, respectively. [a] See Table 12.1 for explanation of variables. [b] Values in parentheses are standard deviations.

are realized only for coffee and beans. Again, this could be associated with multiplication and delivery of improved varieties (in the case of beans) and high-yielding clonal material (in the case of coffee) to farmers. In the southwest, labour elasticities are quite high compared with those of other regions, with matooke having the highest (0.63), followed by millet (0.55) and beans (0.40). Farm size has no significant effect on productivity. Extension has a positive and significant (1%) effect on matooke production, which is the most important crop in the region (both food and cash).

Table 12.5. Determinants of labour use for different crops, south-west Uganda.

Variable[a]	All crops (n = 137)		Banana (n = 137)		Bean (n = 99)		Millet (n = 49)	
	Coefficient	t-value	Coefficient	t-value	Coefficient	t-value	Coefficient	T-value
Constant	20.848		10.497		5.005		4.096	
Ln (A)	1		1		1		1	
Ln (w/p)	−2.455		−2.695		−1.675		−2.242	
Hhsz	−0.037	−1.170	−0.059**	−1.990	−0.019	−0.310	0.071	0.790
Plab	−0.446	−1.630	−0.568**	−2.270	−0.254	−0.450	1.231	1.490
Gender	0.322*	1.960	0.386***	2.610	0.274	0.770	0.157	0.370
Age	0.016	1.060	0.004	0.270	0.074***	3.250	0.065*	1.690
Age_2	−0.00020	−1.02000	−0.00001	−0.09000	−0.00080***	−2.85000	−0.00100*	−1.79000
Edhh	0.007	0.410	0.006	0.380	−0.008	−0.260	0.021	0.430
D	−0.049***	−7.690	−0.036***	−6.080	−0.048***	−5.740	−0.043***	−3.700
FWY	0.002	1.050	−0.006***	−2.720	0.011***	2.900	0.001	0.270
CY	0.005***	4.150	0.008***	6.000	0.004	1.530	0.003	0.440
RY	−0.00100***	−5.0300	−0.00100***	−4.78000	−0.00100	−1.65000	−0.00006	−0.03500
SY	−0.002***	−5.000	−0.002***	−4.410	−0.004***	−4.190	−0.001	−0.760
R²	−0.776		0.561		−0.633		−0.277	
Adjusted R²	−0.948		0.541		−0.860		−0.703	
Durbin-Watson stat	2.128		1.219		1.747		1.860	

*, ** and *** represent 10, 5 and 1% levels of significance, respectively. [a] See Table 12.1 for explanation of variables.

Allocative efficiency

The results on marginal products and allocative inefficiency scores are summarized in Table 12.10. Banana production in Masaka and the south-west has the highest-value marginal products, followed by coffee in Masaka and sweet potatoes in the central region. Value marginal products are lowest in central region, especially for coffee and sweet potatoes. Apart from banana production in Masaka and the south-west, the rest of the crops have their value marginal products well below the going market wage rates, implying that more labour than optimal is used in their production. However, average value products for most crops are close to or even higher than the market wage rates, implying that farmers try to equate wage rates to their average-value products in their labour allocation decisions.

The averages of allocative inefficiency scores were all negative, except for banana production in Masaka and the south-west. The negative scores imply that more labour is employed than optimal, while positive scores imply that less labour is employed than optimal. Since banana is the main crop for the two regions (Masaka and south-west), employing less labour than is optimal implies that farmers are constrained in access to labour (specifically hired labour). The allocative inefficiency scores

Table 12.6. Production function estimates for different crops, 3SLS (three stages least square) (overall sample).

Variable[a]	All crops[b] (n = 560)	Banana (n = 509)	Coffee (n = 174)	Bean (n = 347)	Maize[†] (n = 237)	Sweet potato (n = 207)	Cassava (n = 212)
Constant	8.964***	5.584***	3.197***	3.382***	6.061***	7.001***	4.974***
	(43.56)[c]	(20.22)	(4.18)	(9.97)	(18.37)	(20.29)	(9.15)
Ln(A)	0.486***	0.701***	0.667***	0.568 ***	0.887***	0.886***	0.689***
	(17.90)	(18.30)	(6.01)	(11.88)	(19.74)	(17.55)	(9.15)
Ln(L)	0.514	0.299	0.333	0.432	0.113	0.114	0.311
M		0.00003*					
		(1.82)					
G		–0.000001					
		(–0.05)					
C		0.00001					
		(0.39)					
Farm size	–0.0003	0.0020	–0.0200	–0.0030	–0.0110**	–0.0030	–0.0180***
	(–0.08)	(0.42)	(–1.10)	(–0.83)	(–2.07)	(–0.27)	(–2.71)
Ext	0.011	0.005	0.070**	0.051**	–0.027	0.052**	0.052
	(0.63)	(0.24)	(2.07)	(2.46)	(–0.91)	(1.98)	(1.54)
South-west	0.711***	1.339***	1.558***	0.305***	0.582*	–0.276	0.406
	(9.07)	(13.83)	(5.69)	(3.04)	(1.68)	(–1.52)	(0.86)
Masaka	0.229***	0.754 ***	1.111***	0.339***	0.205	–0.823***	0.3
	(2.87)	(7.806)	(5.49)	(2.96)	(1.28)	(–4.21)	(1.36)
R^2	0.230	0.648	0.373	0.417	0.581	0.287	0.056
Adjusted R^2	0.223	0.643	0.354	0.408	0.572	0.269	0.033
Durbin-Watson stat	0.771	1.338	1.074	1.634	1.744	1.257	1.411

*, ** and *** represent 10, 5 and 1% levels of significance, respectively. [a] See Table 12.1 for explanation of variables. [b] Value of the major crops harvested in a year; physical output used in the individual crop production functions. [c] Values in parentheses are standard deviations.

were negative for all crops in the central region, implying oversupply of labour in farm production. Binswanger and Rosenzweig (1984) attribute the oversupply of labour for own production by small farms to rigidities or search-induced frictions in the casual labour markets. However, imperfections in the commodity markets could also force farmers into subsistence production to meet their subsistence needs if the purchase prices are higher than selling prices. Own production is a secure way of meeting subsistence needs for farmers who face imperfections in either the labour market or the commodity market. Such farmers are likely to have lower marginal products than the going real markets wage rates.

The joint null hypothesis, a = 0, b = 1, is rejected in all cases, implying that farmers in all three regions exhibit allocative inefficiency in terms of farm labour employment (see Table 12.11). The same results have been obtained in the literature for developing countries (Jacoby, 1993; Abdulai and Regmi, 2000).

Table 12.7. Production function estimates for different crops for central Uganda, 3SLS (three stages least square).

Variable[a]	All crops (n = 294)[b]	Banana (n = 246)	Coffee (n = 105)	Bean (n = 183)	Maize (n = 177)	Sweet potato (n = 141)	Cassava (n = 170)
Constant	9.206***	5.391***	3.458***	4.005***	5.540***	6.044***	4.730***
	(36.94)[b]	(12.89)	(4.62)	(10.05)	(9.73)	(9.09)	(8.62)
Ln(A)	0.503***	0.668***	0.690***	0.649***	0.794***	0.722***	0.633***
	(15.49)	(11.64)	(6.29)	(11.64)	(10.16)	(7.22)	(8.11)
Ln(L)	0.497	0.332	0.307	0.351	0.206	0.278	0.367
M		0.0001**					
		(2.01)					
G		0.00004					
		(0.22)					
C		0.00002					
		(0.27)					
Farm size	−0.022**	−0.003	−0.026	−0.010	−0.036***	−0.012	−0.048***
	(−2.50)	(−0.24)	(−1.49)	(−0.90)	(−2.92)	(−0.82)	(−2.81)
Ext	−0.006	−0.027	0.061	0.039	−0.052	0.039	0.058*
	(−0.24)	(−0.88)	(1.49)	(1.51)	(−1.53)	(1.37)	(1.73)
R^2	0.141	0.351	0.092	0.308	0.620	0.060	−0.018
Adjusted R^2	0.133	0.333	0.065	0.297	0.613	0.039	−0.036
Durbin-Watson stat	0.826	1.297	0.843	1.605	1.690	1.345	1.384

*, ** and *** represent 10, 5 and 1% levels of significance, respectively. [a] Value of the major crops harvested in a year; physical output used in the individual crop production functions. [b] Values in parentheses are standard deviations.

Table 12.8. Production function estimates for different crops for Masaka, 3SLS (three stages least square).

Variable[a]	All crops[b] (n = 129)	Banana (n = 126)	Coffee[2] (n = 69)	Bean (n = 65)	Maize[2] (n = 60)	Sweet potato[t2] (n = 66)	Cassava (n = 42)
Constant	8.939***	4.465***	3.688***	4.915***	5.370***	5.383***	6.520***
	(18.67)[c]	(8.95)	(5.50)	(10.34)	(8.29)	(5.03)	(7.27)
Ln(A)	0.445***	0.437	0.567***	0.747	0.761***	0.731***	0.900***
	(6.70)	(6.34)	(5.84)	(11.01)	(7.87)	(4.33)	(6.16)
Ln(L)	0.555	0.563	0.433	0.239	0.253	0.269	0.100
Farm size	0.006	0.006	0.004	−0.002	−0.005	0.015	−0.013*
	(1.34)	(1.35)	(0.17)	(−0.39)	(−0.83)	(0.80)	(−1.71)
Ext	−0.024	0.023	0.116**	0.135**	0.081	0.092	0.013
	(−0.70)	(0.62)	(2.39)	(2.40)	(1.33)	(1.46)	(0.11)
R^2	−0.003	−0.080	0.443	0.370	0.375	0.128	0.284
Adjusted R^2	−0.027	−0.107	0.417	0.339	0.341	0.086	0.228
Durbin-Watson stat	0.685	1.346	2.049	2.017	1.546	1.290	1.099

*, ** and *** represent 10, 5 and 1% levels of significance, respectively. [a] See Table 12.1 for explanation of variables. [b] Value of the major crops harvested in a year; physical output used in the individual crop production functions. [c] Values in parentheses are standard deviations.

Table 12.9. Production function estimates for different crops for south-west Uganda, 3SLS (three stages least square)

Variable[a]	All crops[b] (n = 137) Coefficient (robust standard error)		Banana (n = 137) Coefficient (robust standard error)		Bean (n = 99) Coefficient (robust standard error)		Millet (n = 49) Coefficient (robust standard error)	
Constant	9.015***	(19.10)	4.359***	(9.13)	3.896***	(7.36)	2.418	(1.49)
Ln(A)	0.407***	(6.35)	0.371***	(5.87)	0.597***	(7.76)	0.446**	(2.03)
Ln(L)	0.593		0.629		0.403		0.554	
M			0.00001	(1.56)				
G			−0.000008	(−0.48)				
C			0.00002	(1.05)				
Farm size	0.002	(0.244)	0.005	(0.670)	−0.006	(−0.450)	0.105	(1.530)
Ext	0.123***	(3.830)	0.151***	(5.270)	0.034	(0.690)	−0.063	(−0.850)
R^2	0.362		0.561		0.583		0.414	
Adjusted R^2	0.347		0.541		0.569		0.375	
Durbin-Watson stat	0.673		1.219		1.537		1.365	

*, ** and *** represent 10, 5 and 1% levels of significance. [a] See Table 12.1 for explanation of variables. [b] Value of the major crops harvested in a year; physical output used in the individual crop production functions.

The deviation from the textbook condition:

vmpl = w

is a sign of imperfections in the labour market. The dominance of the negative sign of the allocative inefficiency scores in the central region is a confirmation of binding labour constraints in that region. This is in complete disregard of the fact that the region is in close proximity to major urban centres where the non-farm sector is more developed. However, it is possible that the labour market is segmented and some household members are segregated in the labour market. Udry *et al.* (1995) indicate that many categories of household labour are held captive within the household for reasons of age and gender, as customs prohibit them from working outside their home.

To investigate the labour allocation decisions further in these regions, we compare returns to land and labour on an acre basis for all the crops considered for analysis, and the results are presented in Table 12.12. In the central region, farmers obtain highest returns to labour from bananas and lowest returns from coffee. There are differences in returns to land for the different crops, which implies imperfections in either the land market or differences in land quality or rent for different crops. Prime land is allocated to matooke, followed by sweet potatoes, beans, maize, cassava, and the least productive is allocated to coffee. However, the low returns in coffee could be due to the effect of coffee wilt disease

Table 12.10. Average and value marginal products of labour, selected crops.

Crop	Total sample				Central				Masaka				South-west			
	n	AVP	MVP	Ln(ai)	n	Vapl	Vmpl	Ln(ai)	n	Vapl	Vmpl	Ln(ai)	n	Vapl	Vmpl	Ln(ai)
All crops	560	351.2 (187)[a]	180.5 (96.2)	-0.71 (0.80)	294	271.80 (142.00)	135.2 (71)	-1.24 (0.62)	129	357.7 (139)	198.4 (77)	-0.32 (0.47)	137	525 (195)	311.2 (115)	0.25 (0.40)
Banana	509	585.4 (560)	175.3 (168)	-0.86 (0.89)	246	517 (634)	171.7 (211)	-1.25 (0.84)	126	565.8 (261)	318.6 (147)	0.13 (0.47)	137	581.9 (243)	366.0 (153)	0.412 (0.39)
Coffee	174	336.5 (451)	112 (150)	-1.77 (1.25)	105	158.7 (188)	48.7 (58)	-2.55 (0.77)	69	569.8 (459)	246.5 (199)	-0.39 (0.88)				
Bean	347	331.9 (401)	144 (174)	-1.13 (0.87)	183	314.6 (526)	110.5 (185)	-1.72 (0.71)	65	501.9 (852)	127.1 (216)	-1.45 (1.26)	99	406.2 (419)	163.8 (169)	-0.60 (0.74)
Maize	237	412 (758)	46.5 (86)	-2.87 (1.33)	177	276.4 (516)	56.8 (106)	-2.60 (1.08)	60	563.2 (621)	134.7 (149)	-1.16 (1.15)				
Sweet potato	207	837.3 (1288)	95.7 (147)	-2.03 (1.10)	141	746.2 (750)	207 (208)	-1.18 (0.82)	66	586.9 (809)	157.8 (218)	-1.07 (1.12)				
Cassava	212	442.4 (568)	137.6 (177)	-1.65 (1.08)	170	351.9 (403)	129 (148)	-1.72 (0.90)	42	892.8 (1401)	89.5 (140)	-1.82 (1.12)				
Millet													49	279.1 (150)	154.6 (83)	-0.51 (0.67)

AVP, average value product; MVP, marginal value product.
[a] Values in parentheses are standard deviations.

Table 12.11. Wald test for allocative inefficiency (F-values).

Crop	Overall sample	Central	Masaka	South-west
All crops	660.9	732.0	76.1	49.9
Banana	394.4	296.5	11.05	100.4
Coffee	281.3	573.5	42.8	
Bean	496.1	550.1	97.6	48.2
Maize	673.4	556.6	49.8	
Sweet potato	395.4	157.9	66.0	
Cassava	283.8	314.7	249.5	
Millet				58.4

Table 12.12. Returns to land and labour per acre, selected crops.

	Overall sample		Central		Masaka		South-west	
	Per acre return to		Per acre return to		Per acre return to		Per acre return to	
Crop	Labour	Land	Labour	Land	Labour	Land	Labour	Land
All crops	244765	231380	208401	210554	239734	192461	417962	287239
	(164783)[a]	(155771)	(173426)	(175218)	(13188)	(114953)	(209823)	(144198)
Banana	171755	401870	130952	263266	337859	262115	614824	239325
	(101771)	(238122)	(82660)	(166180)	(217609)	(168825)	(320219)	(226004)
Coffee	56253	112752	26981	60971	129927	170418		
	(53171)	(106576)	(14462)	(32681)	(85262)	(11834)		
Bean	115585	151827	85519	157890	77406	228342	105143	155676
	(67739)	(88979)	(40814)	(75353)	(45655)	(134678)	(55707)	(82480)
Maize	21317	167520	37126	143423	59233	188446		
	(8501)	(66806)	(15798)	(61032)	(29245)	(93042)		
Sweet potato	31359	24293	93710	243620	49361	134194		
	(15630)	(115393)	(46422)	(120683)	(23388)	(63583)		
Cassava	59418	131683	74264	128057	17609	158102		
	(27674)	(61332)	(43136)	(74381)	(5034)	(45194)		
Millet							108131	87047
							(47384)	(38145)

[a] Values in parentheses are standard deviations.

and farmers becoming reluctant to replace old trees with another crop because of the high labour requirements involved in uprooting the trees. As already observed, coffee is grown by farmers who are located far away from market access, and where land rent is relatively low.

Returns to labour in Masaka differ for the different crops, with crops that are highly commercialized having high returns and the returns for subsistence crops (beans, maize and sweet potato) being low. Returns to land from the subsistence crops are close to returns from coffee, implying that land quality and price for land allocated to coffee and the subsistence crops are almost the same. Labour allocation decisions are guided by subsistence needs, where more labour is allocated to maize, sweet potatoes and cassava than is optimal because of the need to satisfy subsistence requirements; this observation is based on returns obtained for coffee and beans. Whereas labour returns are higher in coffee relative to returns from beans, the reverse is true for returns to land, implying that labour intensification is higher in bean production, which leads to higher returns to land and lower returns to labour in beans. Returns to land in matooke are high, implying that prime land is allocated to matooke, while returns are lowest in sweet potato, implying that marginal land is allocated to sweet potatoes in the region.

In the south-west, the highest returns to both land and labour are obtained for bananas, which imply that the best land is allocated to banana production. However, imperfections in the commodity market and/or labour market force farmers to allocate more labour to beans and millet, when they could still earn more from banana production. Returns to labour for beans and millet are close, which shows some element of optimization behaviour for crops that are produced with the same motive. Returns to land are lowest for millet, which confirms our hypothesis that millet is allocated to marginal land that is unsuitable for banana and bean production (poorly drained land in the valley bottoms).

Finally, we present results from the simulation exercise, in which we simulate a 10% increase in wage rate after imposing equal value marginal products of labour (VMPs) across the crops grown in each region. The VMPs are set to be equal to a value that is close enough to the value marginal products obtained for the different crops in each region. The value of VMP for central region was fixed at 151.45 (35% of the market wage rate), that of Masaka was fixed at 200 (74.7% of the market wage rate) while the one for south-west was fixed at 225 (98.8% of the market wage rate). The results obtained are presented in Table 12.13.

The results show a decrease in labour allocated to crops in the central region, with a complete cessation in production of coffee and maize. The amount of labour allocated to beans reduces by 63.7% although output reduces by only 24.4%. This implies that only farmers

Table 12.13. Labour use and output response to improvement in market conditions and a subsequent 10% increase in wage rate, selected crops.

Crop	Central		Masaka		South-west	
	Labour	Output	Labour	Output	Labour	Output
All crops	−39.1	−20.6	−23.9	−13.3	26.8	19.9
Banana	0.1	0.6	42.9	28.9	48.9	37.3
Coffee	−100	−100	9.8	5.7		
Bean	−63.7	−24.4	−79.9	−22.2	−68.9	−30.3
Maize	−100	−100	−68.5	−17.9		
Sweet potato	18.3	6.1	−45.3	−13.2		
Cassava	−38.7	−15.2	−100	−100		
Millet					−100	−100

that use less labour-intensive techniques (e.g. intercropping and high-yielding varieties) remain in bean production. Cassava labour decreases by 38.7% while output decreases by 15.2%. Labour allocated to banana production increases by 0.05% while output increases by 0.60%. Highest increase in labour used is realized for sweet potato production, in which labour use increases by 18.3% and output increases by 6.1%.

The overall response in labour use for Masaka is also negative, but the magnitude is smaller than that of central region. Labour is completely withdrawn from cassava production; use for bean production decreases by 79.9%, in maize production by 68.5% and by 45.3% in sweet potato production. Labour use in bananas and coffee increases by 42.9 and 9.8%, respectively. Output, respectively, increases by 28.9 and 5.7%.

In the south-west, labour use increases in banana production, reduces in bean production and is withdrawn from millet production.

Conclusions

This study was carried out to explain the observed differences in banana productivity between south-west and central Uganda. Results show that land and labour inputs have the expected positive impact on output. Manure application has positive effects on banana productivity. However, relying on organic amendments seems not to be a feasible option for sustainable productivity in agriculture. The null hypothesis for equating the value of marginal products to wage rate is rejected, an indication of imperfections in the labour market. Household characteristics influence farm labour use, which confirms existence of imperfections in the labour market and support for the non-separability hypothesis.

Simulating equal marginal returns to labour for crops grown, followed by an increase in wage rate by 10%, results in an increase in sweet potato production for central Uganda, banana and coffee for Masaka and banana for the south-west. Development of infrastructure (roads) and marketing institutions could foster development in the farm sector, but this has to be complemented by policies that favour investment in the non-farm sector to absorb surplus labour and, in the long-term, stimulate demand for agricultural products.

Acknowledgements

The Rockefeller Foundation funded the study through the National Banana Research Programme (NBRP). INREF funding through the Response Programme at Wageningen University Research, The Netherlands, is highly appreciated. Most of the data used in this study were collected through a collaborative research project between IFPRI and NBRP funded by USAID.

References

Abdulai, A. and Regmi, P.P. (2000) Estimating labour supply of farm households under nonseparability: empirical evidence from Nepal. *Agricultural Economics* 22, 309–320.

Adams, D.W. (1967) Resource allocation in traditional agriculture: comment. *Journal of Farm Economics* 49, 930–932.

Adesina, A.A. and Djato, K.K. (1996) Farm size, relative efficiency and agrarian policy in Côte d'Ivoire: profit function analysis of rice farms. *Agricultural Economics* 14, 93–102.

Aguilar, R. and Bigsten, A. (1993) An analysis of differences in technical efficiency among Kenyan smallholders. *East African Economic Review* 9, 295–306.

Bagamba, F., Ruben, R., Kuyvenhoven, A., Kalyebara, R., Rufino, M., Kikulwe, E. and Tushemereirwe, W.K. (2004) Determinants of resource allocation in low input agriculture: the case of banana production in Uganda. Paper presented at the *American Agricultural Economics Association Annual Meeting*, Denver, Colorado, 1–4 August.

Bardhan, P. (1973) Size, productivity and returns to scale: an analysis of farm-level data in Indian Agriculture. *Journal of Political Economy* 81, 1370–1386.

Barrett, C.B. (1996) On price risk and the inverse farm size–productivity relationship. *Journal of Development Economics* 51, 193–215.

Barrett, C.B., Sherlund, S.M. and Adesina, A.A. (2005) *Shadow Wages, Allocative Inefficiency and Labour Supply in Smallholder Agriculture*. Discussion Paper, Department of Applied Economics and Management, Cornell University, Ithaca, New York.

Benjamin, D. (1992) Household composition, labour markets, and labour demand: testing for separation in agricultural household models. *Econometrica* 60, 287–322.

Benjamin, D. (1995) Can unobserved land quality explain the inverse productivity relationship? *Journal of Development Economics* 46, 51–84.

Binswanger, H.P. and Rosenzweig, M.R. (1984) Contractual arrangements, employment and wages in rural labour markets: a critical review. In: Binswanger, H.P. and Rosenzweig, M.R. (eds) *Contractual Arrangements, Employment and Wages in Rural Labour Markets in Asia*. Yale University Press, New Haven, Connecticut.

Byiringiro, F. and Reardon, T. (1996) Farm productivity in Rwanda: effects of farm size, erosion, and soil conservation investments. *Agricultural Economics* 15, 127–136.

Carter, M.R. (1984) Identification of the inverse relationship between farm size and productivity: an empirical analysis of peasant agricultural production. *Oxford Economic Papers* 36, 131–145.

Chennareddy, V. (1967) Production efficiency in South Indian agriculture. *Journal of Farm Economics* 49, 816–820.

Croppenstedt, A. and Demeke, M. (1997) An empirical study of cereal crop production and technical efficiency of private farmers in Ethiopia: a mixed fixed random coefficients approach. *Applied Economics* 29, 1217–1226.

De Janvry, A., Fafchamps, M. and Sadoulet, E. (1991) Peasant household behaviour with missing markets: some paradoxes explained. *Economic Journal* 101, 1400–1417.

Ellis, F. (1993) *Peasant Economics: Farm Households and Agrarian Development*. Cambridge University Press, Cambridge, UK.

Evenson, R.E. (1978) Time allocation in rural Philippine households. *American Journal of Agricultural Economics* 60, 323–330.

Feder, E. (1967) The latifundia puzzle of Professor Schultz: comment. *Journal of Farm Economics* 49, 507–510.

Gold, C.S., Karamura, E.B., Kiggundu, A., Bagamba, F. and Abera, A.M.K. (1999) Geographical shifts in highland banana production systems in Uganda. *The International Journal of Sustainable Development and World Ecology* 6, 45–59.

Goldman, R.H. and Squire, L. (1982) Technical change, labour use and labour distribution in the Muda irrigation project. *Economic Development and Cultural Change* 30, 754–775.

Gronau, R. (1973) The intrafamily allocation of time: the value of housewives' time. *American Economic Review* 68, 634–651.

Heltberg, R. (1998) Rural market imperfections and the farm size–productivity relationship: evidence from Pakistan. *World Development* 26, 1807–1826.

Heshmati, A. and Mulugetya, Y. (1996) Technical efficiency of the Ugandan Matoke farms. *Applied Economics Letters* 3, 491–494.

Hopper, D.W. (1965) Allocation efficiency in traditional Indian agriculture. *Journal of Farm Economics* 47, 611–624.

Jacoby, H.G. (1993) Shadow wages and peasant family labour supply: an econometric application to the Peruvian Sierra. *Review of Economic Studies* 60, 903–921.

Kanwar, S. (1998) *Wage Labour in Developing Agriculture: Risk, Effort and Economic Development*. Ashgate Publishing Ltd, Andover, UK, 191 pp.

Low, A. (1986) *Agricultural Development in Southern Africa: Farm Household Economics and the Food Crisis*. James Currey, London, 217 pp.

Mbowa, S., Niuewoudt, W.L. and Despins, P. (1999) Efficiency of sugar cane farms in Kwa Zulu Natal. *South African Journal of Economic and Management Sciences* 2, 54–76.

Mishra, K.A. and Goodwin, B.K. (1997) Farm income variability and the supply of off-farm labour. *American Journal of Agricultural Economics* 79, 880–887.

Mubarik, A. and Flinn, J.C. (1989) Profit efficiency among basmati rice producers in Pakistan Punjab. *American Journal of Agricultural Economics* 71, 303–310.

Newman, J.L. and Gertler, P.J. (1994) Family productivity, labour supply, and welfare in a low income country. *Journal of Human Resources* 29, 989–1026.

Olowofeso, O.E. (1999) On maximum likelihood estimation of frontier production functions: empirical application to small-scale rice farmers in Ondo state, Nigeria. *Indian Journal of Applied Economics* 8, 121–136.

Pender, J., Nkonya, E., Jagger, P., Sserunkuuma, D. and Ssali, H. (2004) Strategies to increase agricultural productivity and reduce land degradation: evidence from Uganda. *Agricultural Economics* 31, 181–195.

Randhawa, N.S. and Heady, E.O. (1964) An interregional programming model for agricultural planning in India. *Journal of Farm Economics* 46, 137–149.

Rufino, M. (2003) On-farm analysis of nematode infestation and soil fertility as constraints to the productivity of banana-based production systems in Uganda. MSc thesis, Plant Sciences Department, Wageningen University, The Netherlands, 91 pp.

Sahota, G.S. (1968) Efficiency of resource allocation in Indian agriculture. *American Journal of Agricultural Economics* 50, 584–605.

Schultz, T. (1964) *Transforming Traditional Agriculture*. Yale University, New Haven, Connecticut.

Seyoum, E.T., Battese, G.E. and Fleming, E.M. (1998) Technical efficiency and productivity of maize producers in Eastern Ethiopia: a study of farmers within and outside the Sasakawa Global 2000 Project. *Agricultural Economics* 19, 341–348.

Skoufias, E. (1994) Using shadow wages to estimate labour supply of agricultural households. *American Journal of Agricultural Economics* 76, 215–227.

Smithson, P.C., McIntyre, B.D., Gold, C.S., Ssali, H. and Kashaija, I.N. (2001) Nitrogen and potassium fertilizer versus nematode and weevil effects on yield and foliar nutrient status of banana in Uganda. *Nutrient Cycling in Agroecosystems* 59, 239–250.

Ssali, H., McIntyre, B.D., Gold, C.S., Kashaija, I.N. and Kizito, F. (2003) Effects of mulch and mineral fertilizer on crop, weevil and soil quality parameters in highland banana. *Nutrient Cycling in Agroecosystems* 65, 141–150.

Udry, C.R., Hoddinott, J., Alderman, H. and Haddad, L. (1995) Gender differentials in farm productivity: implications for household efficiency and agricultural policy. *Food Policy* 20, 407–423.

13 Land and Labour Market Participation Decisions under Imperfect Markets: a Case Study in North-east Jiangxi Province, China

SHUYI FENG,[1] NICO HEERINK[2] AND RUERD RUBEN[3]

[1]Development Economics Group, Wageningen University, Wageningen, The Netherlands; e-mail: ShuyiFeng@wur.nl; [2]International Food Policy Research Institute, Beijing Office, China; e-mail: nico.heerink@wur.nl; [3]Centre for International Development Issues, Redbrund University, Nymegen, The Netherlands; email: R.Ruben@maw.ru.nl

Abstract

This chapter analyses the factors influencing household land and labour market decisions in rural China. It shows that, when land and labour market imperfections exist, decisions on land and labour use are made jointly and therefore need to be analysed as such. The insights gained from jointly analysing land and labour market decisions are significantly different from those obtained from studies focusing on each market separately.

The estimation results for 329 farm households collected in three villages in Jiangxi province show that households with both low and high land availability are less likely to participate in off-farm employment. A possible explanation could be that households with small land endowments are not wealthy enough to work off-farm, while households with relatively large land endowments have difficulties in renting out their land and hence prefer to work on-farm.

The average adult age shows an inverted U-shaped relationship with the likelihood of being involved in migration. Differences in estimated turning points suggest that households involved in migration and renting-in of land comprise mainly migrants who return home after a few years of working elsewhere and reinvest in agriculture.

The number of durable assets has a significant positive effect on the probability that households are involved in local off-farm employment.

This result suggests that richer households may have better access to local off-farm jobs than poorer households.

Possession of a land contract and having a migration network has a positive effect on migration. Enjoying more land transfer rights has a significant positive effect on land renting-in. These results confirm that households facing low transaction costs in labour and land market are more likely to participate in these markets.

Introduction

Imperfections in factor markets are a typical feature of the rural economy in many developing countries. Limited access to land and labour markets or even absence of these markets imply that rural households are unable to exchange their land and labour as much as they would like to and have to rely to a large extent – or even fully – on their own resources. Under such circumstances, households' shadow wages and rents become important determinants of production and consumption decisions, and their production and consumption decisions are non-separable (Singh *et al.*, 1986; Hoff *et al.*, 1993).

Rural China is characterized by surplus and underemployed rural labour, while land rental markets are rather thin (Brandt *et al.*, 2002, 2004). Off-farm employment has become an important source of rural incomes since the start of the economic reforms and has accelerated since the mid-1990s (De Brauw *et al.*, 2002). Recent studies also show an increasing incidence of land rental activities in China (Lohmar *et al.*, 2001; Deininger and Jin, 2002; Kung, 2002). However, land and labour markets in rural China are still far from perfect and exhibit high transaction costs (Benjamin and Brandt, 2002; Bowlus and Sicular, 2003; Kuiper, 2005).

Several studies analyse labour migration decisions in rural China either at individual level (Zhao, 1997, 1999b, 2002, 2003; Kung and Lee, 2001; Li and Yao, 2001; Shi *et al.*, 2006) or at the farm household level (Lohmar, 1999; Rozelle *et al.*, 1999b; Zhao, 1999a; De Brauw *et al.*, 2001). Few studies have investigated on-farm labour demand decisions at the household level and tested the separability between household labour demand and supply decisions (Benjamin and Brandt, 2002; Bowlus and Sicular, 2003). Separability was rejected in these studies. The development of land rental markets has recently attracted attention, but empirical analyses of land market participation decisions in rural China are still scarce (Yao, 2000; Lohmar *et al.*, 2001; Deininger and Jin, 2002; Kung, 2002).

The purpose of this chapter is to analyse the factors influencing household land and labour market decisions in rural China, taking into account prevailing factor market imperfections. The insights gained from such an analysis may lead to conclusions and policy recommendations that differ significantly from those obtained from studies focusing on each of these two markets separately.

The effect of land rental market participation on migration decisions is examined by Kung and Lee (2001) and Shi *et al.* (2006), but the land market participation decision is exogenously determined in their studies. Some studies of land markets in rural China include the effect of off-farm employment, especially migration, on land market participation decisions. However, only a few of these studies consider the endogenous character of off-farm employment decisions (Kung, 2002). To our knowledge, no studies have thus far analysed the joint decision making on land and labour market participation under market imperfections in rural China.

The remainder of this chapter is structured as follows. The next section gives a brief description of the study area and data, followed by a description of the land and labour market development and the interaction of land and labour markets in the study area. We then provide an analytical framework of land and labour market participation under market imperfections. In the penultimate section, the results of a multinomial probit analysis explaining household land and labour market participation decisions are presented. We conclude with a summary of the main conclusions and draw policy implications.

Area Selection and Data Description

This chapter uses data from a farm household survey that was held in three villages in north-east Jiangxi province,[1] located in the south-east of China. Agriculture plays an important role in the economy of this province. In 2002, 21.9% of its GDP derived from agriculture, 6.5% higher than the average level for the whole country, and its GDP per capita was 71% of the national average (National Bureau of Statistics of China, 2004).

The villages were selected based on a series of criteria including market access, geographical conditions, off-farm employment and land tenure system in order to display the variety in these conditions in north-east Jiangxi Province and in a much larger hilly area with rice-based production system in south-east China.

Table 13.1 shows the major characteristics of the three villages. Banqiao is the smallest village, and is located in a hilly area. Market access is good in this village, and a major city is located within 10 km. Shangzhu is a middle-sized village, located in a mountain area. The basic infrastructure for transportation is very bad. It takes about 2 h by car from the county capital to the hamlet where the village offices are. Gangyan is the largest village in terms of population. It is located in a flat area, at 20 km from a major market. Road conditions are good.

The farm household survey was carried out in 2000–2001. The questions asked in the survey referred to the entire year 2000. In each village, 23% of the households were interviewed. A stratified random sample was used for selecting the households, with the natural villages

Table 13.1. Summary description of the three villages (from Kuiper *et al.*, 2001).

		Banqiao	Shangzhu	Gangyan
Location				
	Prefecture	Yingtan	Yingtan	Shangrao
	County	Yujiang	Guixi	Yanshan
	Township	Honghu	Tangwan	Wang-er
	Distance	10 km from city	Remote	20 km from city
	Road quality	Poor	Bad	Sand and tarmac
Population				
	Persons	900	2028	3200
	Households	220	472	730
	Hamlets	4	16	7
Income				
	Net income per capita[a] (yuan)	1700	1600	1600
Land (mu[b])				
	Farmland	1700	2759	3880
	Paddy land	1234	2359	3780
	Dry land	500	400	100
	Farmland/capita	1.89	1.36	1.21
	Upland/total land (%)	60–70	97	'Plain'
Agriculture				
	Main crops	Rice, groundnut, fruit trees	Rice, bamboo, fir	Rice, vegetables
	Farm infrastructure condition	Good	Rain-fed or irrigated with conserved water	Good
	Rice yield[c] (kg/ha)	5099	3950	4629
Degradation				
	Soil quality	Worsened	Not clear	Worsened
	Erosion	Dry land	Landslides	None
	Erosion/flood control	No government projects	No government projects	No government projects
	Natural compaction	Yes	Yes, but less	Yes
	Other problems	Acidification	–	Limited flooding
Land tenure				
	Quality/distance classification	4	3	3/4
	Allocation criterion	Family size and labour force	Family size	Family size
	Frequency of adjustment	For some hamlets: never adjusted	Small adjustments	Small: 3–5 years; large: 5–10 years (depends on hamlet)
	Collective management	–	–	Hamlet management of some forests

[a] Net income per capita for Jiangxi province and China were 2135.3 and 2253.4, respectively, in 2000 (National Bureau of Statistics of China, 2002); [b] 1 mu = 1/15 ha; [c] average rice yields for Jiangxi province and China were 5268 and 6272, respectively, in 2000 (National Bureau of Statistics of China, 2002).

within each village forming the strata (Kuiper *et al.*, 2001). In total, 329 farm households were interviewed. Collected information includes, among other things, demographic characteristics, assets, land tenure and participation in factor markets.

Participation in Land and Labour Markets in North-east Jiangxi Province

Land rental market

Since the introduction of the Household Responsibility System (HRS) at the end of the 1970s, land use rights have been assigned to farm households based on family size, labour force or a combination of both. Land transfers were not allowed at first, because policy makers believed that land transfers would lead to a concentration of land with a few households, leaving most households landless. Frequent administrative reallocations of land by village leaders were the only way at that time to correct for demographic changes.

Since the mid-1980s, however, land rentals have been permitted by the government. An overview of land rental market participation in the three villages, subdivided into irrigated and dryland, is presented in Table 13.2. Land rental activities take place mainly for irrigated land. Of all the households in the three villages, 46% rent-in irrigated land, while only 6% rent-in dryland. In Banqiao village, the village with a relatively large area of dryland, 20% of the households rent additional dryland.

Large differences exist between the shares of households that rent-in irrigated land (46%) and those households that rent-out irrigated land (8%). One reason for this large discrepancy may be that households who rent-out land rent it out to more than one household. But part of the discrepancy may also be caused by the fact that some households who rent-out their land have migrated, but still retain their land use rights, and could not be interviewed during the survey. In addition, some households may not report renting-out because they fear losing their land in the next land reallocations.

Table 13.2. Land rental market in three villages in 2000 (from farm household survey).

Village	No. of farm households	Percentage of households involved in:									
		Renting-in		Self-sufficient		Renting-out		Renting-in and -out		Total	
		I	D	I	D	I	D	I	D	I	D
Banqiao	54	54	20	35	76	11	2	0	2	100	100
Shangzhu	108	48	3	46	96	6	1	0	0	100	100
Gangyan	167	41	4	48	93	8	2	2	0	100	100
Total	**329**	**46**	**6**	**45**	**91**	**8**	**2**	**1**	**0**	**100**	**100**

I = Irrigated land; D = Dryland.

Few households rented land in and out at the same time. Of all the households in the three villages, only one rented dryland in and out, while four rented irrigated land in and out simultaneously.

Labour market

Off-farm employment

China's population has recently reached 1.3 bn, with about 60% still living in rural areas. The average size of landholdings is only around 0.5 ha per family, and normally cannot fully employ a family's labour force. An off-farm economy, consisting of jobs in urban centres as well as in rural township and village enterprises and (at a later stage) private enterprises, has begun to emerge since the early 1980s and has accelerated since 1995 (De Brauw et al., 2002).

Local off-farm employment and migration are the two basic off-farm employment categories. Local off-farm employment includes agricultural wage employment, non-agricultural wage employment and self employment. Participation in off-farm employment in the three villages in 2000 is presented in Table 13.3. As many as 82% of the households participated in off-farm employment. Migration is relatively more important than local off-farm employment. Of all the households in the three villages, 21% participated in local off-farm employment and 61% in migration. This difference is caused mainly by the much higher participation in migration than in local off-farm employment in Gangyan village. The overall participation in off-farm employment is also much higher in this village (92%) than in the other two villages (70 and 73%).

Agricultural labour demand

Despite the surplus of rural labour, agricultural labour markets exist in rural China. They provide mechanisms for labour-constrained households to deal with labour shortages, especially during peak agricultural seasons. A distinction can be made between agricultural wage labour and exchange labour. Exchange labour takes place mainly among relatives and

Table 13.3. Participation in off-farm employment in three villages in 2000 (from farm household survey).

Village	No. of farm households	Percentage of households participating in:		
		Off-farm employment	Local off-farm	Migration
Banqiao	54	70	27	43
Shangzhu	108	73	21	52
Gangyan	167	92	19	73
Total	**329**	**82**	**21**	**61**

friends, and does not require payment. Rural labour demand for the production of rice, the most important crop, in the three villages is presented in Table 13.4. Exchange labour is more important than hiring of labour. Only 22% of the households hired additional labour, while 40% of the households used exchange labour in rice production. Both hiring of agricultural labour and exchange of labour are highest in Gangyan village, the village where migration is also highest.

Household labour supply and demand

A summary of the labour market participation of households in the surveyed villages is provided in Table 13.5. Only 15% of the households neither hire-in nor hire-out labour. Out of the 22% of households that hire-in labour, the majority (19%) are involved in off-farm employment. However, most households (63%) are involved in off-farm employment and do not hire-in labour. The share of households that simultaneously hire-in and -out is largest in Gangyan village, while the share of self-sufficient households is smallest in that village. This is consistent with the relatively high incidence of migration and hiring-in labour in this village.

Land and labour markets

The participation in land and labour markets is presented in Table 13.6. Farm households may engage in land and labour markets in 16 different

Table 13.4. Labour demand for rice production in three villages in 2000 (from farm household survey).

Village	No. of farm households	Percentage of households involved in:	
		Hiring	Exchange
Banqiao	54	15	26
Shangzhu	108	15	38
Gangyan	167	29	45
Total	**329**	**22**	**40**

Table 13.5. Labour market participation in three villages in 2000 (from farm household survey).

Village	No. of farm households	Percentage of households involved in:				
		Hiring-in	Self-sufficient	Hiring-out	Hiring-in and -out	Total
Banqiao	54	7	22	63	7	100
Shangzhu	108	3	24	61	12	100
Gangyan	167	1	7	64	28	100
Total	**329**	**2**	**15**	**63**	**19**	**100**

Table 13.6. Percentage of households participating in land and labour markets in 2000 (from farm household survey).

Labour market	Land market				
	Renting-in	Self-sufficient	Renting-out	Rent-in and -out	Total
Hiring-in	1	2	0	0	2
Self-sufficient	8	6	1	0	15
Hiring-out	30	29	3	1	63
Hiring-in and -out	6	9	4	0	19
Total	**46**	**45**	**8**	**1**	**100**

combinations.[2] The hiring-out of labour combined with either renting-in of land (30%) or self-sufficiency in land (29%) are the two most frequent combinations in the research area. As mentioned above, very few households hire-in agricultural labour without being involved in off-farm employment. The share of households simultaneously involved in renting-in and -out of land is also very small. Therefore, the bottom-left area of Table 13.6, which depicts the most frequent land and labour market participation combinations, is most important for our research.

Analytical Framework of Land and Labour Market Participation

The foregoing analysis shows that large differences exist in land and labour market participation between households in north-east Jiangxi province, suggesting the presence of land and labour market imperfections. In this section, we present an analytical framework for analysing farm household behaviour under such imperfections.

A farm household facing land rental and labour hiring decisions is assumed to maximize the utility function:

$$Max \quad U(y, l, Z^h)$$
$$A^{in}, A^{out},$$
$$l^h, l^f, l^o, l$$

(13.1)

where y is the income, l is the leisure and Z^h is farm household characteristics pertaining to consumption, such as wealth; A^{in} and A^{out} are the land rented-in and -out by the household, l^h is hired on-farm labour, l^f is the family labour used for on-farm production and l^o is the off-farm labour.

The utility is maximized subject to the income constraint. The household's total income is the sum of the income from agricultural production, off-farm employment and renting-out of the land, minus the costs for hiring-in labour and renting-in the land:

$$y = f(l^a, A, Z^q) + w^o l^o - w^a l^h - (r + T^A) A^{in} + (r - T^A) A^{out}$$

(13.2)

where w^o is the off-farm wage, w^a is the on-farm wage, r is the market land rent and T^A is the transaction cost involved in the land rental market (we assume that transaction costs for land demand and supply are equal for simplicity), such as tenure security and transfer rights. The price of agricultural product is set equal to one. All other prices are therefore expressed relative to the agricultural product price. $f(l^a, A, Z^q)$ is the agricultural production function, where l^a and A are the labour and land used in agricultural production and Z^g are fixed factors and farm characteristics, such as cattle, skills and experiences. The production function satisfies the standard assumptions. The land used in agricultural production is the sum of the household land endowment ($\bar{A}$), and the land rented-in, minus the land rented-out.

$$A = \bar{A} + A^{in} - A^{out} \qquad (13.3)$$

The labour used in agricultural production is the sum of the family labour for on-farm production, and the labour hired-in.

$$l^a = l^f + l^h \qquad (13.4)$$

The household is subject to a time constraint:

$$\bar{L} = l^f + l^o + l \qquad (13.5)$$

where $\bar{L}$ is the total labour endowment.

The following non-negativity constraints apply:

$$A^{in}, A^{out}, l^h, l^f, l^o, l \geq 0 \qquad (13.6)$$

In order to investigate the interactions of land and labour markets, we examine three cases. First, we analyse how imperfections in labour market affect the household labour allocation decisions when there is no land rental market. Secondly, we relax the assumption of no land rental market and examine how imperfections in the land rental market affect household land allocation decisions. Thirdly, we analyse land and labour markets when there are imperfections in both markets.

Case 1: household labour allocation decisions without land rental market

When there is no land rental market, the household's utility maximization problem becomes:

$$Max \quad U(y, l, Z^h) \qquad (13.1')$$
$$l^h, l^f, l^o, l$$
$$y = f(l^a, \bar{A}, Z^q) + w^o l^o - w^a l^h \qquad (13.2')$$
$$l^a = l^f + l^h \qquad (13.4)$$
$$\bar{L} = l^f + l^o + l \qquad (13.5)$$
$$l^h, l^f, l^o, l \geq 0 \qquad (13.6')$$

Different sub-cases can be distinguished, depending on the functioning of the labour market.

Case 1A: household labour allocation decisions with a perfect labour market

We start with the case of a perfect labour market. In that case, the wage rates for hiring-in agricultural labour and off-farm employment are equal ($w^o = w^a = w$). Substituting (13.4) and (13.5) into (13.2′), household optimal labour allocation can be represented by the following first order condition:

$$f_{l^a} = w \tag{13.7}$$

That is, the household will equate the marginal product of labour in agriculture to the market wage.

Equation (13.7) can be rewritten such that the household labour demand is expressed as a function of a set of exogenous variables, including wage, land and other fixed factors and farm-specific characteristics:

$$l^{i*} = l^{i*}(w, \overline{A}, Z^q) \tag{13.8}$$

where l^{i*} represents the demand for family labour (l^{f*}), hired labour (l^{h*}) and on-farm labour in total (l^{a*}). Equation (13.8) shows that household characteristics pertaining to consumption (Z^h) do not affect household production decisions. So, production decisions are independent of consumption decisions.

Case 1B: household labour allocation decisions with an imperfect labour market

In the analysis so far, the labour market is assumed to be perfect, and household production and consumption decisions are separable. However, farm households are typically facing labour market imperfections. Labour markets can be imperfect for several reasons (Benjamin, 1992). The main reasons are: (i) the household may face constraints in looking for off-farm employment opportunities; (ii) the household may have constraints in hiring-in agricultural labour; and (iii) family and hired labour may not be perfect substitutes in agricultural production. We will explore each of these three circumstances successively and then analyse their consequences for household labour allocation decisions.

Constraints in off-farm employment opportunities

The household can participate in off-farm employment, and earn a wage (w^o) that is higher than the agricultural wage (w^a). In this situation, the household faces an off-farm employment constraint that enables not all its labour to move out of agriculture. The labour market, as a result, is assumed to be cleared by quantity rationing. Following Yao (2000), we assume that the quantity rationing takes the form of imposing a ceiling (l^o_{max}) to the household. Therefore, the household faces an additional constraint:

$$l^o \le l^o_{max} \tag{13.9}$$

Inserting Equations (13.2'), (13.4), (13.5) and (13.9) into Equation (13.1'), the household labour allocation decisions under the off-farm employment constraint give the following first order conditions (assuming an interior solution):

$$f_{l^a} = w^o - \mu^o / U_y \tag{13.10}$$

$$f_{l^a} = w^a \tag{13.11}$$

In Equations (13.10) and (13.11), the left-hand side is the marginal product of agricultural labour and μ^o is the Lagrange multiplier for the constraint on l^o. Denote $w^* = w^o - \mu^o / U_y$, where w^* is only equal to w^o when there is no off-farm employment constraint, and is smaller than w^o when there is off-farm employment constraint. Substituting w^* into Equation (13.10) gives:

$$f_{l^a} = w^* \tag{13.10'}$$

As mentioned above, rural China is characterized by a surplus of rural labour. Therefore, we assume that w^* is smaller than w^a for the constrained household and that the household will not hire-out labour.

In reality, however, some households hire-out labour but also hire-in labour during peak agricultural seasons. We turn to the case of constraints in hiring-in labour now.

Constraints for hiring-in labour

Agricultural production is characterized by seasonality. There are peak seasons and slack seasons in agricultural production. During the slack season, labour is over-supplied. In the peak season, however, labour is scarce compared with the employment opportunities. As in the previous section, we assume that the market for hiring-in labour is cleared by quantity rationing that takes the form of imposing a ceiling (l^h_{max}) to the household:

$$l^h \le l^h_{max} \tag{13.12}$$

Inserting Equations (13.2'), (13.4), (13.5) and (13.12) into Equation (13.1'), the labour allocation decisions of households faced with a hiring-in constraint give us the following first order conditions (assuming an interior solution):

$$f_{l^a} = w^o \tag{13.13}$$

$$f_{l^a} = w^a + \mu^a / U_y \tag{13.14}$$

In Equations (13.13) and (13.14), the left hand side is the marginal product of agricultural labour and μ^a is the Lagrange multiplier for the constraints on l^h. Denote $w^{*\prime} = w^a + \mu^a / U_y$, where $w^{*\prime}$ is equal to w^a when there is no constraint in hiring-in labour, and is greater than w^a

when there is constraint in hiring-in labour. Substituting $w^{*\prime}$ into Equation (13.14), we have

$$f_{l^a} = w^{*\prime} \qquad (13.14')$$

We may assume that $w^{*\prime}$ is greater than w^o for the constrained household so that the household will not hire-in labour.

Varying efficiency of family and hired labour

Agricultural production has several specific features such as spatial dispersion and synchronic timing (Binswanger and Rosenzweig, 1986). Hired labour, as a result, has an incentive to work less efficiently. These incentive problems lead to high monitoring cost of hired labour. Family and hired labour are therefore not perfect substitutes. We assume that one unit of hired labour is perfectly substitutable to β units of family labour, with $\beta < 1$. Accordingly, we replace Equation (13.4) by (13.4'):

$$l^a = l^f + \beta l^h \qquad (13.4')$$

Inserting (13.2'), (13.4') and (13.5) into (13.1'), the household's labour allocation decisions with different efficiency of family and hired labour give the following first order conditions (assuming an interior solution):

$$f_{l^a} = w^o \qquad (13.15)$$

$$f_{l^a} = w^a / \beta \qquad (13.16)$$

In Equations (13.15) and (13.16), the left hand side is the marginal product of agricultural labour. Under a perfect labour market, w^o is equal to w^a. Denote $w^{*\prime\prime} = w^a/\beta$, where $w^{*\prime\prime}$ is equal to w^o (w^a) when hired labour is perfectly substitutable to family labour, and is greater than w^o (w^a) when hired labour is less efficient than family labour. Substituting $w^{*\prime\prime}$ into Equation (13.16), we have

$$f_{l^a} = w^{*\prime\prime} \qquad (13.16')$$

The first order conditions in the case of different efficiency of family and hired labour (Equations (13.15) and (13.16)) are consistent with that in the case of constraints in off-farm employment opportunities (Equations (13.10) and (13.11)) and that in the case of constraints in hire-in labour. In all cases, the household shadow wage does not equal the market wage.

The first order conditions under labour market imperfections can be rewritten as a function of the shadow wage (w^s) and a set of exogenous variables, including land and other fixed assets and farm-specific characteristics:

$$l^{i*} = l^{i*}(\overline{A}, Z^q, w^s) \qquad (13.17)$$

where l^{i*} represents the demand for family labour (l^{f*}), hired labour (l^{h*}) and on-farm labour in total (l^{a*}). As the shadow wage is endogenous, the household production and consumption decisions are non-separable.

Labour market imperfection in rural China is characterized mainly by lack of off-farm employment opportunities. The shadow wage (w^s) can then be expressed as a function of the market wage (w) and the transaction costs involved in the labour market (T^L), such as education and social networks (we assume that transaction costs for labour demand and supply are equal for simplicity). Therefore, the household's status in the labour market can be generalized by:

$f_{l^a} = w + T^L$ for households hiring-in labour,

$w - T^L \leq f_{l^a} \leq w + T^L$ for households self-sufficient in labour use and

$f_{l^a} = w - T^L$ for households hiring-out labour.

Case 2: household land allocation decisions with a land rental market

When there is a land rental market, we can insert Equations (13.2) and (13.3) into (13.1) and the household land allocation decisions are characterized by the following first order conditions (assuming an interior solution):

$$f_A = r + T^A \tag{13.18}$$
$$f_A = r - T^A \tag{13.19}$$

If the land rental market functions perfectly, T^A is equal to 0, so that there are no transaction costs in the land rental market. In that case, the first order condition is $f_A = r$, that is the marginal product of land is equal to the land rent.

However, when $T^A > 0$, the household faces a land rent band. The household's status in the land market is then characterized by:

$f_A = r + T$ for households renting-in land,

$r - T^A \leq f_A \leq r + T^A$ for self-sufficient households and

$f_A = r - T^A$ for households renting-out land.

As can be seen from Equations (13.18) and (13.19), the household will not rent-in and rent-out land simultaneously. In reality, however, some households both rent-in and rent-out in order to consolidate scattered plots, or substitute low-quality plots for high-quality plots (or vice versa).

Case 3: interactions of land and labour markets with imperfections in these markets

In the above, we identified four different alternatives in which a household can operate with respect to the labour market, namely hiring-in, self-sufficient, hiring-out and both hiring-in and -out. We have also identified four different alternatives for households with respect to the land market, namely renting-in, self-sufficient, renting-out and both renting-in and -out. As a result, a household may participate in land and labour markets in 16 different combinations. These 16 different combinations are represented by FOC1–FOC16 (FOC means first order condition) and are presented in Table 13.7. They reflect the variation in household transaction costs and resource endowments that underlie participation in land and labour markets.

Model Specification and Regression Results

Model specification

Based on the first order conditions presented above, we can derive the reduced reform equations for land and labour market participation decisions:

$$I = I(Z^h, Z^q, \overline{L}, \overline{A}, r, T^A, w, T^L) \tag{13.20}$$

where I represents the land and labour allocation decisions on A^{in}, A^{out}, l^h, l^f, l^o and l.

Substituting Equations (13.20) into (13.1), we get the indirect utility function:

$$V = V(Z^h, Z^q, \overline{L}, \overline{A}, r, T^A, w, T^L) \tag{13.21}$$

As the model is non-separable, the functional form of the reduced-form equations and the indirect utility function cannot be derived analytically. To estimate the impact of the exogenous variables at the right-hand side of Equation (13.20) on land and labour allocation decisions, we assume for simplicity that their relationship can be approximated by a linear function. In other words, decisions on the 16 land and labour market participation alternatives by the ith household can be represented by:

$$V_{ij} = \beta'_j x_{ij} + \varepsilon_{ij} \tag{13.22}$$

where V_{ij} is a dummy variable, which equals j if household i chooses alternative j ($j = 1,2,...,16$), and x_{ij} is the set of exogenous variables listed in Equation (13.20) for the ith household.

In our empirical analysis, the household characteristics pertaining to consumption (Z^h) are represented by the number of durable assets[3] in a household, household size, number of dependents[4] in a household and the female:male adult ratio. Household characteristics have a direct effect on consumption preferences: they can have either positive or negative effects on the demand for leisure and consumption goods. In the case of household size and number of dependents, it may be expected that larger households and households with fewer dependents consume more food at a given income level. If household decisions are non-separable, this will put an upward pressure on agricultural production and therefore stimulate hiring-in of labour and renting-in of

Table 13.7. First order conditions for household participation in land and labour markets.

Land and labour markets	Renting-in	Self-sufficient	Renting-out	Renting-in and -out
Hiring-in	FOC1: $f_A = r + T^A$ $f_{ja} = w + T^L$	FOC2: $r - T^A \leq f_A \leq r + T^A$ $f_{ja} = w + T^L$	FOC3: $f_A = r - T^A$ $f_{ja} = w + T^L$	FOC4: Renting-in: good quality, etc. Renting-out: poor quality, etc. $f_{ja} = w + T^L$
Self-sufficient	FOC5: $f_A = r + T^A$ $w - T^L \leq f_{ja} \leq w + T^L$	FOC6: $r - T^A \leq f_A \leq r + T^A$ $w - T^L \leq f_{ja} \leq w + T^L$	FOC7: $f_A = r - T^A$ $w - T^L \leq f_{ja} \leq w + T^L$	FOC8: Renting-in: good quality, etc. Renting-out: poor quality, etc. $w - T^L \leq f_{ja} \leq w + T^L$
Hiring-out	FOC9: $f_A = r + T^A$ $f_{ja} = w - T^L$	FOC10: $r - T^A \leq f_A \leq r + T^A$ $f_{ja} = w - T^L$	FOC11: $f_A = r - T^A$ $f_{ja} = w - T^L$	FOC12: Renting-in: good quality, etc. Renting-out: poor quality, etc. $f_{ja} = w - T^L$
Hiring-in and -out Hiring-in: peak season	FOC13: $f_A = r + T^A$ Hiring-out: slack season Hiring-in: peak season	FOC14: $r - T^A \leq f_A \leq r + T^A$ Hiring-out: slack season Hiring-in: peak season	FOC15: $f_A = r - T^A$ Hiring-out: slack season Hiring-in: peak season	FOC16: Renting-in: good quality, etc. Renting-out: poor quality, etc. Hiring-out: slack season Hiring-in: peak season

land while having a negative impact on off-farm employment and renting-out of land at a given time endowment.

Fixed factors and farm characteristics (Z^q) are represented by the number of cattle in a household at the end of the previous year, the average adult age and the female:male adult ratio. Cattle are very important draught animals for small-scale households in rural China. The use of cattle has a positive impact on agricultural productivity, and hence increases both the shadow rent and the shadow wage. Therefore, land renting-in and labour hiring-in are expected to depend positively on the number of cattle, while land renting-out and off-farm employment are expected to depend negatively on it.

Average adult age is used as a proxy for skills and experiences of a family. A household with older members is more productive in agriculture and, as a result, has a higher shadow rent and wage. So, again, it is expected that land renting-in and labour hiring-in depend positively on it, while land renting-out and off-farm employment depend negatively on it. The square of the average adult age is added to the equation in order to capture possible life cycle effects. The female:male ratio is used to test for differences between females and males in physical strength or other differences in productivity. If males are more productive in agriculture, then a higher value of this ratio leads to less renting-in and more renting-out of land and to less hiring-in of labour and a larger involvement in off-farm employment.

The household time endowment ($\bar{L}$) equals household size minus the number of dependents. In addition, household time endowment may depend on the female:male adult ratio, as taking care of children is usually a female task in Chinese society. Households with a relatively large time endowment face a lower shadow wage and are therefore expected to hire fewer agricultural labourers and to be more involved in off-farm employment. When the labour market is imperfect, they will face a higher shadow rent. As a consequence, renting-in of land is expected to be higher while renting-out of land is expected to be lower.

The household land endowment ($\bar{A}$) is represented by irrigated land contracted per adult. Households with relatively more land have a lower shadow rent and are therefore expected to rent-out more land and rent-in less land. When land markets are imperfect, they will also face a higher shadow wage and are therefore expected to hire-in more labour and participate less in off-farm employment. The square of this variable is added to the equation to capture possible nonlinearities in its impact.

Tenure security and transfer rights are used as proxies for transaction costs in the land rental market (T^A). In our survey, households were asked whether they possessed a land contract. Possession of a land contract is usually considered as having tenure security. In addition, households were asked their opinions about land reallocation.[5] We performed a principle component analysis to create a factor score for subjective tenure security, with higher values indicating greater tenure security. Tenure security is expected to stimulate land market participation because it reduces the transaction costs in land market transactions and hence reduces the rent band (Lohmar, 1999; Lohmar *et al.*, 2001; Kung, 2002).

Another proxy for transaction costs in the land rental markets is the extent to which households can freely enjoy land transfer rights. In our survey, households were asked whether they had the right to transfer land within the village, right to transfer outside the village, inheritance right and mortgage right. The indicator used is the number of transfer rights enjoyed by the household, divided by four. Households were also asked their opinions about land transfer rights.[6] We again performed a factor analysis and created a factor score for subjective transfer rights, with higher values indicating more limitations on transfer rights. More limitations on transfer rights are expected to increase transaction costs and rent band, and hence reduce land market transactions (Li and Yao, 2001).

Average adult education (average years of schooling of adults), presence of a migration network, land endowment and number of durable assets in a household are used as proxies for transaction costs in the labour market (T^L). Farmers with relatively high education levels are generally less constrained by the limited off-farm employment opportunities. Presence of a migration network is a dummy variable that equals one if the household received remittances from family members who are not living together with the household or who participated in migration before the survey year. Having a migration network is expected to meet fewer migration constraints (Zhang and Li, 2001; Zhao, 2003). Households with relatively more land (the most productive asset) and durable assets are also expected to face fewer obstacles in labour market participation, because they have more resources available for paying fixed costs and for obtaining information in gaining access to off-farm employment.

The price bands around the land rent and wage rate partly depend on credit availability. Credit-constrained households mark-up the land rent and wage rate by the shadow value of the credit (Sadoulet and de Janvry, 1995). Credit availability is defined as a dummy variable in the model. Credit-constrained households are more likely to rent-out land and to be involved in off-farm employment and less likely to rent-in land or hire agricultural labour.

The market land rent (r) and labour wage (w) are exogenous in the model and are assumed to be the same for all households living in the same village. They are therefore captured by village dummy variables.

As discussed above, hiring of additional labour takes place only in peak seasons and is done only to a small extent. We therefore leave hiring-in of labour out of our analysis. We subdivide households that are involved in off-farm employment into local off-farm employment and migration.[7] The reason for making this distinction is that local off-farm employment can be combined with working on-farm, whereas this is not possible for migration. Moreover, people involved in local off-farm employment usually live and consume at home. So, the effects of migration and local off-farm employment on agricultural production will differ when household decision making is non-separable. With respect to the land rental market, households are divided into households renting-out, those using only own land and those renting-in.[8] Nine combinations with respect to land and labour market participation can therefore be distinguished. The percentage distribution of the 329 households over these nine combinations is presented in Table 13.8.

Only a small share of the households rent-out land while at the same time being either self-sufficient in labour (1%) or participating in local off-farm employment (2%). We drop these two alternatives because the number of observations is insufficient for our empirical analysis. We therefore confine our analysis to seven land and labour market participation combinations. Accounting for missing observations, the total number of observations that can be used in the analysis is 280. Descriptive statistics of the explanatory variables, subdivided by household groups, are shown in Table 13.9.

For estimating model, we may use either the multinomial logit model or the multinomial probit model. The multinomial logit model has been widely used because of its computational ease. However, it assumes independence of irrelevant alternatives (Judge et al., 1995). The multinomial probit model does not have this limitation, and is therefore used in our empirical analysis.

Estimation results

We chose the household group that is self-sufficient in both land and labour as the basis of the analysis. The estimated coefficients therefore represent the effect of the explanatory variable in question on the probability that households choose a particular combination of participation in the land and/or labour markets over the base situation of being self-sufficient in both markets. The estimation results are presented in Table 13.10. The differences between the mean proportions of each regime and the mean predicted probabilities of each regime are small, indicating a good fit of the model to the data.

Table 13.8. Percentage distribution of participation in land market and off-farm employment in 2000 (from farm household survey).

	Land market			
Labour market	Renting-out	Self-sufficient	Renting-in	Total
On-farm	1	8	9	18
Local off-farm	2	9	11	21
Migration	7	29	26	61
Total	**9**	**45**	**46**	**100**

Note: Totals may not always add up due to rounding errors.

Table 13.9. Descriptive statistics of explanatory variables by household groups. Values are means (standard deviations in parentheses).

Explanatory variable	On-farm and renting-in land (n = 29)	On-farm and self-sufficient in land (n = 24)	Local off-farm and renting-in land (n = 34)	Local off-farm and self-sufficient in land (n = 27)	Migration and renting-in land (n = 69)	Migration and self-sufficient in land (n = 79)	Migration and renting-out land (n = 18)	All households (n = 280)
Household size (n)	4.07 (1.00)	3.38 (1.31)	3.91 (1.11)	4.11 (1.19)	5.04 (1.42)	4.71 (1.74)	4.78 (1.48)	4.46 (1.50)
Dependants (n)	1.59 (1.21)	1.21 (1.25)	1.32 (1.04)	1.70 (0.99)	1.14 (1.10)	1.03 (1.10)	1.56 (0.78)	1.26 (1.10)
Durable assets (n)	6.07 (2.27)	5.08 (2.02)	6.76 (1.58)	7.00 (1.96)	6.62 (1.54)	6.34 (1.83)	6.44 (1.38)	6.40 (1.83)
Cattle (n)	1.24 (1.77)	0.67 (0.48)	0.85 (0.61)	0.56 (0.51)	0.75 (0.53)	0.72 (0.60)	0.39 (0.70)	0.76 (0.80)
Average adult age (years)	38.00 (8.19)	46.31 (11.25)	38.30 (8.24)	35.85 (6.31)	35.37 (4.31)	37.60 (6.00)	38.48 (5.00)	37.81 (7.29)
Adult female:male ratio	1.09 (0.55)	0.96 (0.44)	1.06 (0.55)	0.95 (0.31)	0.92 (0.55)	1.05 (0.62)	1.40 (0.96)	1.03 (0.58)
Irrigated land per adult (mu[a])	2.07 (0.86)	2.61 (1.78)	2.05 (0.68)	2.29 (1.20)	1.60 (0.64)	1.73 (0.59)	2.31 (0.80)	1.94 (0.93)
Subjective security score	−0.004 (1.32)	0.04 (0.89)	0.07 (1.18)	0.07 (0.83)	0.01 (1.14)	−0.07 (0.79)	−0.03 (0.76)	0.00 (1.00)
Subjective transfer score	0.22 (1.09)	−0.25 (1.00)	−0.06 (1.02)	0.28 (0.99)	−0.08 (1.13)	0.14 (0.86)	0.03 (0.80)	0.00 (1.00)
Possession land contract (1 = yes)	0.21 (0.41)	0.29 (0.46)	0.29 (0.46)	0.26 (0.45)	0.33 (0.47)	0.29 (0.46)	0.17 (0.38)	0.28 (0.45)
Land transfer rights	0.57 (0.13)	0.53 (0.17)	0.60 (0.12)	0.58 (0.12)	0.59 (0.12)	0.59 (0.14)	0.57 (0.12)	0.58 (0.13)
Average adult education (years)	3.79 (1.50)	3.72 (2.16)	4.17 (1.63)	4.34 (1.35)	4.61 (1.44)	4.73 (1.70)	3.78 (1.96)	4.35 (1.67)
Migration network (1 = yes)	0.17 (0.38)	0.29 (0.46)	0.15 (0.36)	0.04 (0.19)	0.39 (0.49)	0.44 (0.50)	0.44 (0.51)	0.31 (0.47)
Credit availability (1 = yes)	0.48 (0.51)	0.46 (0.51)	0.38 (0.49)	0.59 (0.50)	0.51 (0.50)	0.34 (0.48)	0.39 (0.50)	0.44 (0.50)
Banqiao dummy (1 = yes)	0.28 (0.45)	0.25 (0.44)	0.26 (0.45)	0.15 (0.36)	0.14 (0.35)	0.11 (0.32)	0.06 (0.24)	0.17 (0.37)
Shangzhu dummy (1 = yes)	0.41 (0.50)	0.63 (0.49)	0.38 (0.49)	0.26 (0.45)	0.30 (0.46)	0.27 (0.44)	0.11 (0.32)	0.33 (0.47)

[a] 1 mu = 1/15 ha.

Table 13.10. Multinomial probit analysis results of land and labour market participation.

			Coefficients (z–score)			
Independent variabole	On-farm and renting-in land	Local off-farm and renting-in land	Local off-farm and self-sufficient in land	Migration and renting-in land	Migration and self-sufficient in land	Migration and renting-out land
Household size (n)	0.01 (0.04)	0.01 (0.04)	-0.02 (-0.05)	0.78 (2.46)**	0.78 (2.44)**	0.79 (2.16)**
Dependants (n)	0.20 (0.60)	-0.11 (-0.35)	0.29 (0.91)	-0.55 (-1.84)*	-0.65 (-2.21)**	-0.55 (-1.42)
Durable assets (n)	0.07 (0.54)	0.29 (2.65)***	0.30 (2.43)**	0.02 (0.18)	-0.01 (-0.05)	0.10 (0.79)
Cattle (n)	0.50 (1.61)	0.24 (0.76)	-0.33 (-0.93)	0.11 (0.35)	0.12 (0.40)	-0.75 (-1.82)*
Average adult age (years)	0.07 (0.32)	0.13 (0.64)	0.01 (0.04)	0.62 (1.99)**	0.37 (1.68)*	1.03 (3.03)***
Average adult age^2	-0.001 (-0.60)	-0.002 (-0.79)	-0.001 (-0.32)	-0.01 (-2.30)**	-0.01 (-1.93)*	-0.01 (-3.18)***
Adult female:male ratio	0.02 (0.06)	-0.01 (-0.02)	-0.66 (-1.63)	-0.48 (-1.53)	-0.10 (-0.32)	0.38 (1.00)
Irrigated land per adult (mu[a])	-0.07 (-0.10)	1.65 (1.79)*	-0.47 (-0.86)	0.73 (0.89)	2.22 (2.30)**	3.29 (2.62)***
Irrigated land per adult^2	-0.10 (-0.79)	-0.43 (-2.47)**	0.01 (0.10)	-0.29 (-1.99)**	-0.59 (-2.94)***	-0.62 (-2.96)***
Subjective security score	-0.05 (-0.29)	0.04 (0.25)	0.08 (0.45)	0.05 (0.32)	0.01 (0.07)	-0.04 (-0.24)
Subjective transfer score	-0.04 (-0.19)	0.04 (0.20)	0.30 (1.49)	-0.10 (-0.56)	0.07 (0.45)	0.01 (0.08)
Possession land contract (1 = yes)	0.45 (0.89)	0.60 (1.32)	0.52 (1.02)	1.02 (2.27)**	0.76 (1.72)*	0.65 (1.19)
Land transfer rights	0.73 (0.58)	2.44 (1.80)*	2.19 (1.53)	1.67 (1.33)	1.84 (1.50)	-0.14 (-0.09)
Average adult education (years)	-0.20 (-1.44)	-0.12 (-0.86)	-0.10 (-0.69)	-0.11 (-0.77)	0.002 (0.02)	-0.35 (-2.16)**
Migration network (1 = yes)	0.03 (0.06)	-0.36 (-0.77)	-0.95 (-1.55)	0.59 (1.43)	0.72 (1.79)*	0.80 (1.72)*
Credit availability (1 = yes)	0.20 (0.51)	-0.02 (-0.05)	0.59 (1.41)	0.42 (1.13)	-0.08 (-0.22)	-0.002 (-0.01)
Banqiao dummy	-0.41 (-0.74)	-0.50 (-0.92)	-0.88 (-1.56)	-1.19 (-2.24)**	-1.34 (-2.50)**	-2.33 (-3.17)***
Shangzhu dummy	-0.79 (-1.40)	-0.60 (-1.07)	-1.09 (-1.84)*	-1.33 (-2.42)**	-1.46 (-2.63)***	-1.90 (-2.93)***
Intercept	0.19 (0.04)	-5.35 (-1.18)	0.17 (0.04)	-11.92 (-1.86)*	-10.32 (-2.11)**	-24.87 (-3.21)***
Share of households participating in different regimes	0.1036	0.1214	0.0964	0.2464	0.2821	0.0643
Mean predicted probability of participation in different regimes	0.1054	0.1214	0.0970	0.2431	0.2810	0.0652

Log likelihood = -364.38; Wald chi2(108) = 391.06; Prob. > chi2 = 0.00.
*, **, and *** indicate statistical significance at the 10, 5, and 1% levels, respectively.
Base outcome is self-sufficiency in land and labour markets.
[a] 1 mu = 1/15 ha.

Household size has a significant positive impact on all three migration combinations, while the number of dependants in a household has a significant negative effect on the probability of migration combined with either land renting-in or land self-sufficiency. These findings confirm the results of earlier studies that larger households and households with fewer dependents tend to migrate (Zhao, 1997, 1999a,b, 2002, 2003; Rozelle *et al.*, 1999a, b; De Brauw *et al.*, 2002).

In terms of our model, it means that the time endowment effect of household size and dependents exceeds the food consumption effect. For local off-farm employment, however, the results indicate that the time endowment effect is counterbalanced by the food consumption effect. The fact that, contrary to migrants, people involved in local off-farm employment generally live and consume food at home seems to play an important role here. The number of durable assets has a significant positive effect on the probability that households are involved in local off-farm employment but not on the probability that households are involved in migration. This result suggests that richer households may have better access to local off-farm jobs than do poorer households.

The number of cattle owned by a household does not significantly affect most land and labour market participation decisions. Only the likelihood that a household has a migrated member and rents-out land is negatively affected. The average adult age shows an inverted U-shaped relationship with the likelihood of being involved in migration. The turning point is at 34 years of age for households that migrate and rent-in land, at 37 for households that migrate and are self-sufficient in land and at 40 for households that migrate and rent-out land.

So, young households are relatively more likely to rent-in additional land when a member migrates, while older households are relatively more likely to rent-out land when a member migrates. A possible explanation for this finding is that households involved in migration and renting-in of land are mainly young migrants who return home after a few years of working elsewhere and reinvest in agriculture, which is a common phenomenon in China (Hare, 1999; Zhao, 2002). High economic and psychological costs of settlement in migration destinations are assumed to cause this phenomenon. The inverted U-shape that we find only partly supports the findings of previous studies that young farmers tend to migrate (Zhao, 1997, 1999a,b, 2002, 2003; Rozelle *et al.*, 1999a, b; De Brauw *et al.*, 2002).

The size of the contacted irrigated land has a significant effect on the participation in migration, in all three land market combinations, and on the probability of being involved in local off-farm employment and renting-in land. Interestingly, the likelihood of belonging to these groups displays an inverted U-shaped relationship with land availability per adult, except for the migration and land renting-in combination. So, households with both low and high land availability are more likely to stay on-farm. This result is consistent with the finding of Li and Yao (2001) that land resources in rural China do not only have a wealth effect, used for financing migration, but also have a substitution effect that holds back migration when the land rental market is imperfect. So, households with small land endowments may not be wealthy enough to be able to migrate, while households with relatively large land endowments may have difficulties in renting-out their land and hence prefer to work on-farm instead of migrating. This result partly contradicts the finding of Deininger and Jin (2002) that households with large land endowments rent-in land while households with small land endowments rent-out land.

Possession of a land contract increases the likelihood that households migrate and either rent-in land or remain self-sufficient in land. Having a migration network has a significant positive effect on the probability of migration combined with either land self-sufficiency or land renting-out. As discussed above, the number of durable assets owned by a household also has a significant positive effect on off-farm labour participation decisions. These results confirm the fact that households facing low transaction costs in the labour market are more likely to participate in off-farm employment. Enjoying more land transfer rights significantly affects the probability of land renting-in combined with local off-farm employment only. Other transaction cost proxies specified in the model do not significantly influence land and labour market participation, except for average adult education which (surprisingly), has a negative effect on the probability of migration and land renting-out. Measurement problems and the inadequacy of these proxies in representing differences in transaction costs may play a role in this.

Finally, the results for the two village dummy variables indicate that there exist significant differences in market wages, land rents or other variables affecting land and labour market participation decisions that differ between these villages.

Conclusions

Exchange of land and engagement in labour markets are two important decisions faced by rural households. In the absence of well-functioning factor markets, decisions on land and labour use are jointly made, implying that production and consumption behaviour should be analysed within a non-separability framework. In this chapter we assess different combinations of land and labour market participation, using data from a survey among 329 farm households in three villages in Jiangxi province. We estimate a multinomial probit model to examine the determinants of the seven most common combinations of land and labour market participation in the three villages.

The empirical results indicate that households with small land resources and those with relatively large land resources are both more likely to stay on-farm. This finding suggests that households with small land endowments may not be wealthy enough to gain access to off-farm employment, while households with relatively large land endowments may have difficulties in renting-out their land and hence prefer to work on-farm instead of migrating. Policies aimed at improving access to credit for households with relatively small land endowments and facilitating the renting-out of land by land-abundant households may therefore play an important role in stimulating the development of rural rental markets and thereby improving the efficiency of agricultural production.

The average age of the adults in a household shows an inverted U-shaped relationship with the likelihood of being involved in migration. The turning point is at 34 years of age for households that migrate and rent-in land, at 37 for households that migrate and are self-sufficient in land and at 40 for households that migrate and rent-out land. A possible explanation for this finding is that households involved in migration and renting-in of land are mainly young migrants who return home after a few years of working elsewhere. Policies that reduce the high economic and psychological costs of settlement in migration destinations are likely to reduce such return migration flows, but may also adversely affect the investments in agriculture made by such return migrants.

A larger household size and fewer dependents have a significant positive impact on migration, but not on local off-farm employment. The time endowment effect is counterbalanced by the food consumption effect for local off-farm employment. The fact that, contrary to migrants, people involved in local off-farm employment generally live and consume food at home seems to play an important role here. The number of durable assets in a household has a significant positive effect on the probability that households are involved in local off-farm employment. This result suggests that richer households may have better access to local off-farm jobs than do poorer households. Policies that improve the access of poorer households to off-farm jobs may therefore provide an important contribution to reducing income gaps and alleviating rural poverty.

We further find that possession of a land contract and having a migration network has a positive effect on migration. Enjoying more land transfer rights has a significant positive effect on land renting-in. These results confirm that households facing low transaction costs in labour and land market are more likely to participate in these markets. Policies that improve tenure security, increase the bundle of land transfer rights and provide more off-farm employment information may therefore reduce transaction costs in both the land and the labour markets and contribute to the further development of rural factor markets.

Additional surveys among different household types are needed to gain more insights into their motives for participating or not participating in land and labour markets and to formulate more specific policy recommendations.

Endnotes

[1] The data were collected for a joint research project of Nanjing Agricultural University, Wageningen University and the Institute of Social Studies, The Hague on economic policy reforms, agricultural incentives and soil degradation in south-east China, financed by the Netherlands Ministry of Development Cooperation (SAIL programme) and the European Union (INCO-DC programme).

[2] The analysis here deals only with irrigated land, the most important category. Furthermore, we consider only hiring of labour against payment.

[3] Durable assets include durable goods such as televisions, fridges, radios, transportation vehicles, etc.

[4] Numbers of dependants in a household are numbers of family members under 16 and over 66 years old.

[5] Opinions about land reallocation include: (i) whether households should receive more land when they increase in size; (ii) whether households should receive less land when they decrease in size; (iii) whether households should return their land when they start working in enterprises; (iv) whether households should be allowed to contract more land when they rent-in land; and (v) whether households should return their land when they rent-out land.

[6] Opinions about land transfer rights include: (i) whether households should be able to transfer their land to other households within the village; (ii) whether households should have inheritance rights; and (iii) whether households should have mortgage rights.

[7] Migrants are family members working off-farm and not living together with other household members. Migration means that the household has at least one family member working as a migrant. Local off-farm employment means that none of the household members involved in off-farm employment works as a migrant.

[8] The four households that both rent-in and rent-out land are treated as renting-out in order to increase the limited number of observations in the renting-out category.

References

Benjamin, D. (1992) Household composition, labour markets, and labour demand: testing for separation in agricultural household models. *Econometrica* 60, 287–322.

Benjamin, D. and Brandt, L. (2002) Property rights, labour markets, and efficiency in a transition economy: the case of rural China. *Canadian Journal of Economics* 35, 689–716.

Binswanger, H.P. and Rosenzweig, M.R. (1986) Behavioural and material determinants of production relations in agriculture. *Journal of Development Studies* 22, 503–539.

Bowlus, A.J. and Sicular, T. (2003) Moving toward markets? Labour allocation in rural China. *Journal of Development Economics* 71, 561–583.

Brandt, L., Huang, J., Li, G. and Rozelle, S. (2002) Land rights in rural China: facts, fictions and issues. *The China Journal* 47, 67–97.

Brandt, L., Rozelle, S. and Turner, M.A. (2004) Local government behaviour and property rights formation in rural China. *Journal of Institutional and Theoretical Economics* 160, 627–662.

De Brauw, A., Taylor, J.E. and Rozelle, S. (2001) Migration and Incomes in Source Communities: a New Economics of Migration Perspective from China. Working paper, Department of Agricultural and Resource Economics, University of California, Davis, California.

De Brauw, A., Huang, J., Rozelle, S., Zhang, L. and Zhang, Y. (2002) The evolution of China's rural labour markets during the reforms. *Journal of Comparative Economics* 30, 329–353.

Deininger, K. and Jin, S. (2002) *Land Rental market as an Alternative to Government Reallocation? Equity and Efficiency Considerations in the Chinese Land Tenure System.* Policy Research Working Paper, World Bank, Washington, DC.

Hare, D. (1999) 'Push' versus 'pull' factors in migration outflows and returns: determinants of migration status and spell duration among China's rural population. *The Journal of Development Studies* 35, 45–72.

Hoff, K., Braverman, A. and Stiglitz, J.E. (1993) *The Economics of Rural Organization: Theory, Practice and Policy.* Oxford University Press, New York.

Judge, G.G., Griffiths, W.E., Hill, R.C., Lütkepohl, H. and Lee, T.C. (1995) *The Theory and Practice of Econometrics*, 2nd edn. John Wiley, New York.

Kuiper, M. (2005) *Village Modelling: a Chinese Recipe for Blending General Equilibrium and Household Modelling.* Development Economics Group, Wageningen University, Wageningen, The Netherlands.

Kuiper, M., Heerink, N., Tan, S., Ren, Q.-E. and Shi, X. (2001) *Report of Village Selection for the Three Village Survey.* Department of Development Economics, Wageningen University, Wageningen, The Netherlands.

Kung, J.K.S. (2002) Off-farm labour markets and the emergence of land rental markets in rural China. *Journal of Comparative Economics* 30, 395–414.

Kung, J.K.S. and Lee, Y.F. (2001) So what if there is income inequality? The distributive consequence of nonfarm employment in rural China. *Economic Development and Cultural Change* 50, 19–46.

Li, J. and Yao, Y. (2001) Egalitarian land distribution and labour migration in rural China. *Land Reform* 1, 81–91.

Lohmar, B. (1999) Land tenure insecurity and labour allocation in rural China. In: *Annual Meeting of the American Agricultural Economics Association*, Nashville, Tennessee.

Lohmar, B., Zhang, Z. and Somwaru, A. (2001) Land rental market development and agricultural production in China. *Annual Meeting of the American Agricultural Economics Association*, Chicago, Illinois.

National Bureau of Statistics of China (2002) *Statistical Yearbook 2001.* China Statistics Press, Beijing.

National Bureau of Statistics of China (2004) *Statistical Yearbook 2003.* China Statistics Press, Beijing.

Rozelle, S., Li, G., Shen, M., Hughart, A. and Giles, J. (1999a) Leaving China's farms: survey results of new paths and remaining hurdles to rural migration. *The China Quarterly* 158, 367–393.

Rozelle, S., Taylor, J.E. and De Brauw, A. (1999b) Migration, remittances, and agricultural productivity in China. *The American Economic Review* 89, 287–291.

Sadoulet, E. and de Janvry, A. (1995) *Quantitative Development Policy Analysis.* The Johns Hopkins University Press, Baltimore, Maryland and London.

Shi, X., Heerink, N. and Qu, F. (2006) Choices between different off-farm employment sub-categories: an empirical analysis for Jiangxi Province, China. *China Economic Review* {In press}

Singh, I., Squire, L. and Strauss, J. (eds) (1986) *Agricultural Household Models: Extensions, Applications and Policy.* The Johns Hopkins University Press, Baltimore, Maryland and London.

Yao, Y. (2000) The development of the land lease market in rural China. *Land Economics* 76, 252–266.

Zhang, X. and Li, G. (2001) *Does Guanxi Matter to Nonfarm Employment?* EPTD Discussion Paper No. 74, Environment and Production Technology Division, International Food Policy Research Institute, Washington, DC.

Zhao, Y. (1997) Labour migration and returns to rural education in China. *American Journal of Agricultural Economics* 79, 1278–1287.

Zhao, Y. (1999a) Leaving the countryside: rural-to-urban migration decisions in China. *The American Economic Review* 89, 281–286.

Zhao, Y. (1999b) Labour migration and earnings differences: the case of rural China. *Economic Development and Cultural Change* 47, 767–782.

Zhao, Y. (2002) Causes and consequences of return migration: recent evidence from China. *Journal of Comparative Economics* 30, 376–394.

Zhao, Y. (2003) The role of migration networks in labour migration: the case of China. *Contemporary Economic Policy* 21, 500–511.

14 Land and Labour Allocation Decisions in the Shift from Subsistence to Commercial Agriculture

MOTI JALETA* AND CORNELIS GARDEBROEK**

*Agricultural Economics and Rural Policy Group, Wageningen University, Hollandsweg 1, 6706 KN, Wageningen, Netherlands; fax: +31 317 484736; e-mails: *Moti.JaletaDebello@wur.nl; moti_jaleta@yahoo.com; **koos.gardebroek@wur.nl*

Abstract

This chapter assesses farm household land and labour allocation decisions under imperfect factor and product markets. A non-separable farm household model is adapted to obtain econometric estimates of land and labour allocation decisions under different market participation regimes. Survey data obtained from a sample of 154 farm households producing both cash and food crops in central and eastern Ethiopia is used for the analysis. Endogenously switching regression estimation results shows that market participation, which is household specific, affects the quantity of land and labour that households allocate to food and cash crop production. There is also a significant regional difference in resource allocation decisions. Farm capital ownership (including motor pump) influences the allocation decisions. Thus, policies improving the efficiency in rural factor and product markets promote farm households' participation in these markets and orient their resource allocation decisions towards more cash crop production and higher farm income.

Introduction

Farm households in developing countries mostly operate under imperfect factor and/or product markets resulting from high transaction costs, shallow or thin markets for factors and/or products, price risks and risk aversion, or less accessibility to market information (Sadoulet and De Janvry, 1995, pp. 149–150). Under such circumstances, production and consumption decisions taken at farm household level are far from separable (Singh *et al.*, 1986; Taylor and Adelman, 2002). Especially when there are high transaction costs to participate in a factor or product market, farm households prefer to be self-sufficient in production and/or consumption of that particular factor or product. In these cases, the value of this particular factor or product in which the households become self-sufficient is evaluated at a household-specific endogenous or shadow price, and this internal price has an implicit effect on the outcome of optimal resource allocation decisions of households in other markets (De Janvry *et al.*, 1991; Skoufias, 1994).

In addition to market failures resulting in endogenous prices for non-tradable factors or products at a household level, markets may exist for other factors or products in which the buying and selling decision prices of households are discontinuous due to high transaction costs prevailing in these markets (Omamo, 1998; Key *et al.*, 2000; Woldehanna, 2000). This discontinuity in decision prices occurs due to the fact that transaction costs put a wedge between market prices at which households are willing to buy and sell the same factor or product, having considered all the searching, negotiation, monitoring and enforcement costs. Note that, for risk-averse farmers, this price wedge may be widened by price risks. Due to price risks, farmers will mark up purchase prices positively, whereas they mark up selling prices negatively (Sadoulet and De Janvry, 1995, p. 150).

Given all these market features, farm households in developing countries earn far less than the potential income they could have attained under perfect markets. For instance, areas around Lake Ziway in central and Lake Haro-Maya in eastern Ethiopia have good potential for cash crop production. However, households in these areas are still engaged in producing both cash and food crops using their limited land and labour resources. Although it is believed that cash crops can help these households earn more profit per unit of resource used, a complete shift of land and labour towards cash crop production is rarely seen, and the share of land allocated to cash crops is still minimal.

Of the total farmland cultivated by the sample households from both the Ziway and Haro-Maya areas covered under this study, only 32.5% was covered by cash crops during the 2002–2003 production period. This lack of a complete or partial shift towards specialized, high-value cash crop production is linked to households' resource use behaviour under market imperfections (De Janvry *et al.*, 1991; Omamo, 1998).

The overall objective of this chapter is to assess farm households' land and labour allocation decisions for cash and food crop production in these two regions of Ethiopia. This analysis contributes to identifying variables that influence farm household decisions in shifting resources from subsistence food production towards market-oriented cash crop production, which is important given current food self-sufficiency policies and poverty reduction strategies for Ethiopia.

Although regions in Ethiopia differ in their natural conditions and human interference, the country as a whole can be considered as a less-favoured area (LFA). It is important to recognize that the indication of a LFA not only refers to natural and biophysical conditions, but also to constraints originating from lack of human interference. Areas with good agricultural potential that are currently used for low-value production are therefore also included in this definition (Kuyvenhoven *et al.*, 2004).

To attain the above-mentioned objective, a theoretical, non-separable farm household model is used as a starting point. This model gives a detailed explanation of households' land and labour allocation decisions between cash and food crop production activities, while taking into account market imperfections due to high transaction costs in the markets. Based on the optimal land and labour allocation decisions derived from the model's first-order conditions, an empirical model is then formulated and estimated. A switching regression model with endogenous switching (Maddala, 1983, pp. 223–228) is used to investigate differences in input allocation between participants and non-participants in land and labour markets. This methodology differs from the 'standard' Heckman two-step procedure that is often mechanically applied in studies like this. The advantage of the switching regression model with endogenous switching is that both regimes are estimated jointly and that one can easily test for differences in impact of variables in both regimes.

The remainder of the chapter is structured as follows. The next section describes the non-separable household model that underlies our analysis. From this model it follows which variables have to be used in the empirical model. We then discuss the specification of the reduced form equations that are based on the theoretical model, then the data used are discussed in the following section. This section also presents some basic statistics on land and labour market participation for food and cash crop production. The section following deals with technica-lities of our estimation procedure; the switching regression model with endogenous switching is discussed here, as well as calculation of price indices used. Estimation results are given in the penultimate section, and finally conclusions and implications are presented.

The Basic Farm Household Model

Since first developed by Singh *et al.* (1986), non-separable farm household models have frequently been used to address research questions related to the complex behaviour of farm households under missing or imperfect markets (De Janvry *et al.*, 1991; Sadoulet and De Janvry, 1995; Key *et al.*, 2000). The theoretical model described in this section is adapted from the work of Woldehanna (2000).

In building up our theoretical farm household model, the following two basic assumptions are made. First, there is at least one imperfect factor or product market for rural farm households. Secondly, due to these market imperfections the production and consumption decisions of peasant households are non-separable (Sadoulet and De Janvry, 1995). Given these assumptions, the optimization problem of households is to maximize utility subject to liquidity, technology, commodity balance and non-negativity constraints:

$$\underset{c,q,x,A,L}{Max}\ U(C_f, C_m, l; z_u) \tag{14.1}$$

Subject to:

$$\sum_i \left[(p_i - d_i^s)s_i - (p_i + d_i^b)b_i \right]$$
$$-p_x X + T + R \geq 0 \qquad i \in (c, f, m, L_c, L_f, A_c, A_f) \tag{14.2}$$

$$G_i(q_i, X_i, L_i, A_i, K_i, W_i, z_{qi}) \geq 0 \qquad i \in (c, f) \tag{14.3}$$

$$q_i + e_i + b_i - s_i - X_i - C_i \geq 0 \qquad i \in (c, f, m, L_c, L_f, A_c, A_f) \tag{14.4}$$

$$C_f,\ C_m,\ l,\ q_c,\ q_f,\ X_f,\ X_c,\ L_f,\ L_c,\ A_c,\ A_f,\ W,\ K \geq 0 \tag{14.5}$$

where U (.) is household utility, which is a function of household consumption of food, C_f, consumption of manufactured goods, C_m, and leisure, l, and household-specific characteristics, z_u, commonly denoted as taste shifters. In the constraint equations (14.2) to (14.5), s_i and b_i are quantities of the ith commodity sold and bought, respectively, at market prices p_i that are adjusted by transaction costs for selling (d_i^s) and buying (d_i^b). Commodities include cash crops (c), food crops (f), manufactured goods (m), labour (L) and land (A). Buying and selling transaction costs are assumed to be different for the same household and the same commodity. p_x is price for variable input X (that comprises seed, fertilizer, herbicides, pesticides and fuel for irrigation). T is net transfers received including remittances,[1] and R is credit available to the household[2]. Produced quantity of crop i is denoted by q_i, K_i is capital employed on the farm, W_i refers to water use for irrigation and z_{qi} represents farm characteristics like soil type or fertility index.

In Equation (14.4), for a given commodity, the sum of home-produced, initially endowed and purchased quantity should not be less than the sum of what the household consumed, sold or used as an input. This commodity balance holds for outputs (food and cash crops), manufactured goods and inputs (land and labour).

The farm household decision prices, both in factor and product markets, incorporate transaction costs associated with the marketing of factors and products. When factors and/or products are non-tradable for a given farm household, decision prices of these factors and/or products are the endogenous shadow prices of these non-traded commodities. Thus, the decision prices are given as (De Janvry *et al.*, 1991; Key *et al.*, 2000).[3]

$$p_i^* = \begin{cases} p_i + d_i^b & \text{Buying price} \\ p_i - d_i^s & \text{Selling price} \\ \tilde{p}_i = \mu_i \big/ \lambda & \text{Self-sufficient (autarkic) price} \end{cases} \tag{14.6}$$

The Lagrangian associated with the constrained maximization problem is given as:

$$\Gamma = U(C_f, C_m, l; z_u) + \lambda \left[\sum_i \left[(p_i - d_i^s)s_i - (p_i + d_i^b)b_i \right] - p_x X + T + R \right]$$

$$+ \sum_i \phi_i \left[G_i \left(q_i, X_i, L_i, A_i, K_i, W_i; z_{qi} \right) \right] + \tag{14.7}$$

$$\sum_i \mu_i (q_i + e_i + b_i - s_i - X_i - C_i)$$

$$i \in (c, f, m, L_c, L_f, A_c, A_f)$$

Note that λ, ϕ and μ_i are the Lagrange multipliers for the liquidity constraint, the production technology constraint and the commodity balance constraints, respectively. These Lagrange multipliers can be interpreted as shadow prices, so that λ stands for the shadow value of liquidity to a household and μ_i is the shadow value for an additional unit of a commodity (e.g. including land or labour).

The first-order Kuhn-Tucker conditions of the above constrained maximization problem give an interior solution for the optimal quantities and the household-specific decision prices for both tradable and non-tradable factors and products. Using these Kuhn-Tucker conditions, land and labour allocation decisions at a household level are analysed. Rewriting the first-order Kuhn-Tucker conditions for land and labour gives:

$$\phi_i \frac{\partial G_i(.)}{\partial Z_i} = \begin{cases} \lambda(p_i + d_i^b) & \text{for households renting in } Z_i \\ \mu_i & \text{for households self-sufficient in } Z_i \\ \lambda(p_i - d_i^s) & \text{for households renting out } Z_i \end{cases} \tag{14.8}$$

where $Z_i \in (L_c, L_f, A_c, A_f)$. From Equation (14.8) one can derive that households equate the marginal revenue of an input with the corresponding valuation for that input. The valuation of an input depends on the status of a household in an input market, i.e. whether the household is a net seller, net buyer or self-sufficient in that market. For households facing high transaction costs in input markets, renting in an input for production of a given crop is feasible only when the marginal revenue product of that input is high enough to compensate the marginal cost, which is the effective renting in price marked up by household liquidity constraint.

In addition to the effective input costs indicated, renting in an input is almost impossible for households badly constrained by liquidity shortage, as the complementary farm inputs used to increase input productivity also increase the liquidity constraint to a household (Woldehanna, 2000). If there is no input transaction in the household, the optimal input allocation is determined by equality of the marginal value product of the input and shadow value of the input to the household.

Reduced Form Equations for Land and Labour Allocation Decisions

The presence of the Lagrange multipliers in the endogenous prices of Equation (14.8) prevents solution of these first-order conditions. Therefore, reduced form equations based on these optimality conditions are specified in this section. These equations are used to estimate parameters involved in farm household land and labour allocation decisions between cash and food crops.

The optimal allocation of inputs between cash and food crops is determined mainly by the marginal revenue products of inputs used for the production of these alternative crops. The marginal revenue product of farm input by itself is a function of the marginal product of the input in use for the production of crop i (which is also a function of other complementary inputs used in the production process) and the household decision prices of the alternative outputs, which is a function of output market-associated transaction costs and household characteristics governing household taste and preferences.

The marginal product of farm input for crop i can be derived from the production technology specified in Equation (14.3). By substituting this marginal product of farm input in Equation (14.8), we get input demand for the production of the ith crop by each household. Considering that farm households are participating in factor and product markets, the optimal demand for farm input (Z_i^*) to produce the ith crop is given in reduced form as:

$$Z_i^* = Z_i(p_f, p_c, p_{A_i}, d_f^b, d_f^s, d_c, d_{Ai}, p_x; K, W, T, R, z_q) \qquad (14.9)$$

Since the equality of marginal revenue product and marginal cost incorporates the effective input and output prices and the marginal product of a given factor is a function of all inputs used, the demand for each input should be derived simultaneously from the system of Kuhn-Tucker's first-order conditions. These simultaneously derived demands for factor inputs are defined in terms of the exogenous factor and product prices, household-specific transaction costs, fixed inputs and farm characteristics.

However, when households are not participating in some of the factor and/or product markets, prices associated with these factors and/or products are shadow prices for these households, and these shadow prices are a function of the observable market prices, namely experienced prices of other netputs for which they do participate in the market or, in the case of non-participation for another netput, the average price and household and farm characteristics (Dutilly-Diane *et al.*, 2003). Therefore, for such households, allocation of farm inputs for the ith crop is expressed in a functional form as;

$$Z_i^* = Z_i(p_c, p_f, p_x, d_c; K, W, T, R, z_q, z_u)$$ (14.10)

Data

The data used in this study were collected in 2003 by conducting a household survey at two research sites in Ethiopia: around Lake Haro-Maya (500 km east of Addis Ababa) and the Lake Ziway area (160 km south of Addis Ababa), both in the Oromia Regional State of Ethiopia. A total sample of 154 farm households was included in the survey, with 78 from Ziway and 76 from Haro-Maya. Farm households were randomly selected from the cash crop-producing households living in these areas. The two study sites were selected intentionally because of their potential in vegetable production and difference in vegetable market destination. Vegetable products from Haro-Maya area are channelled to Djibouti for export, whereas vegetable products from the Ziway area are traded at Addis Ababa (central/domestic market).

The sample households from the two research sites also differ in their market participation status in different markets. Households around Haro-Maya participate less in the land rental market than households around Ziway, both for cash and food crop production (see Table 14.1). This might be due to relatively small holdings around Haro-Maya and the covering of farmland by perennial crops (i.e. *chat*). This plantation of perennial crops around Haro-Maya does not allow the mobilization of land for more productive activities through renting-in or -out.

The proportion of farm households participating in the labour market is higher for the sample from the Ziway area. More households hire agricultural labour for cash crop production, though labour is hired for food crop production too. Participation in off-farm and non-farm work is also higher for the Ziway area. Most of the households working off-farm are engaged in either petty trade or fishing activities by family members (see Table 14.2).

Table 14.1. Household participation in the farmland rental market.

	Sample households renting farmland[a] (*n*)					
	In for		Out for		Autarkic in land for	
	Cash	Food	Cash	Food	Cash	Food
Haro-Maya	11	0	0	1	65	75
Ziway	25	28	12	3	41	47
Total	36	28	12	4	106	122

[a] The sample sizes are 76 for Haro-Maya and 78 for Ziway.

Table 14.2. Labour market participation status of the sample households.

	Hired labour				Working in off- or non-farm jobs	
	For food crops		For cash crops			
	(n)	%	(n)	%	(n)	%
Haro-Maya	49	64.7	39	51.3	20	26.3
Ziway	67	85.9	76	97.4	34	43.5
Total	116	75.3	115	74.7	54	35.1

Of all sample households from Haro-Maya, 32% did not use a water pump. Most of these households dug wells near their plots and used buckets to get the water up from the wells to irrigate their plots. However, more than half of the sample households at both research sites have a motor pump for private usage, i.e. they have either bought it or rented it in for a specific production period (see Table 14.3 for details). From the 154 sample households at both research sites, 20% of the households are net buyers in food crops whereas about 33% are autarkic in food crops. The remaining 47% are net sellers in food crop markets.

Empirical Models

As indicated in Equations (14.9) and (14.10), the quantity of land and labour allocated by each farm household for the production of either cash or food crops is, among other factors, a function of market prices for these resources. However, market prices are observable only when households are participating in the corresponding markets. Therefore, household market participation status plays a crucial role in modelling the household land and labour allocation decision behaviour. If households do not participate in, say, the land market, it is the unobserved shadow price that is relevant for land allocation decisions and not the rental land price. Approximating the unobserved shadow price by a set of household characteristics leads to the specification of a switching regression model with different specification for participants

Table 14.3. Motor pump ownership rights of the sample households.

	Haro-Maya		Ziway	
Type of ownership rights	(n)	%	(n)	%
Does not use motor pump	25	32.9	1	1.3
Goodwill of neighbours or relatives	6	7.9	2	2.6
Exchange for resource (land or labour)	3	4.0	9	11.5
Share with others as a cooperative	0	0.0	25	32.1
Rented	13	17.1	14	18.0
Owned through purchase	29	38.2	27	34.6

and non-participants. Since participation may be affected by self-selection, the appropriate estimation procedure is a switching regression model with endogenous switching (Maddala, 1983, pp. 223–228).

In explaining the switching regression model with endogenous switching, the model is presented in a simplified form. The discussion focuses on land allocation, although the same specification holds for labour allocation. Since the number of sample households that fall into the category of net sellers in land does not allow us to estimate land allocation, we focus on deriving land allocation equations for net buyers and autarkic households in land (see Table 14.1). Similarly, almost all of the sample households are either net buyers or autarkic in labour for the production of both crops (see Table 14.2). Therefore, labour allocation is estimated only for these two groups.

Defining the market participation decision y_j as a dummy variable:

$$y_j = \begin{cases} 1 & \text{if } Z_j\gamma \geq u_j \\ 0 & otherwise \end{cases} \tag{14.11}$$

Where Z_j is a vector of variables explaining whether a given household is a net buyer or autarkic in a factor market, and u_j is error term with zero mean and $\text{var}(u_j) = 1$.

Based on a household's status in land market, household land allocation can be given in two separate equations for a specific crop as:

$$A_j^b = X_{1j}\beta_1 + \varepsilon_{1j} \qquad \text{if } Z_j\gamma \geq u_j \tag{14.12}$$

$$A_j^a = X_{2j}\beta_2 + \varepsilon_{2j} \qquad \text{if } Z_j\gamma < u_j \tag{14.13}$$

Where A_j^b and A_j^a, respectively, are land allocation by net buyers and households autarkic in land for production of a specific crop (either cash or food). X_{1j} and X_{2j} are explanatory variables for land allocation in the two categories, β_1 and β_2 are parameters to be estimated and ε_{1i} and ε_{2i} are error terms with $\text{cov}(\varepsilon_{1j}, \varepsilon_{2j}) = \sigma_{12}$, $\text{cov}(\varepsilon_{1j}, u_j) = \sigma_{1u}$, and $\text{cov}(\varepsilon_{2j}, u_j) = \sigma_{2u}$. The expected values of ε_{1i} and ε_{2i} conditional on the household's market participation status are given as (Maddala, 1983, p. 224):

$$E[\varepsilon_{1j}|u_j \leq Z_j\gamma] = -\sigma_{1u}\frac{\phi(Z_j\gamma)}{\Phi(Z_j\gamma)} \tag{14.14}$$

and

$$E[\varepsilon_{2j}|u_j \geq Z_j\gamma] = \sigma_{2u}\frac{\phi(Z_j\gamma)}{1-\Phi(Z_j\gamma)} \tag{14.15}$$

Then, the self-selection corrected land allocation equations for net buyers and households autarkic in land are given as:

$$A_j^b = X_{1j}\beta_1 - \sigma_{1u}\frac{\phi(Z_j\gamma)}{\Phi(Z_j\gamma)} + \xi_{1j} \tag{14.16}$$

$$A_j^a = X_{2j}\beta_2 + \sigma_{2u}\frac{\phi(Z_j\gamma)}{1-\Phi(Z_j\gamma)} + \xi_{2j} \tag{14.17}$$

Where ξ_{1j} and ξ_{2j} are the new error terms with zero mean, and $\phi(Z_j\gamma)$ and $\Phi(Z_j\gamma)$ are, respectively, the density and cumulative normal distribution function of the probability that household j participates in land rental market as a buyer.

One can estimate either Equation (14.16) and (14.17) separately by using Heckman's two-stage procedure (Heckman, 1979). However, Maddala (1983, p. 227) suggests that it is sometimes more fruitful to estimate the two Equations (14.16) and (14.17) simultaneously by using all the observations in A_j. The advantage of this simultaneous estimation is that one can test which coefficients are different in the two estimation equations. In our case this is relevant, since it indicates which variables have a significantly different impact in both regimes. In other words, we can learn how allocation behaviour changes in changing from non-market participation to participation. By combining the expected values of land allocation in both categories, we get:

$$E(A_j) = E(A_j \big| Z_j\gamma \geq u_j).pr(Z_j\gamma \geq u_j) + E(A_j \big| Z_j\gamma \leq u_j).pr(Z_j\gamma \leq u_j)$$
$$= \beta_1'X_{1j}\Phi_j + \beta_2'X_{2j}(1-\Phi_j) + \phi_j(\sigma_{2u} - \sigma_{1u}) \tag{14.18}$$

When some of the variables in X_{1i} and X_{2i} are the same, as we have in Equations (14.9) and (14.10), we can rewrite Equation (14.18) by subdividing the vector of explanatory variables into subgroups based on whether they appear in both equations or just in one equation alone. Thus, by specifying $X_{1j} = [X_{11j} \ X_{12j}]'$ and $X_{2j} = [X_{12j} \ X_{22j}]'$ where X_{12j} are variables that appear in both regimes, Equation (14.18) can be rewritten as:

$$E(A_j) = \beta_{11}'X_{11j}\Phi_j + (\beta_{12}' - \beta_{21}')X_{12j}\Phi_j + \beta_{21}'X_{12j}$$
$$+ \beta_{22}'X_{22j}(1-\Phi_j) + \phi_j(\sigma_{2u} - \sigma_{1u}) \tag{14.19}$$

when β_{11} and β_{22} are vector of parameters for variables appearing only in X_{11} and X_{22} respectively, and $\beta_{12}' - \beta_{21}'$ measures whether there is a different impact of variables in participation and non-participation regimes.

Estimation results can be obtained using a two-stage estimation procedure (Maddala, 1983, p. 227). First, parameters for market participation (γ) are estimated by a probit ML estimation procedure. From the estimated probit coefficients for market participation ($\hat{\gamma}$), both ϕ_j and Φ_j are computed for each observation. Finally, Equation (14.19) is estimated by regressing A_j on X_j, $\hat{\phi}_j$ and X_j $\hat{\Phi}_j$ using OLS.

Based on the reduced form land and labour allocation Equations (14.9) and (14.10), variables used in estimation are presented in Table 14.4. The household head's age and education are considered, since household heads are usually the centre of farm resource allocation decisions. The number of dependents in a family may influence the household's taste and preference in consumption, and the effect is expected to be higher particularly for households not participating in markets, as the focus of land and labour allocation for these households is to satisfy the household consumption internally.

Table 14.4. Descriptive statistics of variables used in estimations.

Vertical (symbol, units)	Haro-Maya				Ziway			
	Mean	Std. dev.	Min.	Max.	Mean	Std. dev.	Min.	Max.
Household head's age (age, years)	32.53	7.05	21	56	39.37***	10.46	20	62
Household head's education (edu, years)	4.18	3.14	0	12	5.12**	3.02	1	12
Family labour (famlab, AE[a])	2.04	0.79	1	4	2.73***	1.25	1	7.25
Number of dependents in a family (dependt)	1.97	1.19	0	5	2.18	1.87	0	9
Livestock wealth owned by a household (tlu, TLU[b])	1.75	1.06	0	6.05	5.32***	5.60	0	38.6
Exogenous income (exincome, 1000 Birr[c])	1.44	1.70	0	10.9	1.37	1.43	0	6
Value of agricultural materials owned (vagrmtn, 1000 Birr)	0.38	0.27	0.08	1.47	1.12***	1.94	0.05	15.12
Credits available to a household (credit, 1000 Birr)	0.18	0.63	0	5	0.61	1.21	0	6
Dummy for motor pump ownership (pump, 1 = yes, 0 = no)	0.39	0.49	0	1	0.37	0.49	0	1
Distance of the nearest market from homestead (nstmktkm, km)	2.74	1.87	0.02	7	2.93	3.43	0.01	18
Price index for cash crops (prindexcc[d])	-0.31***	0.22	-0.89	0.73	-0.60	0.39	-1.70	0.78
Price index for food crops (prindexfc)	0.01***	0.04	-0.08	0.25	-0.01	0.07	-0.38	0.14
Fertilizer price index for cash crops (fertprindexcc)	0.02**	0.13	-0.22	0.78	-0.02	0.12	-0.57	0.23
Fertilizer price index for food crops (fertprindexfc)	0.01*	0.08	-0.41	0.12	-0.01	0.04	-0.14	0.13
Land rental price for cash crop production (landricpr, Birr)	198.75**	34.16	120	350	184.88	49.05	50	330
Land rental price for food crop production (landrifpr, Birr)	75.64	0	75.64	75.64	75.64	29.17	25	220
Owned land available for cash crop (ownlncc, Qarxi[e])	1.38	1.21	0	8	2.25***	2.16	0	9
Owned land available for food crops (ownlnfc, Qarxi)	2.53	1.16	0	9	9.90***	9.77	0	64

*, ** and *** indicate a subsample mean significantly larger than the other subsample mean at 10 , 5 and 1% significance levels, respectively.

[a] Adult equivalent (8 h work per day per adult); [b] Tropical Livestock Unit; [c] Ethiopian unit of currency ($€1$ = 10 Birr during the study period); [d] price indices are reported in their natural logarithm; [e] a local unit for farmland measurement (1 Qarxi = 0.25 ha or a 1-day farm plot with two-oxen draught power).

Livestock wealth measured in Tropical Livestock Units (TLU) is assumed as a proxy for the wealth status of a household. Exogenous income and credit available to a household usually affect the household's liquidity position and also the demand for farm inputs as well. Value of agricultural tools, motor pump ownership and distance from the nearest local market are considered as a proxy for the household's farm capital, access to irrigation water and transaction costs, respectively. A dummy for regional difference is also included when the model is estimated using data from both regions.

To be specific in line with notations in Equation (14.19), X_{11} represents land price for households participating in land market, $X_{12} = X_{21}$ refers to age and education of household head, livestock wealth, available family labour for agricultural use, farm capital, credit available, exogenous income, distance to local market, price indices for cash crop, food crop and fertilizer. The number of dependents in a household is represented by X_{22}.

Since there are a number of crops grown and various inputs are used in production, aggregating input and output prices is important. Though price variation is not expected much in cross-section data, there is variation observed among the sample households on factors and products marketed. This variation could be due to variations in the nature of markets, quality of input or output marketed, period of year the item is marketed, individual bargaining power in the markets, etc.. The variation in land rental prices is due mainly to the location of farmland and its proxy to the water sources, which largely determine the type of crop grown on the farmland.

Price indices are calculated for two categories of outputs (cash and food crops) and inputs (fertilizer and other chemicals). Cash crops consist of tomato, onion, cabbage and pepper for households around Ziway, and potato, beetroot (red), leek and carrot for households around Haro-Maya.[4] Maize, wheat, teff, haricot bean and sorghum are considered in the food crop category; the first three are dominant for Ziway, but only maize and sorghum for households around Haro-Maya.

Since households participate in different markets to buy and/or sell the same crop, there is no unique price for most of the crops marketed. Thus, average prices for each crop, weighted by the quantity of each crop marketed in different markets at the market-specific prices, are considered. After obtaining the average prices for each crop, Divisia price indices are computed (Higgins, 1986):

$$\ln P_j^k = \frac{1}{2}\sum_{i}^{g}(r_{ij}^k + \bar{r}_{ij})(\ln P_{ij}^K - \overline{\ln P}_{ij}) \qquad (14.20)$$

Where P_j^k is the price index for the jth aggregate for household k, r_{ij} is the share of the ith item in the value of the jth aggregate for the kth household, $\bar{r}_{ij}$ is the average value of the share of the ith item in the jth aggregate on all households, P_{ij} is the natural log of the price of the ith item in the jth aggregate for household k, $\overline{\ln P}_{ij}$ is the average of the natural log of the price of the ith item in the jth aggregate on all households, and g is the number of items in the jth aggregate. The base of the index is the average value of the sample. For households without observation, the average value of the other households with observation is considered (Higgins, 1986).

Estimation Results

Based on Equations (14.11) and (14.19) presented above, estimation results for land and labour market participation equations and the endogenously switching regression model for household land and labour allocation are presented in this section. The first subsection presents estimation results of household farmland allocation, while household labour allocation is presented in the subsequent subsection.

Household farmland allocation

The probit estimation results presented in Table 14.5 show that farm household decisions to rent-in land for cash crop production are strongly influenced by motor pump ownership. Cash crops are mostly produced using irrigation water pumped up with motor pumps from lakes in the region. Family labour availability for agricultural use significantly increases the probability that households will rent land in for cash crop production. Land market participation as a buyer decreases by size of farmland holding. Elderly household heads are usually the ones that have the use-right contract with the government on land. Thus, young household heads obtain land for crop production either by renting-in or arranging sharecropping contracts with elderly household heads.

There is a significant regional effect in explaining the probability of renting-in land. Households around Ziway are more involved in renting land in. Distance from local market significantly decreases a household's probability of being a buyer in land market for food crop production around Ziway. Although the number of significant explanatory variables is limited, the LR test indicates that, overall, the included variables do contribute to explaining land renting-in decisions.

For households renting-in land, Table 14.6 shows that an increase in the land rental price has a significant negative effect on the quantity of land rented in for cash crops. Note that, for food crops, land price has no significant effect on allocation of land for land market participants. In most cases, households with larger farm capital use more land for cash crops in Haro-Maya and for food crops around Ziway. At Haro-Maya, households with higher exogenous income, mostly income from sale of *chat*, operate both cash and food crops on relatively larger farmland areas. Households owning a motor pump allocate more land to cash crop production (see Table 14.6 for details). Since food crops are rain

Table 14.5. Probability of land market participation as a buyer.

| Variable[a] | Land for cash | | | Land for food |
	Pooled (n = 142)	Haro-Maya (n = 76)	Ziway (n = 66)	Ziway (n = 75)
Age	−0.06*	−0.03	−0.47	0.03
	(0.03)[b]	(0.17)	(0.31)	(0.02)
Edu	0.06	−0.16	0.08	−0.17**
	(0.07)	(0.29)	(0.26)	(0.07)
Tlu	−0.06	−2.60	−0.31	0.00
	(0.09)	(2.33)	(0.33)	(0.04)
Famlab	0.69***	−1.02	4.71*	−0.51**
	(0.25)	(1.39)	(2.86)	(0.20)
Dependt				0.22
				(0.14)
Exincome	−0.01	0.38	−1.53	0.02
	(0.13)	(0.77)	(1.25)	(0.14)
Vagrmtn	0.29	−2.23	−0.58	0.08
	(0.55)	(4.38)	(1.49)	(0.17)
Credit	0.20	−1.10	2.35	−0.12
	(0.21)	(1.41)	(2.78)	(0.20)
Pump	1.05**	4.75	7.44	0.30
	(0.41)	(3.96)	(5.79)	(0.44)
Region (1 = Ziway, 0 = Haro-Maya)	1.25**			
	(0.60)			
Nstmktkm	0.05	−1.73	0.75	−0.12*
	(0.07)	(2.04)	(0.52)	(0.07)
Prindexcc	−0.48	−1.70	−2.57	0.70
	(0.60)	(1.71)	(2.18)	(0.58)
Prindexfc	−1.34	5.49	−6.86	4.89
	(3.13)	(15.72)	(8.09)	(3.49)
Fertprindexcc/fc	−0.40	18.87	−6.34	
	(2.16)	(14.57)	(4.63)	
Landricpr (landrifpr)	0.00	0.03	−0.02	0.00
	(0.00)	(0.02)	(0.01)	(0.01)
Ownlnhr/fc	−1.49***	−6.08	−5.92*	−0.05*
	(0.29)	(5.99)	(3.45)	(0.03)
Constant	−0.85	8.24	8.99	1.00
	(1.37)	(13.94)	(5.72)	(1.08)
LR chi2(15)	97.42	48.5	71.72	29.86
Prob. > chi2	0.000	0.000	0.000	0.012
Pseudo R^2	0.61	0.77	0.82	0.30

***, ** and * refer to 1, 5 and 10% significance levels, respectively.
[a] See Table 14.4 for explanation of variables; [b] standard errors are given in parentheses.

dependent, motor pumps are not used in food crop production, which is also reflected in the estimation results.

There is a positive and significant effect of household heads' age and education, as well as livestock wealth, on land allocation for cash crop

Table 14.6. Land allocation for cash and food crop production.

	Cash			Food	
	Pooled	Haro-Maya	Ziway	Ziway	Haro-Maya
Variable[a]	($n = 142$)	($n = 76$)	($n = 66$)	($n = 75$)	($n = 75$)
Age	0.00	0.03	−0.06	1.12***	−0.04*
	(0.02)[b]	(0.02)	(0.06)	(0.18)	(0.02)
Edu	−0.02	0.01	−0.19	−0.88*	0.01
	(0.07)	(0.04)	(0.19)	(0.45)	(0.05)
Tlu	0.05	−0.03	0.01	−0.58**	−0.01
	(0.05)	(0.16)	(0.09)	(0.27)	(0.14)
Famlab	0.52**	0.25	0.41	−8.72***	0.23
	(0.22)	(0.20)	(0.56)	(1.34)	(0.19)
Dependt	0.00	−0.02	0.00	2.92***	−0.01
	(0.14)	(0.10)	(0.29)	(1.03)	(0.12)
Vagrmtn	0.75**	2.07***	0.75	3.48***	0.67
	(0.30)	(0.58)	(0.53)	(0.81)	(0.58)
Credit	−0.29	−0.01	−0.47	−3.85***	−0.13
	(0.29)	(0.31)	(0.64)	(1.21)	(0.21)
Exincome	−0.01	0.18*	−0.19	0.23	0.21**
	(0.15)	(0.11)	(0.38)	(1.13)	(0.09)
Pump	0.94**	0.72**	1.18	3.16	0.19
	(0.46)	(0.29)	(1.06)	(2.06)	(0.30)
Nstmktkm	−0.04	−0.02	−0.01	−1.57***	0.26***
	(0.07)	(0.08)	(0.14)	(0.27)	(0.08)
Prindexcc	−0.63	0.51	−0.76	23.58***	−0.16
	(0.78)	(0.61)	(1.73)	(4.12)	(0.59)
Prindexfc	−6.23	−2.48	−1.52	89.53***	2.94
	(4.44)	(3.77)	(9.83)	(18.31)	(3.34)
Fertprindexcc(fc)	0.26	0.93	1.62	−144.32***	1.71
	(1.54)	(0.85)	(4.49)	(38.69)	(1.78)
Age_phi	0.11*	0.05	0.01	−1.68***	
	(0.06)	(0.19)	(0.11)	(0.31)	
Edu_phi	0.47***	−0.44	0.94**	−1.15	
	(0.15)	(0.31)	(0.35)	(1.01)	
Tlu_phi	0.46***	1.30	0.52**	2.93***	
	(0.13)	(1.92)	(0.21)	(0.63)	
Famlab_phi	−1.53**	−1.99	−0.99	7.95**	
	(0.61)	(1.79)	(0.90)	(3.34)	
Dependt_phi	−0.97**	−0.25	−0.52	−1.98	
	(0.41)	(1.00)	(0.65)	(1.63)	
Vagrmtn_phi	0.80*	−5.49	1.07	−6.11**	
	(0.42)	(5.65)	(0.73)	(2.40)	
Credit_phi	0.32	−0.40	0.42	7.99**	
	(0.60)	(2.14)	(1.04)	(3.20)	
Exincome_phi	0.53*	0.63	0.57	0.08	
	(0.27)	(0.70)	(1.04)	(2.38)	
Nstmktkm_phi	0.57**	0.06	0.41		
	(0.23)	(0.77)	(0.38)		
Prindexcc_phi	0.17	−5.29	2.73	−42.32***	
	(1.47)	(3.58)	(2.68)	(9.44)	

Table 14.6. *Continued.*

	Cash			Food	
Variable[a]	Pooled (n = 142)	Haro-Maya (n = 76)	Ziway (n = 66)	Ziway (n = 75)	Haro-Maya (n = 75)
Prindexfc_phi	26.38***	9.64	17.81	−102.63**	
	(8.98)	(32.35)	(14.64)	(45.31)	
Fertprindexcc(fc)_phi	−8.35*	−11.68	−14.84	198.79**	
	(4.97)	(7.77)	(10.13)	(76.36)	
Landric(rif)pr_phi	−0.02***	0.00	−0.03**	−0.03	
	(0.01)	(0.02)	(0.01)	(0.04)	
$(\sigma_{2u} - \sigma_{1u})$	−3.85**	0.29	−3.97	−36.05***	
	(1.53)	(1.03)	(3.68)	(12.32)	
Constant	0.68	−0.56	4.47	28.39***	1.89**
	(0.98)	(0.70)	(3.11)	(6.16)	(0.91)
F(k, n − 1)	16.46	4.31	5.92	10.02	1.93
Prob. > F	0.000	0.000	0.000	0.000	0.044
R^2	0.80	0.72	0.81	0.84	0.29
Adj. R^2	0.75	0.55	0.67	0.76	0.14

***, ** and * refer to 1, 5 and 10% significance levels, respectively.
[a] See Table 14.4 for explanation of variables; [b] standard errors are given in parentheses.

production for households that participate in the land market. The effect of available family labour on land allocation for cash crop production is stronger for autarkic households in land for cash crop production. Surprisingly, distance to local market has no significant effect on land allocation to cash crops, although the parameters have the expected negative sign. Distance to nearest market has a significant impact on land allocation for food crops. For Ziway the effect is negative, which is counterintuitive. This impact is positive for households around Haro-Maya, who are mainly dependent on their own holding for food crop production. Results further show that sample selection has a significant impact in the land allocation equations for cash crops using pooled data, and for food crops in Ziway.

Household labour allocation

Probit estimation results in Table 14.7 show that, for households around Haro-Maya, labour market participation in hiring labour for cash crop production is positively influenced by exogenous income and negatively by food crop prices. Around Haro-Maya, the probability that households will hire labour for food crop production decreases with the distance from local markets. High cash crop prices also have a significantly reducing effect on the probability of hiring labour for food crop production. Around Haro-Maya, high food crop prices reduce the probability of hiring labour for cash crop production. Around Ziway,

Table 14.7. Probability of labour market participation as a buyer.

	Labour for cash		Labour for food		
Variable[a]	Pooled (n = 154)	Haro-Maya (n = 76)	Pooled (n = 164)	Haro-Maya (n = 76)	Ziway (n = 78)
Age	−0.02	0.00	−0.02	−0.01	−0.06***
	(0.02)[b]	(0.03)	(0.02)	(0.03)	(0.03)
Edu	0.06	0.08	0.03	0.06	−0.14
	(0.05)	(0.06)	(0.04)	(0.07)	(0.09)
Famlab	−0.24	−0.07	0.03	−0.25	0.35
	(0.18)	(0.26)	(0.15)	(0.29)	(0.26)
Tlu	−0.09	−0.29	0.21**	0.20	0.19
	(0.08)	(0.19)	(0.09)	(0.27)	(0.12)
Region	2.71***		−0.10		
	(0.70)		(0.35)		
Nstmktkm	0.06	0.09	−0.08*	−0.42***	0.14
	(0.08)	(0.10)	(0.05)	(0.12)	(0.11)
Vagrmtn	1.00*	0.00	0.37	1.83*	0.56
	(0.59)	(0.71)	(0.30)	(1.03)	(0.49)
Credit	−0.29	−0.78	0.07	0.10	0.13
	(0.20)	(0.51)	(0.17)	(0.31)	(0.27)
Exincome	0.27**	0.53***	0.13	0.02	−0.25
	(0.12)	(0.16)	(0.11)	(0.18)	(0.24)
Prindexcc	−0.12	−0.96	−1.41***	−1.82*	−2.79***
	(0.60)	(0.83)	(0.50)	(0.94)	(1.06)
Prindexfc	−3.34	−13.06**	−1.65	−2.49	0.62
	(4.21)	(6.59)	(2.87)	(4.59)	(4.94)
Fertprindexcc/fc	1.25	1.46	−1.86	−3.05	−10.32
	(1.12)	(1.40)	(2.08)	(3.00)	(8.42)
Constant	0.03	−0.82	0.10	0.69	0.72
	(0.87)	(1.09)	(0.67)	(1.30)	(1.29)
LR chi2(12)	68.35	23.77	37.66	33.13	22.33
Prob. > chi2	0.000	0.014	0.000	0.001	0.022
Pseudo R^2	0.39	0.23	0.22	0.34	0.35

***, ** and * refer to 1, 5 and 10% significance levels, respectively.
[a] See Table 14.4 for explanation of variables; [b] standard errors are given in parentheses.

participation in the labour market to hire labour for food crop production is significantly lower for older household heads.

Labour allocation estimates for cash and food crop production, for both Haro-Maya and Ziway areas, are presented in Table 14.8. The amount of labour allocated to cash crop increases with motor pump ownership. Farm capital also has a positively significant effect on labour allocation to cash crop production around Ziway area. To explain more, the effect of motor pump ownership on labour allocation can be seen from its indirect effect on the expansion of labour-intensive cash crop production. Farm capital also increases both land and labour productivities and helps households employ more of these resources for higher profit.

Table 14.8. Estimates of household labour allocation for both cash and food crop production.

	Labour for cash			Labour for food		
	Pooled	Haro-Maya	Ziway	Pooled	Haro-Maya	Ziway
Variable[a]	($n = 154$)	($n = 76$)	($n = 78$)	($n = 154$)	($n = 76$)	($n = 78$)
Age	−0.54	0.29	1.03	−0.10	1.5	−12.6
	(4.90)[b]	(1.76)	(3.27)	(2.40)	(1.04)	(12.4)
Edu	−29.78*	8.13	10.25	18.50	1.1	−67.1**
	(15.94)	(6.48)	(10.63)	(11.68)	(5.52)	(31.4)
Famlab	58.19	18.06	−24.25	45.54	−9.0	95.0
	(63.01)	(24.48)	(26.24)	(40.35)	(11.81)	(125.9)
Dependt	9.52	−13.63		−24.82	10.2	−91.8
	(41.34)	(14.39)		(22.71)	(12.08)	(61.2)
Tlu	9.83	−20.17	−4.85	65.55*	36.1**	158.6**
	(34.13)	(17.24)	(5.35)	(33.50)	(17.03)	(83.4)
Pump	99.18***	21.94	205.41***			
	(35.91)	(14.63)	(60.84)			
Nstmktkm	−27.40	6.15	8.31	−19.85*	−21.0	134.1**
	(25.78)	(9.91)	(8.53)	(11.43)	(13.03)	(61.6)
Vagrmtn	−354.48	65.24	97.68***	155.80*	104.0	305.3
	(242.46)	(84.02)	(20.12)	(92.28)	(94.71)	(193.9)
Credit	109.52*	−20.05	−47.92*	30.48	119.1**	73.5
	(61.32)	(20.68)	(28.09)	(49.51)	(56.92)	(113.7)
Exincome	−38.19	30.72	2.55	48.55	−26.9**	−63.9
	(54.28)	(27.44)	(20.75)	(36.95)	(13.33)	(95.8)
Prindexcc	168.10	10.97	37.20	−235.77	−46.9	−1561.9*
	(191.24)	(54.73)	(81.54)	(162.64)	(62.13)	(791.8)
Prindexfc	100.16	111.23	494.49	−670.48	−668.2***	5.2
	(752.96)	(238.51)	(429.71)	(556.81)	(242.87)	(2694.0)
Fertprindexcc/fc	17.88	206.13	29.52	9.45	−84.3	−2844.0
	(520.81)	(228.12)	(234.63)	(636.69)	(192.27)	(3562.7)
Age_phi	1.74	0.23		−1.75	−2.9**	8.4
	(5.74)	(2.93)		(3.05)	(1.34)	(12.1)
Edu_phi	40.04**	−15.77		−26.55*	−1.4	59.0*
	(19.18)	(10.85)		(15.07)	(7.44)	(30.9)
Famlab_phi	−69.24	3.14		−38.89	34.2*	−74.7
	(75.01)	(43.81)		(50.29)	(18.21)	(132.0)
Dependt_phi	−19.47	36.59		38.83	−20.6	112.6*
	(47.43)	(25.23)		(26.16)	(17.38)	(64.7)
Tlu_phi	−12.67	16.41		−63.83*	−43.3**	−153.5*
	(34.74)	(26.05)		(33.79)	(19.78)	(83.6)
Nstmktkm_phi	35.22	−9.13		16.85	34.4*	−137.9**
	(27.91)	(18.06)		(12.82)	(17.85)	(62.7)
Vagrmtn_phi	466.08*	−59.78		−124.97	−128.6	−277.3
	(244.13)	(142.43)		(93.81)	(111.88)	(193.0)
Credit_phi	−169.68**	59.35		−19.14	−149.2**	−49.5
	(74.42)	(67.00)		(58.68)	(69.12)	(121.3)
Exincome_phi	36.72	−14.97		−56.86	41.1***	33.6
	(60.06)	(25.25)		(40.55)	(14.18)	(98.2)

Continued

Table 14.8. *Continued.*

	Labour for cash			Labour for food		
Variable[a]	Pooled (n = 154)	Haro-Maya (n = 76)	Ziway (n = 78)	Pooled (n = 154)	Haro-Maya (n = 76)	Ziway (n = 78)
Prindexcc_phi	−174.24	80.58		167.50	70.4	1448.6*
	(226.35)	(95.01)		(176.11)	(94.08)	(794.7)
Prindexfc_phi	297.53	−76.41		843.20	1052.8***	134.1
	(859.97)	(592.81)		(651.10)	(359.67)	(2808.2)
Fertprindex_phi	5.19	−186.95		−6.88	92.8	2808.6
	(648.61)	(344.85)		(752.72)	(219.30)	(3655.4)
$(\sigma_{2u} - \sigma_{1u})$	20.88	58.05		−581.70**	−122.1	−386.4
	(223.50)	(79.30)		(227.38)	(119.81)	(381.7)
Constant	96.46	−20.48	125.75	174.56**	65.5	156.5
	(122.06)	(61.57)	(163.70)	(72.61)	(46.57)	(111.9)
F(k,n−1)	6.64	2.35	6.27	7.21	3.27	3.31
Prob. > F	0.000	0.005	0.000	0.000	0.000	0.000
R^2	0.58	0.56	0.54	0.58	0.62	0.61
Adj. R^2	0.49	0.32	0.45	0.50	0.43	0.43

***, ** and * refer to 1, 5 and 10% significance levels, respectively.
[a] See Table 14.4 for explanation of variables; [b] standard errors are given in parentheses.

Higher cash crop prices have a reducing effect on household labour allocation for food crop production around Ziway. The effect is even higher for households participating in the labour market to hire labour for food crop production. Once households are participating in the labour market to hire labour for cash crop production, the household head's education and farm capital have a positive and significant effect on the size of labour that households allocate to cash crop production. The impact of credit availability on household labour allocation to both food and cash crop production is significantly higher for autarkic households in labour for both cash and food crop production. In food crop production, for both Haro-Maya and Ziway, households with larger livestock wealth allocate more labour to food crop production, and the impact of livestock wealth on allocating labour for food crop is higher for autarkic households in labour for food crops. The impact of sample selection is limited for the labour allocation equations; it is significant only for food crop labour allocation using pooled data.

Conclusions

Farm household resource allocation decisions are complex and difficult to predict, especially when household production and consumption decisions are non-separable. This is a feature of households producing outputs both for own consumption and marketing purposes. Such households are neither completely commercialized nor subsistent in production, but in-between. To assist these households in moving towards more commercially oriented production strategies, it is important to study and understand their behaviour in market participation and resource allocation decisions. This chapter examines these behavioural decisions of farm households in the context of Ethiopia's rural economy, where both cash and food crop production is practised. Based on the estimation results, some general conclusions can be drawn.

Farm households that own large farm capital and have exogenous income sources are allocating more land and labour to cash and food crop productions. The more farm capital employed on a given farm, the more productive land and labour are. This increased productivity of land and labour at household level encourages households to rent-in (hire) more land (labour), as the marginal benefit from renting (hiring) factors from local markets is attractive compared with the market and marketing costs of these resources.

Since cash crops are produced mostly using irrigation, motor pump availability plays a central role. Thus, enabling farm households to use water resources for irrigation and producing relatively high-value vegetable crops for marketing purpose can increase household annual income and their overall level of welfare. Moreover, the production of labour-intensive vegetable crops has an income distribution effect for landless households depending on labour markets for their livelihood. In doing so, the shift from subsistence to commercial farming contributes towards the general objectives of sustainable poverty reduction in rural areas.

Endnotes

[1] The net transfer includes net surplus from livestock marketing and used to finance crop production.

[2] Credit includes the values of variable inputs (like fertilizer, pesticides, etc.) obtained on credit and the potentially available credit for production and consumption purposes.

[3] Note that price mark-ups in this model originate from transaction costs. These markups may also be due to (price) risk effects (e.g. Sadoulet and De Janvry, 1995, pp. 112–126). However, risk is not modelled explicitly in this chapter, since that would make the model too complicated.

[4] Chat is produced as a cash crop around Haro-Maya, but is not included here because chat is a perennial crop and barely competes with other crops, at least for land, in the short term. However, its effect on the allocation of land and labour via household income is included, as is the income from livestock for households around Ziway.

References

De Janvry, A., Fafchamps, M. and Sadoulet, E. (1991) Peasant household behaviour with missing markets: some paradoxes explained. *Economic Journal* 101, 1400–1417.

Dutilly-Diane, C., Sadoulet, E. and De Janvry, A. (2003) *Household Behavior under Market Failures: how Natural Resource Management in Agriculture Promotes Livestock Production in Sahel.* Working Paper 979, University of California, Berkeley, California.

Heckman, J. (1979) Sample selection bias as a specification error. *Econometrica* 47, 153–161.

Higgins, J. (1986) Input demand and output supply on Irish farms: a micro-economic approach. *European Review of Agricultural Economics* 13, 477–493.

Key, N., Sadoulet, E. and De Janvry, A. (2000) Transactions costs and agricultural household supply responsiveness. *American Journal of Agricultural Economics* 82, 245–259.

Kuyvenhoven, A., Ruben, R. and Pender, J. (2004) Development strategies for less-favoured areas. *Food Policy* 29, 295–302.

Maddala, G.S. (1983) *Limited-dependent and Qualitative Variables in Econometrics.* Cambridge University Press, Cambridge, UK.

Omamo, S.W. (1998) Transport costs and smallholder cropping choices: an application to Siaya District, Kenya. *American Journal of Agricultural Economics* 80, 116–123.

Sadoulet, E. and De Janvry, A. (1995) *Quantitative Development Policy Analysis.* The Johns Hopkins University Press, Baltimore, Maryland.

Singh, I., Squire, L. and Strauss, J. (1986) The basic model: theory, empirical results, and policy conclusions. In: Singh, I., Squire, L. and Strauss, J. (eds) *Agricultural Household Models: Extensions, Applications, and Policy.* The Johns Hopkins University Press, Baltimore, Maryland, pp. 17–47.

Skoufias, E. (1994) Using shadow wages to estimate labour supply of agricultural households. *American Journal of Agricultural Economics* 76, 215–227.

Taylor, J.E. and Adelman, I. (2002) Agricultural household models: genesis, evolution, and extensions. *Review of Economics of the Household* 1, 33–58.

Woldehanna, T. (2000) Economic analysis and policy implications of farm and off-farm employment: a case study in the Tigray Region of Northern Ethiopia. PhD dissertation, Wageningen University, The Netherlands.

15 Effects of Deregulation of the Rice Market on Farm Prices in China: a Marketing Channel Model

LE CHEN[*] AND JACK PEERLINGS

Agricultural Economics and Rural Policy Group, Wageningen University, Hollandseweg 1, 6706 KN Wageningen, The Netherlands; Fax: + 31 317 484736; e-mail: [*] Le.Chen@wur.nl

Abstract

The objective of this research is to analyse the effects of market liberalization and deregulation in the rice marketing channel on farm households, using information collected from three selected villages with different market access in Jiangxi Province, China. Intermediates are modelled as a state monopsonist, as oligopsonists or assumed to have no market power (perfect competition). Results indicate that rice producers benefit from market liberalization and deregulation. How much rice producers benefit depends on the degree of remaining market imperfections and the degree of market access.

Introduction

China's policies with respect to agriculture can be divided into two categories: (i) policies aimed at market intervention in order to secure food self-sufficiency; and (ii) policies with respect to market structure, i.e. the way trade is organized. The latter policies refer to the degree that State Grain Trading Companies (SGTCs)[1] have a monopoly position and the barriers of entry set by the government for private intermediates. Agricultural market liberalization (less market intervention) and market deregulation (lesser role of the government in trade) have been implemented over the past two decades.

Since 1978, China's grain marketing channel has undergone series of reforms that have brought the market mechanism into play. Possible effects of market liberalization and deregulation on farm households have been of great concern for China's central government.

The opening of the grain purchase market has linked farm households directly with the market. The government realizes that the participation of private intermediates in the grain purchase market increases competition: increased competition affects farm prices and, therefore, farm households' welfare in less-favoured regions in China. There are a number of studies that have observed the effects of market liberalization and deregulation at the macro level (Rozelle *et al.*, 1997; Wu, 2000).

Although Wu (2000) has analysed vertical market integration focusing on wheat, maize and pig products in China, studies focusing on the grain purchase market are still lacking. Another example of this research is to be found in Huang *et al.* (2004), who found that China's village markets were highly integrated with regional markets. However, they did not take into account differences between villages with respect to market access.

Economic reforms have provided a major stimulus to agricultural development and economic growth since the early 1980s in China. Not all regions, however, benefited equally. Although poverty and food insufficiency declined considerably since that time, around 102 million people in China are estimated to be living in poverty (using the international poverty line of US$1/day) at present (Ravallion and Chen, 2004), while an estimated 140 million people lack secure access to sufficient food for an active and healthy life (FAO, 2000).

Most of these poor and food-insecure people live in rural areas in the west and centre of China, where agriculture is the predominant economic activity. Agriculture in these areas is often constrained by poor and fragile soils that are vulnerable to degradation; township, village and private enterprises that stimulated rural development in the coastal provinces are absent, and market development has lagged behind due to infrastructural, physical, market and other constraints (Sicular, 1995). Jiangxi Province is a typical example of a province where agriculture faces these constraints (Kuiper *et al.*, 2001).

Research objective

The objective of the research is to analyse the effects of market liberalization and deregulation in the rice marketing channel on farm households, using information collected from three selected villages with different market access in Jiangxi Province, China.

The villages are selected such that they are comparable with respect to the crops they produce. Rice is the most important cultivated crop. However, the villages have varying degrees of market access, and this allows a comparison of the effect of market liberalization and deregulation on farm households under different degrees of market access. Market access is defined in terms of distance to the closest consumer market. Although there are other agricultural crops grown also in the study area, it is expected that a focus solely on rice provides useful insights of the effects of market liberalization and deregulation.

Methodology

In this research, marketable supply of rice is assumed to result from utility-maximizing behaviour of farm households given a certain level of rice production. Thus we have a short-term model, since all the rice is already produced and production decisions are not taken into account. For simplicity, we assume that rice is a homogenous product. The consumer price of rice in the nearby consumption centres is an assumed given and not influenced by the marketable supply of the selected villages, because marketable supply of the three villages is small in comparison with overall consumption. However, the farm price is affected by changes in the consumer price. The degree to which the farm price depends on the consumer price is determined by the degree of market access of the selected villages and market imperfections in the rice market channel. We assume that, in a village, the quantity of rice supplied by the farm households equals the quantity of rice purchased by the intermediates. Therefore, the supply response of farm households due to a price change is represented by the change in traded quantity of the intermediates.

The situation of the rice purchase market in 2000 (the government's attempt to control the rice purchase market for a short period after harvest) is compared with the situation in 2003 (the official government approval of the participation of private traders in the rice purchase market). In 2000, in the villages of Banqiao and Shangzhu, intermediates (the SGTC and private traders) are modelled as oligopsonists. In the village Gangyan, the SGTC is modelled as a monopsonist and private traders are modelled as oligopsonists (a limited number of private traders, excluding the SGTC). The monopsony case refers to the situation where the SGTC is the only intermediate in Gangyan for a short period after rice harvest. The oligopsony case represents closely the situation in Gangyan in 2000 when, after the monopsony period of the SGTC, private traders entered the purchase market. Oligopsony is relevant because the small, marketable supply of the selected villages[2] limits the number of intermediates that can be active in the villages.

We assume that, with a large marketable supply, more intermediates are active than with a small marketable supply. Limited market access, through higher transportation costs, restricts the number of intermediates. The situation in all three villages in 2003 is modelled as perfect competition due to the development in transportation and telecommunication[3] that reduced marginal costs of private traders and facilitated market entrance. In order to analyse the impact of market liberalization and deregulation on farm households, the effect of different market structures, i.e. oligopsony versus perfect competition

(Banqiao and Shangzhu), monopsony versus perfect competition and oligopsony versus perfect competition (Gangyan) in combination with consumer price changes, are compared.

Overview

The subsequent section introduces the study area, followed by a brief description of China's grain market reform in 1998, the grain purchase market in 2000–2003 in the study area and the situation after 2003. The theoretical model is then presented, followed by the empirical model and a discussion of data. The model will then be used to analyse the effects of imperfect competitive behaviour of intermediates and consumer price changes on farm households and intermediates, in the penultimate section. Conclusions are drawn in the final part.

Study Area

Jiangxi Province is located in the south-east of China. It is one of the most important provinces for rice production, but is a rather poor province in terms of income per capita (CSY, 2004). Jiangxi Province borders some of the richest provinces in China.

The study area is located in the north-east of Jiangxi Province, where agricultural production has suffered from severe water and soil erosion since the 1980s (Shen and Wu, 2004). Three villages were selected in the study area according to environmental indicators, market access and geographical conditions (Kuiper *et al.*, 2001). They are Banqiao, Shangzhu and Gangyan. Rice production is one of the main agricultural activities in the villages. Gangyan has the best market access; Shangzhu is located in a remote area; Banqiao occupies an intermediate position. The villages are considered to be representative for a large area.

The official survey was conducted in July 2003. A pilot survey was carried out in March 2003 to test the designed questionnaires. During the official survey, farm households, private traders, owners of processing factories, wholesalers and retailers were interviewed.[4]

China's Grain Market Reforms since 1998

This section discusses the policy reforms in China's grain market since 1998.[5] The policy reform of 1998 is first described, followed by a description of the grain market in the study area in 2000–2003.

The grain purchase market in 1998

A series of policies announced in 1998 were intended to reduce the budget burden caused by subsidization of the SGTCs. Under the new regime, the SGTCs had to reduce overhead costs by reducing the number of employees (SC, 1998a). It was intended that the SGTCs monopolize the purchase market while private companies were allowed to participate in grain trade in the wholesale and retail markets (SC, 1998b, c). The idea was that, being a monopolist, the SGTCs could set higher selling prices to make enough profit so that the subsidies could be reduced.

However, since the 1980s, despite being frequently suppressed by the government, China's informal free market had grown vigorously (Sicular, 1995). It became very expensive to prevent private traders from trading with farm households. With the help of local governments, some SGTCs monopolized the purchase market in a short period after harvest season. Although the SGTC was supposed to purchase at the price fixed by the government, it often reduced the purchase price by downgrading the grain sold by farm households. Meanwhile, farm households realized that the SGTC could control the market only during a limited period, so that they had opportunity, after this period, to sell to private traders from whom they might receive a higher price. Therefore, the amount of grain purchased by the SGTCs dropped (Huang, 1998).

The existence of private traders made selling at higher prices by the SGTCs impossible. The SGTCs were burdened by overhead costs that were higher than that of the private traders. Hence, SGTCs selling prices at which profit could be made were much higher than the selling prices of the private traders. Since selling at prices that were lower than the purchase prices is not profitable, the SGTCs kept the grain and asked the government for storage subsidy (Huang, 1998). As a result, the fiscal deficit increased instead of being diminished.

Grain purchase market in the study area in 2000–2003

The unsuccessful reform in 1998 was abolished at the end of 1999. Realizing that the SGTCs were no longer dominant in the grain market, in 2000 China's government stopped procuring low-quality grain (especially early rice) that was proving difficult to resell. For high-quality grain (such as late rice and one-season rice), the SGTCs still had to purchase at the protective price fixed by the government. Meanwhile, private traders were allowed to enter the grain purchase market officially but could trade only designated varieties (SC, 2000). All private traders were required to register at the grain administrations at the county level.

In 2001, the government established the state grain reserve system,[6] which took over the functions of the SGTCs for national food self-sufficiency and grain price stabilization. By 2002, grain procurement was abolished in most grain-surplus regions, while it still existed in most grain-deficit regions.[7] In 2003, grain production declined to the lowest point since 1990, which brought a rapid increase in grain prices (PDO, 2003). Learning from its past experiences, China's government did not reverse its policies.

Our survey observed that the number of private traders increased in 2000–2003 (Chen, 2003). However, private traders and the SGTCs faced an unbalanced competition environment in 2000–2003. Information presented below is summarized from interviews held with private traders and SGTCs in the study area.[8]

Private traders

The majority of the private traders who purchase rice directly from farm households are farmers. These farmers are engaged in normal agricultural production and conduct the rice purchase business during harvest season. In the study area, a private trader does not purchase rice until he receives an order from his buyer. A private trader not only purchases rice from the village where he lives but also from nearby villages. He visits the farm households he knows in advance and informs them about the date of purchase so that they are prepared when he comes. On the selected day, the private trader purchases the rice and brings it to the truck he has hired. When the truck is full, he transfers the rice directly to his buyer.

The private trader pays farm households in two ways. If the private trader knows the farm household well, e.g. his neighbour, he makes an oral agreement to pay after he has sold the rice. If not, he has to pay immediately after the transaction is concluded using money received from his buyer beforehand. The prepayment requires close contacts between the private trader and his buyer. In other words, certain transaction costs, such as networking costs and search and information costs,[9] are made before potential private traders can enter the rice purchase market.

State grain trading companies

In order to compete with private traders, each SGTC has its own strategy. The SGTCs from Banqiao and Shangzhu follow the same strategy as private traders. For example, every employee of the SGTC is assigned to purchase a certain amount of rice. To create incentives for its employees, the SGTC allows price differences between the price farm households receive from the employees and that which employees receive from the SGTC. While supported by the local government, the SGTC from Gangyan still behaves as monopsonists in the local purchase market for a limited period, e.g. 1 month following the harvest season. During that period, transactions of private traders are prohibited. However, the cost of enforcement is too high[10] for the local government to maintain prohibition for a long time. Nevertheless, farm households could still sell their grain to private traders after the prohibition period. This had negative effects on the amount of grain sold to the SGTCs by the farm households in the prohibition period. The large enforcement costs and the minimal effects of the prohibition led to the abolishment of this strategy by the SGTC in Gangyan in 2003. All three SGTCs also purchase rice from private traders, or from the SGTCs from other regions where the purchase price is lower.

Competition between private traders and the SGTCs

In order to improve efficiency and reduce losses of the SGTCs, government policies favoured the SGTCs in 2000–2003. For example, the SGTCs are allowed to use state-owned storage facilities and fleets of trucks, in contrast with the private traders. In addition, the SGTCs benefit from their social networks and the marketing information system provided by the government. Compared to the razor-thin profit of grain trade, search and information costs are high and are often an entrance barrier for potential private traders. Private traders are burdened by cumbersome licensing procedures, e.g. a private trader must possess a certain amount of financial resources that are often far beyond the entire wealth of an average farm household. Besides, different kinds of local taxes leave little profit for a licensed private trader. This leads to many unregistered private traders. Since they are not recognized by the government they have limited access to legal means for enforcing contracts. These constraints have increased the risk of their trade.

On the other hand, the need to perform policy-oriented business has prevented the SGTCs from competing freely with private traders in 2000. To avoid losses caused by purchasing rice at the protective prices fixed by the government, the SGTCs have used their administrative power to downgrade the rice sold by farm households, which has resulted in the payment of lower prices. This has led to farm households selling rice to the private traders. Moreover, the lower overhead costs of private traders[11] made them strong competitors for the SGTCs in 2000–2003. The SGTCs have higher buying costs, e.g. they are overstaffed, have storage costs, capital depreciation costs, pension costs of retirees, etc. Furthermore, stocking rice requires a low humidity, which means that more time needs to be spent on drying rice before selling to SGTCs. Since private traders do not store the rice, the non-strict humidity requirement has attracted the sale of rice by farm households.

According to all the SGTCs interviewed, by adopting the same strategy as the private traders and making full use of their former network, the SGTCs have reduced their transaction costs to the same level as private traders in 2000–2003. The development in road construction in the three villages (observed from survey) and the development in telecommunication (such as special price offers to farmers in rural areas) had reduced the search and information costs of the private traders in 2000–2003. As a result, the competition between the SGTCs and the private traders had become more intense.

Grain market reforms since 2003

In 2004 the regulation of grain circulation[12] was established to facilitate the reform started in 2000 (SC, 2004a). At the same time, the State Grain Administration announced the required qualifications for private traders to enter the purchase market (SGA, 2004). The criteria indicated that private traders whose trade volume was < 50 tons per annum did not have to register, which left room for small-scale private traders to enter the newly opened grain purchase market (SGA, 2004). In 2005, after almost 5 years' preparation, a system of monitoring and forecasting of market prices was introduced in July. Nevertheless, this system only serves the central and local government instead of the private traders in the markets. At the same time, the grain information centre releasing information on grain demand and supply, grain quality and grain price was established. However, this information is accessible only when a fee is paid. It is therefore very expensive for private traders (farm households) to benefit from this information system.

In summary, although several reform withdrawals have slowed down the pace of China's agricultural market reform, there is a clear movement towards market liberalization and deregulation (Tian and Zhang, 2003).

Theoretical Model

Competition in 2000

Three villages are selected in the study area and each village has its SGTC. Information obtained from interviews with the SGTCs in the study area shows that the SGTCs adopt different competitive strategies. We therefore model the competition between the SGTC and private traders in two ways.

Competition in Banqiao and Shangzhu

As mentioned above, the SGTCs in Banqiao and Shangzhu adopt the same strategy as private traders in order to survive competition. Search and information costs are high due to the underdevelopment of telecommunication and transportation, which act as an entry barrier for private traders entering the purchase market. Therefore, we assume a limited number of private traders in the market (oligopsony) competing with the SGTCs on quantity (Cournot) in 2000. In this case, we assume the SGTC reduces its marginal costs and equalizes it to the marginal costs of the private traders. The profit function of intermediate i (SGTC and private traders) ($\forall i = 1,..., n$) is:

$$\pi_i = (p - c_i - w)q_i - f_i \tag{15.1}$$

where π_i is the profit of intermediate i, p the consumer price of rice (price that intermediates receive),[13] c_i the marginal other costs of intermediate i (e.g. transportation costs, search and information costs, etc.), w the farm price of rice (price that intermediates pay to farm households), q_i the quantity of rice purchased by intermediate i, f_i the fixed costs of intermediate i – for example, the costs of cellphones for the private traders and the overhead costs for the SGTC, such as salaries for its employees. We assume that c_i and p are constant. The total quantity purchased by the intermediates (Q^d) equals the total quantity supplied by farm households (Q^s), i.e. $Q^d = Q^s = \sum_{i=1}^{n} q_i = Q$. We assume that the total quantity supplied by farm households (Q^s) depends on farm price (w) so that the total quantity purchased by the intermediates (Q^d) also depends on farm price (w).

Therefore, the first order condition of intermediate i is given by:

$$\frac{\partial \pi_i}{\partial q_i} = p - c_i - w - \frac{\partial w}{\partial Q}\frac{\partial Q}{\partial q_i}q_i = 0 \tag{15.2}$$

This can be written as:

$$p = c_i + w + w\frac{\partial w}{\partial Q}\frac{\partial Q}{\partial q_i}\frac{q_i}{w} \tag{15.3}$$

where $\dfrac{\partial Q}{\partial q_i}$ is private trader i's beliefs about how total demand would change when its own demand changes. Here, it is assumed that it equals 1, which indicates intermediates play Cournot. Therefore equation (15.3) can be written as:

$$p = c_i + w + \frac{w}{\varepsilon_i} \tag{15.4}$$

Where ε_i is price elasticity that individual intermediate i faces $(\varepsilon_i = \dfrac{\partial Q}{\partial w}\dfrac{w}{q_i})$. In other words, it shows the percentage change in demand of intermediate i when the farm price changes by 1%. Equation (15.4) shows that intermediate i receives a mark-up (extra profit above marginal costs of production) equal to: $\dfrac{w}{\varepsilon_i}$. From this equation, we see that the mark-up of an intermediate depends on the price he pays to farm households (farm price) and the price elasticity he faces. This mark-up decreases when the price elasticity that intermediate i faces becomes larger, namely a larger effect of changes in farm price on the rice demand of intermediate i leads to a smaller mark-up. When the price elasticity that intermediate i faces becomes infinite, the mark-up of intermediate i becomes zero. In that case the farm price is constant and the intermediates play perfect competition.

Competition in Gangyan

The SGTC in Gangyan behaves as a monopsonist for a short period after harvest, and private traders only enter the market afterwards (see above). In this case, we assume that competition between the SGTC and private traders comprises two stages. In the first stage, with the help of the local government, the SGTC is the only buyer (monopsony) in the purchase market 1 month after the harvest and decides the farm price (w_{SGTC}) it offers. Farm households know that they could sell their rice to private traders after the 1-month period but they do not know the price offered by private traders. In the second stage, private traders enter the purchase market. The strict control is eased but not eliminated. To a certain extent, private traders still incur risks of being prosecuted. Moreover, they bear the costs of searching for information.

Therefore, we assume there are a limited number of private traders (oligopsony). We know that the SGTC stores rice after purchasing and therefore has storage costs; moreover, search and information costs are higher. Thus marginal other costs of the SGTC are higher than those of private traders, i.e. $c_{SGTC} > c_{PT}$. Because of the high marginal other costs, the SGTC cannot increase the price it pays to farmers (w_{SGTC}) to the level of the price private traders pay (w_{PT}), and thus it does not attract any rice supply from farm households in the second stage. This indicates that the amount of rice purchased by the private traders as a whole (q_{PT}) equals total rice supply minus rice purchased by the SGTC in the first period, i.e. $q_{PT} = Q - q_{SGTC}$.

We solve this by looking first at the second stage. The profit function of private trader $i(\forall i = 1, \ldots, n)$ is:

$$\pi_{PTi} = (p - c_{PTi} - w_{PT})q_{PTi} - f_{PTi} \tag{15.5}$$

where π_{PTi} is the profit of private trader i, p the consumer price of rice (price that private traders receive), c_{PTi} the marginal other costs of private trader i (e.g. transportation costs, search and information costs, etc.), w_{PT} the farm price of rice offered by private trader i, f_{PTi} the fixed costs of private trader i (e.g. the costs of cellphones), q_{PTi} the quantity of rice purchased by private trader i and we have $q_{PT} = \sum_{1}^{n} q_{PTi} = \sum_{1}^{r} x_{PTh}$, in which x_{PTh} is the quantity of rice supplied to private trader by farm household $h(\forall h = 1, \ldots, r)$. We assume that c_{PTi} is constant.

Private trader i maximizes their profit. The first order condition of Equation (15.5) is:

$$\frac{\partial \pi_{PTi}}{\partial q_{PTi}} = p - c_{PTi} - w_{PT} - \frac{\partial w_{PT}}{\partial q_{PT}}\frac{\partial q_{PT}}{\partial q_{PTi}} q_{PTi} = 0 \tag{15.6}$$

This can be written as:

$$p = c_{PTi} + w_{PT} + w_{PT}\frac{\partial w_{PT}}{\partial q_{PT}}\frac{\partial q_{PT}}{\partial q_{PTi}}\frac{q_{PTi}}{w_{PT}} \tag{15.7}$$

where $\frac{\partial q_{PT}}{\partial q_{PTi}}$ is private trader i's beliefs about how total demand of all the private traders in the second stage would change when its own demand changes. Here, we assume that equals 1, which indicates intermediates play Cournot. Therefore, equation (4.7) can be written as:

$$p = c_{PTi} + w_{PT} + w_{PT}\frac{\partial w_{PT}}{\partial q_{PT}}\frac{q_{PTi}}{w_{PT}}) \tag{15.8}$$

or:

$$p = c_{PTi} + w_{PT} + \frac{w_{PT}}{\varepsilon_{PTi}} \tag{15.9}$$

where ε_{PTi} represents the price elasticity that private trader i faces $(\varepsilon_{PTi} = \frac{\partial q_{PT}}{\partial w_{PT}}\frac{w_{PT}}{q_{PTi}})$. Equation (15.9) shows that private trader i receives a

mark-up equal to: $\frac{w_{PT}}{\varepsilon_{PTi}}$. This indicates the mark-up of private trader i

depends on the price he pays to farm households (farm price) and the price elasticity he faces. This mark-up increases when the price elasticity that private trader i faces becomes smaller.

From Equation (15.9), the price paid by private traders (w_{PT}) can be derived. The amount of rice supplied in the second stage (q_{PT}) can be derived by writing aggregated rice supply of farm households as a function of the farm price offered by private traders, i.e.

$$q_{PT} = \sum_1^r x_{PTh}(w_{PT}).$$

Now we look at the first stage. In the first stage, the SGTC is the only buyer of rice in a village. Since we know the aggregated rice supply to private traders (q_{PT}), the residual supply to the SGTC is derived as: $q_{SGTC} = Q(w_{SGTC}, w_{PT}) - q_{PT}$. Therefore, we have $q_{SGTC} = q_{SGTC}(w_{SGTC}, w_{PT}, q_{PT})$. We let $x_{PTh} = x_{PT}(x_{SGTCh})$ be the reaction function of household h that depicts the supply decision of household h to private traders given its supply to the SGTC in the previous stage. The aggregated rice supply

to private traders can be written as $\sum_1^r x_{PTh} = x_{PT}(\sum_1^r x_{SGTCh})$, which is equivalent to $q_{PT} = q_{PT}(q_{SGTC})$. In the end, the residual supply function can be written as $q_{SGTC} = q_{SGTC}(w_{SGTC}, w_{PT}, q_{PT}(q_{SGTC}))$. The inversed residual supply function is then $w_{SGTC} = w_{SGTC}(w_{PT}, q_{SGTC}, q_{PT}(q_{SGTC}))$.

The profit function of the SGTC is:

$$\pi_{SGTC} = (p - c_{SGTC} - w_{SGTC})q_{SGTC} \tag{15.10}$$

where π_{SGTC} is the profit of the SGTC, c_{SGTC} is the marginal other costs of the SGTC (e.g. transportation costs and storage costs,[14] etc.), w_{SGTC} the farm price offered by the SGTC, and q_{SGTC} the rice purchased by the SGTC. We assume c_{SGTC} to be constant.

The first order condition of Equation (15.10) is:

$$\frac{\partial \pi_{SGTC}}{\partial q_{SGTC}} = p - c_{SGTC} - w_{SGTC} - (\frac{\partial w_{SGTC}}{\partial q_{SGTC}} + \frac{\partial w_{SGTC}}{\partial q_{PT}}\frac{\partial q_{PT}}{\partial q_{SGTC}})q_{SGTC} = 0 \tag{15.11}$$

This can be written as:

$$p = c_{SGTC} + w_{SGTC} + w_{SGTC}(\frac{\partial w_{SGTC}}{\partial q_{SGTC}}\frac{q_{SGTC}}{w_{SGTC}}$$

$$+ \frac{\dfrac{\partial q_{PT}}{\partial Q}\dfrac{Q}{q_{PT}}}{\dfrac{\partial q_{PT}}{\partial w_{SGTC}}\dfrac{w_{SGTC}}{q_{PT}} \cdot \dfrac{\partial q_{SGTC}}{\partial Q}\dfrac{Q}{q_{SGTC}}}) \tag{15.12}$$

or:

$$p = w_{SGTC}(1 + \frac{1}{\varepsilon_{SGTC}} + \frac{\varepsilon_Q^{PT}}{\varepsilon_{SGTC}^{PT} \cdot \varepsilon_Q^{SGTC}}) + c_{SGTC} \tag{15.13}$$

where $\varepsilon_{SGTC} = \dfrac{\partial q_{SGTC}}{\partial w_{SGTC}}\dfrac{w_{SGTC}}{q_{SGTC}}$ represents the price elasticity that the SGTC

faces (total marketable rice supply of farm households in this stage

equals the demand from the SGTC). $\varepsilon_{SGTC}^{PT} = \dfrac{\partial q_{PT}}{\partial w_{SGTC}}\dfrac{w_{SGTC}}{q_{PT}}$ represents the

cross-price elasticity of marketable rice supply to private traders.

$\varepsilon_Q^{SGTC} = \dfrac{\partial q_{SGTC}}{\partial Q}\dfrac{Q}{q_{SGTC}}$ represents the elasticity of supply to the SGTC

(q_{SGTC}) with respect to the total marketable rice supply (Q).

$\varepsilon_{PT}^Q = \dfrac{\partial q_{PT}}{\partial Q}\dfrac{Q}{q_{PT}}$ represents the elasticity of supply to the private traders

(q_{PT}) with respect to the total marketable rice supply (Q).

Equation (15.13) shows that the mark-up of the SGTC equals

$(\dfrac{1}{\varepsilon_{SGTC}} + \dfrac{\varepsilon_Q^{PT}}{\varepsilon_{SGTC}^{PT}\varepsilon_Q^{SGTC}})w_{SGTC}$, among which $\dfrac{w_{SGTC}}{\varepsilon_{SGTC}^s}$ is monopsony mark-up.

The second component of the mark-up includes the cross-price elasticity of supply to private traders (ε_{SGTC}^{PT}), which is expected to have a negative sign since an increase in w_{SGTC} results in a decrease in q_{PT}. The signs of ε_Q^{SGTC} and ε_Q^{PT} are expected to be positive since an increase in total marketable rice supply leads to an increase in supply to the SGTC or to the private traders. Hence, the second component has a negative sign, which counterbalances the monopsony mark-up. This indicates that the expected monopsonistic market power of the SGTC in Gangyan is offset by the participation of the private traders in the second stage.

Competition in 2003

China's government officially allowed private traders to enter the rice purchase market in 2000. In 2000–2003, the number of private traders increased (Chen, 2003). The SGTC is no longer subsidized by the government and has to compete with private traders in the purchase market. During interviews with private traders in the study area, we found the trading amount of an individual private trader to be very low, indicating that the market share of an individual private trader is also small, so that the farm price is determined in a competitive market.

Due to the rapid development of China's telecommunication system, most of the private traders interviewed possess a cellphone, which has increased the accessibility of private traders to market information. We also found the difference between the buying and selling price of private traders to be very thin, again an indication of a perfect competitive market. According to Baumol *et al.* (1982), when entry costs are sufficiently low, the threat of potential entry may already yield an efficient outcome. Given the large number of potential traders (farm households) in each selected village, we assume the rice purchase market in 2003 in the study area to be perfectly competitive.

We assume that there are n private traders and the SGTC in the purchase market in one village. Thus we have $n + 1$ intermediates competing in the market. The profit function of intermediate i equals:

$$\pi_i = (p - c_i - w)q_i - f_i \tag{15.14}$$

where π_i is the profit of intermediate i (the SGTC or private trader i), p the consumer price received by intermediate i, c_i the marginal other costs of intermediate i, w the farm price, q_i the amount of rice purchased by intermediate i and f_i the fixed costs of intermediate i (e.g. costs of cellphones or overhead costs).

In a perfectly competitive market, intermediate i is a price-taker. Therefore, w is given. The first order condition of Equation (15.14) is:

$$p - c_i - w = 0 \tag{15.15}$$

or:

$$p = c_i + w \tag{15.16}$$

Therefore, intermediate i receives zero profit, i.e. $\pi_i = 0$. The market share of intermediate i equals $\dfrac{1}{n+1}$.

Empirical Model and Data

In this section the theoretical model is presented. It consists of the first order conditions of the oligopsonists (Equations 15.4 and 15.9) and monopsonist (Equation 15.13) in the purchase market of 2000; and the first order condition of the intermediates in the purchase market of 2003 (Equation 15.16).

Elasticities

The elasticities in Equations (15.4), (15.9) and (15.13) are estimated using data from the SERENA project (Chen and Peerlings, 2005 unpublished results) and presented in Table 15.1. All elasticities have the expected sign.

Other data

In Banqiao and Shangzhu, marginal other costs of private traders consist of transport costs, search and information costs (e.g. costs of visiting farm households and costs of making telephone calls) and networking costs (e.g. cigarettes, drinks and meals) in 2000. As discussed in above, the SGTCs in Banqiao and Shangzhu adopted similar strategies to those of private traders in 2000, namely they paid their employees to go to the village to purchase rice from farm households. In addition, the SGTCs store rice they purchase so they bear storage costs. Therefore, the marginal other costs of the SGTCs include transport costs, costs they pay their employees to go to the villages to purchase rice and storage costs. Although the marginal other costs of the SGTCs and the private traders have different composition, we assume that the marginal other costs of the SGTC and the private traders have the same value.

For Gangyan in 2000, we assume that the marginal other costs of the SGTC are higher than those of private traders because these includes monitoring costs. Therefore, the marginal other costs of the SGTC in

Table 15.1. Estimated elasticities from the SERENA project.

Banqiao ε_i	Shangzhu ε_i	Gangyan				
		ε_{PTi}	ε_{SGTC}	ε_{SGTC}^{PT}	ε_Q^{SGTC}	ε_Q^{PT}
38.666	33.764	34.148	0.275	−0.127	2.941	1.348

Gangyan include transport, storage and monitoring costs. In 2003, since the rice purchase market was perfectly competitive, we assumed marginal other costs of intermediates (the SGTC and private traders) in Banqiao and Shangzhu to be lower than those of intermediates in 2000 (due to developments in the telecommunication system and the improvement of roads). For Gangyan in 2003, since the SGTC had to reduce its marginal other costs in order to compete with private traders, we assumed that the marginal other costs of the SGTC and the private traders had the same value.[15] We also assumed the marginal other costs of intermediates in Gangyan in 2003 to be lower than those of private traders in 2000 (see Table 15.2 for the composition of the marginal other costs of the SGTCs and the private traders in 2000 and in 2003).

Market access is measured in terms of transport costs. Transport costs differ for the villages depending on the distance from the village to the next buyer of the intermediates. Transport costs are a measure of market access; as mentioned above, Gangyan has the best market access. Shangzhu is located in a mountainous area and is the most remote village; Banqiao occupies an intermediate position. Transport costs are reflected in the marginal other costs.

Farm households' marketable rice supply in 2000 in the three selected villages was taken from a household survey carried out in the same villages in 2000 and 2001[16]: 191 tons in Banqiao and 80 in Shangzhu. The marketable rice supply of Gangyan in 2000 consisted of supply to the SGTC (141 tons) and supply to private traders (429 tons). Farm prices were taken from the SERENA project. The farm price equalled 1073 Yuan/ton in Banqiao, while in Shangzhu the farm price equalled 977 Yuan/ton. In Gangyan the farm price offered by the SGTC was 960 Yuan/ton[17] and the farm price offered by private traders was 1024 Yuan/ton.

Table 15.2. Composition of marginal other costs.

	2000 (imperfect competition)		2003 (perfect competition)	
	SGTC	Private trader	SGTC	Private trader
Marginal other costs Banqiao and Shangzhu	$c_{SGTC} =$ Transport costs Employee costs	c_{PTi} Transport costs Search and information costs	$c_{SGTC} =$ Transport costs Employee costs	c_{PTi} Transport costs Search and information costs
	Storage costs	Networking costs	Storage costs	Networking costs
Marginal other costs Gangyan	$c_{SGTC} >$ Transport costs Monitoring costs	c_{PTi} Transport costs Search and information costs	$c_{SGTC} =$ Transport costs Employee costs	c_{PTi} Transport costs Search and information costs
	Storage costs	Networking costs	Storage costs	Networking costs

During the survey conducted in July 2003, price margins of private traders in the three villages were obtained, namely 50 Yuan/ton for the private traders in Banqiao, 58 Yuan/ton for Shangzhu and 50 Yuan/ton for Gangyan (Chen, 2003). Therefore, the consumer price (price that private traders receive) in 2000 was calculated by the summation over farm price and the price margin: this equalled 1123 Yuan/ton in Banqiao, 1035 Yuan/ton in Shangzhu and 1074 Yuan/ton in Gangyan (see Table 15.3).

Simulations and Results

Simulations

Three simulations were performed, all comparing the situation in 2000 with that in 2003, but different assumptions are made with respect to changes in the consumer price. Since we model the SGTC and private traders differently in Gangyan, we also look at the differences between both.

Simulation 1

Comparison between 2000 and 2003 while keeping the consumer price constant. First, we examine the changes for the intermediates (the SGTC and private traders) from 2000 to 2003. In the case of perfect competition (situation in 2003), there are several differences compared with the case of monopsony and oligopsony (situation in 2000):

- Both first-order conditions of monopsony and oligopsony are changed such that the mark-ups become zero. It is expected that this will lead to an increase in farm prices, an increase in total marketable supply, zero profits for the intermediates and an increase in profit for the farm households.
- The marginal other costs of all the intermediates in all the villages were lower in 2003 than in 2000 due to increased competition (SGTC), the development of telecommunication and the improvement

Table 15.3. Other data used for the simulations (from SERENA, 2001; RESPONSE, 2003).

	Banqiao intermediate	Shangzhu intermediate	Gangyan	
			SGTC	Private traders
Consumer price (Yuan/ton)	1123	1035	1074	
Farm price (Yuan/ton)	1073	977	960	1024
Rice production (ton)	265	201	692	
Marketable rice supply (ton)	191	80	141	429
Marginal other costs 2000 (Yuan/ton)	23	28	86	20
Marginal other costs 2003 (Yuan/ton)	15	20	12	

in road conditions. We expect an increase in farm prices and marketable supply to the intermediates.

• Since rice is a normal good in this chapter, there is a negative correlation between farm price and rice consumption, i.e. it is expected that an increase in farm prices will lead to a decrease in rice consumption and therefore to an increase in total marketable rice supply. Secondly, we compare the differences between the SGTC and private traders of Gangyan in 2000. Notice that different mark-ups and marginal other costs result in different farm prices between the SGTC and private traders (see Equations 15.9 and 15.13). Which mark-up is larger is an empirical issue, so that it is not clear beforehand whether the SGTC or private traders offer the higher farm price.

Simulation 2

Comparison between 2000 and 2003 when the consumer price decreases by 10%. The liberalization and deregulation of China's rice market has introduced price uncertainty. In this simulation the effects of a lower consumer price are examined. We expect the price decrease lowers the gain for farm households from more competition among the intermediates. There will be a smaller increase in farm prices and rice supply to the SGTC and private traders. In 2003, no intermediates made profits because of the perfect competition assumption.

Simulation 3

Comparison between 2000 and 2003 when the consumer price increases by 10% following international trade liberalization resulting from WTO accession.[18] The consumer price has been increasing since 2001, however the speed of this price increase slowed down in 2005 (PDO, 2005). We expect that the effects have the same sign as in the first simulation, but that they will be larger.

Results

The first three columns of Table 15.4 report the results of the comparison of the rice marketing channel in 2000 with 2003 (simulation 1) for the three selected villages. Since the intermediates faced perfect competition in all three villages in 2003, they have zero mark-ups. In 2000, intermediates had a higher mark-up (29.9 Yuan/ton) in Shangzhu than in Banqiao (27.1 Yuan/ton). This is because of the smaller price elasticity that an intermediate faces in Shangzhu than in Banqiao that leads to a higher mark-up. In Gangyan, the mark-up of the SGTC (27.2 Yuan/ton) is lower than the mark-up of private traders (30.0 Yuan/ton). This is because the monopsony market power of the SGTC is offset by the potential competition of private traders after the monopsony period. The slightly higher mark-up of private traders in Gangyan than in Banqiao and Shangzhu indicates the greater market power of private traders in Gangyan.

In Banqiao and Shangzhu, the farm price increased by 3.3 and 3.9%, respectively. The higher price increase in Shangzhu is due to the loss of its higher mark-up. In Gangyan, the farm price for the SGTC and private traders increased by 10.5 and 3.7%, respectively. The larger price increase of the SGTC is due mainly due to the reduction of its marginal other costs.

The traded quantity for individual intermediates in Banqiao and Shangzhu decreased by 20.5 and 27.5%, respectively. However, because more intermediates participated in the purchase market in 2003, the total quantity supplied increased by 0.6 and 1.0% in Banqiao and Shangzhu, respectively. The increase in total quantity supplied is also due to the increase in farm price. The larger increase in Shangzhu represents the larger market power of intermediates. In Gangyan, the trading quantity for the SGTC increased by 8.9% and decreased by 17.4% for private traders. The large increase in rice supplied to the SGTC is due to the large increase in farm price. Although rice supplied to private traders decreased by 17.4%, the total rice supplied in Gangyan increased by 3.3% due to the increase in the number of private traders.

In all the villages, the loss in profit for the intermediates is smaller than the gain of farm households. Therefore, the overall profit increase is positive. The lower gain in overall profit in Shangzhu is due to its

Table 15.4. Simulation results, comparison between 2000 and 2003.

	Simulation 1 (consumer price constant)			Simulation 2 (consumer price −10%)			Simulation 3 (consumer price +10%)		
	Banqiao	Shangzhu	Gangyan	Banqiao	Shangzhu	Gangyan	Banqiao	Shangzhu	Gangyan
Change in trade (%)									
Δq_{SGTC}	−20.5	−27.5	8.9	−22.0	−29.5	−0.5	−18.9	−25.5	18.3
Δq_{PTi}			−17.4			−21.9			−12.8
Δ_Q	0.6	1.0	3.3	−1.4	−1.8	−3.2	2.6	3.8	9.8
Change in price (%)									
Δ_p	0	0	0	−10	−10	−10	10	10	10
Δw_{SGTC}	3.3	3.9	10.5	−7.2	−6.7	−0.7	13.7	14.5	21.7
Δw_{PT}			3.7			−6.8			14.2
Change in mark-up per unit (Yuan/ton)									
SGTC	−27.1	−29.9	−27.2	−27.1	−29.9	−27.2	−27.1	−29.9	−27.2
Private Trader			−30.0			−30.0			−30.0
Total profit change (1000 Yuan)									
$\Delta\pi_{SGTC} + \Delta\pi_{PT}$	−5.2	−2.4	−16.7	−5.2	−2.4	−16.7	−5.2	−2.4	−16.7
$\Delta\pi_{farm}$	6.3	2.8	70.4	−13.7	−4.8	−46.7	26.5	10.6	183.5
$\Delta\pi_{SGTC} + \Delta\pi_{PT} + \Delta\pi_{farm}$	1.1	0.4	53.7	−18.9	−7.2	−63.4	21.3	8.2	166.8

relatively small trading quantity. In Gangyan, the loss in profit of private traders is the sum of profit loss of each individual private trader. Therefore, it is much higher than the profit loss of the SGTC.

The middle three columns of Table 15.4 report the results of the comparison of 2000 with 2003, when the consumer price had decreased by 10% (simulation 2). According to Equations (15.4), (15.9) and (15.13), the mark-ups of the intermediates do not include the consumer price, so the change of mark-ups for the intermediates remains the same as in simulation 1. Farm prices decreased by 7.2 and 6.7% for Banqiao and Shangzhu, respectively. In Gangyan, farm prices decreased by 0.7% for the SGTC and 6.8% for private traders. This implies a decrease in marketable rice supply and farm profits. As expected, profit deficit of the intermediates remains the same as in simulation 1 because they have zero profit in the case of perfect competition. Profit of farm households decreased for all villages. Therefore, the overall profit change was negative for all villages.

The last three columns of Table 15.4 present the results of simulation 3, a comparison of 2000 and 2003 when the consumer price increases 10%. The results look similar to those for simulation 1, but with a larger increase or smaller decrease in farm price, farm profits and marketable rice supply. As mentioned above, the change of intermediates' profit stays the same as in simulation 1. Consequently, the overall profit increased more in all three villages.

Simulation 2 shows that the decrease of the consumer price (10%) is less than fully transmitted to farm households. Simulation 3 shows that the increase of the consumer price (10%) is more than fully transmitted to farm households. In simulation 2, the negative effects of a consumer price decrease are partly offset by the positive effects of an increase in competition from 2000 to 2003, which leads to a decrease in farm prices of < 10% for all villages.

Conclusions and Policy Implications

The objective of the chapter is to analyse the effects of market liberalization and deregulation in the rice marketing channel on farm households, using information collected from three selected villages with different market access in Jiangxi Province, China. Market access in this chapter is measured by transport costs. Results indicate that rice producers benefit from market liberalization and deregulation, but how much depends on the degree of market imperfections existing before market liberalization and deregulation and the degree of market access.

For small villages in terms of production and for villages far from consumer markets (limited market access), it is to be expected that the number of intermediates in a liberalized restructured market will be small. Therefore, the gains from market liberalization and deregulation will be smaller than for farm households in villages close to consumer markets and with a relatively large production. The competition with private traders forces the SGTC to reduce its marginal other costs, which reduces the subsidies going to the SGTC. This indicates that the commercialization of the SGTCs has partly reached its objective – to relieve the government of the budget burden caused by subsidies.

The results of our study are obviously subject to some qualifications. First, the model is a short-term model because we take production as given, and the effects on rice production of a farm price change are not taken into account. Secondly, we looked only at the effects on farm profits, which is not the same as utility of farm households; neither were the effects on the supply and consumption of other goods taken into account. Thirdly, it is a major assumption that the rice market was perfectly competitive in 2003. Fourthly, there is still uncertainty about the data, especially the number of intermediates. Finally, there is uncertainty about the data used, especially the number of intermediates.

Using a full household production model that includes production (instead of just marketable supply) decisions and a more extended consumption block may enrich the analysis. The model presented here can serve as a building block in this type of extended analysis.

Some policy recommendations can be drawn from this study. Positive effects from liberalizing the rice purchase market can be enhanced by improving market access. Public investment in the physical infrastructure (roads and transport systems) might be a means for this. Moreover, to reduce search and information costs, market information availability could be improved, e.g. by improving the accessibility of the information provided by the government, namely to reduce or remove the costs of the private traders for obtaining information on rice prices.

Endnotes

[1] The commercialization of state grain bureaux involved with the grain trade began in 1992; since then, they have been termed state grain trading companies (SGTC).

[2] According to the data collected for the SERENA project (Kuiper *et al.*, 2001) the total rice supply of the selected villages was 841 tons, which is relatively small compared with the total rice supply of the whole region (northern part of Jiangxi Province).

[3] From 2000 to 2003 the length of railways and highways in China increased 6 and 29%, respectively, and the number of cellphones increased 219% (CSY, 2004).

[4] For more information on the survey, see Chen and Peerlings (2005, unpublished data).

[5] To understand fully how China's grain market functions nowadays, it is important to look at the history of policy reforms in China's grain market (see, e.g. Zhong, 2001).

[6] The purpose of setting up the grain reserve system is to avoid the fluctuation in grain prices with an early-warning and immediate-market monitoring system.

[7] Grain-surplus regions mean that regions produce more grain than they consume, so that they export grain to grain-deficient regions.

[8] For more information about the survey, see Chen and Peerlings (2005, unpublished data).

9   Search and information costs of private traders comprise costs incurred during the search for farm households (who are willing to sell) and buyers (who are willing to purchase), and also the costs of maintaining such personal contacts.

10 Normally, the local government monitors grain transactions by setting up checkpoints at the entrances to a village. In addition, the time of inspection is also important due to the fact that private traders might transport rice out of the village in the middle of the night. Therefore, the cost of enforcement is mainly in terms of labour.

11 Overhead costs are fixed costs that include costs of interest, utilities and taxes.

12 According to the regulation, grain prices are to be determined by market supply and demand. The central government intensifies its supervision over the grain marketing channel (SC, 2004b).

13 According to the regulation, grain prices are to be determined by market supply and demand. The central government intensifies its supervision over the grain marketing channel (SC, 2004b).

14 The SGTC might be able to obtain a higher selling price later in the season. However, the model does not consider this aspect.

15 Due to government investments in infrastructure (e.g. road construction) and the boom in the private transportation business, transport costs have been decreasing since 1996 (Luo and Crook, 1997).

16 The household survey was conducted as part of the SERENA project (Kuiper *et al.*, 2001).

17 The farm price offered by the SGTC in Gangyan was originally 1068 Yuan/ton. However, as mentioned in the section covering the grain purchase market in 1998, above, the SGTC often reduces the farm price by downgrading the rice sold by farm households. Therefore, the price announced by the SGTC is not the price actually paid. We assume in our analysis that downgrading by the SGTC decreases the farm price by 10%.

18 Huang *et al.* (2004) conclude that the effect of China's accession to the WTO on rice prices is positive because of rising rice exports.

References

Baumol, W., Panzar, J. and Willig, R. (1982) *Contestable Markets and the Theory of Industry Structure.* Harcourt Brace Jovanovich, San Diego, California.

Chen, L. (2003) Summary of information and data collected in surveys conducted in March and July 2003 for project 'Agricultural Policy and Rural Food Security in Southeast China: the Mediating Role of Markets' from RESPONSE program. WUR, The Netherlands and IFPRI, Washington, DC.

Chen, L. and Peerlings, J. (2005) Survey description and estimation of elasticities. Background document RESPONSE program, Wageningen University and Research Centre (WUR), The Netherlands and International Food Policy Research Institutue (IFPRI), Washington, DC.

CSY (2004) *China Statistical Yearbook.* China Statistical Press, Beijing.

FAO (2000) *The State of Food Insecurity in the World 2000.* FAO, Rome.

Huang, J. (1998) *Investigating the Practice of Current Reform Policies in the Grain Circulation System.* Internal discussion paper, Centre for Chinese Agricultural Policy, Chinese Academy of Sciences, Beijing.

Huang, J., Rozelle, S. and Chang, M. (2004) Tracking distortions in agriculture: China and its accession to the World Trade Organization. *The World Bank Economic Review* 18, 59–84.

Kuiper, M., Heerink, N., Tan, S., Ren, Q. and Shi, X. (2001) *Report of Village Selection for Three Villages Survey.* SERENA-project working paper, Nanjing Agricultural University, Nanjing, China and Wageningen University, Wageningen, The Netherlands.

Luo, X. and Crook, F.W. (1997) *The Emergence of Private Rice Marketing in South China.* International Agricultural and Trade Reports, China: Situation and Outlook series (WRS-97-3), USDA, Washington, DC.

PDO (2003) Grain price hikes do not indicate supply shortage. *People's Daily Online,* 8 November (http://english.peopledaily.com.cn/200311/08/eng20031108_127899.shtml).

PDO (2005) Grain production rises; price fall hurts farmers. *People's Daily Online,* 26 September (http://www.chinadaily.com.cn/english/doc/2005–09/26/content_480757.htm).

Ravallion, M. and Chen, S. (2004) *China's (uneven) Progress against Poverty.* World Bank Policy Research Working Paper 3408, Washington, DC.

Rozelle, S., Park, A., Huang, J. and Jin, H. (1997) Liberalization and rural market integration in China. *American Journal of Agricultural Economics* 79, 635–642.

SC (1998a) *The Resolution of Further Deepening Grain Circulation System Reform.* State Council P.R. of China, Document No. [1998] 15.

SC (1998b) *The Notification of Distributing Suggestions on Advancing Grain Circulation System Reform.* State Council PR of China, Document No. [1998] 35.

SC (1998c) *Grain Procurement Regulations.* State Council P.R. of China, Document No. [1998] 244.

SC (2000) *The Notification of Further Improving Relevant Policy Measures with Respect to Grain Production and Circulation.* State Council PR of China, Document No. [2000] 12.

SC (2004a) *The Opinion on Enhancing the Income of Farm Households.* State Council P.R. of China, Document No. [2004] 1.

SC (2004b) *The Administrative Ordinance of Grain Circulation.* State Council P.R. of China, Document No. [2004] 407.

SGA (2004) *Provisional Measures of Auditing Qualifications for Grain Purchasing.* State Grain Administration P.R. of China, Document No. [2004] 121.

Shen, D. and Wu, J. (2004) *Mountain–River–Lake Integrated Water Resources Development Program, Jiangxi, People's Republic of China. Water and Poverty – The Realities.* Asian Development Bank, Philippines.

Sicular, T. (1995) Redefining state, plan and market: China's reforms in agricultural commerce. *The China Quarterly* 143, 1020–1046.

Tian, W. and Zhang, L. (2003) Recent economic and agricultural policy development in China. Paper prepared for the *Roles of Agriculture Project International Conference,* 20–22 October, Agricultural and Development Economics Division (ESA), FAO, Rome.

Wu, L. (2000) *Studies on the Integration of Agricultural Product Markets.* China Agriculture Press, Beijing.

Zhong, F. (2001) *The Political Economy of the Chinese Grain Marketing System.* Technical Report ACIAR Project 9721 Working Papers, APSEM, the Australian National University, Canberra, Australia.

16 Consequences of the Abolition of Multi-fibre Arrangement Import Quotas on the Apparel Industry of Bangladesh: a Computable General Equilibrium Analysis

NAZNEEN AHMED[1] AND JACK PEERLINGS[2]

[1]*Bangladesh Institute of Development Studies (BIDS), E-17, Agargaon, Sher-e-Bangla Nagar, Dhaka 1207, Bangladesh; e-mail: nahmed@sdnbd.org;* [2]*Agricultural Economics and Rural Policy Group, Wageningen University, Hollandseweg 1, 6706 KN Wageningen, The Netherlands; e-mail: jack.peerlings@wur.nl*

Abstract

This chapter explores and analyses the changes and challenges faced by the apparel industry of Bangladesh following the abolition of import quotas under the Multi-fibre Arrangement (MFA). Being the most important export industry of Bangladesh, changes related to the apparel industry affect the economy as a whole. This chapter applies the global computable general equilibrium model and data of the Global Trade Analysis Project (GTAP). We find that abolition of MFA import quotas has had negative consequences for the Bangladeshi economy. Depending on various scenarios, the country's apparel exports decreased by 9.6% to 15.1%, while production of apparel decreased by 8.9% to 13.9%. There are also negative consequences for the rural income and poverty levels.

Introduction

Apparel is the main export industry of Bangladesh. Apparel exports grew from US$1.0 million in 1978 to US$5 billion in 2004, comprising 75% of total export earning and 80% of manufacturing export earning of Bangladesh (Bangladesh Bureau of Statistics, 1982; Bangladesh Bank, 2005). In 2004 the apparel industry had a 9.5% share in the total GDP and a share of 29.7% in manufacturing GDP (Bangladesh Bank, 2005). For the last two decades the apparel industry has been the main source of growth of export and formal employment of unskilled workers. At present, nearly 1.9 million workers are employed in this industry and 90% of them are female (Mlachila and Yang, 2004; Razzaque, 2005). Around 75% of these female apparel workers are migrants from rural areas, coming from the poorest rural households (Afsar, 2001).

The Multi-fibre Arrangement (MFA) import quotas and availability of cheap labour were the two main reasons behind the rapid growth of the apparel industry in Bangladesh. Relatively less restrictive MFA import quotas for Bangladesh compared with those for traditional apparel exporters (Korea, Hong Kong, Japan, China, etc.) ensured a market for Bangladesh apparel in the USA and stimulated the growth of this industry (Bhattacharya and Rahman, 2005). The apparel industry has been exposed to the challenges of the import quota-free world since 1 January 2005, as the MFA import quotas were fully abolished on 31 December 2004.

Starting in 1974, MFA governed the trade in textile and apparel until the end of the Uruguay round (31 December 1994). On 1 January 1995 the WTO's Agreement on Textile and Clothing (ATC) replaced the MFA. The ATC was a 10 year-long transitional trade regime devised to integrate fully the trade of textile and apparel into the WTO rules in four phases, mainly by phasing out the MFA import quotas. Thus, since 1 January 2005, for textile and apparel trade, the free trade rules under WTO apply. This has posed challenges for the apparel industry of Bangladesh.

Different studies (e.g. Yang *et al.*, 1997; Islam, 2001; Lips *et al.*, 2003; Mlachila and Yang, 2004; Nordas, 2004; Razzaque, 2005) have looked into the possible impacts of MFA import quota abolition. In general, these studies have concluded that both the world trade in apparel and welfare would increase and that consumer price would decrease; however, impacts will vary across countries and some countries may face a welfare loss.

Given the importance of apparel for the Bangladesh economy, abolition of the MFA import quotas affects not only the apparel industry but also the economy as a whole. Such impacts will occur through both industry linkages and changes in income, investment and savings of various actors. MFA import quota abolition also includes textile, an important input for the apparel industry. Moreover there will be changes in the relative competitiveness of Bangladesh in relation to its competitors in the international market, and explicit knowledge of these changes is important in deciding upon policies needed to cope with the changed situation. A multi-country computable general equilibrium (CGE) model is the best possible way to analyse economy-wide effects of international trade policy changes and changes in the relative position of a single country. Therefore, this chapter applies the Global Trade Analysis Project (GTAP, Hertel, 1997) multi-country CGE model and data and aims to analyse the possible consequences of MFA import quota abolition on Bangladesh using GTAP.

The chapter is organized as follows. The next section provides a brief overview of the Bangladesh apparel industry and the challenges faced; the following part describes the model and data, and we then provide three different simulations; analysis of the results and comparison with the outcomes of other studies follows, including a discussion of the effects on the poor; the final section provides some conclusions.

Overview of the Apparel Industry

Production chain of apparel

To obtain an overview of the Bangladesh apparel industry, it is important to understand the production chain of apparel. In general, apparel commodities follow the production chain illustrated in Fig. 16.1.

Up to the fabric stage, the product is considered to be textile and then, through the process of cutting and sewing, it becomes apparel, and there are various intermediate stages between textile and apparel. In general, there are two broad categories of apparel products – woven apparel and knitted apparel. These two categories use different types of yarn, fabric, machineries, manufacturing technology and even labour. Woven apparel uses mostly female workers and knitted apparel uses mostly male workers: this is because of differences in skill requirements. More skilled workers are required in knitted apparel, and female skilled workers are relatively scarce in Bangladesh.

Emergence and growth

The apparel industry of Bangladesh has grown at a very fast pace in terms of export value and share in total export (see Table 16.1). As a result, the composition of export has changed. The jute-dominant (both raw and

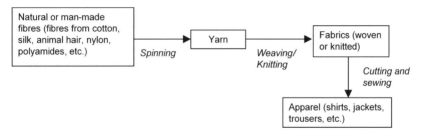

Fig. 16.1. General production chain of apparel.

Table 16.1. Exports from Bangladesh, in US$ million; fiscal year starts 1 July (from CPD, 2001; Bangladesh Bank, 2005).

Item	1980–1981	1990–1991	1999–2000	2003–2004
Primary goods	209	306	469	485.8
	(29.4)[a]	(17.8)	(8.2)	(6.4)
Manufactured goods	501	1411	5283	7117.2
	(70.6)	(82.2)	(91.8)	(93.6)
Jute goods	367	290	266	246.5
	(51.7)	(16.9)	(4.6)	(3.2)
Leather and leather goods	57	136	195	211.4
	(8.0)	(7.9)	(3.4)	(2.8)
Woven apparel	3	736	3083	3538.1
	(0.4)	(42.9)	(53.6)	(46.5)
Knitted apparel	0	131	1270	2148
	(0.0)	(7.6)	(22.1)	(28.3)
Chemical products	11	40	94	na
	(1.5)	(2.3)	(1.6)	
Other manufactured goods	63	78	375	973.2
	(8.9)	(4.5)	(6.5)	(12.8)
Total export	710	1717	5752	7603
	(100)	(100)	(100)	(100)

na, not available.
[a] Values for percentage share in total exports given in parentheses.

processed jute) export bundle of the early 1980s has shifted to an apparel-dominant export bundle. While the apparel industry contributed only 0.4% to total export earning in the fiscal year 1980–1981, it contributed 75.7% in 1999–2000 and 74.8% in 2003–2004. Apparel export growth has acted as the main force behind the total export growth of Bangladesh.

Availability of workers on a low wage is a major competitive advantage of the apparel industry of Bangladesh. As a consequence of the rapid growth of the apparel industry, employment in this labour-intensive industry has also grown rapidly. While the industry employed 0.1 million people in 1985, it employed nearly 1.9 million in 2003 (Razzaque, 2005). It will be noted from Table 16.2 that, in 2003, the wage per worker in Bangladesh was 157% lower than that in China. Comparing Bangladesh's 1997 data with those for India in 1998 shows that the wage per worker in Bangladesh was 20–30% lower than that in India. Workers in the apparel industry are mostly young unmarried females, migrating (first time) from rural areas and maintaining a close link with the rural economy. There is an almost unlimited supply of such people, which makes it easy to employ them at a low wage.

Internationally, the apparel industry of Bangladesh flourished under the umbrella of the MFA. Relatively less restrictive import quotas for Bangladesh under MFA ensured a market for Bangladesh apparel in the USA and stimulated growth of the apparel industry (Bhattacharya and Rahman, 2000).

Table 16.2. Wage and productivity in the apparel industries of some selected countries (from Mlachila and Yang, 2004; World Bank 2005).

Country	Year	Wages and salary per employee (US$)	Value added per employee (US$)
Bangladesh	2003	2,500	700
	(1997)	(900)	(400)
China	2003	7,000	1,800
	(2001)	(5,000)	(1,600)
Hong Kong SAR	1999	27,600	14,800
India	1998	2,600	700
Indonesia	1999	2,500	600
Sri Lanka	1998	2,500	700

Another important stimulator of the growth of apparel in Bangladesh is the tariff and import quota-free access in the European Union (EU) under the Generalised System of Preference (GSP) and Everything but Arms (EBA) initiatives that contributed to the expansion of apparel export to the EU, provided that Bangladesh met the Rules of Origin (ROO) requirement. The GSP scheme allows EU importers to claim full tariff drawback on imports when they import from Bangladesh (Bhattacharya and Rahman, 2000). Table 16.3 shows that Bangladesh is relatively less restricted in both the EU and the USA than other exporting countries in Asia.

MFA import quota abolition and the apparel industry

Currently, abolition of MFA import quota is the main challenge faced by the apparel industry of Bangladesh. MFA import quota abolition has increased competition, and various supply-side structural rigidities are making it difficult to compete. Thus, the challenge originates as a result

Table 16.3. Export Tax Equivalents (ETEs)[a] of import quotas in selected countries, 2003 (from Mlachila and Yang, 2004).

Country/Economic unit	Textile		Apparel	
	USA	EU	USA	EU
Bangladesh	0.0	0.0	7.6	0.0
India	3.0	1.0	20.0	20.0
China	20.0	1.0	36.0	54.0
Pakistan	9.8	9.4	10.3	9.2

[a] Export tax equivalents (ETE) of quotas are expressed as a percentage of f.o.b. price (net of quota price); for detail on ETEs please see the web document by Joseph Francois and Dean Spinanger at http://www.gtap.agecon.purdue.edu/resources/download/723.pdf (accessed 9 April 2005).

of preference erosion following implementation of the WTO rules. This situation is illustrated in Fig. 16.2.

Figure 16.2 illustrates the effects on production and welfare of import quota abolition for a small exporting country. To simplify the analysis we assume that a country produces only for the export market. In the initial situation P_m is the world market price, Q the quota the country faces and P_s is the shadow price of supply. The shadow price of supply is the marginal cost of production, at which price the country is willing to produce exactly the quota level Q. Producers in this country receive the market price P_m, while their marginal cost of production equals P_s, and therefore the unit quota price equals $(P_m - P_s)$.

Suppose now that the import quotas are abolished and this country no longer faces a quantitative restriction on output. There are two possible situations. First, the new world market price is P_{s1}, which is higher than the initial marginal cost of production P_s. Note that, in a quota-free world, the world market price equals the new marginal cost of production. Moreover, the world market price is lower since world production increases because it is no longer constrained. Under this situation, although the world market price has decreased, the country increases its production to Q_1.

Under the alternative situation, the new world market price P_{s2} is lower than the initial marginal cost of production P_s. The country in that case lowers its production to Q_2. The more a country is constrained by

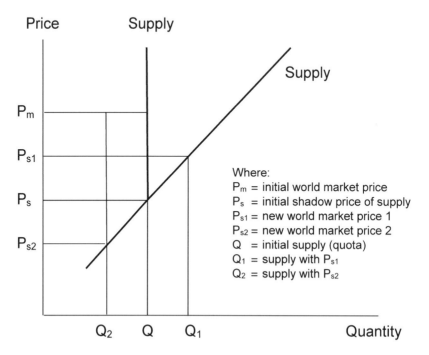

Where:
P_m = initial world market price
P_s = initial shadow price of supply
P_{s1} = new world market price 1
P_{s2} = new world market price 2
Q = initial supply (quota)
Q_1 = supply with P_{s1}
Q_2 = supply with P_{s2}

Fig. 16.2. Effects of import quota abolition on an exporting country.

the quota the lower the shadow price of supply and the more likely it will increase its production after quota abolition. Countries that are barely constrained by the quota have a relatively high shadow price of supply. After quota abolition it is more likely that they will reduce production.

In the first case, the producer surplus decreases with (a + b) and increases with e. In the second situation, producer surplus decreases with (a + b + c + d + f + g). Here, it is assumed that the initial quota rent (a + b + c + d) was received by the producers. This is, of course, not always true (the government may sell the quota rights). If the quota rent is received by buying houses or by other middlemen, then (a + b) actually constitutes a loss on the part of those middlemen. The welfare effects also depend on the effect of trade liberalization on input prices. If, for example, the price of textile as an input in apparel production falls, the supply curves shift outwards and there is an extra positive welfare effect for the producers.

In Bangladesh, after receiving the quota from importers, the Export Promotion Bureau (EPB) allocates the quotas to private traders. After this the quotas can be traded in an open market. Thus, at least a part of the quota rent goes to these private traders. Abolition of the import quotas increases competition and decreases the world market price. Table 16.3 shows that the Export Tax Equivalents (ETEs) of import quotas for Bangladesh were relatively low. This means that removal of import quotas might increase the export of apparel from Bangladesh's competitors, such as China and India, while Bangladeshi exports might decrease.

Competitiveness of Bangladeshi apparel

Impacts of MFA import quota abolition on a country will ultimately be determined by its relative cost of production. The f.o.b. (without quota cost) price of apparel can be considered as an indicator of this cost. Table 16.4 presents the f.o.b. price for one 180 g T-shirt (under the HS code 6109) in selected Asian countries as calculated by the World Bank (2005). The f.o.b. price in Bangladesh is higher than that in either India or China, but lower than that in Nepal.

Figure 16.3 shows that 78% of the total production cost is for materials, which mainly include imported cotton fabrics (93% of total material cost). The 12% administrative cost includes profits, which comprises 39% of the total administrative cost. Labour is utilized mostly at the sewing/assembly stage, and therefore 67% of sewing cost is spent on labour. The higher cost of production of Bangladesh's apparel products implies that, after import quota abolition, when all countries will be treated equal, the competitive position of Bangladesh may deteriorate.

Table 16.4. F.o.b. price comparison (from World Bank, 2005).

Country	F.o.b. price[a] of one 180 g T-shirt (US$)
Bangladesh	1.3
China	0.9
India	1.2–1.5
Nepal	2

[a] Price net of quota cost.

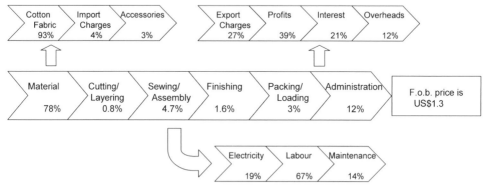

Fig. 16.3. Cost structure of producing one 180 g T-shirt in Bangladesh.

Bangladesh apparel faces various structural bottlenecks, which increase the cost of production. These bottlenecks include: (i) heavy dependence on imported inputs (especially in case of woven apparel); (ii) product and market concentration of apparel export; (iii) low labour productivity offsetting the advantage of low labour cost; and (iv) infrastructure bottlenecks (in terms of electricity supply, efficiency of ports, telecommunication facilities, corruption, high cost of finance, etc.). Compared with woven apparel, knitted apparel is less import dependent, has a higher labour productivity and produces higher-quality products. While 85% of the fabric used in woven apparel is imported, this comprises only 25% of the yarn and fabrics used in knitted apparel (World Bank, 2005). The reason for this is the higher investment cost for setting up factories producing fabrics for woven apparel, because of the type of machines and manufacturing technology required.

Model and Data

We apply version 5.1 of the GTAP database, which uses the year 1997 as the base (Dimaranan and McDougall, 2002). Some of the previous studies also applied GTAP to analyse the impacts of MFA import quota abolition, and some of those also concentrate on Bangladesh. However, these studies (excepting that of Lips *et al.*, 2003) have not considered (in their model) the tariff drawback incentive of the EU. To encounter the tariff drawback issue, this study follows the approach used by Lips *et al.* (2003).

Lips *et al.* (2003) assume full employment for all countries, where as unemployment among unskilled labour is evident in many developing countries. Therefore, the current study has incorporated unemployment in unskilled labour for the South Asian countries (i.e. SAFTA member countries) as well as for China. Thus, the wage is fixed and any number of unskilled labour can be employed at this wage, which implies welfare decreases if less labour is employed after a policy change (negative endowment effect, see Appendix 16.1). With a fixed labour amount and a labour price decrease, labour will be employed elsewhere in the economy and the endowment effect in this case is zero. In addition, the database of the model has been updated to incorporate preferential access of Bangladesh in the EU and Canada, which is absent in Lips *et al.* (2003). As mentioned earlier, Bangladesh receives tariff and import quota-free entry in the EU under the GSP scheme of the EU.[1] The other reality is the tariff and import quota-free entry of Bangladesh's textile and apparel in Canada.[2] For the used region and commodity aggregation see Appendix 16.2.

Scenarios

Using the updated data, the effects of the following three scenarios were determined: (i) S1: abolition of the MFA import quotas; (ii) S2: S1 plus EU enlargement (no import tariffs between the EU and Central and Eastern European Countries (CEEC), a preferential trade agreement between Turkey and the EU and China's WTO accession; and (iii) S3: S2 plus a worldwide import tariff reduction of 36% for textile and apparel.

Scenario S1 considers the implementation of the ATC. It refers to a complete abolition of the import quotas on textiles and apparel. The import quotas are modelled as export tariff equivalents (ETEs) in GTAP (Francois and Spinanger, 2002). In S1, abolishing import quotas means that the ETEs are eliminated. In addition to abolition of MFA import quotas, there are other issues that might potentially affect Bangladesh's apparel export. These are the enlargement of the EU, the preferential trade agreement between Turkey and the EU and China's accession to the WTO (scenario S2). WTO accession for China implies tariff reduction, liberalization of agricultural and service trade. Accession of China to the WTO implies an import tariff reduction for textile and apparel exported by China in order to respect the Most Favoured Nation (MFN) clause. Here, such import tariff reductions for two products have been considered – textile and apparel.

EU enlargement may affect Bangladesh's export, as the EU is the most important export market for Bangladesh and there are similarities between the apparel export items of Bangladesh and those of the CEECs

(Ahmed, 2001). Turkey also exports apparel to the EU and receives a preferential treatment from the EU. Enlargement of the EU and the preferential trade agreement between itself and Turkey may negatively affect the existing preferential position of Bangladesh within the EU. The geographical location of Bangladesh as compared with that of Turkey may lead to a relatively higher cost for the products of Bangladesh compared with that from Turkey, even if they have similar production cost.

The last scenario (S3) includes a worldwide import tariff reduction of 36% for textile and apparel, which is a possible outcome of the ongoing trade liberalization negotiations in the WTO. This final scenario also includes all elements of the other two scenarios.

Results

The results of the different scenarios are summarized in this section. As the main interest of this chapter is to look into the impacts of various scenarios on the apparel industry of Bangladesh compared with that of her main competitors, India and China, results addressing this matter receive prime attention. A summary picture of the results under different scenarios is presented in Table 16.5; detailed results can be obtained from the authors.

Results of scenario 1

When MFA import quotas are abolished the export price of apparel decreases. As noted in Fig. 16.2, the magnitude of the price decrease depends on how restrictive the initial import quotas were. The export price of apparel from India and China decreases much more than that of Bangladesh, as import quotas were more restrictive for India and China than for Bangladesh. For the same reason, exports of apparel from India and China increase while export of apparel from Bangladesh decreases by 9.6%. Although in percentages, the increase of Chinese export is less than that of India, the very large export base of China refers to a much larger export growth in value terms. Sri Lanka also faces an export decrease (–0.2%) while other South Asian countries export more (+0.9%).

In the two main markets of apparel, i.e. the EU and the USA, only India and China experience an export increase while all other regions (excepting Sri Lanka and the rest of South Asia in the EU) experience an export decrease. Even the Central American countries experience a 54% decrease in their export to the USA. Bangladesh's export decrease to the EU is higher, as the preferences under GSP become less important as other countries acquire import quota-free entry to this market. Production of apparel decreases in Bangladesh by 8.9%, whereas it increases by 111.6

Table 16.5. Summary of simulation results.

	S1	S2	S3
Change in production of apparel (%)			
Bangladesh	−8.9	−11.7	−13.9
India	111.6	108.4	119.9
China	32.2	30.3	42.9
Change in export of apparel (%)			
Bangladesh (base value US$4058.8)	−9.6	−12.5	−15.1
India (base value US$4171.8)	283.8	276.9	305.8
China (base value US$33659.3)	67.5	63.6	90.8
World (base value US$141897.8)	12.8	15.9	26.7
Market share in world apparel export (%)			
Bangladesh (base value 2.9%)	2.3	2.2	1.9
India (base value 2.9%)	10.0	9.6	9.4
China (base value 23.7%)	35.2	33.5	35.7
Change in aggregate export price of apparel (%)			
Bangladesh	−3.5	−4.5	−5.3
India	−20.7	−21.2	−20.9
China	−10.9	−11.3	−11.8
Change in employment of unskilled workers (%)			
Bangladesh	−1.3	−1.7	−0.6
India	1.8	1.7	1.9
China	1.4	1.3	2.2
Change in real GDP (%)			
Bangladesh	−0.7	−0.9	−0.5
India	1.1	1.1	1.2
China	0.9	0.9	1.4
Change in welfare (equivalent variation, US$ million)			
Bangladesh	−694.0	−884.3	−758.6
India	4,239.0	3,892.8	4,619.8
China	6,420.7	5,319.4	10,432.5
World	15,853.6	17,299.0	25,894.9

and 32.2% in India and China, respectively. Production of textile for all three countries increases. In terms of production, textile production is larger than that for apparel in Bangladesh, but this industry experiences a 0.9% production growth while apparel production falls by 8.9%.

Considering exports and imports of all Bangladesh's industries, it is observed that exports of all industries other than apparel increase while imports of all industries decrease. There is a 10.8% increase in the textile export while export of apparel falls by 9.6%. Export of another important product of Bangladesh, leather, increases by 23.9%. The aggregate export of Bangladesh decreases by 1.4% and aggregate import decreases by 5.7%. These results reflect the fixed balance of trade assumption. Exports of apparel fall, the exchange rate increases (which implies a devaluation of the Taka, the currency of Bangladesh), other exports increase and imports fall.

With the abolition of the MFA import quotas, ultimately Bangladesh faces a 0.7% drop in GDP and a 1.5% loss in per capita utility. India and China both gain in terms of GDP (1.1 and 0.9%, respectively) and per capita household utility (1.2 and 0.7%, respectively). The overall welfare loss of Bangladesh is US$694 million. Both India and China achieve welfare gains (US$4239 million and US$6420 million, respectively). The welfare loss of Bangladesh arises mainly from the negative endowment effect (less labour employed) and the terms of trade effect. The US$249 million negative endowment effect for Bangladesh is generated mostly by a 1.3% decrease in the unskilled labour employment. This decrease in unskilled labour employment is likely to affect mostly the poor households as the main owners of unskilled labour.

The terms of trade effect is negative for Bangladesh and all import quota-restricted exporters as the world price of apparel decreases. However, in India and China, such negative impacts are compensated by the increase in allocative efficiency and positive endowment effects (more labour employed). The negative allocative efficiency for Bangladesh occurs due to the fact that, as a result of ATC the production of the subsidized textile industry increases, leading to the allocation of resources in an economically inefficient industry.

Table 16.5 shows that world apparel export increases by 12.8%, but the share of Bangladesh in world apparel export decreases from 2.9% at base to 2.3%; the shares of both India and China increase. Ultimately, world welfare increases by US$15,853 million.

Results of scenario 2

MFA import quota abolition (S1), along with eastern EU enlargement, the preferential trade agreement between the EU and Turkey and WTO accession of China, results in an export price reduction for apparel of 4.5, 21.2 and 11.3% for Bangladesh, India and China (see Table 16.5), respectively. As in S1, the price of apparel from India and China decreases much more than that of Bangladesh. Export of apparel by India increases by 276.9% and that from China by 63.6%. Export of Bangladesh decreases by 12.5%. Compared with S1, the percentage increase of Indian and Chinese apparel export is lower in S2, as the CEEC and Turkey receive trade preferences from the EU. For the same reason, the relative position of Bangladesh deteriorates more than in S1, and the percentage decrease of apparel export is larger in S2.

While the CEEC and Turkey faced a decrease in aggregate apparel export (by 17.1 and 15.4%, respectively) in S1, they now experience a large increase in their export (by 68.2 and 55.1%, respectively). Sri Lanka also faces a larger export decrease than before (by 3.0% instead of 0.2%) and also all the other South Asian countries face an export decrease (by 4.1% instead of 0.9%). Apparel exports of Turkey and the CEEC to the EU increase (by 94.9 and 80.8%, respectively) but decrease to the USA (by 38.3 and 43.7%, respectively). However, export of apparel from Bangladesh to the EU decreases by a higher percentage point than export to the USA (by 15.8 and 1.0%, respectively). Compared with those in S1, Bangladesh's exports to the EU decrease more in the current scenario (by 15.8% instead of 10.1%), but the opposite is true for exports to the USA (−1.0% instead of −6.1%). India and China also experience a lower increase in their apparel export to the EU than in S1 (for India, 84.3 instead of 105.2% and, for China, 73.7 instead of 94.3%).

Production of apparel decreases in Bangladesh by 11.7%, whereas it increases by 108.4 and 30.3% in India and China, respectively. Production of textile by all three countries increases. Compared with that in S1, in S2 the production decrease in Bangladesh is larger and the production rise in India and China is smaller. Considering exports and imports of all industries of Bangladesh, it is observed again that exports of all industries other than apparel increase while imports of all industries decrease. There is a 12.3% increase in textile export, while export of leather increases by 32.5%. The aggregate export of Bangladesh decreases by 1.9% and aggregate import decreases by 7.5%. These results reflect the fixed balance of trade assumption. Exports of apparel fall, the exchange rate increases, other exports increase and imports fall.

Implementation of S2 results in a 0.9% GDP loss and 1.9% per capita household utility loss for Bangladesh. India and China both gain in terms of GDP and per capita household utility, although the increase is slightly less than that in S1. The CEEC and Turkey also experience GDP (0.7 and 0.1%, respectively) and utility (1.8 and 0.4%, respectively) gains. The overall welfare loss of Bangladesh is US$884 million, which is higher than that in S1 (US$694.0 million). Welfare gains of both India and China are also lower than in S1. Where the CEEC and Turkey were facing a welfare loss in S1, they are facing a welfare gain in the current scenario. The welfare loss of Bangladesh arises mainly from the negative endowment effect (less labour employed) and terms of trade effect. Further investigation reveals that the US$330 million negative endowment effect for Bangladesh is generated mostly by a 1.7% decrease in the unskilled labour employment. As in S1, a decrease in unskilled labour employment is likely to affect poor households most.

As in S1, the terms of trade effect is negative for Bangladesh and all import quota-restricted exporters as the world price of apparel decreases. Again, as in S1 the negative terms of trade effects for India and China are compensated by the increase in allocative efficiency and the positive endowment effects. In Bangladesh, however, allocation of more resources in the inefficient textile industry results in negative allocative efficiency. Welfare of the world increases by US$17299 million.

Table 16.5 shows that world apparel export increases in this scenario by 15.9%, but the share of Bangladesh in world apparel export decreases from 2.9% in the base to 2.2%, which is lower than that in S1. Shares of India and China both increase, but less so than in S1.

Results of scenario 3

In this scenario, S2 is accompanied by a reduction in worldwide import tariffs of textile and apparel. Scenario 3 results in an export price reduction for apparel of 5.3, 20.9 and 11.8% for Bangladesh, India and China, respectively (see Table 16.5). As in S2, the price of apparel from India and China decreases much more than that of Bangladesh. Export of apparel of India increases by almost 305.8% and that of China by 90.8%, with a decrease for Bangladesh of 15.1%. Compared with the first two scenarios, the percentage increase of Indian and Chinese apparel export is highest in S3, as these two countries were greatly restricted in the base. For Bangladesh, the percentage decrease of apparel export is highest in S3, and export to the EU falls by 24.8%.

As Bangladesh enjoyed tariff-free entry to the EU market (GSP scheme), the relative position of Bangladesh deteriorates when the EU reduces import tariffs for all countries, including the competitors of Bangladesh. The reduction of the USA import tariff leads to an increase of exports of Bangladesh to the USA with 12% even after the MFA import quota abolition. India and China also experience an increase in their apparel export to the USA (808.0 and 240.3%, respectively).

Production of apparel decreases in Bangladesh by 13.9%, whereas it increases by 119.9% and 42.9% in India and China, respectively (see Table 16.5). Taken together, import quota abolition and import tariff reduction result in a higher production decrease for Bangladesh compared with that in the other two scenarios. India and China, however, increase production most compared with that in the other two scenarios. Considering exports and imports of all industries of Bangladesh, it is observed again that exports of all industries other than apparel increase while imports of all industries decrease. There is a 13.8% increase in textile export, and export of leather increases by 39.0%. The aggregate export of Bangladesh decreases by 2.4% and the aggregate import decreases by 8.2%. These results reflect the fixed balance of trade assumption. Exports of apparel fall, the exchange rate increases, other exports increase and imports fall.

Scenario 3 results in a 0.5% GDP loss and 1.6% per capita household utility loss for Bangladesh. India and China both gain in terms of GDP (1.2 and 1.4%, respectively) and per capita household utility (1.3 and 1.2%, respectively), and these increases are higher than those in the other two scenarios. The overall welfare loss for Bangladesh is US$ 758.6 million. The welfare loss for Bangladesh arises mainly out of the negative endowment effect (less labour employed) and terms of trade effect. The US$ 119.4 million endowment effect for Bangladesh is generated mostly by a 0.6% decrease in the unskilled labour employment. The decrease in labour employment is lower in S3 compared with that in the other two scenarios but, nevertheless, this loss will negatively affect the poor households who own most unskilled labour. Negative terms of trade impacts for India and China are compensated by the increase in allocative efficiency and positive endowment effects.

Table 16.5 shows that world apparel export increases in S3 by 25.7%, which is higher than that in the other two scenarios. The share of Bangladesh in world apparel export decreases from 2.9% in the base to 1.9%. Shares of India and China both increase, but less so than in S2. Ultimately, world welfare increases by US$25895 million in this scenario.

Discussion

Comparison of findings from various studies

Possible consequences of MFA import quota abolition have been analysed in a number of studies. Yang *et al.* (1997) have mentioned that MFA import quota abolition will result in a large increase in textile and apparel export from South Asia. However, they also mentioned that there is a possibility that textile and apparel exporters, with large import quotas relative to their production potential, may suffer from the MFA import quota abolition because of the increased competition from more efficient producers and the deterioration in their terms of trade. The CGE analysis of Islam (2001) also suggests that MFA import quota abolition would strengthen the competitive positions of the Association of Southeast Nations (ASEAN), China, and South Asia in apparel at the expense of industrially developed regions such as the EU, North America and Japan, and of marginally competitors from Latin America and the rest of the world. It is predicted that Bangladesh will not only face more competition from India and China, but that there also may be a shortage in fabric supply in the post-MFA period, as exporters of fabrics might use their fabrics for own increased production and export of apparel.

One study by Mlachila and Yang (2004) has estimated that, depending on different assumptions, MFA import quota abolition may result in a decrease in exports of apparel from Bangladesh by 6.2 to 17.7%. They have also predicted a 2.3% GDP loss for Bangladesh. Spinanger (2004, cited in Razzaque, 2005) has looked into the impacts of MFA import quota abolition and greater access of China to the world market and has noted that, taken together, these two scenarios may cause apparel export from Bangladesh to decrease by 7.9% and result in a GDP loss of 0.1%. The study by Nordas (2004) does not find an absolute decrease in total export from Bangladesh, but finds a decrease in the market share of Bangladesh's apparel in the USA. His study has forecasted a 2-percentage point decrease in the market share of Bangladesh's apparel in the USA (from the current share of 4 to 2%), with a 1-percentage point increase in the EU (from the current share of 3 to 4%). The market share of China's apparel in the EU has been forecast to increase from 18 to 29% and in the USA from 16 to 50%.

This current study has also found a decrease in the market share of Bangladesh's apparel and an increase in the market share of China's apparel in the world market as a whole.

The study of Lips *et al.* (2003), the framework followed in the current study, has estimated that elimination of the MFA import quotas will lead to a decrease in the Bangladesh apparel export of 13%. Comparing findings of some of the above-mentioned studies (see Table 16.6) shows that these studies mostly predict a decrease in apparel export from Bangladesh, although the results vary under different assumptions and scenarios.

Effects on the poor

GTAP does not permit investigation of the poverty effects at the disaggregated household level. Winters (2005) has noted that, in order to discover the impacts of trade liberalization on poverty, one has to look into (among others) the income sources and expenditure patterns of the poor. Following this notion, some indicators are reported in Table 16.7 to establish estimations of possible poverty consequences of MFA import quota abolition.

It is observed that factor return decreases for both land and unskilled labour. Although the GTAP database does not include disaggregated

Table 16.6. Possible consequences of MFA abolition on Bangladesh: comparison of different study findings,[a] percentage change from the base.

Indicator	Current study	Spinanger (2000)	Lips *et al.* (2003)	Mlachila and Yang (2004)
Export of apparel	−9.6 to −15.1	+1.5 to −7.9	−13.0	−6.2 to −17.7
GDP	−0.5 to −0.9	−0.14 to −0.54	nm	−0.3 to −3.7
Total export	−1.4 to −2.4	+2.8 to −0.1	nm	−3.0 to −14.2

nm, not mentioned.
[a] All the above studies use GTAP database 5.

Table 16.7. Poverty implications for Bangladesh, percentage change from the base.

Indicator	S1	S2	S3
Return for land	−1.9	−3.1	−2.5
Return for unskilled labour	−2.4	−3.5	−5.0
Private consumer price for rice	−2.4	−3.6	−4.2
Private consumer price for food	−1.7	−2.6	−3.0
Private consumer price for grains	−1.7	−2.6	−3.0
Regional household income	−3.8	−5.3	−6.4

household data, the 1999–2000 social accounting matrix of Bangladesh (Arndt *et al.*, 2002) points out that poor households (agricultural and non-agricultural) in Bangladesh own around 36% of the total land area. Moreover, around 75% of unskilled labours belong to poor households living in rural and urban areas. Using this information, we may conclude that income of the poor is decreasing in all scenarios. The current study has shown that employment of unskilled labour is decreasing. However, Table 16.7 shows that prices of various food items are decreasing, which may be beneficial for the poor. According to Winters *et al.* (2004), however, it may be questioned whether lower consumer prices cascade to the poor or not. What is more obvious from the results is that overall income is decreasing for all households, which indicates that, not only is the income of the poor decreasing but also the income transfer capacity of the rich people is decreasing, and this may contribute to the increase in poverty.

Apparel employs mostly unskilled workers, and one-quarter of these workers send remittances to rural households from where they have emerged (Afsar, 2001). We may infer that the remittance flow may decrease with the decrease in income of unskilled workers. Thus, there is a possibility that rural household income will decrease, especially for the households from where apparel workers originate.

Conclusions

The current study analyses various implications of MFA import quota abolition on Bangladesh, using GTAP. The study considered import quota abolition under three scenarios. The scenarios considered were: (i) abolition of the MFA import quotas; (ii) EU enlargement, preferential trade agreement between Turkey and the EU and China's WTO accession; and (iii) a worldwide import tariff reduction of 36% for textile and apparel.

Results show that abolition of MFA import quotas will decrease the world market price of apparel, while increasing the world export of apparel and world welfare in all scenarios. However, for some regions, apparel export and welfare will fall. For Bangladesh, production and export of apparel decrease in every scenario. In the base, Bangladesh was enjoying import tariff and import quota preferences in some important markets like the EU and Canada. Additionally, in the USA the import quotas were less restrictive for Bangladesh as compared with her competitors. As a result of abolition of the MFA import quotas, and in other cases reduction of import tariffs (worldwide or region specific), the preferential position of Bangladesh ultimately deteriorates, and this decreases the export and production of apparel.

Depending on various scenarios, apparel export of Bangladesh decreases by 9.6–15.1%, while production of apparel decreases by 8.9–13.9%. India and China are found to acquire the largest gains from liberalization of world trade in textile and apparel. Abolition of the very restrictive import quotas for them leads to a large increase in their apparel production and exports. The study has also indicated that MFA import quota abolition has negative impacts on poverty and rural household income in Bangladesh.

Results of the current study depend on the assumptions made in the GTAP model. For example, GTAP is a static CGE model. Therefore, this study cannot analyse the dynamic effects for different regions from MFA import quota abolition and import tariff reduction. The level of aggregation in GTAP of commodities (industries) and households is high. This implies that, for individual commodities (industries) and households, results could be different. GTAP also assumes competitive markets (e.g. no market power) and does not take into account transaction costs and institutional constraints. Moreover it was not possible to incorporate all realities in the model. For example, in updating the base there could be a possible overvaluation of EU's tariff preference for Bangladesh, as the EU's Rules of Origin (ROO) requirement often constrains Bangladesh's apparel export to avail tariff-free access in the EU. Therefore, not considering the ROO constraint, the magnitude of the deterioration of the relative position of Bangladesh in the EU under different simulations could be less.

In addition, the GTAP model and data did not allow one to distinguish between woven apparel and knitted apparel, although this could be relevant for Bangladesh. Nevertheless, global CGE models like GTAP are the best possible way to cover interactions between domestic and international trade activities for different regions of the world as a whole.

Looking at the actual export performance of Bangladesh in the USA after abolition of the MFA import quota (US Department of Commerce, Office of Textiles and Apparel, 2005), it is observed that there has been a 20.4% growth in apparel export from Bangladesh to the USA during the period January–August 2005 as compared with the same period in 2004. However, exports of some apparel products fell, e.g. cotton underwear, coats for women, etc., but it is too early yet to draw conclusions.

Even with the limitations of GTAP the study indicates that, in the present import quota-free world, Bangladesh's apparel is exposed to greater competition. This requires Bangladesh to improve the competitiveness of the apparel industry by removing structural weaknesses like poor infrastructure. Moreover, product diversification can reduce the risk of losing markets by increasing competition.

Appendix 16.1: The GTAP Model

GTAP is a comparative static, global CGE model. It assumes that all markets are perfectly competitive, and constant returns to scale prevail in all production and trade activities. In the model, firms maximize their profits and households maximise their utility.

Income and expenditure

Each region has a single representative regional household. The regional household income is generated through: (i) factor income; (ii) output flow from producers to the regional household; and (iii) tax flow from the private household, producers and government (including export and import taxes) net of subsidies. The regional household allocates expenditure over private household expenditure, government expenditure and savings according to a Cobb Douglas per capita utility function. Thus, each component of final demand maintains a constant share of total regional income.

The constrained optimizing behaviour of the private household is represented in GTAP by a CDE (Constant Difference of Elasticity) implicit expenditure function. Both private households and government spend their incomes on consumption of both domestic and imported commodities and pay taxes. Here, consumption is a CES composition of domestically produced goods and imports.

Production technology

A nested production technology is considered, with the assumption that every industry produces a single output and constant returns to scale prevail in all markets. The model employs the Armington assumption, which enables the distinguishing of imports by their origin and explains intra-industry trade of similar products. Producers sell their products to domestic private households and government as consumption goods, to the rest of the world (exports), to other domestic producers (intermediate inputs) and to the savings sector (as investment goods).

Welfare measurement

In GTAP, welfare effects are measured using the equivalent variation. GTAP decomposes the welfare effects into: (i) changes in allocative efficiency; (ii) changes in the country's terms of trade; (iii) changes in factor endowments; and (iv) changes in technology. A welfare change may also occur from changes in foreign income flows. Allocative efficiency refers to the efficient allocation of scarce resources. The terms of trade effect increases with a relative increase in the price of exports as compared with that of imports. The factor endowment effect is a measure of how much countries gain due to an increase in employment of land, labour or capital. A welfare gain from technological change occurs as a result of increased productivity.

Appendix 16.2: Region and Commodity Aggregation

The current study follows the commodity and region aggregation of Lips *et al.* (2003):

Aggregated regions	Comprising regions
Bangladesh	Bangladesh
India	India
Sri Lanka	Sri Lanka
rSAFTA	Rest of South Asian Free Trade Association (Bhutan, Maldives, Nepal, Pakistan)
China	China and Hong Kong
hASIA	Japan, Korea, Singapore, Taiwan
oASIA	Indonesia, Malaysia, Philippines, Thailand, Vietnam
EU	EU-15
CEEC	Hungary, Poland, rest of Central and Eastern European Countries
Turkey	Turkey
USA	USA
Canada	Canada
cAMERIKA	Mexico, Central America and Caribbean
ROW	Rest of the world

Aggregated commodities	Comprising commodities
Rice	Paddy rice and processed rice
Grains	Non-rice grains
Fibres	Plant-based fibres
rAGR	Rest of agriculture (oilseeds, sugarbeet, cattle, pigs and poultry, milk)
Food	Processed food without processed rice
Textiles	Textile
Apparel	Apparel
Leather products	Leather products
Extract	Fishing, forestry, coal, oil, gas, minerals
LiMANF	Labour-intensive manufactures
CiMANF	Capital-intensive manufactures
Services	Services

Endnotes

¹ The Rules of Origin requirements are not incorporated, as the modelling framework does not permit this.

² WTO document No. WT/COMTD/N/15/Add.1(2003). Although Canada gives this preference to all LDCs, the aggregation of the current study does not permit incorporation for others. Moreover, this study considers only tariff and import quota-free entry of textile and apparel, as only few other categories of products are covered by this concession.

References

Afsar, R. (2001) Sociological implications of female labour migration in Bangladesh. In: Sobhan, R. and Khundker, N. (eds) *Globalisation and Gender: Changing Patterns of Women's Employment in Bangladesh*. Centre for Policy Dialogue and University Press Limited, Dhaka, pp. 91–165.

Ahmed, N. (2001) *Trade Diversion Due to the Europe Agreements: Should Bangladesh Care?* Research Report 171, Bangladesh Institute of Development Studies, Dhaka.

Arndt, C., Dorosh, P., Fontana, M. and Zohir, S. with El-Said, M. and Lungren, C. (2002) Opportunities and challenges. In: *Agriculture and Garments: a General Equilibrium Analysis of the Bangladesh Economy*. TMD Discussion Paper 107, International Food Policy Research Institute, Washington, DC.

Bangladesh Bank (2005) http://www.bangladesh-bank.org/econdata/exprtrec.html (accessed 17 February 2005).

Bangladesh Bureau of Statistics (1982) *Statistical Yearbook of Bangladesh*, Dhaka.

Bhattacharya, D. and Rahman, M. (2000) *Experience with Implementation of WTO-ATC and Implications for Bangladesh*. CPD Occasional Paper Series No. 7, Centre for Policy Dialogue, Dhaka.

Bhattacharya, D. and Rahman, M. (2005) Bangladesh's apparel sector: growth trends and the post-MFA challenges. Paper presented at the seminar on *Growth of Garments Industry in Bangladesh: Economic and Social Dimensions*, organized by BIDS and OXFAM, 21–22 January, Dhaka.

CPD (2001) *Policy Brief on Industry and Trade*. CPD Task Force Report, Centre for Policy Dialogue, Dhaka.

Dimaranan, B.V. and McDougall, R.A. (2002) *The GTAP 5 Data Base*. Center for Global Trade Analysis, Purdue University, West Lafayette, Indiana.

Francois, J. and Spinanger, D. (eds) (2002) *ATC Export Tax Equivalents*. Center for Global Trade Analysis, Purdue University, West Lafayette, Indiana.

Hertel, T.W. (1997) *Global Trade Analysis: Modeling and Applications*. Cambridge University Press, New York.

Islam, S. (2001) *The Textile and Clothing Industry of Bangladesh in a Changing World Economy*. Centre For Policy Dialogue and the University Press Limited, Dhaka.

Lips, M., Tabeau, A., Van Tongeren, F., Ahmed, N. and Herok, C. (2003) Textile and wearing apparel sector liberalization – consequences for the Bangladesh economy. Paper presented at the *6th Conference on Global Economic Analysis*, The Hague, The Netherlands.

Mlachila, M. and Yang, Y. (2004) The End of Textile Quotas: a Case Study of the Impact on Bangladesh. IMF Working Paper, WP/04/108, International Monetary Fund, Washington, DC.

Nordas, H.K. (2004) *The Global Textile and Clothing Industry post the Agreement on Textiles and Clothing.* WTO Discussion Paper, World Trade Organization, Geneva, Switzerland.

Razzaque, A. (2005) *Sustaining RMG Export Growth after MFA Phase-out: an Analysis of Relevant Issues with Reference to Trade and Human Development.* Study conducted for the Ministry of Commerce, Government of Bangladesh and United Nations Development Programme (mimeo).

Spinanger, D. (2000) *The WTO and Textile and Clothing in a Global Perspective: what's in it for Bagladesh?* Centre for Policy Dialogue, Dhaka.

Winters, L.A. (2005) The European agricultural trade policies and poverty. *European Review of Agricultural Economics* 32, 319–346.

Winters, L.A., McCulloch, N. and McKay, A. (2004) Trade liberalization and poverty: the evidence so far. *Journal of Economic Literature* XLII March, 72–115.

World Bank (2005) *Bangladesh: Growth and Export Competitiveness.* Dhaka.

Yang, Y., Martin, W. and Yanagishima, K. (1997) Evaluating the benefits of abolishing the MFA in the Uruguay Round Package. In: Hertel, T.W. (ed.) *Global Trade Analysis: Modeling and Applications.* Cambridge University Press, Cambridge, UK, pp. 271–298.

V Strategies and Policy Priorities

17 Poverty Targeting with Heterogeneous Endowments: a Micro-simulation Analysis of a Less-favoured Ethiopian Village

MARIJKE KUIPER[1] AND RUERD RUBEN[2]

[1]International Trade and Development, Public Issues Division, Agricultural Economics Research Institute (LEI) – Wageningen UR, PO Box 39703, 2502 LS, The Hague, The Netherlands; telephone: +31 70 335 8194; fax: +31 70 335 8196; e-mail: marijke.kuiper@wur.nl; [2]Centre for International Development Issues Nijmegen (CIDIN), Radboud University Nijmegen, PO Box 9104, 6500 HE Nijmegen, The Netherlands; e-mail: R.Ruben@maw.ru.nl

Abstract

Spatially targeted programmes for poverty reduction in less-favoured areas (LFAs) are typically constrained by a large heterogeneity amongst households in terms of the quantity and quality of available resources. The objective of this chapter is to explore, in a stylized manner, the role of heterogeneous household endowments for: (i) policies aimed at poverty reduction; (ii) within-village income inequality; and (iii) resource degradation.

Using a micro-simulation model, we analyse for each household in a remote Ethiopian village three sets of policies commonly put forward to reduce poverty: technology improvement, infrastructure investment and off-farm employment through migration or cash for work programmes.

In the analysis of single policies, migration was found to produce the largest decrease in poverty headcount. Because of self-selection, cash-for-work (CFW) programmes performed best in terms of reaching the poorest of the poor. This policy also results in the largest reduction of within-village income inequality, while a reduction in price band increases income inequality. Richer households buy more consumer goods and thus benefit more from reduced consumption good prices.

Only in the case of technology improvements was a trade-off between poverty reduction and soil erosion found. Price band and non-farm policies, however, lead to (sometimes considerable) reductions in erosion, while also having a better performance in terms of poverty reduction than technology improvements.

Analysing the relation between assets and income, ownership of oxen was found to be crucial: households with no oxen are below the US$1/day poverty line, households above the US$2/day poverty line own oxen. Oxen ownership does not fully determine income: in addition to oxen, sufficient land (in terms of quantity and quality) is needed to escape poverty.

Analysing combinations of policies, we find that combining policies helps poorer households overcome the limitations of their asset endowments. This complementarity of policies is less important for better-endowed households. As a result, combining complementary policies helps in targeting the poorest households, reducing income inequalities. Combining a CFW programme with a reduction in price bands yields most in terms of poverty reduction and income inequality.

In terms of the effect on soil erosion, the combination of a reduction in fertilizer prices with improved technologies yielded unexpected interaction effects. Changed relative prices of inputs affect the choice of technology such that erosion levels increase, as opposed to decrease, as was expected based on the impact of single policies.

Introduction

Rural households living in less-favoured areas (LFAs) represent, globally, around one-third of the chronic poor that are difficult to reach with standard programmes for poverty alleviation (Ruben and Pender, 2004; Hazell *et al.*, 2005). Persistent poverty in LFAs is usually attributed to a fragile natural resource base (compared with population density) and poor market linkages, leading to structural low factor returns and reducing the effectiveness of agrarian policies. Whereas broad spatial targeting, based on access and distance criteria, can be effective for reducing absolute poverty due to adverse geographical conditions (Ravallion and Jalan, 1996), a large part of the variation in household income can be attributed to within-village differences in initial resource endowments (Jayne *et al.*, 2003; Elbers *et al.*, 2004).

Heterogeneity in wealth and asset stocks, education levels, engagement in social networks and risk attitudes will result in a selection of activity portfolios with different potential for escaping poverty (Barrett *et al.*, 2004). Spatially targeted programmes for poverty reduction in LFAs are thus typically constrained by the large heterogeneity amongst households, in terms of the quantity and quality of available resources and related differences in transaction costs and risk behaviour, which tends to inhibit a 'one-size-fits-all' policy.

Most of the poverty prevailing in LFAs has been characterized as 'asset poverty', where rural households possess limited resources (land, labour or cattle) to cope with adverse events and are extremely vulnerable to shocks that could lead to an irreversible breakdown of their asset base (Carter and Barrett, 2006). Under these conditions, farmers typically engage in low-return activities and try to diversify their activity portfolio in order to be able to deal with unexpected income shortfalls. However, asset inequality leads to the selection of low-risk portfolios and the maintenance of reserves by poorer households that generally offer lower returns (Zimmerman and Carter, 2003). Overcoming such poverty traps is possible only when minimum-asset thresholds can be reached that enable households to become engaged in better-remunerated activities and to engineer a self-sustaining pathway to growth (Lybbert *et al.*, 2004).

Complementarities between social and physical assets, and between private and public assets, may be responsible for different patterns of income growth (Escobal, 2005). In the absence of certain minimum levels of public investment, returns to private investment in agricultural production or education tend to be structurally low. Moreover, differences in the distribution of assets will also affect the rate of returns to investments, thus reinforcing the tendency towards income inequality (Van de Walle and Gunewardena, 2001). It is therefore of critical importance to identify the right mix and sequence of policy interventions for LFAs that enable the rural poor to overcome binding thresholds and benefit from the opportunities provided. Programmes for raising agricultural productivity of food and cash crops can thus be expected to be more effective for poverty alleviation when, simultaneously, price bands are decreased (Ruben *et al.*, 2000). In a similar vein, poor households might become more engaged in off-farm employment when price bands on local food and factor markets are reduced (De Janvry *et al.*, 1991).[1]

In the current debate on poverty targeting towards LFAs, specific attention is given to two key factors that determine the effectiveness of policy interventions: (i) the role of *initial conditions*; and (ii) the impact of *asset complementarities*. Both factors have a decisive impact on farm households' returns to investments, but could also influence income distribution within villages. Heterogeneous asset endowments are likely to reinforce income inequality and could easily lead to biased policies, where the better-off households capture the lion's share of the benefits (Lipton, 2005). Policies focusing on improved agricultural technologies are likely to benefit mainly land-based households (thus excluding the landless[2]), and tend to replicate existing inequality in asset ownership.

Otherwise, programmes of infrastructure provision that reduce price bands will reach net-buying and net-selling households (but leave self-sufficient producers unaffected), whereas programmes for creating local employment through food-for-work activities have the potential to involve a large number of intended beneficiaries while relying on self-selection mechanisms to guarantee preferential access for the poorest households.

Development strategies for poverty alleviation in LFAs should also consider the pattern of interaction between distribution and growth.

Cross-country evidence consistently shows that inequality of income and unequal distribution of assets lead to a reduction in growth rates (Birdsall and Londoño, 1998; Deininger and Squire, 1998; Keefer and Knack, 2002). Within LFAs, it is likely that similar mechanisms are in force. Policies aimed at reducing both poverty levels and income equality may thus be expected to contribute most to sustained economic development in LFAs. Single-policy incentives are usually not capable of influencing both processes. Complementary interventions that enable rural households to intensify their resource use and, at the same time, reduce market transaction costs and risks, might offer better perspectives for pursuing such a win-win strategy.

Targeting poor households thus requires insight into minimum asset thresholds and complementarities among assets and policies. Income distribution is also a matter of concern, with a view to future development. Furthermore, poverty in LFAs is often attributed to a fragile natural resource base, which may limit the scope for reducing poverty by intensifying agricultural production. The objective of this chapter is to explore, in a stylized manner, the role of heterogeneous household endowments when assessing the impact of policies in terms of: (i) poverty reduction; (ii) within-village income inequality; and (iii) resource degradation.

Using a micro-simulation model, we analyse for each household in a remote Ethiopian village three key elements of policies commonly put forward to reduce poverty: (i) technology improvement; (ii) infrastructure investment; and (iii) off-farm employment through migration or local government programmes. With a simulation approach we can analyse policies separately, as well as in combination. This allows us to gauge whether combined policies can overcome limitations posed by household endowments. By using a micro-simulation approach we maintain household heterogeneity, allowing us to assess the impact of policies on income inequality. By using technical data on input and output data we are able to assess the impact of agricultural practices on erosion levels (a primary concern in the case study village).

We start by introducing the micro-simulation model and the underlying data. We then discuss results of having single policies on household income, followed by an analysis of the role of household assets in the poverty impacts of policies. We then proceed by analysing the impact of combined policies on poverty, income distribution and erosion. The last section concludes.

A Stylized Micro-simulation Farm Household Model

The key characteristic of micro-simulation models is the use of actual survey observations. This allows one to include household heterogeneity missed by models using representative households. The origins of micro-simulation methods are traced to Orcutt (1957). Despite these early origins of micro-simulation models, representative agent models still dominate the literature. Apart from the micro-simulation models developed in the 1970s for analysing the distributive effects of taxes and welfare schemes, the idea of using individual observations in simulation models did not catch on until poverty studies in the late 1990s (Cogneau and Robilliard, 2000).

Focusing on heterogeneous resource endowments and income distribution, household heterogeneity is central to our analysis and we thus employ a micro-simulation model. We formulate a standard farm household model in the tradition of Sing *et al.* (1986). Households maximize utility:

$$U_h = \sum_j \mu_j \ln QC_{hj}, \qquad\qquad \forall\ h, j \in C \qquad (17.1)$$

where U_h is household utility, μ_j is the budget share of good j, QC_{hj} is the quantity consumed of good j and J is the set of all commodities in the model of which C is a subset of consumed commodities (Appendix I contains a list of commodities included in the case study model and their set membership). We use a Cobb-Douglas utility function for reasons of tractability.

Utility maximization is, first of all, constraint by the available production technologies, described by Leontief technologies that make full use of technical data available for the case study area. For each activity (a) we define input use and output supplied in relation to an activity level ($QA_{h,a,tech}$):

$$QI_{h,j} = \sum_a \sum_{tech} \alpha_{a,j,tech} QA_{h,a,tech} \qquad\qquad \forall\ h, j \in I \qquad (17.2)$$

$$QO_{h,j} = \sum_a \sum_{tech} \beta_{a,j,tech} QA_{h,a,tech} \qquad\qquad \forall\ h, j \in O. \qquad (17.3)$$

In the case of crop activities, $QA_{h,a,tech}$ refers to the cultivated area; in the case of livestock activities, to the number of animals. Activities can be performed with different technologies (*tech*). For crop activities, technology consists of a combination of soil types, technology (choices regarding the use of fertilizer and type of traction) and levels of intensity. In the case of livestock activities there is only a choice in energy intake. The input ($\alpha_{a,j,tech}$) and output ($\beta_{a,j,tech}$) coefficients determine input demanded ($QI_{h,j}$) and output supplied ($QO_{h,j}$) of each crop and livestock activity.

Next to crop output, each crop technology has an associated level of erosion. Given the activities chosen by the households, we can thus compute the total amount of erosion at village level (E):

$$E = \sum_{h} \sum_{a} \sum_{tech} \varepsilon_{a,j,tech} QA_{h,a,tech} \qquad\qquad (17.4)$$

to assess the impact of policies on soil erosion, in addition to analysing the income effects of policies.

Next to crop and livestock activities, households can also engage in off-farm activities (*off*), in the case study consisting of CFW or migration. These activities require only labour as input and are limited by an upper bound (*offav*$_{h,labor,off}$):

$$QIoff_{h,labor} = \sum_{off} QOoff_{h,off} \qquad\qquad \forall\, h \qquad (17.5)$$

$$QOoff_{h,labor,off} \leq offav_{h,labor,off} \qquad\qquad \forall\, h,,off. \qquad (17.6)$$

This specification allows us to account for segmented labour markets (with different wages earned in CFW and migration) that limit the access of households to off-farm employment.

Household utility maximization is further constrained by commodity balances. How these affect household decision making depends on the level of tradability. Goods that are household non-tradable (*HNT*) cannot be bought or sold. Household demand thus needs to satisfy household supply:

$$QC_{h,j} + QI_{h,j} + QIoff_{h,j} \leq QO_{h,j} + \omega_{h,j} \qquad\qquad \forall\, h, j \in HNT \qquad (17.7)$$

where, in addition to the variables defined before, $QC_{h,j}$ is household consumption and $\omega_{h,j}$ are household endowments. We include the amount of labour used in off-farm activities in the commodity balance for household non-tradables since, in the case study, households cannot hire-in labour. The available labour is thus determined by household endowments (additional labour cannot be produced). Through the off-farm activities defined above, households can hire-out labour to earn off-farm income (CFW or migration). This is accounted for in the cash constraint below.

In the case of goods that can be bought or sold, two different types of commodities are distinguished. There are goods that are traded only locally, for which there are no transaction costs driving a wedge between buying and selling prices. These are household tradables and village non-tradables (*VNT*) with the following commodity balance:

$$QC_{h,j} + QI_{h,j} + QMS_{h,j} \leq QO_{h,j} + \omega_{h,j} \qquad\qquad \forall\, h, j \in VNT \qquad (17.8)$$

where $QMS_{h,j}$ is the marketed surplus of village non-tradables.

There are also goods that are traded outside of the village for which the households do incur transaction costs. Buying prices of these village tradables (*VT*) then exceed selling prices, and we need to distinguish sold purchased ($QP_{h,j}$) from sold commodities ($QS_{h,j}$) in the commodity balance:

$$QC_{h,j} + QI_{h,j} + QS_{h,j} \leq QO_{h,j} + QP_{h,j} + \omega_{h,j} \qquad\qquad \forall\, h, j \in VT \qquad (17.9)$$

Since household tradables can be bought and sold by the household, the commodity balances do not affect the utility maximization. Instead, it is the associated cash constraint that restricts utility maximization:

$$\sum_{j \in vnt} p_j QMS_{h,j} + \sum_{j \in vt} ps_j QS_{h,j} + \sum_{off} Y_{h,off} \leq \sum_{j \in vt} pp_j QP_{h,j} \qquad \forall\, h \qquad (17.10)$$

with

$$pp_j > ps_j \qquad\qquad\qquad\qquad \forall\, j \in VT$$

Village non-tradables have a single price, p_j. The village tradables have a price $(ps_{h,j})$ received when selling and a price $(pp_{h,j})$ paid when buying.

In addition to crop and livestock activities, households can earn income by engaging in off-farm activities $(Y_{h,off})$:

$$Y_{h,off} \leq \sum_j w_{j,off} QOoff_{h,j,off} \qquad\qquad \forall\, h, off \qquad (17.11)$$

where $w_{j,off}$ are wages differing by type of off-farm employment.

The stylized household model limits the variation across households to variation in household resource endowments. All households are thus assumed to have access to the same technologies and holding identical preferences (as can be seen from the absence of a household index in the production and utility functions). Restricting heterogeneity to household endowments implies that our assessment of the impact of household heterogeneity on policy impacts can be taken as a lower bound estimate. Introducing other sources of heterogeneity, like differences in household preferences, would lead to an even more diverse impact of policies. We furthermore use a static household model in which households are not saving or investing in new assets.[3]

Despite the static nature of the model, we can explore the impact of asset thresholds by relying on the variation in endowments across households. This gives an indication of the asset accumulation needed by poor households in order to gain access to higher-yielding activities. It furthermore allows us to explore whether (combinations) of policies can overcome the restrictions of limited assets. Finally, we do include the possibility of locally traded goods, renting of land and animal traction within a village by defining village non-tradables (*VNT*). We do not, however, account for price formation in local markets. We faced a trade-off between the use of technical data to model production and its impact on soil erosion versus well-behaved functional forms needed to arrive at a local market equilibrium. Given the importance of fragile natural resources in LFAs, we gave precedence to analysing the impact of agricultural activities on soil erosion, forcing us to ignore interactions among households in local markets.

A Village in the Northern Highlands of Ethiopia

In order to study the impact of heterogeneous household endowments, we rely on census data covering all 200 households in an Ethiopian village (the appendix, below, contains a summary of the census data used in the simulations). The village is located in the eastern part of Tigray. Tigray is a poor region in Ethiopia with limited agricultural potential due to poor soils and erratic rainfall. The village has an average annual rainfall of 450 mm and lacks irrigation structures, thus limiting agricultural opportunities. Fertile land is limited to a small valley bottom set amidst steep slopes and high plateaux. As a result of the steep slopes and loss of vegetative cover, erosion (including gully erosion) is a severe problem in the village. In addition to limited natural endowments, the village has no roads to the nearest town. As a result of these unfavourable conditions, many households in the village cannot satisfy their own food requirements (Meijerink, 2002).

In addition to data on household endowments, we need data on consumption and technologies in order to implement the model numerically. We use secondary data from Ethiopia on household expenditure patterns for rural households in the lowest five quintiles (Diao *et al.*, 2004) to compute budget shares (μ_j). This resulted in the following expenditure pattern by category: cereals, 0.42; non-cereals, 0.21; purchased food, 0.15; milk and meat, 0.05; and leisure, 0.18. We use technical data of production systems in Tigray to derive input and output coefficients. The output coefficients include the effect of production on soil erosion, computed through the Revised Universal Soil Loss Equation (Hengsdijk, 2003).

In the stylized farm household models, each household can choose from five crops (barley, millet, sorghum, wheat and pulses) and three types of livestock (cows, sheep and goats), each producing two types of output (milk and meat). For each activity, households can choose between different technologies, as explained in the mathematical exposition of the model.

Although endowments play a central role, they do not completely bind households. Household endowments of land and oxen can be rented in or out (against fixed prices, in the absence of a village market in the model). Household labour and livestock endowments (goats, sheep and cows) are assumed to be non-tradable. Household labour can be used to a limited extent for off-farm employment (see model description). These assumptions reflect the findings of a Rapid Diagnostic Appraisal (RDA) held in the village (Meijerink, 2002). The assumptions on tradability of production factors make the availability of off-farm employment a primary driver of model results. This reflects the importance attached by villagers to finding employment in CFW or migration as a main coping strategy.

Livestock products can be sold locally without a price band between buying and selling prices. Crop outputs are sold and purchased outside the village with a price band representing transaction costs of buying or selling. This assumption is based on the RDA, in which households indicated the selling of high-value cereals to buy cheaper cereals as a coping strategy (Meijerink, 2002). There is assumed not to be a local market for crop output, since all households are facing the same periodicity in crop production. Finally, there is a CFW programme in place providing a base income to all households in the village.

To analyse poverty we need to define a poverty line. Given the stylized nature of the model, we employ a relative poverty line obtained from the RDA. According to a wealth ranking by villagers, 74% of the households are considered poor (Meijerink, 2002, p. 33). We use income per adult equivalent (to account for differences in household size and composition) from the base run to determine a poverty line such that 74% of the households are considered very poor. This results in a poverty line of 760 birr (the currency unit of Ethiopia) per capita per year, translating to US$2.04/day.[4] The household classification by villagers thus appears to correspond rather well to the international poverty line of US$2/day. If we would use a US$1/day poverty line, 45% of the households would be classified as poor.

Analysing the Poverty Impact of Single Policies

We start by analysing the poverty impact of three sets of policy elements often suggested for reducing poverty in LFAs:

1. Technology: improving technologies by raising production from 70 to 100% of the maximum attainable production level:

- Fertilizer: raise only the productivity of technologies by the use of fertilizer.
- Non-fertilizer: raise only the productivity of technologies that do not use fertilizer.
- All technologies: raise productivity of all technologies.

2. Prices: infrastructure investments are assumed to reduce the price band (i.e. the difference between buying and selling outside the village) by 50%.

- Inputs: reduce only price band of agricultural inputs (fertilizer).
- Consumption goods: reduce only price band of consumption goods.
- All goods: reduce price band of all goods.

3. Off-farm: increased off-farm employment through:

- Migration: allow households with an adult male to have up to one migrant.
- Cash for work (CFW): double the availability of local employment through the CFW programme from 30 to 60 days per household.

These policy shocks applied to the model are extreme, designed to explore the potential impact on different households. In the migration scenario, for example, we allow any household with an adult male to migrate.[5] In practice, there will be barriers in terms of (social) capital that will limit the access to migration for poor households. We furthermore run a number of sub-scenarios, such as simply reducing the price band for inputs, in order to decompose the impact of different policy components.

Poverty impact of single policies

Analysing the impact of single policies (see Table 17.1), we find that policies promoting the use of fertilizer, through technology improvement or by lowering the fertilizer price, reach only some of the households. In the base run, 36% of households use fertilizer. By reaching 66 and 68% of the households, the impact of policies targeting fertilizer still stretches far beyond the households initially using fertilizer. Not all households

Table 17.1. Poverty, income, income inequality and erosion by policy scenario.

	Reach[a] (%)	Poverty[b]			Income inequality (Gini)	Erosion[c] (% change)
		P0	P1	P2		
Base run	na	74	40	26	0.40	na
Single policies						
Technology						
Fertilizer	66	72	39	26	0.40	−14
Non-fertilizer	100	68	37	24	0.39	3
All technologies	100	68	37	24	0.39	3
Price band						
Inputs	68	71	39	26	0.40	−17
Consumption goods	100	66	35	22	0.41	−1
All goods	100	66	34	22	0.41	−18
Off-farm						
Migration	40	64	34	22	0.39	−28
CFW	100	70	33	20	0.36	−18
Combined policies						
Technology (1c) and prices (2c)	100	64	32	20	0.40	1
Technology (1c) and migration (3a)	100	63	31	20	0.38	−27
Technology (1c) and CFW (3b)	100	65	30	17	0.35	−14
Prices (2c) and migration (3a)	100	56	29	19	0.41	−41
Prices (2c) and CFW (3b)	100	57	27	16	0.37	−34

na, not available
[a] Households affected by the policy; [b] Foster-Greer-Thorbecke poverty measures: P0, the poverty headcount; P1, the poverty gap; P2, the severity of poverty; [c] change in erosion is computed with respect to erosion levels in the base run.

benefit from fertilizer policies, due to the cash constraint households face. In contrast, improvement of non-fertilizer technologies reaches all households in the village. This is due to the presence of a land rental market. In the absence of a land market the landless households (8% of the village) would not benefit from improved crop technologies.

The impact of promoting migration is limited to households with an adult male that can migrate (40% of the households). Although migration has a limited reach, it does lead to the largest reduction in the poverty head count. This is due to the assumption that all households with an adult male can participate in migration. No additional investments are needed that may, in practice, limit the access of poor households to migration.

In terms of reaching the poorest of the poor, the CFW programme outperforms all other policies, reducing the poverty gap by seven percentage points and the severity of poverty by six percentage points. This focus on the poorest households seems to come at the expense of reducing the poverty headcount. Only the policies targeted at fertilizer are less effective in reducing the poverty headcount. Since the poor are net buyers of food, reducing the price band for consumption goods is more important for poverty reduction than the price of agricultural inputs. The poorest households are not those using the fertilizer-based technologies. This is also reflected in the poverty impact of improved fertilizer technologies, which is less than that of the poverty impact of improving technologies that make no use of fertilizer.

Single policies and income inequality

To assess the income inequality within the village, we computed the Gini coefficient. In the base run we find a Gini coefficient of 0.40. This is less than the 0.59 reported in Jayne *et al.* (2003) for Ethiopia as a whole. This is not surprising, since we compare the incomes within a single village, which one would expect to be more homogeneous than the income distribution across a country with widely varying natural resource endowments.

The CFW programme performs best in terms of reducing income inequality, thus reducing the Gini to 0.36. This is not surprising, since the poverty indicators showed that the CFW programme targets the poorest of the poor. This is also reflected in the income increases from the CFW programme, which are 49% for the lowest-income quintile and only 6% for the highest-income quintile. This finding underscores the importance of the CFW programmes for sustaining the living of the poorest households in the village.

A different pattern appears with a reduction in the price band of consumption goods. Compared with the other policies, it does perform rather well in terms of the poverty indicators, since the poor are net buyers of food. It does also, however, increase the Gini coefficient to

0.41. Richer households have more money to spend on consumption goods and thus benefit more from the price reduction. Reductions in the price band, for example infrastructure investments, thus seem to reinforce income inequality, with better households capturing a larger share of the benefits.

Single policies and soil erosion

A key question when combating poverty in LFAs is a possible trade-off between rising incomes and maintaining the fragile natural resource base. In the case study village we focus on erosion, a key natural resource issue in the village.

All policies except for the improved non-fertilizer technologies result in a decrease in erosion (see Table 17.1). Non-fertilizer technologies have higher erosion levels, and improved technologies result in more intensive land use. Improved fertilizer technologies reduce soil erosion because more households start using fertilizer. In terms of poverty impacts, improved non-fertilizer technologies outperformed fertilizer technologies. In focusing only on technology improvements, there seems to be a trade-off between combating poverty and maintaining the resource base.

Broadening the scope towards price band and off-farm policies, the trade-off between poverty and soil erosion disappears. These policies all reduce erosion levels while having a stronger impact on poverty than technology improvement. These policies affect erosion through two pathways: access to fertilizer and reduced need to derive cash income from agriculture. Access to fertilizer is improved through a reduced fertilizer price (price band policy) or by cash income from migration or CFW. Improved access to fertilizer increases the number of households using less-erosive fertilizer technologies. A reduction in the price of consumption goods or alternative non-farm income (migration or CFW) reduces the pressure on households to derive (cash) income from agriculture. Reducing the intensity of agriculture also reduces soil erosion. Price band and, especially, off-farm policies thus seem to offer more scope for both erosion and poverty reduction than improved (non-fertilizer) technologies.

Asset Poverty and its Implications for Policy Impacts

The discussion of poverty has, so far, centred around the common approach of defining poverty in terms of household income. As discussed in the introduction, assets play a central role in poverty, which has rendered a recent literature analysing poverty in terms of assets (Carter and Barrett, 2006). This literature finds that households that are structurally poor do not manage to progress beyond a minimum asset threshold needed to grow out of poverty.

With our static model, we are not able to examine changes in assets and the impact on poverty over time. We can, however, examine the relation between assets and income across households, which provides a cross-sectional look at the level of assets needed to progress beyond the poverty line. Figure 17.1 presents the income per adult equivalent in combination with the asset endowments by income quintile.

The most striking result is that the lowest two income quintiles do not own any oxen. Comparing this with the distribution of income across households in the top part of Fig. 17.1, we find that lack of oxen results in household income being < US$1/day. The importance of oxen ownership is also apparent at the top part of the income distribution. Households in the top income quintile (which roughly corresponds to households above the US$2/day poverty line) own oxen. Oxen and other livestock are the only assets that show a consistent increase when moving from the lowest to the highest income quintile. Land and labour endowments do not show a clear pattern. This may be a result of land reforms that have occurred in the recent past.

Looking in more detail at the relation between household assets and income levels, we find that oxen ownership does not totally determine income status. There are six households above the US$2/day poverty

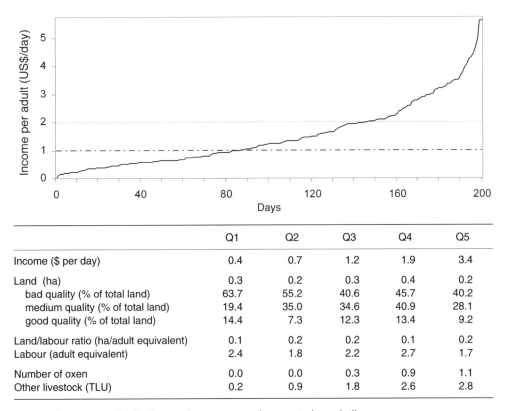

	Q1	Q2	Q3	Q4	Q5
Income ($ per day)	0.4	0.7	1.2	1.9	3.4
Land (ha)	0.3	0.2	0.3	0.4	0.2
bad quality (% of total land)	63.7	55.2	40.6	45.7	40.2
medium quality (% of total land)	19.4	35.0	34.6	40.9	28.1
good quality (% of total land)	14.4	7.3	12.3	13.4	9.2
Land/labour ratio (ha/adult equivalent)	0.1	0.2	0.2	0.1	0.2
Labour (adult equivalent)	2.4	1.8	2.2	2.7	1.7
Number of oxen	0.0	0.0	0.3	0.9	1.1
Other livestock (TLU)	0.2	0.9	1.8	2.6	2.8

Fig. 17.1. Income distribution and resource endowments by quintile.

line that do not own oxen. In the presence of local markets for oxen services, they are able to compensate for a lack of oxen by exceptionally good land endowments (both in terms of size and quality). There are also 30 households that do own oxen but are still below the US$2/day poverty line. The majority of these households have a very low land:labour ratio, some have exceptionally bad land and a few lack any livestock. This suggests that possession of oxen alone is not enough to escape poverty, despite the fact that we allow for local renting of land.

Analysing the asset endowments suggests that oxen ownership does separate the very poor (those below the US$1/day poverty line) from the relatively rich (above the US$2/day poverty line). In between the two poverty lines we find about 30% of the village households. One-half of these households have oxen but limited land, and the other half have land but lack oxen. Oxen ownership thus does not provide a clear threshold separating the poor from the non-poor. Land endowments are complementary to oxen ownership and may even compensate for a lack of oxen (assuming oxen can be rented-in).

Figure 17.1 suggests the presence of an asset threshold around the US$2/day poverty line. Incomes of households above this poverty line show a much stronger increase when moving to the richer households than decrease when moving to the poorer households. This suggests that there is a minimum amount of assets for households for engaging in productive activities.

The relative flatness of the income line just below the US$2/day poverty line indicates that the reductions in the poverty head counts in Table 17.1 present a flattered picture of the poverty impact of policies. The improved technologies, for example, lead to minor changes in income. Because of the shape of the income distribution, such a minor change in income still results in a 2–4% reduction of the poverty head count. In other words, a relatively large number of households are lifted from just below to just above the poverty line.

Analysing the Poverty Impact of Combined Policies

The analysis of single policies indicates that heterogeneous assets affect the impact of policies. A reduction in price bands, for example was found to worsen income inequality by disproportionally benefiting richer households. Combined policies are more effective if complementarities between policies overcome asset limitations. With the improved fertilizer technology, for example, households' cash constraint appears to play a central role. Combining improved technologies with reductions in price band and/or increases in off-farm income may put new technologies within the reach of the poorest households.

In order to analyse complementarities between policies, we run combinations of the policies analysed above. To limit the number of possible combinations we combine a change in all technologies (1c), in prices of all goods (2c) with the two off-farm income scenarios. We do maintain the migration and CFW scenarios because of their rather different distributional impacts. This leaves us with five different scenarios, the results of which are reported in Table 17.1.

Poverty impact of combined policies

Comparing poverty indicators of single and combined policies, there appear to be complementarities between the policies in terms of combating poverty. Any combination of two policies outperforms all single policies for all three poverty measures. Although no policy combination strictly dominates all other combinations in terms of poverty reduction, the combination of a price band reduction with the CFW programme performs best in terms of reaching the poorest of the poor and is almost at equal stance with the prices and migration combination in terms of reducing the poverty headcount.

Combined policies and income inequality

To analyse the presence of interaction effects of policies, we examine whether a combination of policies results in income increases that exceed the summed income of separate policies. When comparing the incomes there are three possible outcomes: (i) competing policies (income of combined policies is less than the sum of separate policies); (ii) no interaction (income of combined policies is equal to the sum of separate policies); and (iii) complementary policies (income of combined policies exceeds the sum of separate policies). We present the results by income quintile which, because of the close link between income and assets found above, indicates whether combined policies help overcome the asset restrictions of poor households (Fig. 17.2).

Overall, we find most policy combinations to be complementary. The only exception is the combination of improved technologies with a reduction in price bands. Reduced price bands reduces the costs of purchasing food. A set of households in the middle quintiles reduce their own agricultural production (renting-out land) and rely more on the market for food. With a reduction in agricultural production, the benefits of improved technology are also reduced. A price band reduction may thus change the (relative) attractiveness of new technologies because of a greater reliance on the market for food consumption.

The most striking finding in Fig. 17.2 is that, for the poorest income quintile, all policy combinations are complementary. Moreover, the effect is consistently stronger for the poorest income quintile than for any of the other quintiles. This strong positive impact of combined policies on the poorest households drives the reductions in Gini coefficients found for combined policies. Complementarity of policies

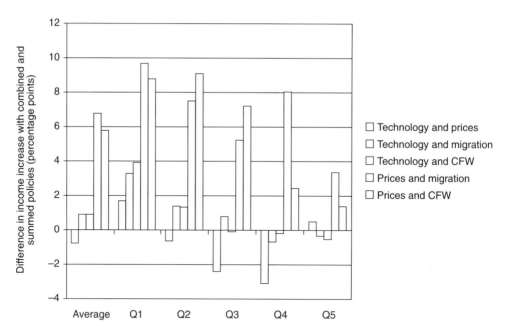

Fig. 17.2. Comparison of the income effects of combined policies with the sum of separate policies. Table entries are the increase in income (%) from combined policies minus the sum of the increase in income (%) from separate policies.

can thus help in overcoming limited assets. This is more important for asset-poor households and thus aids in targeting policies at the poorest households.

Combined policies and soil erosion

When analysing single policies, we found reduced price bands reduced erosion by reducing the price of fertilizer, while improved technologies increased erosion levels by stimulating non-fertilizer technologies. Given the reduction percentages with the single policies one may expect the price band effect (−17%) to dominate the technology effect (+3%). Their combined effect, however, is a 1% increase in erosion. The improved non-fertilizer technologies reduce the shift to using fertilizer when its price is reduced. More important, however, is the combination of policies that changes the choice of fertilizer technologies. With improved technology and a reduced fertilizer price, households opt for a fertilizer technology without the use of oxen (prices of oxen are fixed). Technologies without oxen are more erosive than those with oxen, and thus the amount of erosion increases.

Changing technologies as well as relative prices of inputs thus affects technology choice in such a way that the combination of policies leads to a different result than the summing of their separate effects would suggest.

Combining technology improvements with off-farm policies leads to fewer interactions between policies. Although reductions in erosion from off-farm employment are tempered by the improved technologies, strong reductions in erosion remain. Combining price band and off-farm technologies also does not generate surprising results. The reductions in erosion from combined policies are only somewhat smaller than the summed erosion reductions of single policies.

Conclusions

In this chapter we have explored the implications of heterogeneous household asset endowments for policies aimed at reducing poverty in LFAs. Analysing the poverty impact, we found that combining policies helps poorer households overcome the limitations of their asset endowments. This complementarity of policies is less important for better-endowed households. As a result, combining complementary policies helps in targeting the poorest households, thus reducing income inequalities.

Analysing household endowments in relation to the poverty, we found oxen ownership to be a central asset for households in attempting to rise above the poverty line. Although we found a clear difference between the poorest and richest households in terms of oxen ownership (households with no oxen are below the US$1/day poverty line, households above the US$2/day poverty line own oxen), this does not fully determine the income-earning opportunities of the households. Oxen and land ownership are found to be complementary to some extent. Households in between the US$1 and US$2/day poverty line are found to have oxen but a small land:labour ratio, or to have a favourable land:labour ratio but lack oxen. In addition to oxen, sufficient land (in terms of quantity and quality) is thus needed to escape poverty.

The importance of assets has implications for the distributive consequences of policies. We found a reduction in price bands to reduce poverty while at the same time increasing income inequality. Many households thus benefit, but richer households benefit disproportionally. The opposite holds for the CFW programme which, through self-selection, targets the poorest households. Combining the CFW programme with a price band reduction was found to benefit the poorest households most while still reducing income inequality in the village. This combination of policies was also found to result in the strongest reduction in erosion, thus providing a win-win option for combating poverty and reducing natural resource degradation in the village.

The simple stylized micro-simulation model we employed in this chapter is insightful because of its tractability. Because of the simple nature of the model and the rather extreme scenarios, the findings provide direction only for further analysis and cannot be relied upon for policy advice. The model does not account for limited access to new technologies or migration that is likely to exist for the poorest households. These scenarios, therefore, overestimate the impact of technology or migration on poverty reduction, as well as the associated reductions in income inequality.

The model also lacks risk, which Zimmerman and Carter (2003) found to be an important impediment for poor households to engage in more profitable activities. The absence of risk in our model also implies that we overestimate the poverty impacts of single policies. In terms of combined policies, we may be underestimating the poverty impact. Combining an improved technology with a CFW programme may put the technologies within the reach of the poor households, not only by providing the cash needed for inputs (we find increased adoption of improved technologies employing oxen), but also by allowing households to take more risk by providing a minimum income.

With our static model we are unable to analyse the dynamics of asset accumulation. By analysing individual households, however, we have a cross-sectional perspective on the role of assets. Given the lack of availability of panel data for most developing countries, such a cross-sectional analysis of the role of assets provides an important first look at the role of assets for poverty reduction policies.

Our model accounts only for heterogeneity in assets. The analysis could be improved by including differences in household preferences, risk aversion and differences in access to new technologies and off-farm employment. A key limitation of the current model is the absence of village markets from the modelling exercise. We faced a trade-off between analysing impact of agricultural activities on natural resources – a major issue for LFAs – and including village markets in the model. The absence of village markets in the current analysis implies first of all that the total demand for village non-tradables does not equal the supply. The second implication is that we cannot analyse the indirect impact of policies such as, for example, increased employment opportunities if more labour-intensive technologies are adopted by some of the households.

Appendix I

The mathematical model description is formulated in general terms. By assigning commodities to different sets, different assumptions – on, for example, the presence of markets or the consumption of leisure – can easily be implemented. Table 17.A1 presents an overview of the commodities (set J) included in the model and their set memberships. Table 17.A2 presents the descriptives of the census data used in the simulations.

Table 17.A1. Set membership of commodities.

Description	Output (O)	Input (I)	Consumed (C)	Household non-tradable (HNT)	Village non-tradable (VNT)	Village tradable (VT)
Barley (kg)	X					X
Millet (kg)	X					X
Pulses (kg)	X					X
Sorghum (kg)	X					X
Wheat (kg)	X					X
Poor-quality land (ha)		X			X	
Medium-quality land (ha)		X			X	
Good-quality land (ha)		X			X	
Labour (man/days)		X	X	X		
Oxen (days)		X			X	
Nitrogen fertilizer (kg/ha)		X				X
Milk (kg)	X				X	
Meat (kg)	X				X	
Cows and other animals (TLU)		X		X		
Sheep (TLU)		X		X		
Goats (TLU)		X		X		
Purchased non-food			X			X

Table 17.A2. Descriptives of household endowments (*n* = 200).

	Mean	Median	Minimum	Maximum
Labour (adult equivalents)	2.2	1.8	0.5	6.7
Land (ha)	0.3	0.3	0.0	0.8
Poor-quality land (% total land)	49.1	50.0	0.0	100.0
Medium-quality land (% total land)	31.6	33.3	0.0	100.0
Good-quality land (% total land)	11.3	0.0	0.0	57.1
Oxen (TLU)	0.7	0.0	0.0	3.0
Cows (TLU)	1.3	0.9	0.0	7.2
Sheep (TLU)	0.3	0.0	0.0	2.0
Goats (TLU)	0.1	0.0	0.0	1.4

Endnotes

1   Remittances can ultimately be used for hiring-in labour and purchasing food.
2   Depending on the labour intensity of the selected technologies, the landless may benefit, nevertheless, in an indirect way through increased employment opportunities.
3   Introducing saving behaviour in a static model amounts to including a fixed saving propensity. This would reduce the available income for consumption but would not qualitatively alter the results of the model.
4   We use a purchasing parity exchange rate of 1.02 birr per dollar, derived from Alan Heston, Robert Summers and Bettina Aten, Penn World Table Version 6.1, Center for International Comparisons at the University of Pennsylvania (CICUP), October 2002.
5   We lack data on the education levels of household members. Analysing available data on whether children are sent to school (assuming that this could serve as a proxy of the education of the parents), we found all households with an adult male sent their children to school. We therefore assumed that all households with an adult male could engage in migration if the opportunity presented itself.

References

Barrett, C.B., Bezuneh, M., Clay, D.C. and Reardon, T. (2004) *Heterogeneous Constraints, Incentives and Income Diversification in Rural Africa*. Mimeo. Cornell University, Ithaca, New York.

Birdsall, N. and Londoño, J.L. (1998) No trade-off: efficient growth via more equal human capital accumulation. In: Birdsall, N., Graham, C. and Sabot, R. (eds) *Beyond Tradeoffs: Market Reforms and Equitable Growth in Latin America*. Inter-American Development Bank, Washington, DC.

Carter, M.R. and Barrett, C.B. (2006) The economics of poverty traps and persistent poverty: an asset-based approach. *The Journal of Development Studies* 42, 178–199.

Cogneau, D. and Robilliard, A. (2000) *Growth, Distribution and Poverty in Madagascar: Learning from a Micro-simulation Model in a General Equilibrium Framework*. TMD Discussion Paper 61, IFPRI, Washington, DC.

Deininger, K. and Squire, L. (1998) New ways of looking at old issues: inequality and growth. *Journal of Development Economics* 57, 257–285.

De Janvry, A., Fafchamps, M. and Sadoulet, E. (1991) Peasant household behaviour with missing markets: some paradoxes explained. *Economic Journal* 101, 1400–1417.

Diao, X., Gautam, M., Keough, J., Puetz, D., Chamberlin, J., Rodgers, C., You, L. and Yu, B. (2004) *Growth Options and Investment Strategies in Ethiopian Agriculture – a Spatial, Economy-wide Model Analysis for 2004–2015*. Mimeo. IFPRI, Washington, DC.

Elbers, C., Fujii, T., Lanjouw, P., Ozler, B. and Yin, W. (2004) *Poverty Alleviation through Geographic Targeting: how much does Aggregation Help?* Mimeo. Vrije Universiteit, Amsterdam.

Escobal, J. (2005) The role of public infrastructure in market development in rural Peru. PhD thesis, Wageningen University, Wageningen, The Netherlands.

Hazell, P., Ruben, R., Kuyvenhoven, A. and Jansen, H.G.P. (2005) *Investing in Poor People in Less-favoured Areas*. IFPRI 2020 Policy Paper, IFPRI, Washington, DC.

Hengsdijk, H. (2003) *An Introduction to the Technical Coefficient Generator for Land Use Systems in Tigray*. No. 241, Plant Research International, Wageningen, The Netherlands.

Jayne, T.S., Yamano, T., Weber, M.T., Tschirley, D., Benfica, R., Chapoto, A. and Zulu, B. (2003) Smallholder income and land distribution in Africa: implications for poverty reduction strategies. *Food Policy* 28, 253–275.

Keefer, P. and Knack, S. (2002) Polarization, politics and property rights: links between inequality and growth. *Public Choice,* 111, 127–154.

Lipton, M. (2005) Can small farms survive, prosper, or be the key channel to cut mass poverty? Paper presented at *FAO Symposium on Agricultural Commercialisation and the Small Farmer*, Rome, 4–5 May.

Lybbert, T.J., Barrett, C.B., Desta, S. and Layne Coppock, D. (2004) Stochastic wealth dynamics and risk management among a poor population. *The Economic Journal* 114, 750–777.

Meijerink, G.W. (2002) *Rural Livelihoods and Soil Conservation in Eastern Tigray: a Rapid Diagnostic Appraisal Report from Gobo Deguat and Teghane.* Policies for sustainable land management in the Ethiopian Highlands Working Paper 2002-2006, Wageningen University, Wageningen, The Netherlands.

Orcutt, G.H. (1957) A new type of socio-economic system. *The Review of Economics and Statistics* 39, 116–123.

Ravallion, M. and Jalan, J. (1996) Growth differences due to spatial externalities. *Economic Letters* 53, 227–232.

Ruben, R. and Pender, J. (2004) Rural diversity and heterogeneity in less-favoured areas: the quest for policy targeting. *Food Policy* 29, 303–320.

Ruben, R., Kuyvenhoven, A. and Kruseman, G. (2000) Bio-economic models for eco-regional development: policy instruments for sustainable agricultural intensification. In: Lee, D.R. and Barrett, C.B. (eds) *Trade-offs and Synergies: Agricultural Intensification, Economic Development and the Environment.* CAB International, Wallingford, UK, pp. 115–134.

Singh, I., Squire, L. and Strauss, J. (eds) (1986) *Agricultural Household Models. Extensions, Applications and Policy.* Johns Hopkins University Press, Baltimore and London.

Van de Walle, D. and Gunewardena, D. (2001) Sources of ethnic inequality in Vietnam. *Journal of Development Economics* 65, 177–207.

Zimmerman, F.J. and Carter, M.R. (2003) Asset smoothing, consumption smoothing and the reproduction of inequality under risk and subsistence constraints. *Journal of Development Economics* 71, 233–260.

18 Less-favoured Areas: Looking beyond Agriculture towards Ecosystem Services

LESLIE LIPPER,[*] PRABHU PINGALI[**] AND MONIKA ZUREK[***]

Agricultural and Development Economics Division, Food and Agriculture Organization of the UN, Viale delle Terme di Caracalla, 00100 Rome, Italy; telephone: +39 06 570 54217; fax: +39 06 570 55522;
[*]Leslie.lipper@fao.org"; [**]prabhu.pingali@fao.org; [***]monika.zurek@fao.org

Introduction: What are Less-favoured Areas and Why do we Need to Care?

In the context of poverty alleviation, the discussion on less-favoured areas (LFAs) has gained in momentum over recent years. About 40% of the rural population in developing countries are estimated to live in these areas (Kuyvenhoven *et al.*, 2004), which face quite a number of diverse biophysical and socio-economic constraints to sustaining livelihoods: (i) growing population numbers; (ii) limited infrastructure and market access; (iii) land tenure problems; and (iv) increasing degradation problems due to poor management of soils prone to erosion, steep slopes or low rainfall quantities are some of the limitations for agricultural production that have led, in many areas, to growing numbers of poor people.

In addition, policy makers as well as the national and international research and extension systems have neglected these regions in recent decades (Kuyvenhoven *et al.*, 2004), thus aggravating some of the problems. As Pender and Hazell (2000) describe, for a long time development strategies in many countries emphasized the importance of investing in highly productive areas, as returns to investments would be greatest there. Improved food production and therefore increasing food security, together with economic growth, would stimulate the migration out of less-favoured areas, thus reducing pressures on fragile resources and population numbers.

Increasing evidence shows today that upward trends in population and poverty numbers have not changed (Pender and Hazell, 2000;

Kuyvenhoven *et al.*, 2004), while the resource situation has worsened in a number of cases. In addition, investments in favourable areas have not always had the desired effect, as diminishing returns to investments and increasing environmental problems in many intensively used agricultural areas around the world demonstrate. All these developments have put LFAs back on the agenda of policy makers and researchers alike (see, for example, the CGIAR TAC report on research priorities for marginal lands (CGIAR TAC, 1999)).

Varying definitions have been proposed to describe less-favoured or marginal areas. Part of the difficulty of clearly defining what a less-favoured or marginal area is stems from their heterogeneity and the diversity of encountered problems. Furthermore, less-favoured lands can be defined based on a variety of different characteristics, such as their potential or constraints for agricultural production or the encountered socio-economic conditions. Pender and Hazell (2000) give a short, but simple, definition by describing these areas as 'less favoured either by nature or by man'. The CGIAR TAC report on marginal lands (CGIAR TAC, 1999) provided a useful overview of various terms used in this context.

The definition adopted in this chapter is the one on marginal lands from the CGIAR report. Marginal lands are defined as lands with 'limitations which, in aggregate, are severe for sustained application of a given use. Increased inputs to maintain productivity or benefits will be only marginally justified. Options for diversification without the use of inputs are "limited"' (CGIAR TAC, 1999). Important here is the term 'for a given use': an area might be marginal or less favoured for use as a crop production area under a specific production system, e.g. due to either water scarcity or lack of market access. The same area, nevertheless, could become more favourable if either new water-saving technologies or new marketing routes became available.

In order to devise new development options for LFAs, taking a wider approach that looks beyond agriculture is crucial. As most of these areas have important limitations for agricultural production per se, targeting these bottlenecks is definitely one way forward. Nevertheless, in recent years the use of land to produce not just food or fibre, but also provide other ecosystem services such as carbon sequestration or biodiversity conservation, has gained in importance. These options also need to be explored.

The objective of this chapter is to contribute to the discussion on new policy options for LFAs by presenting a set of conceptual ideas in directing policy strategy development for alleviation of the constraints faced by land users in these regions. In the following section, using drylands as an example, we describe first a typology of dryland crop production systems around the world based on differences in resource endowments of land and labour. We then discuss the various pathways that drylands development has taken, based on opportunity costs for land and labour, as well as conditions external to the agricultural sector.

We then develop concepts for the incorporation of environmental services into drylands development strategies, looking at the way in which supply environmental services can contribute to livelihoods by either impacting agricultural productivity or providing an alternative source of income. The potential returns to providing environmental services are also explored in this section, with a discussion on current sources of demand. The chapter concludes with a discussion of the way forward: the type of information, policies and institutions that will be required for successful incorporation of environmental services into drylands development strategies.

Agricultural Development Pathways for Marginal Lands: the Case of the Drylands

Drylands, defined as water-scarce lands (MA, 2005), serve as a good example of less-favoured areas, their problems and possible solutions. Their main biophysical constraint is lack of water, resulting from low precipitation and high evapotranspiration levels, which can restrict crop and livestock production severely. In addition, the quality of soils found in drylands can vary tremendously. The degree of aridity varies as well across drylands, which has resulted in classifying them in four sub-categories along an increasing gradient of moisture deficit: dry sub-humid, semi-arid, arid and hyper-arid. These four dryland types cover about 41% of the Earth's surface (about 6 bn ha) and are inhabited by about 2 bn people (MA, 2005).

Drylands can be found on all continents and, as Fig. 18.1 shows, they are located in developing countries, such as the Sahel region in

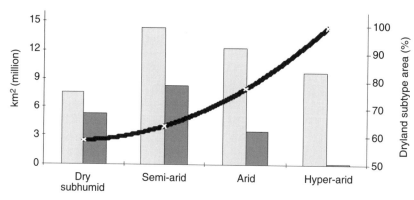

Fig. 18.1. Dryland subtypes in developing and industrialized countries (from Millennium Ecosystem Assessment (MA, 2005).

Africa or Rajasthan in India, as well as in industrialized ones, like central Australia or California in the USA. Nevertheless, the MA (2005) calculated that the lion's share of global drylands – about 72% – are in developing countries. This figure also depicts the trend that, the drier it gets, the more likely one is to find this area in a developing country. For example, no industrialized country has hyper-arid areas.

Figure 18.2 indicates that the vast majority of dryland inhabitants (roughly 90%) live in developing countries (MA, 2005). In many cases they are the poorest of the poor and display the lowest levels of human well-being. The MA compared indicators such as infant mortality and GDP across different ecosystem types and found that dryland populations had the highest infant mortality and the lowest GDP levels (MA, 2005). In addition to water stress, dryland areas in developing countries also face a number of socio-economic constraints, such as increasing population pressure, poor infrastructure and market access, lack of proper land tenure systems and poor governance systems (MA, 2005). Thus, drylands are a good example of less-favoured lands in which biophysical constraints and a number of socio-economic conditions working together render them marginal areas.

Not all dryland environments, however, are low-productivity, subsistence systems. Quite a few examples exist where human modification to the existing constraints has resulted in the conversion of these areas to profitable crop or livestock production systems. However, the profitability of such conversion depends on suitable technologies, economic incentives and supportive institutional set-ups. Moreover, what happens outside the agriculture sector is as important, if not more important, with regard to what happens in the agriculture sector for determining the status and future of the drylands.

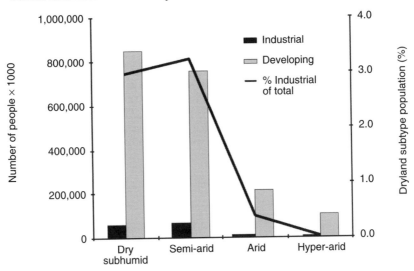

Fig. 18.2. Developing and industrialized countries' populations in different dryland types (from Millennium Ecosystem Assessment (MA, 2005).

We identify four categories of drylands farming systems that can be used to distinguish different drylands development pathways in an induced innovation-type framework (Hayami and Ruttan, 1985; Fig. 18.3). Many traditional subsistence farming systems, often with low productivity, can be found in areas with low opportunity costs for land and labour. Examples include small-scale subsistence farming systems in many sub-Saharan African countries based on the production of traditional staple crops such as sorghum, maize or cassava. The incentives for increasing productivity in these systems are minimal, since low population densities and poor market infrastructure conditions imply that the farmers face an inelastic demand for their output.

Economic growth and trade integration triggers a movement away from subsistence systems towards commercial farming in areas of low population density, provided investments in irrigation and market infrastructure are made. Increasing economic development outside of a LFA is likely to change opportunity costs of labour by providing new job opportunities with the resulting out-migration of labour. Rising labour costs will increase the incentives for farmers to look for labour-saving technologies. Such mechanized, large-scale production systems are observed in cereal production areas in the Argentinean, Australian or US drylands. These areas are also conducive to the development of extensive pastoral systems for livestock rearing, such as beef cattle production systems in Argentina.

Where labour is abundant but land is the more constraining factor, intensive cereal systems have developed that rely more on the use of high-yielding varieties and fertilizers to increase productivity while saving land. Typical examples of this kind of development are the intensively managed rice–wheat production systems in the Indian Punjab and the intensive rice production systems in South-east Asia. Intensive livestock production, generally associated with stall feeding, is

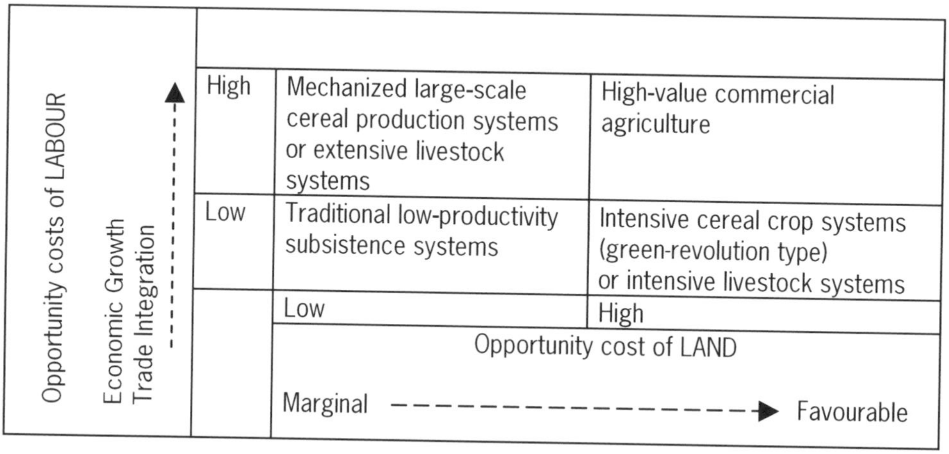

Fig. 18.3. Dryland farming system types.

also common in these systems. Sustainable development of crop and livestock production systems in these environments depends on good access to input and output markets, as well as a functioning R&D system that can provide adapted varieties and develop solutions for better resource use and agronomic management. If such conditions do not exist, one would observe high levels of degradation, with population growth and the associated increase in land use intensity.

Inevitably, situations also exist in which both opportunity costs for land and labour are relatively high. This can be the case in areas with high population density and dynamic, well-functioning manufacturing and/or services sectors that provide off-farm labour opportunities. Systems found in drylands in which these situations exist include the intensively managed fruit and vegetable areas around the Mediterranean (e.g. Israel, Egypt, Spain). As most of these production systems are highly intensified commercial systems, they require access to input and output markets, supporting R&D systems and the appropriate physical infrastructure. Many of these areas are also associated with various environmental problems resulting from inappropriate fertilizer, pesticide or water use.

Though this typology is highly stylized, it nevertheless provides a historical sketch of the pathways followed in the transition out of subsistence agriculture in dryland environments. The pathway chosen depends on two conditions: (i) the land and labour endowments of particular societies; and (ii) the dynamics outside of the agricultural sector that influence their opportunity costs. The above discussion indicates that drylands are not automatically destined for destitution, which is also the case for other LFAs. Demographic, economic and institutional factors are the ultimate determinants of the pathways out of poverty for the marginal environments.

New Opportunities for Dryland Systems: the Provision of Ecosystem Services

In this section we broaden the range of sustainable land use strategies that we consider for drylands to include the potential for generating ecosystem services both as a complement to and substitute for agricultural production. Ecosystem services are defined as all benefits that humans receive from ecosystems (Costanza *et al.*, 1997; Daily, 1997; MA, 2003).

The Millennium Ecosystem Assessment (MA) distinguishes four different categories of services (MA, 2003): (i) provisioning services, which include food production; (ii) regulating services (e.g. climate regulation, nutrient cycling); (iii) supporting services (e.g. biodiversity); and (iv) cultural services (e.g. amenity values). An ecosystem perspective, as opposed to a conventional agricultural development perspective, recognizes that land management can produce more than just food or fibre, and distinguishes between services that sustain agricultural production and those that are public goods in their own right.

Provisioning services in the form of agricultural production are one of the most important ecosystem services generated in drylands, having clearly defined economic value as compared with other environmental services. Supporting and regulating services, such as soil fertility and water flow management, while important complementary services to agriculture production, tend to be undervalued. In cases where the provision of environmental services involves a trade-off with agricultural production, e.g. conserving biodiversity by maintaining forest areas, poor societies have generally chosen in favour of production.

The following two questions are pursued here: (i) can the provision of environmental services be integrated into the development pathways for dryland environments?; and (ii) can payments for environmental services provide an additional source of livelihood for poor households?

Supply of ecosystem services

The ability of dryland systems to supply ecosystem services is determined by the quantity and quality of the land resource base. Population density, agro-ecological conditions, level of market integration and primary technology employed in agriculture are all important determinants of the current returns to land and labour in drylands, and also the potential role that the provision of public environmental services may play in improving these returns.

In land-abundant areas, including areas where rising off-farm employment opportunities have drawn populations out of rural areas, the potential for setting aside land for non-agricultural uses is high. Conversion of agricultural lands to forests contributes to carbon sequestration, watershed protection and biodiversity conservation. An example of this kind of conversion is China's Sloping Lands Programme, in which the Chinese government has the goal of converting 14.6 million ha of cropland on slopes into forest, to stop soil erosion and improve water retention.[1] Given the low opportunity cost of land, the trade-off with food and fibre production is small in these areas, particularly where transport infrastructure is a limiting factor for competitive agricultural production.

On the other hand, in land-scarce environments, the trade-off between agricultural and non-agricultural services is high. In such environments, ecosystem services would have to be complementary to, rather than a substitute for, food and fibre production. This requires the adoption of agricultural production systems that generate environmental services. Conservation tillage, agroforestry systems and silvo-pastoral systems are some of the many examples of agricultural production systems that can generate external environmental benefits in the form of carbon sequestration, biodiversity conservation and watershed protection (see Dutilly-Diane *et al.*, 2004 for a discussion of rangelands management programmes that can generate environmental services in drylands areas). The service is generated by the land user but the benefit is realized off-site. The beneficiaries may include local residents, consumers in global markets or even future generations. These types of environmental services are generally in the form of public goods, with low rivalry in consumption and high exclusion costs.

One example of how the switch to an agroforestry system can be facilitated by payments for carbon sequestration can be found in Mexico, where farmers receive payments from the International Federation of Automobiles as carbon offsets.[2] Another example is the Integrated Silvo-pastoral Approaches to Ecosystem Management Project in Central America, in which small- and medium-sized farmers receive compensation for planting trees, fodder shrubs and living fences around pastures (FAO Livestock Policy Brief No. 3). Improving the supply of environmental services can thus be an important component of agricultural development strategies in drylands areas, depending on the existing natural resource endowment and the type of farming system in place (Dutilly-Diane *et al.*, 2004; Zilberman *et al.*, 2006a).

It is important to recognize that not every country or region has the potential to realize an economic benefit from supplying environmental services, and environmental conditions are an important determinant of the returns. Factors such as soil quality, topography and climate are critical determinants of the productivity of ecosystem service provision and, together, can be considered as 'land quality', which is an input to the production of ecosystem services. The way in which land quality affects the productivity of ecosystem service provision varies by the type of services, e.g. steep topography can result in highly productive watershed protection but very unproductive agriculture. In other cases, land quality has a similar productivity effect on more than simply ecosystem service, e.g. soil fertility is important in determining both agricultural productivity and the productivity of soil and above-ground carbon sequestration.

The differential impact of land quality on the productivity of agricultural versus public environmental goods has an important impact on the degree to which the production of the two goods is complementary (Zilberman *et al.*, 2006a).

Converting lands from agricultural to public environmental service production on a site with land quality that is very poor for agricultural productivity but good for public environmental service provision has a lower trade-off than on land of excellent quality for agriculture. This variability has important implications for the role of supplying public environmental goods in developing strategies for improving drylands.

One key strategy for improving agricultural productivity in drylands has been to focus on improving land quality. This strategy has often been implemented using capital inputs to enhance land quality. The application of fertilizers and other soil amendments, terracing and irrigation are examples. These strategies are essentially attempts to increase the returns to land for agricultural producers. The experience with reliance primarily on capital inputs to increase the productivity of the land has shown that several types of problems can arise (Lipper, 2000): one example is a low return to input use, e.g. low input efficiency due to degraded or poor natural resource base. Another problem is the lack of financial sustainability of applying capital inputs in systems with low input efficiency.

The provision of public environmental goods that are complementary to agricultural production and which contribute to dimensions of land quality important for agricultural productivity thus emerges as a potentially important component of strategies aimed at improving the productivity of land in agricultural production (Zilberman *et al.*, 2006a, b). The benefit of combining payments for the provision of public environmental goods such as soil carbon sequestration or watershed protection for the adoption of agricultural practices that can eventually lead to increased agricultural productivity is quite attractive.[3] In some cases, however, the adoption of the new agricultural practice could lead to a decrease in the returns to agriculture, in which case the payment for the public good component must be sufficient, at the very least, to compensate for such losses.

In some cases, improving the quality of land to enhance agricultural productivity is not economically feasible, even with the potential of additional funds and improvement of natural resources through the provision of public environmental services. The land quality may be highly productive in the provision of public environmental goods, however. An example here might include crop production on steep slopes where erosion has resulted in poor soil fertility. Cultivation in this area generates very low returns to the farmer, and at the same time generates negative impacts on the functioning of the watershed. Converting this land from agricultural production to watershed protection (via planting of trees) will generate a higher return to both land and labour (Zilberman *et al.*, 2006b). This strategy, implemented in conjunction with growth in non-agricultural sources of employment, could be the most viable option for areas with highly degraded resources.

Aside from environmental factors, the supply of environmental services and, particularly, their successful incorporation into the livelihoods of drylands inhabitants, is dependent on other enabling conditions such as property rights, food security and low transactions costs. These have been discussed in several publications on payments for environmental services and poverty (Cacho *et al.*, 2002; Landell-Mills and Porras, 2002; Smith and Scherr, 2002; Lipper and Cavatassi, 2003; Dutilly-Diane *et al.*, 2004; Pagiola *et al.*, 2005; Wunder, 2005; Zilberman *et al.*, 2006a, b). One key issue is the widespread lack of formal property rights among most low-income land users, and the problems this entails with receipt of payments for land use changes. Another is the importance of food security risk as a determinant of land use choice and the necessity of including this concern in national and local planning. High transactions costs associated with environmental service payments to small producers seriously reduce the potential benefit of such payments.

Sources of demand for environmental services

Even where all conditions are in place for the supply of environmental services to support drylands development, there is still the issue of economic incentives that motivate farmers to supply ecosystem services, especially those that have larger off-site than on-site benefits. If producers don't receive payments that cover – at least – their costs of supplying the environmental good, they will have no incentive to participate in PES programmes unless they generate some other form of benefit such as increased agricultural productivity (Lipper and Cavatassi, 2003). Thus the demand for, and potential payment level to farmers for providing, environmental services is critical in determining the feasibility of their incorporation into drylands development strategies.

Two recent developments have created conditions for incorporating a wide range of environmental service provision into land use strategies in drylands: (i) there is increasing willingness to pay for environmental services on the part of external beneficiaries; and (ii) there is increasing recognition of the importance of environmental services in sustainable agricultural management, and the high costs of their depletion. However, the concept of paying for environmental services is still relatively new, since it is only recently that governments, international agencies and individuals have begun to recognize the important role that farmers, ranchers, foresters and any land-users could play in improving environmental management. Payment mechanisms are still being developed and much still remains to be done to fully realize their potential.

The source of payments and funds for environmental services depends upon the beneficiary of the service. One important criterion for assessing potential sources of demand is the scale at which benefits are realized, e.g. global versus local benefits from environmental services.

Climate change mitigation through carbon sequestration and biodiversity conservation (including agricultural biodiversity conservation) are the two main services that fall into the first category. In both cases, the environmental service has potential benefits for the entire global population as well as for future generations. In contrast to the benefits from environmental services for watershed management such as improvements in water flow, soil erosion and water quality are usually realized at a local level.

Another important criterion for classifying demand is whether they are from the public or private sector (Zilberman *et al.*, 2006a). In some cases, PES programmes are more supply driven, with the public sector seeking to utilize funds more efficiently (Pagiola, 2005). In other cases they are more demand driven, and these cases are where private sector participation is most likely.

In the following section we describe some of the emerging sources of payments for environmental service by sector and scale.

Public sector funds

At the international level, a major source of demand for global environmental services in the public sector is the Global Environment Facilty (GEF), which was established as a funding mechanism for several multilateral environmental agreements, including the Convention on Biological Diversity and the UN Framework Convention on Climate Change. The GEF funds projects that generate global environmental goods such as climate change mitigation, biodiversity conservation and the management of international water bodies (GEF website). The GEF is funded by contributions from donor countries: in 2002, 32 donor countries pledged US$3 bn to fund operations between 2002 and 2006 (GEF website). The GEF funds the 'incremental' or additional costs associated with transforming a project with national benefits into one with global environmental benefits, such as climate change mitigation and biodiversity conservation.

Another source of international public sector funds is the Biocarbon Fund, recently established by the World Bank with a capitalization of US$53.8 million for the first phase. The Fund is a purchaser of climate change mitigation services in the form of carbon sequestration and substitution (World Bank, 2002).

Other public sector funds for environmental services are managed at the national level. The Conservation Reserve Program in the USA and various payment schemes for multifunctionality in Europe are models of government-financed agricultural PES programmes. Some large, developing countries such as Brazil and China have established public sector funds for purchasing environmental services as well, with Proambiente in the former and the Sloping Lands Conversion Programme in the latter.

The public sector can be an important purchaser of environmental services at the local level as well. One of the most famous examples is in New York, where the city of New York increased water fees by 9% in order to make payments to farmers in the watersheds feeding the city water supply to adopt farming practices that would generate less water pollution (Mayrand and Paquin, 2004). In Brazil, some states have implemented a programme of biodiversity conservation using sales tax revenues for funding habitat conservation (Grieg-Gran, 2000).

Demand from the private sector

Private firms are already purchasing ES that result in higher profits by reducing production or environmental regulation compliance costs or increasing the sales value of their products on the market (Zilberman *et al.*, 2006a). In some cases, the company is interested in obtaining goodwill rather than the environmental service itself. In the case of payments for biodiversity offsets for example, mining companies are now agreeing to fund biodiversity conservation projects to offset the potentially negative impacts their activities have elsewhere. Among the main benefits from this are improved relationships with local communities, reduced opposition to planning permission and reduced costs of operating (Den Kate, 2005).

Demand for environmental services from the private sector is being generated where payments are the least-cost means of meeting environmental regulations. The demand for climate change mitigation services is a good example, with a combination of legally binding commitments to reduce carbon emissions and the potentially lower cost of emission reduction credits from carbon sequestration relative to other possible means driving the demand for this service (Graff-Zivin and Lipper, 2006).

Several exchanges of carbon offsets are being set up, but the one of most relevance for developing countries is the Clean Development Mechanism of the Kyoto Protocol. This mechanism allows developed countries with binding emissions reductions to offset their carbon emissions with the purchase of carbon emission reduction credits (CERs) from developing countries. The activities that are allowed as sources of carbon emission reductions are restricted. Some which could be beneficial to agricultural producers in drylands areas, such as soil carbon sequestration and avoided deforestation, are not allowed in the first commitment period ending in 2012. Discussions on whether these should be included in the future are currently under way.

Watershed services are another example where the private sector has provided payments. In the Cauca Valle in Colombia, for example, farmers' associations are paying for watershed management practices that improve the supply of irrigation water, using funds from water user fees (Mayrand and Paquin, 2004).

NGOs as purchasers

Some of the most effective ES funds are managed by NGOs that represent groups with specific environmental interests. The Nature Conservancy has invested millions of dollars in various programmes that buy or lease land and purchase development rights and other assets in order to provide ES (Zilberman *et al.*, 2006a). The World Wildlife Fund has an active programme developing PES in both developed and developing countries as a means of attaining sustainable agricultural development and poverty reduction objectives (Zilberman *et al.*, 2006a). Conservation International has recently started a new programme of conservation incentive agreements based on an equitable exchange of natural resource conservation for economic and social benefits; this programme involves compensation to local landowners for maintaining conservation activities (Zurita, 2005).

The Way Forward

In the previous section we asked two questions: (i) can the provision of environmental services be integrated into the development pathways for drylands environments?; and (ii) can payments for environmental services provide an additional source of livelihoods for poor households? Our analysis indicates the answer to both questions is yes – conditional upon several factors, including environmental conditions, as well as institutional and policy factors. In this section we discuss the policy and institutional changes necessary to meet the two objectives, and the role of various public sector actors in promoting these changes.

Perhaps the most important requirement for incorporating environmental services into drylands development strategies is the incorporation of their potential value into mainstream agricultural and economic development strategies. Consideration of the comparative advantages a country has in supplying such services and their potential to contribute to overall development objectives needs a broad perspective, and cannot be formulated solely by the environmental sector. To achieve this type of broad perspective across government sectors is no small or easy task. An important first step is thus information dissemination about the potential of environmental services and a dialogue between various ministries and agencies on if and how they could be integrated into development plans.

At an international level, inclusion of the potential role of environmental services in major development strategies such as the TerrAfrica Program supported by the Global Environmental Facility to promote sustainable land management in Africa, as well as in country-level poverty reduction strategy papers (PRSPS) as part of World Bank lending programmes, are important examples.

Information on the potential demand and supply of environmental services from a given country or region is needed to alert policy makers and planners to the possibilities that environmental services might provide, as well as to give a realistic appraisal of how well they can be integrated into overall development strategies. Ultimately, the information is also necessary to design an effective strategy for their incorporation into drylands development.

For example, a rough analysis of a country's potential to supply environmental services could be obtained from spatially referenced information on various dimensions of land quality as related to agricultural production and environmental services. In many cases, this information already exists, but needs to be analysed for the purpose of assessing supply potential. In other cases, data at the appropriate scale may need to be collected. The Millenium Ecosystem Assessment gives a good overview of available data and major gaps (MA, 2005, Chapter 2).

Assessing the potential demand for the environmental services a country can provide is equally important. For local level environmental services, such as watershed management, some analysis of the potential benefits (in terms of improved water quality, hydro-electric or irrigation operations) is needed. For global services such as biodiversity conservation and climate change mitigation, some assessment of the requirements for participation among major funders and current level of payments for services should be carried out. Much of this information is also already available or partially available but, as in the case of the supply side information, it needs to be analysed to derive an assessment of demand for a specific location. The Ecosystem Marketplace (http://www.ecosystemmarketplace.com) is one important source, as well as studies from international research and technical agencies including CGIAR centres, FAO, the World Bank, the International Institute for Environment and Development and others already involved in conducting some of this work, but more needs to be done and better coordination of information is necessary.

In addition to information gathering, analysis and dissemination, a proactive policy strategy is needed to incorporate environmental services into drylands development strategies and support to livelihoods. Governments, from the local to international level, have an important role to play in terms of creating demand for environmental services and establishing an enabling policy and institutional environment to support livelihoods.

The most straightforward means by which governments create demand for environmental services is by enabling public funds to be spent on environmental services. For example, the SLCP in China was initiated by the central government in 1999 with the stated environmental goals of reducing water and soil erosion, and the Ministry of Finance manages its funding (Bennet and Xu, 2005). In Brazil, several state-level governments decided to allocate some of their sales tax revenues to support biodiversity conservation through the establishment of the ICMS Ecológico (Grieg-Gran, 2000).

Of course, public sector funding for environmental objectives is likely to be quite limited in many developing countries. However, environmental service provision may be a least-cost means of achieving a development goal that is, or will be, funded by the public sector. For example, reducing siltation in major waterways in China under the SLCP provides significant economic benefits to the country in terms of hydroelectric power and improved navigability, and obtaining these benefits through an alternative means, such as dredging, is likely to be more expensive and less effective. Environmental services could be a cost-effective means of improving drinking water quality or in natural disaster preparedness – two important public policy objectives in many developing countries. Thus, the demand for environmental services from the public sector could arise from either environmental or broader development objectives, and exploring the potential of environmental services to be a cost-effective means of meeting broader development goals is an important task that governments need to undertake in partnership with national and international research and technical institutions to successfully incorporate environmental services into drylands development strategies.

The enactment and implementation of environmental regulations that allow for market-based implementation mechanisms is another important role governments and national and international agencies can play in creating demand for environmental services. Private sector entities looking for low-cost means of meeting regulatory requirements is an important basis for several environmental service programmes, such as the Clean Development Mechanism of the Kyoto Protocol. The demand for biodiversity conservation is highly conditioned by regulations governing the conservation and use of biodiversity at the national and international levels. National-level policy makers are important players in the development of international-level environmental agreements, and their support for environmental regulation and the use of market-based approaches affects the level of demand for services.

Multi-lateral environmental agreements such as the UN Convention to Combat Drought and Desertification and the Convention on Biological Diversity, together with other international agencies such as FAO, UNEP, IUCN and the World Bank, have an important role to play in designing the incentive mechanisms to obtain the desired environmental objectives in a way that facilitates the participation of and potential benefits for low-income countries and people.

On the supply side, policies to support potential suppliers are needed to realize the potential of environmental services for drylands development. One of the most important of these is a supportive land tenure policy environment. Lack of formal tenure is one of the biggest barriers to participation in environmental service provision, particularly among the poor. Public sector recognition and clarification of informal property rights to land and water is likely to be a very important requirement for payments for environmental services becoming a significant contribution to livelihoods in many dryland areas. Dutilly-Diane *et al.* (2004) and Pagiola *et al.* (2005) discuss options that could or have been used in cases where tenure is informal. Potential conflicts with currently existing policies and regulations governing land use are another important area for consideration. For example, in some countries productive use of the land has been required to maintain tenure, and this has generally been interpreted to mean agricultural production. Changes in these types of regulations in recognizing environmental services other than agricultural production as a productive use would be necessary to support the integration of environmental services into a development strategy.

One final step in the integration of environmental services in drylands development and in livelihoods is the establishment of institutions to facilitate the exchange of environmental services for payments. Payments for environmental services are a new type of exchange, and institutions are needed to support the process, including the development of projects, certification of product delivery and transfer of payments. An important issue is the need to establish institutions and rules that reduce transactions costs of participation for both buyers and sellers. In many of the existing cases NGOs are involved in facilitating exchanges, usually interacting with national or international public sector agencies. Transactions costs are high and, if they remain so, are likely to render environmental services unimportant; however, these costs could decrease as markets and rules of exchange become more established. Countries committed to incorporating environmental services into their drylands development strategies will need to assess their current institutional capacity, together with the new demands that payments for environmental services will place on the system, and develop a strategy to meet these demands most cost-effectively.

Drylands areas are not condemned to destitution, but rather have several potential pathways towards achieving sustainable development, depending on their specific environments and socio-economic conditions. Incorporating environmental services into drylands development strategies is an important means of broadening the set of potential options for a sustainable development pathway, as is the recognition that going beyond agriculture to develop solutions is necessary.

Acknowledgement

The authors gratefully acknowledge the assistance of Camille Barchers in preparing case study information.

Endnotes

[1] The central government started China's Sloping Land Conversion Program (SLCP) in 1999, with the goal of converting 14.67 million ha of cropland to forests by 2010 (the largest conversion programme in the developing world). Within this area, 4.4 million ha are also on slopes > 25°. The project is taking place across 25 provinces in China and involves the participation of tens of millions of rural households. The total budget for the programme is > US$40 bn. The cash subsidy is US$36/ha/year. Subsidies are given in grain and cash and apply for 8 years if forests are planted and for 5 or 2 years if harvestable forests and grasses are planted. If farmers plant trees in wasteland (baseline classified as 'barren') sites, US$91/ha is given. These funds are also exempt from income taxes normally paid by farmers. The farmers will be the main beneficiaries of the potential environmental services from these activities, which are watershed protection, increased productivity and soil carbon sequestration (Bennet and Xu, 2005).

> [2] The Scolel-Te carbon project in Mexico involved 400 small-scale farmers across 20 different communities. The farmers switched from swidden (an area cleared for cultivation by slashing and burning of vegetation) agriculture to agroforestry systems that combined crops with fruit and timber trees and enriched fallow lands. The expected environmental service of these activities was forest carbon sequestration. The International Federation of Automobiles buys the carbon 'offsets' (approximately 17,000tC from US$10/tC to US$12/tC). They purchase these carbon offsets through the Econergy International Corporation (US) and Future Forests (UK) (De Jong *et al.*, 2000).
>
> [3] Payments for external environmental benefits for the adoption of agricultural systems that have higher private returns than existing systems is controversial, to the extent to which the environmental services generated are 'additional', e.g. would they have been provided even without payments for the externality? Presumably, agricultural producers will choose production systems with the highest private returns; however, barriers to adoption such as capital constraints, property rights or lack of information can prevent adoption. The recognition of socio-economic barriers as a part of the baseline scenario for calculation of incremental environmental benefits has been recognized in some PES mechanisms, however, including the CDM and GEF.

References

Bennett, M.T. and Xu, J. (2005) China's Sloping Land Conversion Program: institutional innovation or business as usual? Paper presented at the *CIFOR/ZEF-Bonn Workshop on Payments for Environmental Services (PES) – Methods and Design in Developing and Developed Countries*, Tititsee, Germany.

Cacho, O., Marshall, G. and Milne, M. (2002) *Smallholder Agroforestry Projects: Potential for Carbon Sequestration and Poverty Alleviation*. FAO ES Technical Series Working Paper, FAO, Rome, 76 pp. (http://www.fao.org/es/esa/en/pubs_wp.htm).

CGIAR TAC (1999) Report of the study on CGIAR research priorities for marginal lands, the framework for prioritizing land types in agricultural research, the rural poverty and land degradation: a reality check for the CGIAR. Consultative Group on International Agricultural Research (CGIAR), Technical Advisory Committee, Washington, DC (http://www.fao.org/Wairdocs/TAC/X5784E/x5784e00.htm).

Costanza, R., Dárge, R., De Groot, R.S., Farber, S., Grasso, M., Hannon, B., Limburg, K., Naeem, S., O'Neill, R.V., Paruelo, J., Raskin, R.G., Sutton, P. and Van den Belt, M. (1997) The value of the world's ecosystem services and natural capital. *Nature* 387, 253–260.

Daily, G.C. (1997) Introduction: what are ecosystem services? In: Daily, G.C. (ed.) *Nature's Services: Societal Dependence on Natural Ecosystems*. Island Press, Washington, pp. 1–10.

De Jong, B.H.J., Tipper, R. and Montoya-Gomez, G. (2000) An economic analysis of the potential for carbon sequestration by forests: evidence from southern Mexico. *Ecological Economics* 33, 313–327.

Den Kate, K. (2005) Presentation at the *CIFOR/ZEF-Bonn Workshop on Payments for Environmental Services (PES) – Methods and Design in Developing and Developed Countries*, Tititsee, Germany.

Dutilly-Diane, C., McCarthy, N., Turkelboom, F., Bruggeman, A., Tiedeman, J., Street, K. and Serra, G. (2004) Could payments for environmental services improve rangelands management in Central Asia, West Asia and North Africa? Paper presented at the

2004 Meetings of the International Association for the Study of Common Property (IASCAP), Oaxaca, Mexico, 9–13 August.

FAO (2006) Livestock Policy Brief Number 03, FAO Sector Analysis and Policy Branch, Animal Production and Health Division (http://www.fao.org/ag/AGAinfo/resources/documents/pol-briefs/03/EN/AGA04_EN_05.pdf).

Graff-Zivin, J. and Lipper, L. (2006) Poverty, risk and soil carbon sequestration. Paper presented at the *International Conference on Economics of Poverty, Environment and Natural Resource Use*, 17–19 May 2006, Wageningen University, Wageningen, The Netherlands.

Grieg-Gran, M. (2000) *Fiscal Incentives for Biodiversity Conservation: The ICMS Ecológico in Brazil*. Discussion Paper 00–01, International Institute for Environment and Development, London.

Hayami, Y. and Ruttan, V.W. (1985) *Agricultural Development: an International Perspective*, revised edn. The Johns Hopkins University Press, Baltimore, Maryland.

Kuyvenhoven, A., Pender, J. and Ruben, R. (2004) Editorial: development strategies for less-favoured areas. *Food Policy* 29, 295–302.

Landell-Mills and Porras, I.T. (2002) *Silver Bullet or Fools' Gold? A Global Review of Markets for Forest Environmental Services and their Impact on the Poor*. International Institute for Environment and Development (IIED), London.

Lipper, L. (2000) Dirt Poor: poverty, farmers and soil resource investment. In: *Two Essays on Socio-economic Aspects of Soil Degradation*. ESA working paper ESA/149, FAO, Rome.

Lipper, L. and Cavatassi, R. (2003) Land use change, poverty and carbon sequestration. *Environmental Management* 33, 374–387.

MA (2003) Ecosystems and human well-being: a framework for assessment. In: *Millennium Ecosystem Assessment*, Island Press, Washington, DC/Covelo, London.

MA (2005) Current state and trends: dryland system. In: *Millennium Ecosystem Assessment*. Vol. 1, Island Press, Washington, DC/Covelo, London, Chapter 22.

Mayrand, K. and Paquin, M. (2004) *Payments for Environmental Services: a Survey and Assessment of Current Schemes*. Commission for Environmental Cooperation, Montreal, Canada.

Pagiola, S. (2005) Presentation at the *CIFOR/ZEF-Bonn Workshop on Payments for Environmental Services (PES) – Methods and Design in Developing and Developed Countries*, Tititsee, Germany.

Pagiola, S., Arcenas, A. and Platais, G. (2005) Can payments for environmental services help reduce poverty? An exploration of the issues and the evidence to date from Latin America. *World Development* 33, 237–253.

Pender, J. and Hazell, P. (2000) *Promoting Sustainable Development in Less-favoured Areas: Overview*. 2020 Focus 4, Brief 1, November 2000, IFPRI, Washington, DC.

Pingali, P. (2001) Population and technological change in agriculture. In: Smelser, N.J., Boltes, P.B. (eds) (2001). *International Encyclopaedia of the Social and Behavioural Sciences*. Elsevier, Amsterdam.

Smith, J. and Scherr, S.J. (2002) *Forest Carbon and Local Livelihoods: Assessment of Opportunities and Policy Recommendations*. Occasional Paper 37, CIFOR, Bogor, Indonesia.

World Bank (2002) Biocarbon Fund. http://biocarbonfund.org/

Wunder, S. (2005) Payments for environmental services: some nuts and bolts. Occasional Paper 42, CIFOR, Bogor, Indonesia.

Zilberman, D., Lipper, L. and McCarthy, N. (2006a) *Putting Payments for Environmental Services in the Context of Economic Development*. (In press)

Zilberman, D., Lipper, L. and McCarthy, N. (2006b) When are payments for environmental services beneficial to the poor? Paper presented at the *International Conference on Economics of Poverty, Environment and Natural Resource Use*, 17–19 May 2006, Wageningen University, The Netherlands.

Zurita, P. (2005) Presentation at the *CIFOR/ZEF-Bonn Workshop on Payments for Environmental Services (PES) – Methods and Design in Developing and Developed Countries*, Tititsee, Germany.

19 Livelihood Strategies, Policies and Sustainable Poverty Reduction in LFAs: a Dynamic Perspective

ANDREW DORWARD

Agricultural Development Economics, Centre for Environmental Policy, Imperial College, London, Wye Campus, Wye, Ashford, Kent TN25 5AH, UK; e-mail: A.Dorward@imperial.ac.uk; website: http://www.imperial.ac.uk/people/A.Dorward

Introduction

Investigation of the prospects for sustainable poverty reduction in LFAs requires micro-analysis of the potential for economic and livelihood growth within LFAs, but it also needs this to be set within an analysis of economic and livelihood opportunities elsewhere in national (and indeed, international) economies. This chapter examines broad livelihood development options for LFAs in the context of dynamic processes of technical, economic, institutional and social change both within and outside LFAs. We address this issue using two different analytical frameworks: the first, a categorization of techno-economic conditions in different areas and the second a simple categorization of rural people's livelihood strategies. We therefore begin the chapter by setting out each of these frameworks, and then bring them together to relate micro-analysis of changing livelihood opportunities in LFAs to wider processes of economic development. We conclude with a brief discussion of the implications of this for poverty reduction policies in LFAs.

Techno-economic Conditions in Different Areas

Hazell *et al.* (2005) suggest that less favoured areas (LFAs) are best defined in terms of poor market access and low natural resource productivity potential. However, these characteristics are not independent, static or exogenous. Thus, areas with good natural resource potential are more likely to attract population growth and infrastructural investment and, conversely, infrastructural investment or population growth that improve market access often stimulate investments in natural resource productivity.[1] We describe some of the interactive dynamics of this relationship in Fig. 19.1.

The two axes in Fig. 19.1 map out market access and agricultural productivity potential as two important variables determining levels and patterns of economic development in an area. We consider two particular dimensions of economic development: density of economic activity and the percentage of household incomes derived from agriculture. Density of economic activity itself is related to household incomes, to thick markets and (inversely) to general economic risk, as higher economic density is generally related to higher incomes and wealth, with increased market volumes market players and with lower market access and production risk (Kydd and Dorward, 2004; Dorward *et al.*, 2005). Production risk is also often associated with low agricultural productivity potential (for example, in rain-fed agriculture in semi-arid areas).

These relationships between market access, potential productivity and economic development are represented in Fig. 19.1 using 'isoquant' curves to show gradients on each of the two dimensions of economic development discussed above. Thus, in the bottom left-hand corner of the graph, poor market access and low productivity potential lead to a generally poor economy with thin markets and high levels of risk. Greater market access and/or productivity potential will generally lead to increasing economic activity and incomes, thicker markets and lower risks. However, at very high levels of market access (for example, in

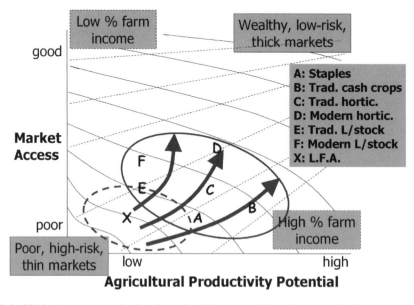

Fig. 19.1. Market access, agricultural productivity potential and economic change.

cities) economic activity is driven only by increasing market access as, in these circumstances, agricultural productivity potential has very little influence on economic activities (indeed, urban investments reduce agricultural productivity potential by building over agricultural land).

Figure 19.1 also shows a gradient of the declining importance of agriculture in incomes from areas with high agricultural productivity potential and poor market access (in the bottom right-corner of the diagram) to areas with good market access and poor agricultural productivity potential (although, again, agricultural productivity potential becomes much less important at high levels of market access, leading to flattening of the isoquant curve).

Having mapped out stylized ways in which market access and agricultural productivity potential influence economic welfare, market density, risk and the relative economic importance of agriculture, we now use this analysis to consider possible development pathways for areas with different combinations of market access and agricultural productivity potential. In particular, we ask how development may occur in LFAs, located in the dashed circle in the bottom left corner of Fig. 19.1.

Three broad development pathways may be envisaged: (i) dramatic investments in market access infrastructure may shift a locality upwards in the figure; (ii) a sudden shift in productivity potential may shift a locality along to the right (e.g. investment in irrigation increasing agriculture productivity); or (iii) incremental technical or institutional change or small infrastructure investments will cause small shifts upwards and to the right.[2] This last pathway is probably the most common, particularly if one includes within it stagnant localities that get 'stuck', with very little and even negative change (for example, loss of productive potential from land degradation).

Possibilities for progress with this pathway will vary according to specific conditions, but alternative 'sub-pathways' may be envisaged involving increasing intensification of livestock in areas with greater agro-ecological constraints, or increasingly productive cultivation of food crops, horticultural crops or other cash crops. Indicative pathways are shown in Fig. 19.1. In very broad terms, pathways through extensive and intensive livestock production systems (E and F) may be least demanding of productivity potential, with extensive grazing systems in more remote areas and more intensive systems in more accessible areas. Among crops, staple food production (A) may be least demanding of access and productivity, traditional cash crops (B) may be more demanding of both and then horticultural crops for domestic and export markets (C and D) are most demanding of access to markets.[3]

This analysis suggests that development pathways for LFAs depend upon current market access and productivity potential, upon external investments in market access and productive infrastructure and upon incremental technical, institutional and infrastructural change. These, however, interact with and depend upon the livelihood decisions and activities of local people, which we consider below.

Livelihood Strategies of LFA Inhabitants

Sustainable poverty reduction in LFAs involves changes in the livelihoods of LFA inhabitants and, as noted above, livelihood changes interact with market access, productivity potential, external investments and institutional and technical change. An understanding of policy options for the promotion of sustainable poverty reduction therefore requires an understanding of LFA inhabitants' livelihood strategies.

Dorward *et al.* (in press) suggest that poor people follow three broad types of livelihood strategy in pursuit of aspirations for security and for advancement:

- 'Hanging in', where assets are held and activities are engaged in to maintain livelihood levels, often in the face of adverse socio-economic circumstances.
- 'Stepping up', where investments are made in assets to increase the productivity or scale of current activities.
- 'Stepping out', where current activities are seen as a means to accumulate assets which, in time, can then provide a base or 'launch pad' for moving into new activities with higher and/or more stable returns.

These three strategies are not mutually incompatible, and are commonly pursued together.

Simple though it may be, this categorization of three strategies is well suited to analysis of the livelihoods of poor people living in LFAs. 'Hanging in' is often a major pre-occupation of the poor everywhere as they engage in low-return/low-risk activities in a daily struggle to survive. 'Stepping up' strategies involve attempts to raise incomes from current natural resource-based activities such as livestock keeping or production. Assisting people with 'hanging in' but, particularly, 'stepping up' strategies is the major focus of most of the chapters in this book. However, though hanging in and stepping up strategies may be very important to poor inhabitants of LFAs, they are unlikely to be the major means by which they escape from poverty – the history of development in richer countries and contemporary patterns of development and migration in today's poorer countries both suggest that the major strategy that people follow to escape from poverty in LFAs is 'stepping out'. This takes two forms: (i) switching from agricultural to non-agricultural activities within LFAs; and, more importantly (ii) migration out of LFAs to urban areas or to richer rural areas. These strategies are illustrated in Fig. 19.2 in terms of people 'stepping out' or 'stepping up'.

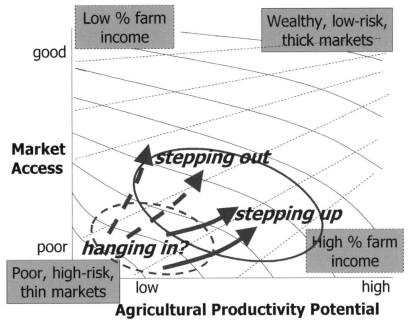

Fig. 19.2. 'Hanging in', 'stepping up' and 'stepping out' in LFAs.

Policies for Sustainable Poverty Reduction in LFAs

The above analysis suggests four broad ways in which the livelihoods of poor LFA inhabitants may be improved. We briefly discuss these and policy measures for encouraging them:

1. More secure 'hanging in': this describes livelihood improvements that reduce people's vulnerability to adverse shocks, for example climatic, health or price shocks. A wide variety of policy approaches may assist people to hang in: these include, for example, social service and social protection measures, livestock health services, irrigation development, market interventions to reduce risks of high consumer prices or low produce price, and a variety of agricultural innovations aimed at reducing production risks and/or protecting and conserving natural resources.

2. Local 'stepping up': this describes scaling-up or intensification of farming activities in LFAs. Policy measures to promote this can include a wide variety of research, extension, market and other agricultural development services. Attempts to expand production must ensure that this is matched by access to markets for new produce.

3. Local 'stepping out': this describes diversification out of agriculture and increased income from and reliance on non-agricultural activities. An important prerequisite for this is prior economic growth in the area, and this is most commonly generated by local agricultural growth (local stepping up) or by remittances or savings from migrant stepping out (see below).

4. Migrant 'stepping out': here, people move out of the area to find employment elsewhere – in urban areas or in more productive agricultural areas. Prerequisites for this most important strategy are growth in the economy elsewhere, with increasing labour demand, together with local people having accumulated the capital needed for a new start elsewhere. An important source of such savings may be income from local agriculture or prior migration by others. Migrant 'stepping out' will also be assisted by improved communications and by education.

Conclusions

Policies for sustainable poverty reduction in LFAs need to be built around and work from LFA's market access and agricultural productivity potential, and should support the inhabitants of these areas in their pursuit of hanging in, stepping up and local or migrant stepping out. This requires policies that reduce vulnerability and encourage growth in both farm and non-farm opportunities in LFAs themselves and elsewhere in the economy. It is important that policies for LFAs should not focus exclusively on agricultural development and resource conservation but place such work, important as it is, in the context of wider processes of economic, social and livelihood change.

Endnotes

[1] The argument presented here is developed with specific regard to natural resources affecting agro-ecological productivity. However the general argument may be applied, with appropriate adjustments, to other types of natural resources, such as exploitation of mineral resources or tourism opportunities.

[2] Other pathways might follow investments in mineral extraction or tourism facilities.

[3] This broad brush categorization of possible pathways hides much important variability within categories (for example there are stable and traditional cash crops that are more and less demanding of market access and water or soils) and the divisions between categories is often blurred (when does a food crop become a cash crop?).

References

Dorward, A.R., Kydd, J.G., Morrison, J.A. and Poulton, C. (2005) Institutions, markets and economic coordination: linking development policy to theory and praxis. *Development and Change* 36, 1–25.

Dorward, A., Anderson, S., Nava, Y., Pattison, J., Paz, R., Rushton, J. and Sanchez Vera, E. (2007) Hanging in, stepping up and stepping out: livelihood aspirations and strategies of the poor. *World Development* (in press).

Hazell, P., Ruben, R., Kuyvenhoven, A. and Jansen, H.G.P. (2005) *Investing in Poor People in Less-Favoured Areas*. IFPRI 2020 Policy Paper, IFPRI, Washington, DC.

Kydd, J.G. and Dorward, A.R. (2004) Implications of market and coordination failures for rural development in least developed countries. *Journal of International Development* 16, 951–970.

Index